P9-ECV-692

GLENCOE/McGRAW-HILL

SCIENCE PROBE I

GLENCOE / McGRAW-HILL

SCIENCE PROBE I

Adapted by Gary E. Sokolis and Susan S. Thee

This product is adapted from Nelson Canada's *Science Probe 9.*

Author Team

Peter Beckett
Jean Bullard
Julie Czerneda
Nancy Flood
Peter Freeman
Eric S. Grace
Alan J. Hirsch
Barry McCammon
Gerry Sieben
Tom Stokes
Mary Kay Winter
Allen Wootton

Pedagogical Consultant
Sandy M. Wohl

New York, New York Columbus, Ohio Woodland Hills, California Peoria, Illinois

Glencoe/McGraw-Hill 
A Division of The ***McGraw-Hill*** *Companies*

Library of Congress Cataloging-in-Publication Data
Thee, Susan S. (Susan Santana), 1950–
Science probe I / Susan S. Thee, Gary Sokolis.
p. cm.
Includes index.
Summary: A collection of experiments concentrating on earth, space, physics, chemistry, and life science.
ISBN 0-538-66900-4
1. Science—Experiments—Juvenile literature. [1. Science—Experiments. 2. Experiments.] I. Sokolis, Gary, 1941– .
II. Title.
Q182.3.T47T44 1996 96-26783
500—dc20 CIP
AC

Send inquiries to:
Glencoe/McGraw-Hill
8787 Orion Place
Columbus, OH 43240-4027

ISBN: 0-538-66900-4
Printed in the United States of America.

8 9 10 11 12 071 06 05 04 03 02 01

ACKNOWLEDGEMENTS

Science Probe I is the result of the efforts of a great many people. The publisher would like to thank those who assisted in the development of the original product, *Science Probe 9*, Nelson Edition, from which this edition is adapted.

Focus Group Members
Anand Atal, Burnaby South Secondary; Patricia Dairon, Delta Secondary; Dr. Gaalen Erickson, University of British Columbia; Lloyd Esralson, Burnaby North Secondary; Roger Fox, Premier's Advisory Council on Science and Technology; Lon Mandrake, Seaquam Secondary; Tony Williams, Dr. Charles Best Junior Secondary; Sandy Wohl, Hugh Boyd Junior Secondary.

Teacher Reviewers
John Eby, Nanaimo District Senior Secondary (Unit III); Tom Hierck, Trafalgar Junior High (Unit II): Glen Holmes, Claremont Secondary (Unit IV); Shane Hood, Skeena Junior Secondary (Unit VI); Douglas Mann, Central Junior Secondary (Unit I); W. A. Vliegenthart, Correlieu Secondary (Unit V).

Content Reviewers
Dr. Tom Dickinson, University College of the Cariboo (Unit VI); Dr. R. L. Evans, University of British Columbia (Unit V); Carlo Giovanella, University of British Columbia (Unit III); Dr. Mel Heit, Simon Fraser University (Unit I); Dr. Peter Matthews, University of British Columbia (Unit V); Dr. Nancy Ricker, Capilano College (Unit II); Dr. J. B. Tatum, University of Victoria (Unit IV).

We would also like to thank those people who provided information or reviewed portions of this book in manuscript form:

ABB Robotics (Unit V), Barry L. Ackerman (Unit VI), Judy Benger (Unit V), Canadian Diabetes Association, B.C.-Yukon (Unit II), Andrew Cowan (Unit II), Daniel David (Unit II), Marcy Dayan (Unit V), Fisheries and Oceans Canada (Unit VI), Gene Hemsworth, Professional Association of Diving Instructors (Unit II), Michael Jessen (Unit I), Chris Kolmel (Unit I), Marchmont Public School (Unit II), Geoff Meggs (Unit VI), Janice Morrison (Unit V), Barry Mowatt (Unit I), Eva Novy (Unit IV), Stefani Paine (Unit VI), Jill Philipchuk (Unit II), Physiotherapy Department, Kootenay Lake District Hospital (Unit V), Cheryl Power (Unit VI), Radarsat International (Unit VI), Dr. W. E. Ricker (Unit VI), Delia Roberts (Unit II), Jeff Roberts (Unit I), Dennis Semanitus (Unit I), Société d'énergie de la Baie James (Unit VI), Space Telescope Science Institute (Unit IV), Ed Stockerl (Unit I), Dr. Ken Tapping (Unit IV), Ellie Topp (Unit II), Rod Waterlow (Unit III), Dr. Andrew Woodsleth (Unit IV), Morris Zallen (Unit VI).

We would especially like to thank the staff and students of Burnaby South Secondary School in Burnaby and Hugh Boyd Junior Secondary in Richmond for taking part in photograph sessions.

TABLE OF CONTENTS

BETHLEHEM

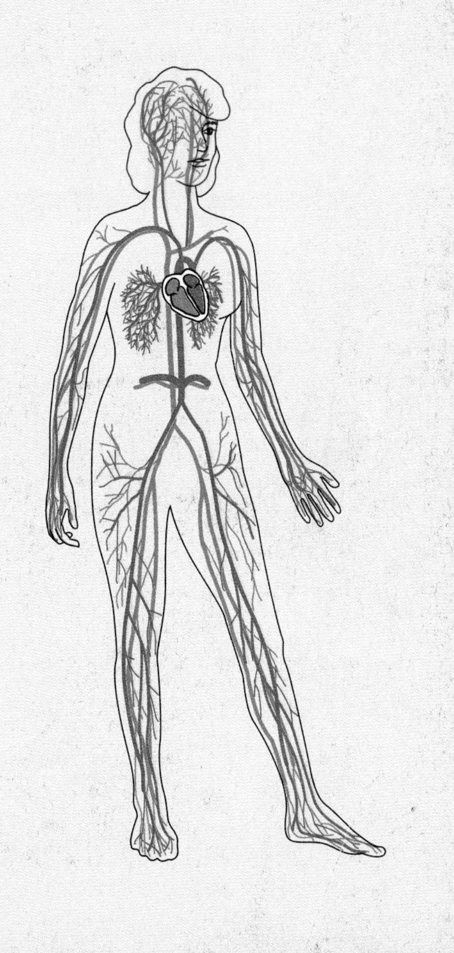

FEATURES OF SCIENCE PROBE I

■ *Science Probe I* has an introductory chapter about the nature of science followed by six units, each representing a major area of science. These units can be studied in any sequence.

■ The text includes a variety of activities designed to foster the development of thinking skills and the processes and skills of investigation. The diverse formats include step-by-step experiments, reading activities, informal discussion, and student-designed investigations. All activities are designated by the same type of heading. The first activity in each chapter is designed to help students consider what they already know about the topic of the chapter and what they may want to find out.

■ Activities that are strong examples of the use of collaborative learning are marked with a Group Activity icon. (In addition to the activities that specifically require collaboration, many of the other activites, which are not marked with icons, can be conducted in pairs or in larger groups, as the teacher chooses.) Similarly, activities that require students to integrate other subject areas with science learning are marked with a Cross-Curricular icon.

■ Within an activity, the word CAUTION! followed by a few words or sentences indicates that the activity contains potentially dangerous procedures or materials. More information on safety can be found in *Safety Rules* on page ix.

■ Enrichment activities in the margin under the heading *Extension* are intended to challenge students or to reinforce new concepts. Under the heading *Did You Know?* are interesting facts and scientific discoveries.

■ Special features, *Profile*, *Career*, *Computer Application*, and *Science in Our World*, examine topics related to the chapter.

■ Terms considered essential for understanding new concepts are printed in **boldface**, defined in the text, included in the *Vocabulary* at the end of the chapter, and listed in the *Glossary*.

■ At the end of each numbered section, questions are listed under the heading *Review*. These questions allow students to check their understanding of content as they progress through the chapter.

■ Each chapter concludes with a five-part chapter review. *Key Ideas* lists the main ideas of the chapter. *Vocabulary* lists the boldface words and suggests two activities using these key terms. *Connections* require skills such as comparison, evaluation, prediction, and problem solving. *Explorations* suggest projects, research, and group activities. *Reflections* allow students to consider how the chapter has changed their ideas and to reflect on what they have learned.

■ The *Glossary* includes definitions of all boldface terms, along with other important scientific terms.

■ The *Appendices* provide information of general use and an activity, Fish Anatomy and Dissection.

■ The *Index* helps students to locate important topics and to find connections between units.

■ The *Science Probe I Teacher's Guide* provides teaching strategies, lesson plans, practical tips, answers to questions, additional questions and activities, sample tests, blackline masters (some for overhead projection, some for student use), and other useful information for planning and teaching each unit.

SAFETY RULES

Your school laboratory, like your kitchen, need not be dangerous. In both places, understanding how to use materials and equipment and following proper procedures will help you avoid accidents.

The activities in this textbook have been tested and are safe, as long as they are done with proper care. Take special note of the instructions accompanying the word CAUTION! whenever it appears in an activity. These instructions will help you understand how to use electricity, laboratory equipment, and chemicals safely.

Follow the safety rules listed below. Your teacher will give you specific information about other safety rules for your classroom. You will also be told about the location and proper use of all safety equipment.

1. Give your teacher your complete attention when listening to instructions.
2. Learn the location and proper use of the safety equipment available to you, such as safety goggles, protective aprons, heat-resistant gloves, fire extinguishers, fire blankets, eyewash fountains, and showers. Find out the location of the nearest fire alarm.
3. Inform your teacher of any allergies, medical conditions, or other physical problems you may have. Tell your teacher if you wear contact lenses.
4. Read through the entire activity before you start. Before beginning any step, make sure you understand what to do. If there is anything you do not understand, ask your teacher to explain.
5. Do not begin an activity until you are instructed to do so.
6. If you are designing your own experiment, obtain your teacher's approval before carrying out the experiment.
7. Clear the laboratory bench of all materials except those you are using in the activity.
8. Follow your teacher's instructions regarding the No Crowding Zone.
9. Wear protective clothing (a lab apron or a lab coat) and closed shoes during activities involving heating substances or using chemicals. Long hair should be tied back.
10. Wear safety goggles when using hazardous or unidentified materials and when heating materials.
11. Do not taste or touch any material unless you are asked to do so by your teacher.
12. Do not chew gum, eat, or drink in the laboratory.
13. Do not rock or lean on lab stools.
14. Do not run or play games in the laboratory.
15. Do not throw any objects, including paper, chalk, pens, and liquids.
16. Carry lab equipment, containers of chemicals, and glassware carefully.
17. Read and make sure you understand all safety labels.
18. Label all containers.
19. When taking something from a bottle or other container, double-check the label to be sure you are taking exactly what you need.
20. If any part of your body comes in contact with a chemical or specimen, wash the area immediately and thoroughly with water. If your eyes are affected, do not touch them but wash them immediately and continuously for at least 15 minutes and inform your teacher.

21. Handle all chemicals carefully. When you are instructed to smell a chemical in the laboratory, follow the procedure shown here. Only this technique should be used to smell chemicals in the laboratory. Never put your nose close to a chemical.

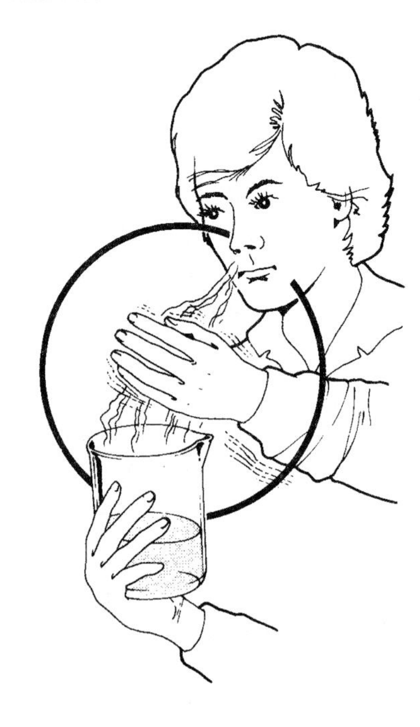

22. Hold the container(s) away from your face when pouring liquids.

23. Place test tubes in a rack before pouring liquids into them. If you must hold the test tube, tilt it away from you before pouring liquids in.

24. Clean up any spilled water, chemicals, or other materials immediately, following instructions given by your teacher.

25. Do not return unused chemicals to original containers. Do not pour them down the drain. Dispose of chemicals as instructed by your teacher.

26. Special care must be taken when dissecting an organism. A dissection must be performed cautiously and patiently. Be sure to follow the instructions in the text and also those that your teacher gives you. Each time you dissect, you should do the following:

- Make sure that the area you are working in is well ventilated.
- Wear safety goggles and an apron at all times.
- Wear disposable gloves when performing a dissection to prevent any chemicals from coming in contact with your skin.
- Gently rinse your specimen under running water to wash away excess preservatives.
- Wash all splashes of the preservative solution off your skin and clothing immediately. If you get any chemical in your eyes, rinse thoroughly for at least 15 minutes and inform your teacher.
- Position your specimen so that it is not directly beneath your face and nose.
- Familiarize yourself with the safe and proper use of all dissecting instruments.
- Report dull or damaged equipment immediately. Dull blades will slip and may cause injury.
- Use the dissection tools carefully. Be sure to follow your teacher's instructions when using a knife or razor blade. In most cases, you will be asked to cut away from yourself and away from others.
- Always cut gradually through layers of tissue.
- Always thoroughly wash your hands and lower arms with soap and warm water after completing your dissection work.
- Dispose of any waste material in the container provided by your teacher.

27. Whenever possible, use electric hot plates for heating materials. Use flames only when instructed to do so. Read the special Bunsen burner safety procedures listed under safety rule number 47.

28. When heating materials, always wear safety goggles and use hand protection if required.

29. When heating glass containers, make sure you use clean Pyrex or Kimax. Do not use broken or cracked glassware. Always keep the open end pointed away from yourself and others. Never allow a container to boil dry.

30. When heating a test tube over a flame, use a test tube holder. Holding the test tube at an angle and facing away from anyone, move it gently through the flame so that the heat is distributed evenly.

31. Handle hot objects carefully. Hot plates can take up to 60 minutes to cool off completely. Hot and cold hot plate burners can look the same. If you burn yourself, immediately apply cold water or ice.

32. Keep water and wet hands away from electrical cords, plugs, and sockets.

33. Always unplug electrical cords by pulling on the plug, not the cord. Report any frayed cords or damaged outlets to your teacher.

34. Make sure electrical cords are not placed where anyone can trip over them.

35. When cutting with a knife or razor blade, follow your teacher's instructions. In most cases, you will be asked to cut away from yourself and away from others.

36. When walking with a pair of scissors or any pointed object, keep the pointed surface facing the floor away from yourself and away from others.

37. Watch for sharp or jagged edges on all equipment.

38. Place broken or waste glass only in specially marked containers.

39. Follow your teacher's instruction when disposing of waste materials.

40. Report to your teacher all accidents (no matter how minor), broken equipment, damaged or defective facilities, and suspicious-looking chemicals.

41. Be sure all equipment is shut off when not in use. Be ready to shut off equipment quickly if it breaks down or if an accident occurs.

42. Clean all equipment before putting it away.

43. Put away all equipment and chemicals after use.

44. Wash your hands thoroughly using soap and warm water after working in the science laboratory. This practice is especially important when you handle chemicals, biological specimens, or microorganisms.

45. Do not take any equipment, materials, or chemicals out of the laboratory.

46. Do not practice laboratory experiments at home unless directed to do so.

47. If a Bunsen burner is used in your science classroom, make sure you follow the procedures listed below. (NOTE: Hot plates should be used in preference to Bunsen burners whenever possible.)

- Do not wear scarves or ties, long necklaces, or earphones suspended around your neck. Tie back long hair and roll back or secure loose sleeves before you light a Bunsen burner.
- Obtain instructions from your teacher on the proper method of lighting and using the Bunsen burner.
- Do not heat a flammable material (for example, alcohol) over a Bunsen burner.
- Be sure there are no flammable materials nearby before you light a Bunsen burner.
- Do not leave a lighted Bunsen burner unattended.
- Always turn off the gas at the valve, not at the base of the Bunsen burner.

48. If a fire occurs, make sure you follow the procedures listed below.

- Do not panic. Remain calm.
- Notify your teacher immediately. Act quickly to provide help in an emergency.
- Shut off all gas supplies at the desk valves if it is safe to do so.
- Pull the fire alarm.
- Follow your teacher's instructions if your assistance is required.
- If your clothing is on fire, roll on the floor to smother the flames. If another student's

clothing is on fire, use a fire blanket to smother the flames.
- Avoid breathing fumes.
- Do not throw water on a chemical fire.
- If the fire is not quickly and easily put out, leave the building in a calm manner.

49. Labelling and placarding assists shippers, carriers, fire departments, police, emergency response personnel, and others in complying with and enforcing the regulations governing the safe transport of hazardous materials by highway, rail, water, and air. The labelling of hazardous material is specific to the hazard class of the material. The placards represent the hazard class(es) of the material(s) contained within the freight container, motor vehicle, or rail car.

Become familiar with the warning labels that are placed on containers of potentially dangerous materials. You should be able to identify and understand each of the labels shown here.

Hazardous Material Warning Placards

The warnings on labels of household products were developed to indicate exactly why and to what degree a product is dangerous. Pay careful attention to any warning labels on the products or materials that you handle.

TO THE STUDENT

This book is about science in your world. You will find information here on how to dispose of hazardous household chemicals, how your body uses the food you eat, what makes volcanoes erupt, why Venus is the brightest planet we can see, what a machine is, how human activities can harm fish, and much, much more. In addition, this book contains activities and other suggestions to help you learn and understand science. A few of the many things you might do during science classes are described below.

Consider What You Already Know

At the beginning of each chapter in this book, you will find an activity that helps you consider what you already know about a particular topic. For instance, Activity 10A in Chapter 10 asks you to consider what you already know about physical fitness. As you read Chapter 10, you may find that you want to change or add to some of your original ideas. You may also find that you know a lot more about the subject than you thought you did.

Write in Your Learning Journal

Throughout this book, you will find references to "your learning journal." A learning journal is a notebook where you write your ideas and comments about your learning as you read and investigate. Writing down your ideas will help make them clear to you and to others, should you choose to share your writing. As well as helping you think, your learning journal will be a record of your learning in science.

Read Actively

Some of the activities in this book ask you to do something as you read. For instance, in Activity 18F in Chapter 18, you are asked to write down information as you read about energy resources.

What makes volcanoes erupt? See Chapter 11.

There are also reading activities that ask you to complete a task while you read. For example, you might be asked to fill in a table, make a flowchart, or draw diagrams. These reading activities will help you learn by allowing you to organize the ideas in a useful way.

Make a Mind Map

In school, at home, and in the workplace, you use your ability to think. There are many ways you can record your thinking, including learning journals, illustrations, flowcharts, poems, and stories. Another way to record your thinking is by using a mind map.

A mind map is drawn like a spider's web. You start a mind map by printing an idea in the center of a page. Next, you think of other ideas that have something to do with the main idea in the center. You print these ideas around the main idea and connect them to the main idea with lines. On the lines, you then print a few words that describe why or how these ideas are connected to the main idea.

Here is the beginning of a mind map that a student drew about machines.

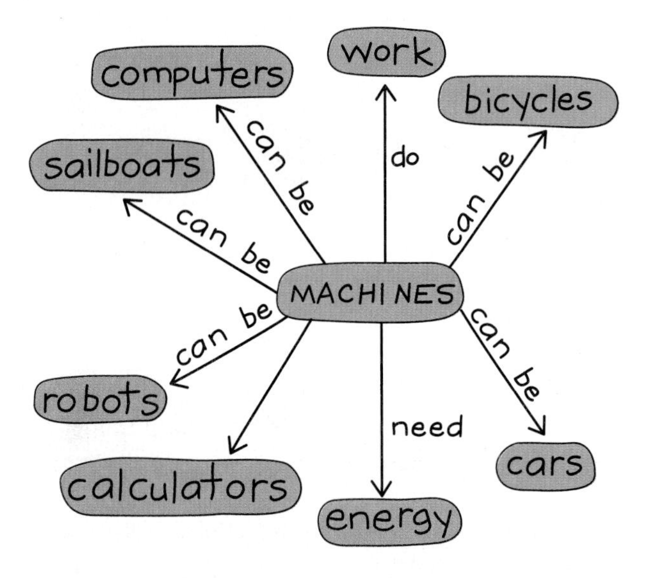

Here is the same mind map about machines with more details added.

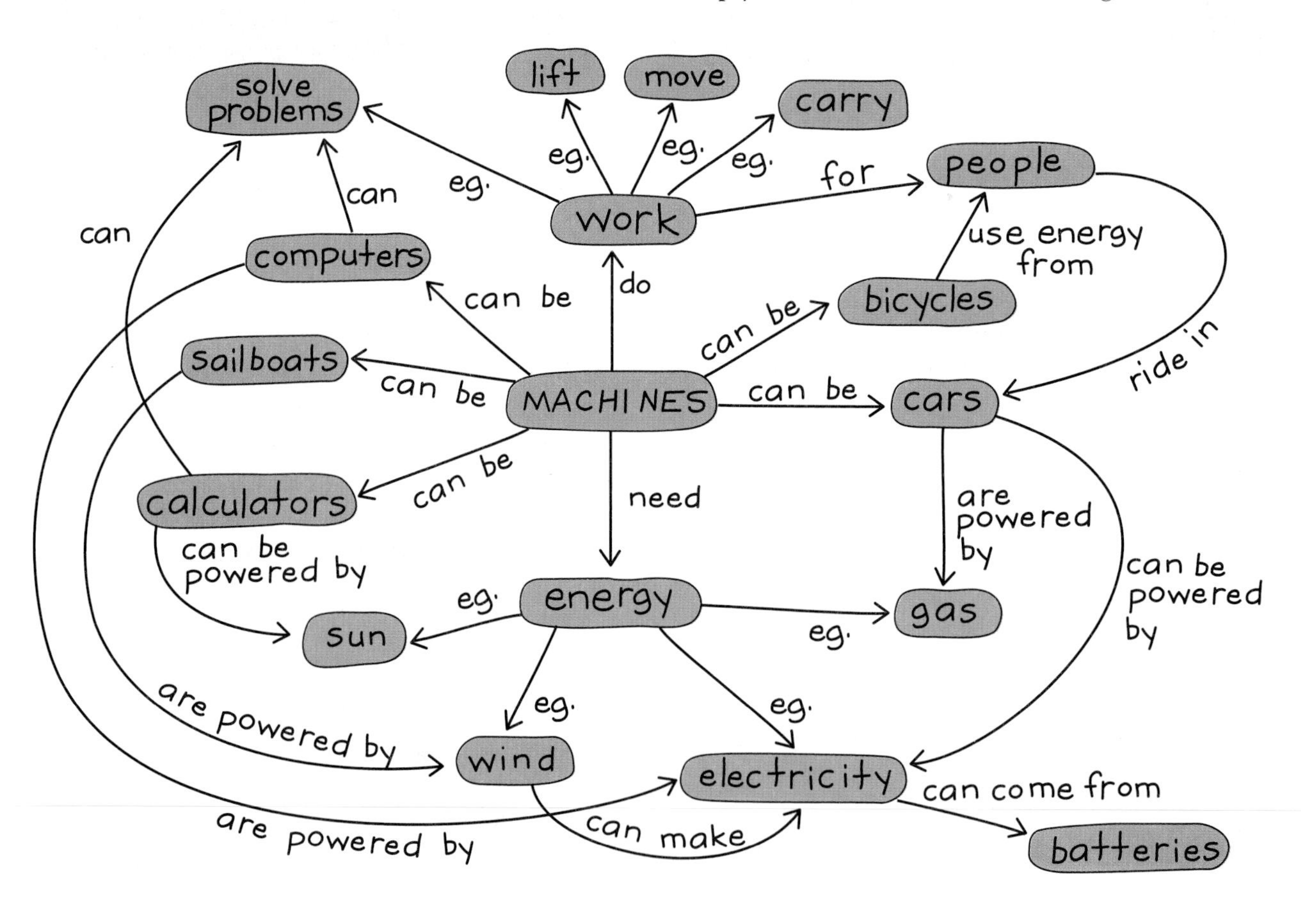

You can see that the student has provided more description for each of the new ideas by printing other ideas around them and connecting the ideas with lines and a few describing words.

Mind mapping is a good way to "brainstorm." You can quickly transfer many ideas from your brain to your paper. When you find an idea that appears in more than one location on your mind map, you can connect it with a line. By the time you have finished mapping, you usually have a giant spider web of ideas and lines. You might use colors and different styles of printing for the different levels of a mind map. You might also add your own drawings and designs to make your mind map attractive.

In this book, you will have many opportunities to practice mind mapping. In fact, at the end of every chapter, you will be asked to draw a mind map of your new ideas. Making a mind map will help you sort information into categories and see

Create a mind map of ideas about sports.

relationships between ideas. The ideas that you include in your mind maps may come from your classroom learning, from reading books, magazines, or newspapers, and from experiences you have had.

Try making a mind map now. In your notebook, print the word "sports" as the main idea for a mind map. Follow the suggestions in this section and your teacher's directions to create a spiderweb of your ideas about sports. Share your ideas with your classmates. Were the same ideas about sports recorded by several students in your class?

Look Back at Your Learning

This book provides many opportunities to review and reconsider your learning and your thinking. The Reflections section at the end of each chapter contains questions that ask you to think about what you have learned or to add to your own ideas. For example, in Chapter 16, you are asked: "Now that you have finished this chapter, have your ideas about the formation of the solar system and the universe changed? If so, describe how they have changed."

Share Your Ideas

In many of the activities in this book you are asked to share your ideas with others, to express what you think you know, and to hear other ideas in return. For example, Activity 23A in Chapter 23 asks you to work with other students to study the water cycle in your area. You are then asked to share what you found out with the rest of the class. If you are afraid to do this because you might be "wrong," then you will miss the opportunity to see whether or not you can explain your thinking to others. So share your ideas. Then listen to the responses and the ideas of others and compare them with your own. You will not only learn more about science, you will also learn how to learn. You will develop skills that will help you long after you have finished this science course.

CHAPTER 1

SCIENCE, TECHNOLOGY, SOCIETY, AND YOU

The boy and the girl in the photograph are ready to do a scientific experiment. They have collected the safety equipment and the materials they will need for the experiment. But before they begin, they have some questions about what they will be doing. Asking questions is an important part of science.

Suppose you decide to use some money you have received for your birthday to buy a new headset for your tape deck. You already know a lot about headsets. But before making a purchase, you plan to ask questions about the price, quality, and design of the headsets that are available. The answers to these questions will help you decide which headset to buy.

Science is about asking questions and then looking for answers in an organized way. Scientists look for the answers to questions by using various process skills. In this chapter, you will review the process skills that you have already learned about. You will also find out about some other process skills. Finally, you will explore how science is part of your everyday life.

ACTIVITY 1A / LOOKING FORWARD

This activity will help you focus on some of the topics you will be studying in science this year.

MATERIALS

a large sheet of paper
writing instruments

SELECTED SCIENCE TOPICS

Changes in Matter
The Sun and the Planets
Chemicals in the Home
The Stars You See in the Sky
Your Body Needs Food
Why Is Energy Important?
Factors Affecting Fitness
Temperature, Thermal Energy, and Heat
When the Earth Shakes
The Soil Ecosystem
The Origin of Continents
The Ocean

PROCEDURE

1. Working with a small group, write "Science" in the middle of a large sheet of paper (Figure 1.1). This word will form the center of your mind map.
2. Write the phrases from the list entitled "Selected Science Topics" around the word "Science." Draw a line from each phrase to "Science."
3. Think about what you have learned about science in earlier grades, as well as what you have learned from newspapers, magazines, books, and other sources of information. Use this information to add two or three subtopics around each topic. For example, you might write "cleaners" and "pesticides" around "Chemicals in the Home." Draw a line from each subtopic to the appropriate topic. Write one or two connecting words on each line to describe how each subtopic is related to the topic.
4. When all the members of your class have finished their mind maps, display them around the classroom.

DISCUSSION

1. How was your group's mind map similar to those of other groups in your class? How was it different?
2. Are there any topics you do not know much about? If so, list those topics in your learning journal.
3. Look at what some of your classmates said about the topics in your list. Write down any information your classmates included on their mind maps that might help you understand the topics more easily. ❖

FIGURE 1.1 ▶
You can use a mind map to focus on what you already know about the science topics you are going to study this year.

1.1 WHAT IS SCIENCE?

You may recall that **science** is both a body of knowledge and a process for gaining more knowledge about the world in which we live. Since science represents a vast body of knowledge, it is impossible for any individual to know everything about science. For this reason, scientists usually specialize in a particular area, or branch, of science.

Just like the branches of a tree, a branch of science is made up of many smaller branches. For instance, one of the main branches of science is biology, the study of living things. A biologist has a good general knowledge of biology but will probably specialize in a particular area of biology, like ecology—the study of the environment. In other words, an ecologist is a biological scientist whose area of expertise is the study of the environment.

Until the middle part of the 20th century, almost every new invention was developed by an individual working alone. Today this is no longer so. Most recent inventions are the result of teams of specialists working together in university, government, or corporate research laboratories. Can you think of a reason for this change?

ACTIVITY 1B / SOME BRANCHES OF SCIENCE

In this activity you will define some branches of science and put them into categories.

MATERIALS

a copy of the sheet entitled "Some Branches of Science"
a dictionary

PROCEDURE

1. On the sheet entitled "Some Branches of Science," use a pencil to lightly number all the terms in the column "Branch of Science" from 1 to 20.
2. Beside each term in the column "Is the Study of," lightly write a letter from A to T. Use your knowledge of science to match as many lettered items in the second column as you can to the numbered terms in the first column so that true statements are formed. For example, the first statement would read, "Biology is the study of life." Use a dictionary to help you match any branch of science that is unfamiliar to you.
3. Compare your completed matched set with those of other students.
4. Using the completed matched set, group the statements into three or more categories. Write the categories in your notebook. The first statement in each category should also act as the title for the entire group. Be prepared to defend your grouping during a class discussion.
5. Refer to the Table of Contents on page vi. Beside each branch of science, write the numbers of the unit and chapter where you might find more information about that branch of science.

DISCUSSION

1. (a) List the branches of science that were most familiar to you.
   (b) How many branches of science were unfamiliar to you?
2. Share your categories with your classmates. Why did you group the branches of science the way you did?
3. Check the Table of Contents on page vi. Which branches of science will you discover more about in this textbook?
4. Write the names of the following careers in your notebook. List the branches of science that people preparing for these careers would study during their training. You may need to list more than one branch of science for some of the careers. ➡

pharmacist	fish and wildlife officer	seismic observer
dietitian	photographer	heart surgeon

5. Suggest the name of a career for each of the descriptions below. Description (a) has been done for you. ❖

Description	Career
(a) a scientist who studies living things	a biologist
(b) a scientist who studies how elements and compounds combine and behave	
(c) a scientist who specializes in the study of animals	
(d) a heart specialist	
(e) an earthquake specialist	
(f) a person who is trained to make weather forecasts	

EXTENSION

■ Choose one of the branches of science that interests you. Use illustrations and labels to prepare a display that will explain the branch of science to someone who knows nothing about it. Include names of any careers connected to this branch of science. Check encyclopedias under the headings "Biology," "Chemistry," "Physics," and so on to learn more about branches of science.

PROCESS SKILLS

Most of us are curious about what goes on around us. We are constantly looking for knowledge that will explain events that we do not understand or that we wish to know more about. Like scientists, we gain knowledge by asking questions, predicting possible answers, performing experiments, examining results to look for causes and effects, and sharing our ideas with others.

Even after scientists have answered a question, they keep on investigating, thinking, and sharing ideas. Often the result is a change in an old explanation, or the formation of a new one. What scientists believe to be true today may turn out in the future to be incomplete or incorrect (Figure 1.2).

Scientists explain unknown events, or revise current explanations, by following a **process**, a set of actions done in a specific order. In the process of doing science, many skills are used. These skills are sometimes known as thinking skills, or **process skills**. A process skill is a procedure, such as observing or measuring, that is used in scientific research (Figure 1.3). You can use process skills to find answers in other subjects as well.

FIGURE 1.2 ▶
The ideas, measurements, and calculations of Copernicus and the investigators who followed him convinced later scientists to revise their view of the solar system to our current sun-centered (heliocentric) view.

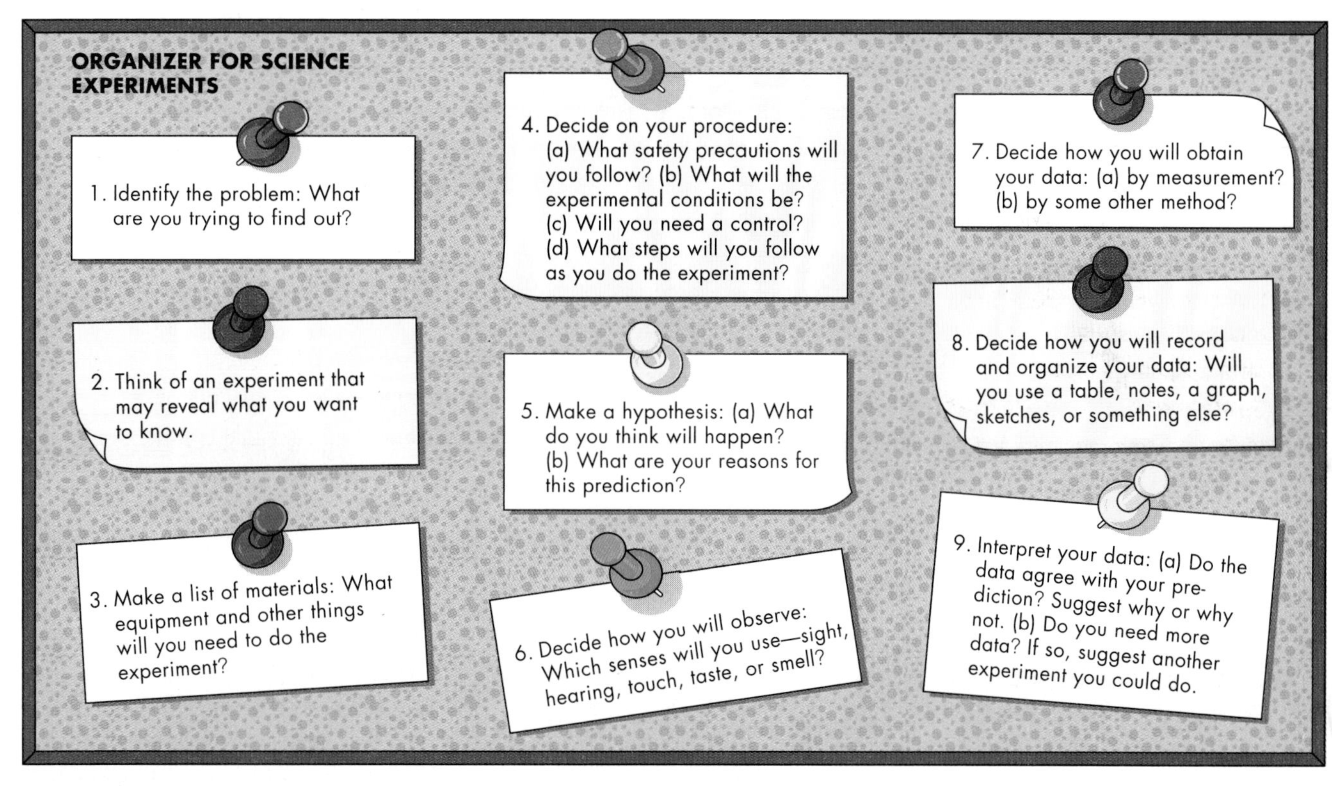

ORGANIZER FOR SCIENCE EXPERIMENTS
1. Identify the problem: What are you trying to find out?
2. Think of an experiment that may reveal what you want to know.
3. Make a list of materials: What equipment and other things will you need to do the experiment?
4. Decide on your procedure: (a) What safety precautions will you follow? (b) What will the experimental conditions be? (c) Will you need a control? (d) What steps will you follow as you do the experiment?
5. Make a hypothesis: (a) What do you think will happen? (b) What are your reasons for this prediction?
6. Decide how you will observe: Which senses will you use—sight, hearing, touch, taste, or smell?
7. Decide how you will obtain your data: (a) by measurement? (b) by some other method?
8. Decide how you will record and organize your data: Will you use a table, notes, a graph, sketches, or something else?
9. Interpret your data: (a) Do the data agree with your prediction? Suggest why or why not. (b) Do you need more data? If so, suggest another experiment you could do.

FIGURE 1.3 ▲
You use many process skills when you design a science experiment.

ACTIVITY 1C / WHAT ARE PROCESS SKILLS?

This activity will help you recall process skills you already know and apply them to a case study.

PART 1 Remembering Process Skills

PROCEDURE

1. On a sheet of notebook paper, write the name and number of each process skill listed in the first column of "Process Skills and Their Definitions" (see page 6).
2. Match each process skill to its definition in the second column by writing the letter of the definition beside the skill.
3. In your learning journal, write the title "Reviewing Process Skills." List each process skill with its correct definition.

PART 2 Identifying Process Skills

PROCEDURE

1. Keeping in mind your knowledge of process skills, carefully read the Stretch Power case study on page 6.
2. List all the process skills you can think of that the students used in their experiment. Beside each process skill, list examples from the case study where the students used the skill.

Process Skills and Their Definitions

Process skills	Definitions
1. Observing	**(a)** using notes, tables, graphs, sketches, reports, or mind maps to record data
2. Classifying	**(b)** stating an explanation that results from an observation
3. Estimating	**(c)** planning and carrying out a series of activities that help solve a problem or find an answer to a question
4. Measuring	**(d)** watching or examining carefully by using your five senses
5. Predicting	**(e)** employing an object, a design, or an idea that helps you understand something that cannot be observed directly
6. Making inferences	**(f)** roughly determining the measurement or nature of something
7. Stating a hypothesis	**(g)** grouping together objects or living things according to the ways they are alike
8. Doing experiments	**(h)** using observations made in the past to describe what you think might happen
9. Using models	**(i)** examining carefully the information obtained from experiments
10. Recording information	**(j)** giving exact information about characteristics such as height, length, mass, or temperature
11. Analyzing data	**(k)** putting ideas and information together so that they can be discussed and more easily understood
12. Organizing data	**(l)** giving a possible explanation of how something happens in nature

Stretch Power: A Case Study

Anand and Janis were discussing what they might do for their science presentation. Anand told Janis that he had had trouble with another science presentation. He was supposed to measure how far a rubber band would stretch when objects of different masses were attached to it, but he had to replace the rubber band because it kept breaking. Each new rubber band caused the results to vary from his previous results. Janis asked if the rubber bands had all been the same thickness. Anand said that although all the rubber bands were the same length, they were of different thicknesses. As their discussion continued, they became curious about the differences in Anand's results. In the end, the lab partners decided to do their presentation on how thickness affects the "stretch power" of rubber bands.

Janis and Anand decided to test whether thicker bands travel farther than thinner ones when they are stretched and then released. They titled their presentation "How the Stretch Power of Rubber Bands Is Affected by the Thickness of the Band." Here are the steps the students followed when carrying out their investigation.

1. They collected the materials needed for their investigation: a meter stick, a 5 m tape measure, a desk, safety goggles, and three rubber bands (all 8.5 cm long), 2, 4, and 6 mm thick.

2. The students moved the desk to an area away from other students and asked their classmates to stay away from the area until they had completed their testing. The partners reminded each other to put on their safety goggles.

3. Before the students began their experiment, they each predicted which rubber band would travel farthest. Anand predicted that the 6 mm thick rubber band would travel farther than the thinner rubber bands. Janis predicted that the thinnest rubber band would travel farthest.

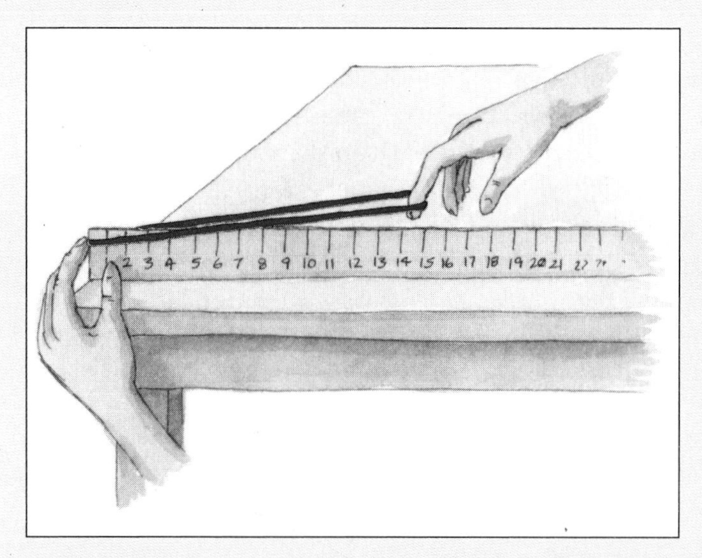

FIGURE 1.4
Why is it important to stretch each rubber band exactly the same amount?

4. The meter stick was held on its edge so that the 0 cm line was at the very end of the flat desktop. Then the thinnest rubber band was placed over the ruler so that one end just caught the end of the ruler. Janis pulled back the free end of the 2 mm thick rubber band with her index finger until it exactly reached the 15 cm line (see Figure 1.4). Making sure the rubber band was flat against the ruler, Janis released the band.

5. Anand carefully measured the distance from the point where the rubber band first hit the floor to the end of the desktop and then recorded the flight distance of the first trial. The 2 mm thick rubber band was tested four more times, and the flight distance was recorded each time. An average flight distance was determined for the thinnest rubber band.

6. Using exactly the same procedure, the students tested the 4 mm thick band next, and then the widest band. The results were carefully recorded for all of the flight distances.

7. The lab partners collected all their records and materials and returned the desk to its place.

8. Janis and Anand examined their data and compared the average flight distance of each of the rubber bands. Using a different color for each of the rubber bands, they plotted the flight distances on a graph.

9. The partners made a successful presentation using a large poster that stated the title of their investigation, the materials they used, their predictions, their data, their calculations, and the conclusions they reached.

DISCUSSION

1. Compare your list with the lists made by other members of the class. Add any examples you may have missed to your list.
2. Did you have trouble identifying any of the skills? If so, go back and review those skills.
3. List the students' activities that are related to safety precautions. ❖

MORE ABOUT PROCESS SKILLS

How often have you heard the expression "That's not fair"? At home, in the classroom, in the schoolyard, or during a sports event, people want things to be fair for everyone. Scientists are concerned that experiments used to test hypotheses are "fair." Many possible factors, or **variables**, can affect the results of an experiment. To ensure that an experiment is fair, all variables except one are kept the same. Such an experiment is called a **fair test**.

Figure 1.5 shows four examples of how two types of flooring could be tested to see which type is bouncier. These examples show several variables that could affect the results of the tests, including the type of ball that was dropped, the height from which each ball was dropped, the way each ball is dropped, and the type of flooring. For the test to be fair, all variables should be kept the same except for the type of flooring. The variables that are kept the same are said to be **controlled variables**. A fair test in science is called a **controlled experiment**.

FIGURE 1.5 ▶
The basketball team at your school has requested that the floor for your school's new gymnasium be as bouncy as possible. Your science class has been given samples of wood and rubber flooring to test for bounce. Look carefully at each of the tests in this figure. Is that the test you would choose? Is the test fair or unfair? Explain why.

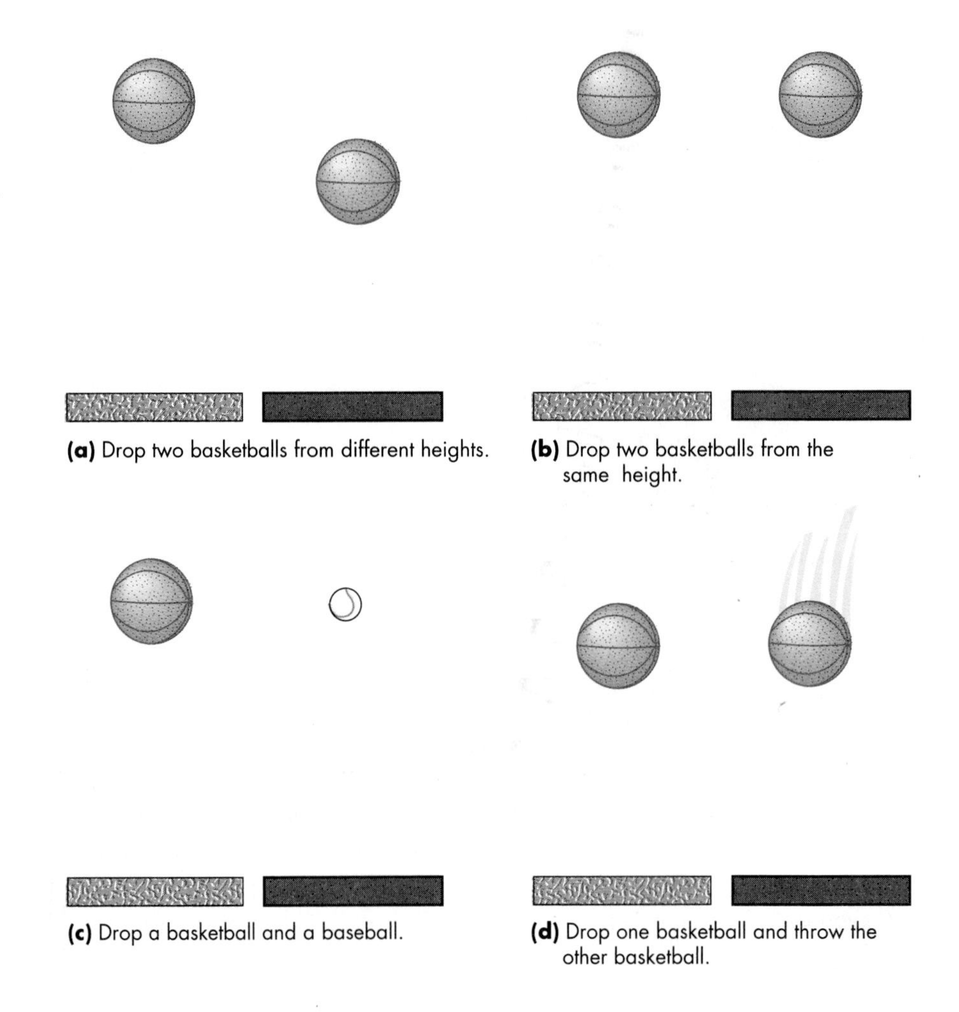

(a) Drop two basketballs from different heights.

(b) Drop two basketballs from the same height.

(c) Drop a basketball and a baseball.

(d) Drop one basketball and throw the other basketball.

DEFINING OPERATIONALLY

It is common for two or more objects or products to be compared. Sometimes people say that one product is the "best," or "better" than other products. If no other data are provided to support the claim, no one knows why the product is best or how much better it is than other products.

Scientists avoid these types of comparisons by using a method of comparison called **defining operationally**. Defining operationally means using observable or measurable characteristics to describe an object or event. Suppose that you want to compare Toothpaste A with Toothpaste B. Instead of simply stating that "A is better than B," you might say that Toothpaste A results in whiter-looking teeth—a characteristic that can be observed. Or you might say that using Toothpaste A results in 60 percent fewer cavities—a characteristic that can be measured.

EXTENSION

■ Find an advertisement for a product or an event in a newspaper or magazine that implies you should buy the product instead of some other product or attend the event instead of another event. Paste the advertisement in your notebook. Read the advertisement carefully. Number each claim made about the product or event and write the numbers in your notebook. Decide whether each numbered claim is either observable, measurable, or neither, and write the appropriate term beside each number. Write a short statement that shows why you think the advertisement has or has not defined the product or event operationally.

ACTIVITY 1D / "DEFINING IT OPERATIONALLY"

This activity helps you define objects operationally.

PROCEDURE

1. Copy Table 1.1 into your notebook.
2. In column two, rewrite each statement in column one so that the objects or events are defined operationally.

Statement of comparison	Defining it operationally
Detergent X is better than detergent Y.	Detergent X washes 20 percent more dishes than Detergent Y.
Gum A is better than Gum B.	
Margarine is better than butter.	
Cold cream is better than soap.	
This ice cream is best.	
Swimming is better exercise than walking.	

TABLE 1.1 ***Sample Data Table for Activity 1D***

DISCUSSION

1. In a small group, compare your statements with those of other group members.
2. Rewrite any statements that are not based on observed or measured data. ❖

QUALITATIVE AND QUANTITATIVE PROPERTIES OF AN OBJECT

When you describe an object in observable and measurable terms, you are expressing two kinds of properties, or characteristics, of the object. A **qualitative property** is a property that supplies information about the appearance, smell, taste, feel, or sound of an object. For example, a description of the qualitative properties of an apple might include: it is red, it tastes sweet, it feels smooth, and it makes a crunching sound when you bite into it. A property that answers the question "How much?" or is the result of exact measurement is called a **quantitative property**. For example, quantitative properties of an apple might include: it weighs 300 g, the thickest part of the apple has a circumference of 28 cm, and it has a volume of 375 cm^3. (A list of the quantities, units, and symbols that are used in this book can be found in Appendix A on page 523.) The next activity allows you to practice using qualitative and quantitative properties to describe an object.

ACTIVITY 1E / DESCRIBING A BURNING CANDLE

In this activity you will describe a burning candle.

MATERIALS

safety goggles
apron
ruler
candle
candle-holder

■ **Do not use your sense of taste in this activity.**

■ **Never leave a lighted candle unattended.**

PROCEDURE

1. Divide a sheet of note paper into two equal columns. Title the first column "Qualitative Properties" and the second column "Quantitative Properties." Title your table "A Description of a Candle."
2. Use your senses to observe as many properties of the candle as you can and note them in the appropriate column in your table.
3. Using the ruler, note as many properties of the candle as you can and record them in your table.
4. Put on your safety goggles and your apron.
5. Place the candle securely in the holder and set it in the middle of the lab table. Make sure the area around the candle-holder is clear, then carefully light the candle.
6. Repeat Steps 2 and 3.

DISCUSSION

1. Compare your list of properties with the lists made by other members of the class. Make corrections and additions to your list if necessary.
2. Did you list properties that you inferred? For example, if you said that the candle "tasted like wax," you would have inferred that property because you did not taste the candle. Remove from your list any properties you inferred.
3. What tools besides the ruler could you have used to measure the candle? List those tools and give an example of a property you could have recorded with each tool. ❖

USING THE SKILLS OF SCIENCE

Science is more than a collection of facts about the world. It is also a way of thinking about the world and investigating things that puzzle us. Scientists admit that there are many things in the world they do not understand. They also realize that ideas about the world that seem correct today may turn out to be incorrect tomorrow. Scientists take the time to observe things around them, suggest possible explanations for their observations, and then conduct experiments to test these explanations. When you notice something that puzzles you and take the time to think through possible solutions, you are doing exactly what scientists do. The skills of science help you discover answers to events that puzzle you (Figure 1.6).

FIGURE 1.6
Suggest what might be puzzling each of the students in these cartoons. How might science skills help the students discover answers to their questions?
What have you done in the past few days that illustrates how science is part of your life? Share this experience with other members of the class.

(a)

(b)

▶ REVIEW 1.1

1. Define each of the following terms: "science," "process," and "process skills."
2. Write the name of the branch of science that deals with each of the following topics. Beside each name, write the term for the scientists who specialize in that area.
   life
   the environment
   the Earth
   elements and compounds
   chemistry of life
   matter and energy
   space
   earthquakes
   weather
   motion
3. Explain why you agree or disagree with the following statement. "Science is constant. It never changes."
4. Use two or three sentences that show you understand the meaning of the terms "variable," "fair test," "controlled," and "controlled experiment."

5. Write two statements that compare the performance of two brands of batteries to show you understand the skill of defining operationally. The second statement should use the skill of defining operationally, whereas the first one should not.

6. Define the terms "qualitative property" and "quantitative property." Supply an example with each definition to show you understand the difference between these two kinds of properties.

■ Lorne Whitehead, a physicist, has invented the Light Pipe, which brings sunlight into places without windows through the imaginative use of prisms. Today, his light-without-heat invention is used in hospitals and chemical laboratories.

■ 1.2 SCIENCE, TECHNOLOGY, AND SOCIETY

Science, technology, and society are important to us in our daily lives. **Technology** is the use of scientific knowledge to solve a problem. Our **society** is made up of many individuals who must work together. Science, technology, and society are all connected to one another. When you are faced with a problem, science skills can help you understand the problem and technology may help you deal with it. For example, suppose you want to save half a cut onion for future use. You decide to put it in the refrigerator to keep it fresh. However, you remember that the strong onion odor can flavor other foods, like milk, if they are stored together in the refrigerator. To solve the problem, you place the onion in a container designed to seal in the onion's odor and to keep it fresh.

A member of society has a responsibility to learn about and use science and technology to make informed personal decisions. For instance, suppose that you get tar on the bottom of your running shoe and track it onto the kitchen floor. You can scrape the tar off the floor with a knife, or you can use a strong cleaner, developed by science and technology, to clean the tar off. But what if your younger sister has an allergy to the cleaner? And how will you dispose of any unused cleaner later? What effect will it have on the environment? When science skills identify a problem, and technology is used to solve the problem, every effort must be taken to ensure that other members of society are not negatively affected by the solution. In this way, science, technology, and society are interconnected.

Members of society must often make personal decisions about an **issue**, a topic about which people have different points of view. Many issues are neither easy to describe nor easy to understand, because people have different personal, cultural, religious, political, or economic points of view about the same issue. For example, if the number of salmon in Washington were steadily decreasing, should the quota (the number of salmon caught over a particular period of time) for sport or commercial fishing be reduced? A sport fisher who enjoys salmon fishing four days a year might not see how lowering the quota from four to two salmon a day would solve the problem. A commercial fisher whose family's only source of income is based on the number of salmon caught might feel that reducing the quota is unfair. An ecologist might say that reducing the quota is the only way to save the supply of salmon for the future.

People with different viewpoints on the same issue often experience strong emotions such as anger or frustration. These emotions sometimes cause individuals to make decisions about an issue that they might not have made had they approached it in the same way as you use science skills to solve a problem. Science can allow people to understand the issue, and technology can help them react to it. When you must make a decision about an issue, it is important to identify the issue, collect as much information about it as you can, analyze the information you have collected, and then make a decision based on the information you have collected and analyzed (Figure 1.7).

■ Wendy Murphy, a medical research technician, realized during the devastating Mexico City earthquake of 1985 that there were no suitable evacuation stretchers to transport tiny babies to hospitals for emergency treatment. In 1989 she invented the Weevac 6, a rescue stretcher that can transport six infants at a time.

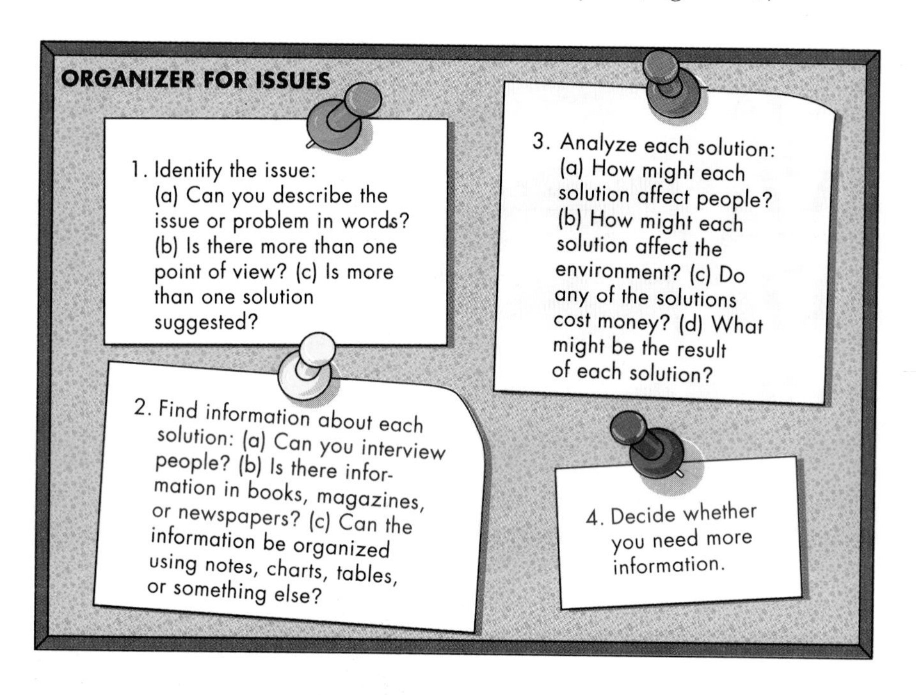

ORGANIZER FOR ISSUES
1. Identify the issue: (a) Can you describe the issue or problem in words? (b) Is there more than one point of view? (c) Is more than one solution suggested?
2. Find information about each solution: (a) Can you interview people? (b) Is there information in books, magazines, or newspapers? (c) Can the information be organized using notes, charts, tables, or something else?
3. Analyze each solution: (a) How might each solution affect people? (b) How might each solution affect the environment? (c) Do any of the solutions cost money? (d) What might be the result of each solution?
4. Decide whether you need more information.

■ Describe one or two issues about which you have strong emotions. Briefly state the facts that you know about the issue, as well as any other points of view that are different from your viewpoint. State how you feel about the issue. Share your ideas with other members of your class.

FIGURE 1.7 ◀
When investigating an issue that affects people, think about following these steps.

ACTIVITY 1F / "THE ISSUE IS . . ."

This activity allows you to analyze an issue.

PROCEDURE

1. Work with four or five classmates in a small discussion group to do this activity.
2. Read the Background and What Was Said on page 14 and then decide what you think the issue is. Write the issue at the top of a sheet of notebook paper.
3. Are there any conflicting points of view? If so, record what they are and who expresses them.
4. Are there any differences in the statements of facts? If so, record what they are and how the statements differ. ➡

5. What will your group choose to do? You may decide that you need to gather more information before you can make a decision. Prepare a report that summarizes the conflicting opinions about the issue and then states your group's decision and the reasons for it.

6. Select a spokesperson to present your report to the rest of the class.

Background

Scientists have commonly used animals in experiments to test new surgical techniques or new drugs for people. Some people are now questioning whether scientists should use animals in this way. Imagine that a rare but powerful virus has been discovered that can lead to the death of many people, particularly the very young and the very old. A medical research team at a local university is working on a cure for this virus. The team of scientists is seeking funding to develop a new drug that must first be tested on rhesus monkeys. During the experiments, some of the monkeys will probably die. A committee examining the research team's request listens to presentations from local residents before making a decision about whether to grant the funding.

What Was Said

Derek Yin, representing the medical research team at the university: This is a very serious situation. We believe we have found a vaccine that will cure this infection. We cannot test the vaccine on humans. The rhesus monkey reacts to the virus in ways that are similar to the ways that humans react. We must continue this research without delay.

Cynthia Illingsworth, president of AHRT (Animals Have Rights Too): This funding should not be granted. Hundreds of laboratory rats and mice have already died to develop this vaccine. Now they want to test it on monkeys, and who knows how many of them will die. Laboratory animals are being treated inhumanely.

Steve Perkins, local resident whose three-year-old daughter is seriously ill with the viral infection: I would like to see the funding granted. It may not be developed in time to save my daughter, but if a cure isn't discovered soon, a lot more children are going to die. I love animals, but I love my children more.

Indira Singh, laboratory-animal technician: I've worked for five years at the university and for a veterinarian for three years before that. I'm responsible for the day-to-day welfare of the laboratory animals. We are trained in providing humane, responsible care for the animals. Most of us have genuine affection for our animals. We can and do report cases of animal abuse.

Hans Pikkert, spokesperson for research methods that do not use animals: Considerable progress has been made in the development of non-animal research methods in many areas. However, this is one area that still requires the use of animals if a cure for this infection is going to be found.

DISCUSSION

1. How was your group's decision different from that of other groups? How was it the same as other group decisions?

2. Did personal emotions affect your decision in any way? Explain your answer.

3. Write a letter to the editor of a local newspaper that either supports or opposes the use of animals in research. ❖

ENVIRONMENTAL CHILDREN'S ORGANIZATION

Severn Cullis-Suzuki, one of the founders of the Environmental Children's Organization (ECO), believes that teenagers can affect decisions made about the environment. ECO was founded after Severn and her family visited a small native village in the Brazilian rain forest. There Severn saw how the forest was being destroyed by burning, mining, and road construction.

Members of ECO at the Earth Summit, the United Nations Conference on Environment and Development, in Rio de Janeiro, Brazil. From left to right, Vanessa Suttie, Severn Cullis-Suzuki, Michelle Quigg, Morgan Geisler, and Severn's sister, Sarika Cullis-Suzuki.

When she got back home, she told her friends what she had seen. They decided to form ECO to do something about the threat to the environment. "We wanted people to start paying attention to what was happening. We are kids, and the environment is our future."

When the members of ECO heard that the Earth Summit, the United Nations Conference on Environment and Development, was being held in Rio de Janeiro, Brazil, in June 1992, they were determined to attend and make their voices heard. They worked for over a year to raise enough money to send Severn, Morgan Geisler, Vanessa Suttie, Michelle Quigg, and Michelle's mother to the Earth Summit.

ECO obtained one of the 600 booths for nongovernmental organizations at the summit. The group members handed out materials and talked to delegates about their concerns about the environment. People listened. They spoke at the Earth Parliament for children, women, native people, and elders. Finally, the ECO members were invited to speak before government leaders from around the world.

How has this experience affected the ECO members?

"I think about science and the environment all the time now. The things I do every day obviously affect the environment," says Vanessa.

"I've become very concerned about the environment," states Morgan. "When I do choose a career, I'll want to consider what its effect on the environment will be."

"We've learned to question things a lot more," Severn comments. "Kids can make a difference, if they try!"

What concerns about the environment do you have? How do you think you can make a difference?

REVIEW 1.2

1. Explain why you agree or disagree with the statement "Scientists have all the right answers."
2. Write one or two statements to show how science, technology, and society are interconnected.
3. Define the term "issue."
4. Explain why people's points of view might differ on the same issue.
5. Suggest the steps you might take before making a personal decision about an issue in which there are several points of view or possible solutions to the issue.

CHAPTER REVIEW

KEY IDEAS

■ Science is both a body of knowledge and a process for gaining more knowledge about the world we live in.

■ Because science includes such a vast body of knowledge, a scientist generally specializes in a particular branch of science.

■ Process skills are used by scientists and other people to look for answers to questions that puzzle them.

■ All variables except one are controlled when a fair test, or controlled experiment, is conducted.

■ Scientists use a method of comparison called defining operationally. This method uses observable or measurable characteristics to describe objects or events.

■ Science, technology, and society are all connected.

■ Members of society can use science to help them understand an issue, and technology to help them react to it.

VOCABULARY

science
process
process skill
variable
fair test
controlled variable
controlled experiment
defining operationally
qualitative property
quantitative property
technology
society
issue

V1. Imagine that you are the host of a science quiz program. Use the words in the vocabulary list to compose six questions that could be used on the quiz program. Supply the answers to the questions as well.

V2. Use any six vocabulary words to create at least two cartoons or illustrations. You may use captions or sentence bubbles to help explain the meaning of your illustration. Your illustration should show the meanings of the words you use.

CONNECTIONS

C1. Choose one of the questions below and design an experiment to find out the answers.
(a) Does wearing nail polish make nails long and strong?
(b) Why doesn't gum stick to teeth?

Draw a line down the right-hand side of a page in your notebook, making a narrow margin. After you have designed your experiment, indicate in the margin which process skills you would use in your experiment. Also indicate what safety precautions you would observe in carrying out your experiment.

C2. Explain why qualitative and quantitative properties are important when you use the method of defining operationally to compare two objects.

C3. Outline a plan for a controlled experiment to test the strength of three brands of paper towels when they are wet. Be sure to state which variables you would control and which variables would not be controlled. State reasons why you think your experiment is a fair test.

C4. State an appropriate unit to measure each of the following:
(a) the volume of juice in a bottle
(b) the amount of work done in a particular task
(c) the area of a farm
(d the volume of a car trunk
(e) the amount of electricity used in a home for a month
(f) the distance around an apple orchard

C5. Outline the steps you might take to find out more about a scientific issue you were interested in.

EXPLORATIONS

E1. Describe three situations in your life in which you have used science, technology, or both to make a personal decision or solve a problem.

E2. Give at least one reason why you agree or disagree with the person's point of view in each statement that follows.
(a) "It takes too much time to store cans and then take them to a recycling depot," said Akira Apartment-Dweller.
(b) "I don't care what you say about this powder being a pollutant, I don't want ants on my patio," said Grace Gardener.
(c) "I prefer to buy frozen orange juice and mix it up at home, rather than buying a new jug every time," said Jean Jucier.
(d) "I used to think smoking was cool, but the facts are clear. Smoking harms my health and the health of those around me, " said Nikki Non-Smoker.

E3. At one time, most people believed that the Earth was the center of the solar system. Name another belief that has changed because of new scientific discoveries.

E4. Find an article in a newspaper or a magazine that discusses an issue. Analyze the facts and the points of view presented in the article. After your analysis, state what you think might be the best way to resolve the issue.

E5. Prepare a presentation or report describing the contributions of two American scientists in different branches of science.

REFLECTIONS

R1. Choose one of the following topics and write a report on it. Your report should be at least three-quarters of a page but not more than one page. Be sure that your report is clear and that it outlines your position.
(a) Science Is Important in My Everyday Life
(b) Science and Technology Will Play an Important Part in the Survival of the Planet

R2. Recall an issue in which you were personally involved in the past. What decision did you make about the issue? Why did you have the point of view on the issue you did? State why you would or would not make the same decision today.

INVESTIGATING MATTER

What do all the things in this picture have in common? If you said they are all examples of matter, you are correct. Another factor they have in common is that they will change over time. The children will grow into adults, the balloons will go flat, and the food in the picnic basket will soon go through a series of chemical reactions as it is digested. How might other examples of matter shown in this picture change over time?

This unit is about the branch of science called chemistry. The science of chemistry includes investigations of matter, how matter changes, and how matter is used in industry and everyday life.

You will first examine how matter can change. Then you will learn how scientists explain these changes and how they use a form of chemical shorthand to communicate their ideas. Next, you will gain some practical knowledge about how changes such as cooking and rusting can be controlled. Finally, you will learn about the matter in household products. The way in which people use matter is an issue that affects all societies on our planet. As you study the science of chemistry in this unit, you will be building a foundation of knowledge and skills that you will find useful for the rest of your life.

UNIT I

CHAPTER 2

CHANGES IN MATTER

Change is everywhere around you. Your classroom becomes noisier or quieter. You work harder or not so hard. Your performance in math improves or gets worse. You become happier or sadder. Outside, the weather shifts, and it gets darker. Meanwhile, taxes increase and technology advances. These are changes in energy, feelings, and abstract ideas.

Matter is constantly changing too. For example, wood rots, snow melts, cookies crumble, paper yellows, buildings collapse, fireworks explode, concrete hardens, cake becomes light and fluffy in the oven—the list is almost endless.

In this chapter, you will learn about changes in matter. You will investigate changes in various materials in your classroom, identify kinds of changes, and explain what happens as matter changes.

ACTIVITY 2A / WHAT CHANGES?

In your everyday life, you are familiar with dozens of changes to matter. In a small group, brainstorm as many changes as you can. Make a list like the one on the right, using a different noun and a different action word for each example.

Concrete hardens.
Wood rots.
Snow melts.
Cookies crumble.
Paper yellows.
Buildings collapse.
Fireworks explode.

After you have completed your list, classify the changes on the list into two categories—those in which you think new materials are produced, and those in which you think no new materials are produced. ❖

2.1 CLASSIFYING MATTER AND CHANGES IN MATTER

Matter is anything that has mass and volume, including all solids, liquids, and gases. To understand changes in matter, you need to know what the sample of matter was like before the change and what it is like after the change. You also need to be able to describe and classify matter. There are many ways in which matter could be classified, but scientists have found it useful to classify matter according to its properties, or characteristics. **Properties** are ways of describing things so that we can identify them.

Any kind of matter can be classified as a mixture or a pure substance.

A mixture**, such as a chocolate chip cookie, **contains two or more substances. The properties of this kind of mixture vary in different samples.

A pure substance**, such as table salt, **has properties that are always the same.

Mixtures are further classified as mechanical mixtures, suspensions, or solutions.

***In a mechanical mixture,** such as granola, you can easily see two or more parts of the mixture.*

***In a suspension,** such as tomato juice, solid pieces are scattered throughout a liquid.*

***In a solution,** such as sea water, the substances are so completely mixed that they look like one substance.*

Pure substances are further classified as either elements or compounds.

***An element,** such as gold, cannot be broken down into any simpler substances.*

***A compound,** such as sugar, is a pure substance that is made up of two or more elements that are chemically combined.*

Figure 2.1 shows the relationship of the five types of matter. In this chapter and in the following chapters, you will learn more about two types of matter—elements and compounds—and how they change.

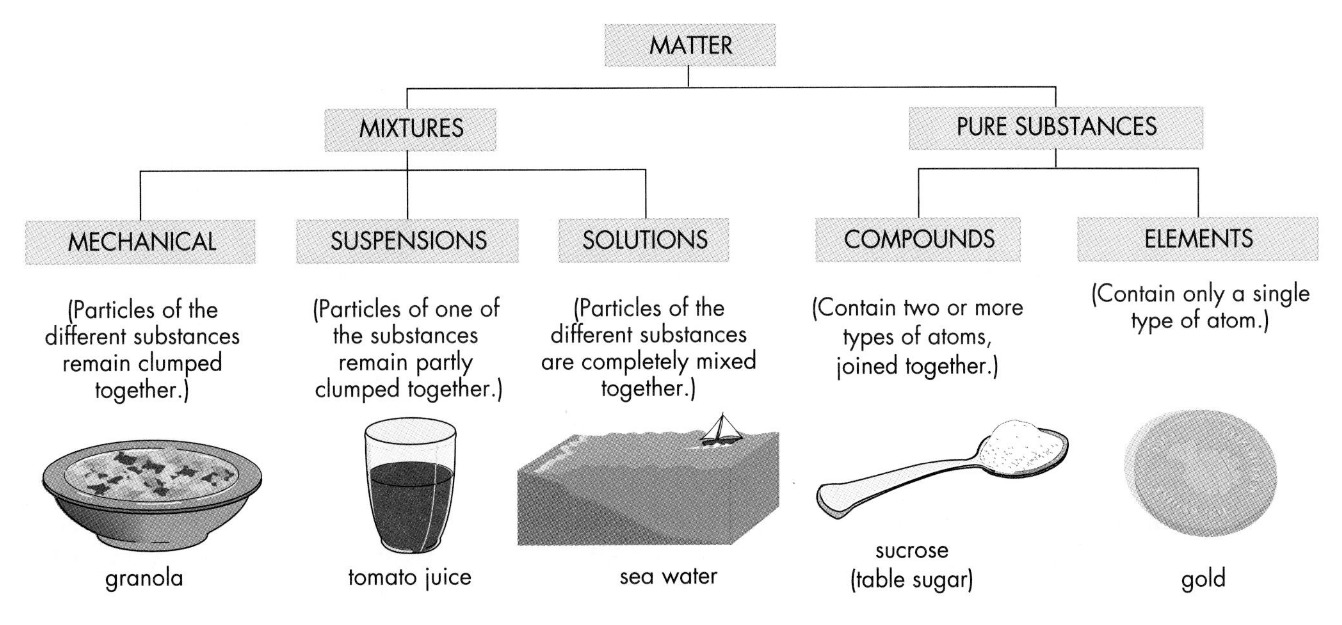

FIGURE 2.1
Classification of matter

PHYSICAL CHANGES AND CHEMICAL CHANGES

When matter changes, what happens to it? After some changes, matter may still have the same properties. When you tear paper into small pieces, for example, each piece still has the properties of the original paper. When a balloon bursts, the rubber is still the same color and is still just as soft and stretchy as it was in the balloon (Figure 2.2). When ice melts, the water is still the same substance—it can easily be changed back into ice if it is cooled. These are **physical changes,** changes in which no new substance is produced.

FIGURE 2.2
The skins of a burst balloon and an inflated balloon have the same properties.

During some changes, a new substance may appear. For example, if you leave your bicycle out in the rain, a reddish-brown crumbly substance may form on it. This substance—rust—is quite different from the shiny, hard metal that was used to make the new bicycle. Changes such as this, in which new substances are produced, are called **chemical changes** (Figure 2.3). If one or more new substances are produced—substance(s) with properties that are different from the properties of the starting material(s)—then a chemical change has occurred. Chemical changes may involve energy; heat or light may be given off during the change. Most chemical changes, such as the one shown in Figure 2.3, cannot be easily reversed. Examples of physical and chemical changes are listed in Table 2.1.

Physical changes	Chemical changes
Water freezes.	A candle burns.
Ice melts.	The metal on a car starts to rust.
Sugar dissolves in water.	A piece of wood burns in a fireplace.
Frost forms on windows.	A firecracker explodes.
Detergent removes grease from a dirty pot.	Concrete becomes hard after it is poured.
The burner on an electric stove glows red.	An egg becomes hard when it is cooked.
Coffee changes color when cream is added.	

TABLE 2.1 ***Examples of Changes in Matter***

FIGURE 2.3
What happens as a log burns? Can you "unburn" a log?

ACTIVITY 2B / OBSERVING CHANGES IN MATTER

In this activity, you may observe physical changes, chemical changes, or situations where nothing appears to change.

MATERIALS

Part I
safety goggles
apron
gloves
small piece of steel wool
two pieces of copper wire (2 cm long)
tongs
Bunsen burner
dilute hydrochloric acid
four test tubes in rack
magnesium ribbon (2 cm strip)
sodium carbonate solution
calcium chloride solution
copper sulfate crystals
Pyrex test tube
test tube holder
water in dropper bottle

PART 1 Student Experiments

PROCEDURE

1. In your notebook, draw two data tables like Tables 2.2 and 2.3. Use them to record the properties of the starting materials and any changes that occur.
2. Put on your safety goggles, apron, and gloves.
3. With the steel wool, polish a piece of copper wire. Examine it carefully and record the properties of the wire in your table. Pick up the wire with tongs and hold it in the hottest part of a Bunsen burner flame for 5 seconds. Remove the wire, let it cool, and examine it again. Record your observations in your data table.

Procedure step	Starting materials	Properties
3.	Copper wire	
4.	Dilute hydrochloric acid	
5.	Magnesium ribbon	
6.	Sodium carbonate solution	
	Calcium chloride solution	
8.	Copper sulfate crystals	

TABLE 2.2 ***Properties of Starting Materials in Activity 2B***

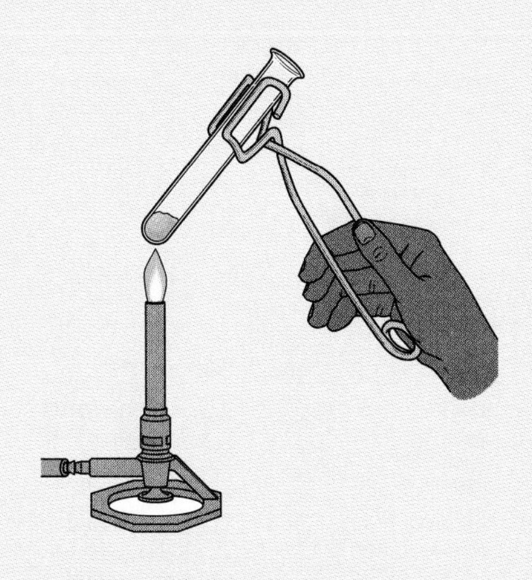

FIGURE 2.4
Heating copper sulfate

Procedure step	Description of change	Type of change (physical or chemical)	Reasons for classification as physical or chemical
3.			
4.			
etc.			

TABLE 2.3 ***Sample Data Table for Activity 2B***

■ **Dilute hydrochloric acid is very corrosive and can cause damage to eyes and skin. If you get any in your eyes or on your skin, immediately rinse with cold water for 15 or 20 minutes and inform your teacher.**

■ **Copper sulfate is poisonous. When heating the test tube, hold it pointing away from yourself and everyone else.**

4. Carefully pour dilute hydrochloric acid into a test tube to a depth of about 4 cm. Polish the second piece of copper wire as you did in step 3 and carefully place it in the dilute hydrochloric acid. Leave it in the acid as you continue with Steps 5 to 7. Then observe the wire carefully and describe any changes in your data table.
5. With the steel wool, polish a piece of magnesium ribbon. Examine it carefully and record the properties of the magnesium ribbon in your data table. Pour dilute hydrochloric acid into a test tube to a depth of about 4 cm. Add the magnesium ribbon to the test tube; describe any changes.
6. Carefully pour sodium carbonate solution into one test tube and calcium chloride solution into another, both to a depth of about 2 cm. Examine each solution and record the properties of each in your data table.
7. Slowly and carefully pour one solution into the other. Describe any changes that you observe in your data table.
8. Place approximately 2 g of copper sulfate crystals in a Pyrex test tube. Hold the test tube as shown in Figure 2.4 and heat the crystals for 10 s. Allow the tube to cool, then heat it again. Replace the test tube in the rack and allow it to cool completely. Record your observations in your data table.
9. Add a few drops of water to the crystals. Record your observations in your data table.
10. Dispose of the materials according to your teacher's instructions.
11. Wash your hands thoroughly after you have completed the activity.

MATERIALS

Part II
ammonium chloride in Pyrex test tube
test tube holder
glass wool
Bunsen burner

DISCUSSION

1. Why did you have to wear safety goggles, an apron, and gloves to carry out these experiments? ➡

■ **Your teacher will carry out Part II of the investigation if the classroom has good ventilation.**

2. Using the steel wool in Steps 3, 4, and 5 removed the evidence of a change. Why was it better to polish the copper wire and magnesium ribbon before studying them?.
3. (a) List all the solid substances used or produced in this experiment (If you do not know the name of the substance, include only a description in your list.)
   (b) Repeat part (a) for all liquids and all gases.
4. In Step 6, solutions of sodium carbonate and calcium chloride are used. Suggest a reason why these solutions are used rather than solid sodium carbonate and solid calcium chloride.

PART II Teacher Demonstration

PROCEDURE

1. Your teacher will place one scoop of ammonium chloride in a Pyrex test tube. The test tube will be plugged with glass wool to slow the escape of any gas (Figure 2.5). The bottom end of the test tube will be heated gently over a flame for 3 seconds only and then allowed to cool.
2. Record your observations in your notebook.

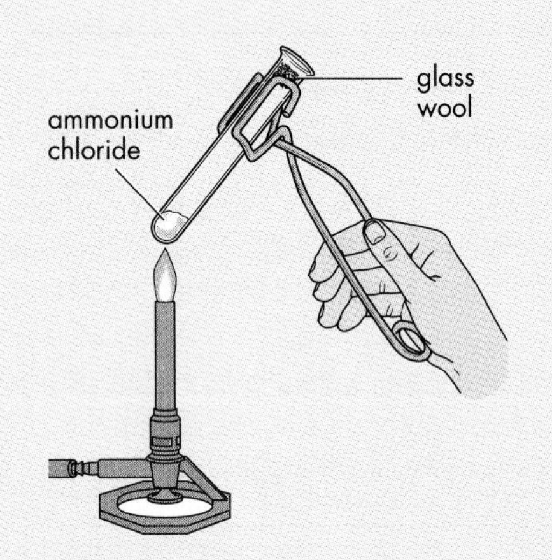

FIGURE 2.5
Is there evidence of a change as the test tube cools?

DISCUSSION

1. What are some characteristics of physical changes? Give an example of a physical change with each characteristic.
2. What are some characteristics of chemical changes? ❖

DISTINGUISHING BETWEEN A PHYSICAL CHANGE AND A CHEMICAL CHANGE

It is not always easy to tell the difference between a physical change and a chemical change. Some chemical changes may produce flames, sparks, and noise (Figure 2.6), but many do not. The chemical changes that take place when you bake a cake, polish brass, or clean the oven are much less spectacular. However, certain clues may tell you that a chemical change has occurred:

- A new color may appear.
- Heat or light may be given off.
- Bubbles of gas may be formed.
- Solid material may form in a liquid.
- The change may be difficult to reverse.

Any of these clues could also be part of a physical change. You need to consider several clues in order to determine which type of change has taken place.

FIGURE 2.6
This welder is using a chemical change to create a physical change. How does his oxyacetylene welding torch work? What are the two changes?

► REVIEW 2.1

1. Give two examples (not shown in Figure 2.1) of each of the following:
   (a) an element
   (b) a compound
   (c) a solution
   (d) a mechanical mixture
2. Use a reference book to find out what gases are in the air. Then decide whether air is a pure substance or a mixture. Explain your answer.
3. Suggest five clues you would consider before deciding whether a change that you observe is a physical change or a chemical change.
4. Which of the following cause physical changes? Which cause chemical changes? Defend your answers.
   (a) making a pot of tea
   (b) bleaching a stain
   (c) cutting the lawn
   (d) using silver polish to polish a silver spoon
5. In Figure 2.7, you can see evidence of both physical and chemical changes.
   (a) Describe the physical change.
   (b) Describe the chemical change.

FIGURE 2.7
A burning candle sheds light on changes.

2.2 CHEMICAL REACTIONS

Another term for chemical change is **chemical reaction**. There are many kinds of chemical reactions. They may be spectacular and explosive, like the reactions in a fireworks display (Figure 2.8). Or they may be quite slow, like the formation of rust.

In every chemical reaction, a new substance is produced. A **reactant** is any substance that you start with. A **product** in a chemical reaction is any substance produced in the action. There may be one or more reactants and one or more products in a chemical reaction. The products have properties that are different from the properties of the reactants.

FIGURE 2.8
Chemical reactions put on a show.

WORD EQUATIONS

Scientists use **word equations** to show the reactants and products of a chemical reaction. Figure 2.9 shows the word equation for the chemical reaction in which iron rusts. As you can see in this figure, a word equation gives the names of all the reactants separated by "plus" signs. Following these, an arrow points to the names of all the products. The names of the products are also separated by a "plus" sign. The equation is read as follows: "Iron plus oxygen plus water produces rust."

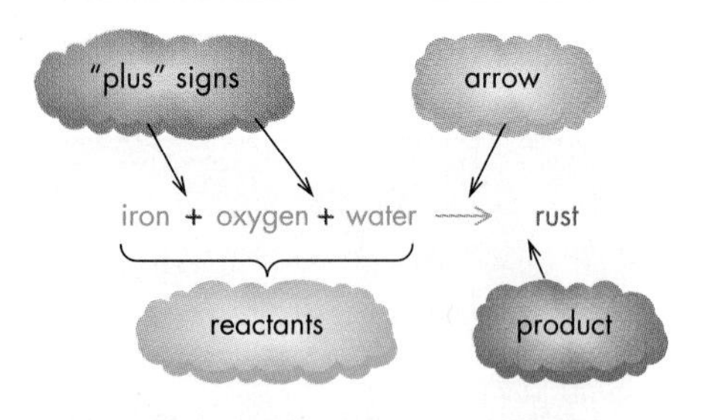

FIGURE 2.9
This word equation describes the chemical reaction that produces rust.

▶ INSTANT PRACTICE

The three chemical reactions that you observed in Activity 2B are described by the following word equations. Read each equation aloud, and identify the reactants and the products.

copper + oxygen → copper oxide
magnesium + hydrochloric acid →
hydrogen + magnesium chloride
sodium carbonate + calcium chloride →
sodium chloride + calcium carbonate

In the three chemical reactions above, elements or compounds react to produce a new substance. In another type of chemical reaction, a compound breaks down to produce two or more products. For example, when hydrogen peroxide breaks down, it forms water and oxygen.

hydrogen peroxide → water + oxygen

▶ INSTANT PRACTICE

Write word equations for the reactions described below.

1. Aluminum reacts with oxygen. The product of the reaction, aluminum oxide, forms a thin layer on the surface of the aluminum and keeps the remaining aluminum from reacting.
2. In our bodies, we use glucose (a type of sugar) and oxygen to produce carbon dioxide and water.

SOME COMMON CHEMICAL REACTIONS

Chemical reactions are going on all around you. When you cook a hamburger, bake cookies, or allow glue to harden, you are carrying out a chemical reaction. When you eat, the glucose in the food reacts with the oxygen in the air you breathe. This reaction produces the energy you need to run, study, and stay alive.

As you have already learned, rusting of iron is another type of chemical reaction. Rusting is an example of **corrosion**, or the eating away of metal through a chemical reaction.

Corrosion poses problems in many aspects of life, from bridge construction to car manufacturing to the production of computer chips.

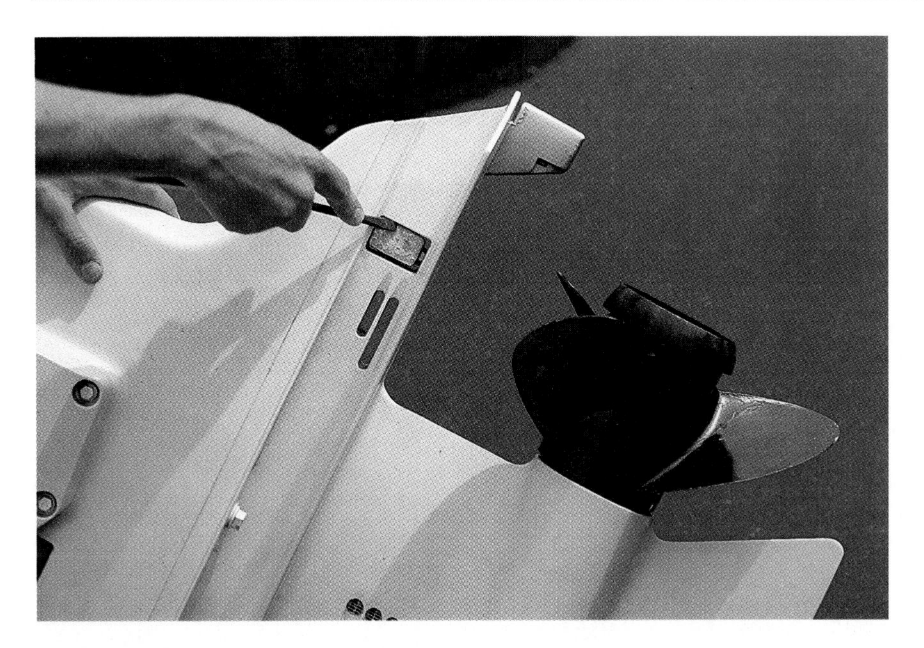

FIGURE 2.10
A zinc bar helps protect the other metal parts on this outboard engine. Why does it have to be cleaned periodically?

EXTENSION

■ A chemical reaction takes place in cakes and cookies when they are baked. Find out what is in baking powder and baking soda. Are they pure substances or mechanical mixtures? What are the reactants in the chemical reactions in which they are involved, and what are the products? Does anything escape from a cake or cookies during baking? How do you think the chemical reactions affect the texture of cakes and other baked goods?

Corrosion is a constant problem with boats and ships. Without some protection, the metal parts corrode quickly in water. This is an especially serious problem for ships in salt water.

Paint and other surface coverings can be used to slow corrosion. Corrosion can also be stopped by fastening **reactive** metal to a ship or boat. A metal that is reactive reacts more easily than other metals. When a reactive metal is attached to a ship, that metal corrodes instead of the more important, less reactive parts of the ship. Zinc and magnesium are two reactive metals used to halt corrosion (Figure 2.10).

In some metals, corrosion produces a protective surface covering. Aluminum is one example. When the surface of a piece of aluminum is scratched, it is shiny, but it soon becomes dull. Some of the aluminum has combined with oxygen from the air to produce aluminum oxide. The aluminum oxide sticks to the surface of the aluminum and protects the rest of it from further reaction. As a result, objects made of aluminum keep their original strength under certain conditions, in contrast to iron objects, which are weakened by rusting. Aluminum can therefore be used in making outdoor furniture, snow shovels, and other items exposed to weather (Figure 2.11). However, pure aluminum does corrode rapidly in salt water.

Figure 2.11
The aluminum oxide on the surface of an aluminum shovel prevents the shovel from rusting.

Another common type of chemical reaction is combustion, which is another name for burning. **Combustion** is a chemical reaction in which oxygen is one of the reactants and in which heat is produced. Many substances, such as wood, kerosene, natural gas, and diesel fuel,

DID YOU KNOW

■ Sunlight causes important reactions in human skin. Ultraviolet light from the sun causes the skin to produce a dark pigment called melanin. This reaction darkens the color of, or tans, the skin.

can undergo combustion. These substances burn easily in air, which is only about 20 percent oxygen (Figure 2.12). In pure oxygen, they burn much more intensely.

FIGURE 2.12 ▶
If firefighters can stop air from reaching a fire, combustion stops. That is impossible for fires like this one, but small fires can be smothered more easily. Knowing about the role of air in combustion, how would you put out a fire in clothing?

ACTIVITY 2C / IDENTIFYING THE PRODUCTS WHEN WAX IS BURNED

In this activity, you will find out what products are produced when wax burns.

MATERIALS

safety goggles
apron
matches
candle
glass plate (or small candle-holder)
two large jars
cobalt chloride paper
limewater

PROCEDURE

1. Put on your safety goggles and apron.
2. Light the candle. Fasten it to the glass plate by dripping some wax onto the plate, then placing the candle in the pool of molten wax.
3. Place a clean, dry jar upside down over the candle (Figure 2.13).
4. Observe what happens to the candle. Then remove the jar and observe its inside surface. Test the surface with blue cobalt chloride paper. (Note: Cobalt chloride paper turns pink in the presence of water.) Record all your observations in your notebook.
5. Set the second clean, dry jar right side up. Place the candle and its stand in the jar (Figure 2.14).

FIGURE 2.13
Step 3

CAUTION!

■ **Limewater (calcium hydroxide) is a skin irritant. Avoid contact with your skin and eyes. Wash all splashes off your skin and clothing thoroughly. If you get any in your eyes, rinse with cold water for at least 15 minutes and inform your teacher.**

6. Light the candle again. Allow it to burn for about 2 min, then blow it out.

7. Immediately add several drops of limewater to the jar and gently swirl it around. (Note: Limewater is a clear, colorless solution that turns milky in the presence of carbon dioxide.) Observe the drops and record your observations in your notebook.

8. Wash your hands thoroughly after you have completed the activity.

FIGURE 2.14
Step 5

DISCUSSION

1. What evidence do you have that a chemical reaction occurred?

2. (a) Describe what you observed on the walls of the upside-down jar. What substance was present? How do you know?
(b) Describe what you observed in the second jar. What substance was present? How do you know?

3. (a) What are two of the products of combustion of candle wax?
(b) Write a word equation for this chemical reaction.
(c) You may have observed evidence for a third product, a black solid. What do you think this substance is? ❖

► REVIEW 2.2

1. Give three examples of chemical reactions that are not given in the text.

2. When a candle burns, the wax first undergoes a physical change as the solid wax melts. Then it undergoes a chemical change as the wax burns.
(a) Describe the physical change that occurs.
(b) Why do you think this change is classified as a physical change?
(c) What are the products of the chemical change?
(d) How could you identify each of the products?

3. In the chemical test for carbon dioxide, limewater combines with the carbon dioxide to produce calcium carbonate. Water is also produced. Write a word equation for this reaction.

4. Write word equations for the following reactions.
(a) Zinc and sulfuric acid react, producing zinc sulfate and hydrogen.
(b) Sodium and chlorine react, producing sodium chloride.

5. Ordinary glass is a solution of sodium silicate and calcium silicate. It is made by heating together sand, limestone, and sodium carbonate. During this process, two chemical reactions occur. Write a word equation for each of them, as they are described below.
(a) Calcium carbonate (limestone) and silicon dioxide (sand) react to form calcium silicate and carbon dioxide.
(b) Sodium carbonate and silicon dioxide react to form sodium silicate and carbon dioxide.

HAIRSTYLIST

Teth Bruyere is a hairstylist. She works in a hairdressing salon. In a recent interview, she had this to say about her work.

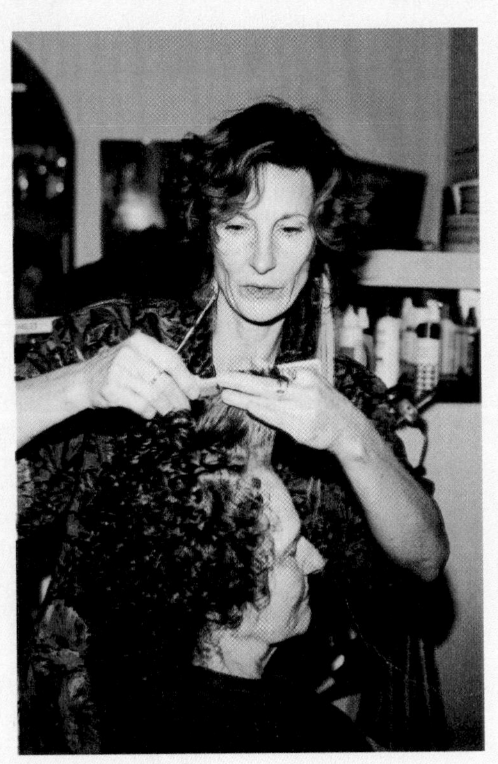

I started my hairdressing training when I was nine years old. My mother was taking a course on weekends and evenings, and I sat through 1600 hours of hairstyling technique! I've been doing it ever since.

Some people will come in and say, "Just make my hair look better." I have total creative freedom. Some people know exactly what they want me to do, and I have to listen. Every time clients sit down in my chair, they put incredible faith in me. Then they leave feeling wonderful. That's the whole point of hairdressing— to make people feel good.

The chemistry of hairstyling is really quite interesting. To dye hair, a chemical is used to open up the outer layer of hair. Color is deposited right into the cortex, the main part of the hair shaft. Like artists, hairstylists need to know a lot about pigment and color.

Our direct contact with chemicals is minimal, but we are careful anyway. We wear gloves and avoid breathing the fumes. You wouldn't have wanted to be a hairdresser in the 1960s. Then they used metallic-based dyes. Today's dyes are much safer.

Besides chemical dyes, we also use hennas, which are natural, vegetable-based colors that coat the hair. Natural rinses also give a sheen and a little bit of a tinge to the hair. Chamomile tea can be used on blonds, marigold tea on redheads, and rosemary tea on brunettes.

Do I ever make mistakes? Sure, everybody does, but nothing too horrible. Those mistakes are all supposed to happen while we're in school. But even not being able to please a client can be devastating.

In this job, you have to love people. You have to be flexible, open, and always willing to learn. You have to like to hear the sound of your own voice and be a good listener. It is perfect for me.

Recent evidence suggests that some chemical hair dyes, especially those that darken hair, may cause cancer. If you were going to have your hair dyed, what questions would you ask the hairstylist? What alternatives would you consider?

2.3 MASS AND CHEMICAL CHANGE

From the beginning of civilization, people have asked many questions about the chemical changes they have observed. They have wondered what happens when matter changes. Does the amount of matter also change? The earliest chemists carried out many experiments to try to answer these questions. These early scientists learned a lot, but they were limited by the instruments they had to work with. Scientists could not really answer these questions because they could not make accurate measurements. Your classroom balances can make far more accurate measurements than the instruments of the early chemists (Figure 2.15).

(a)

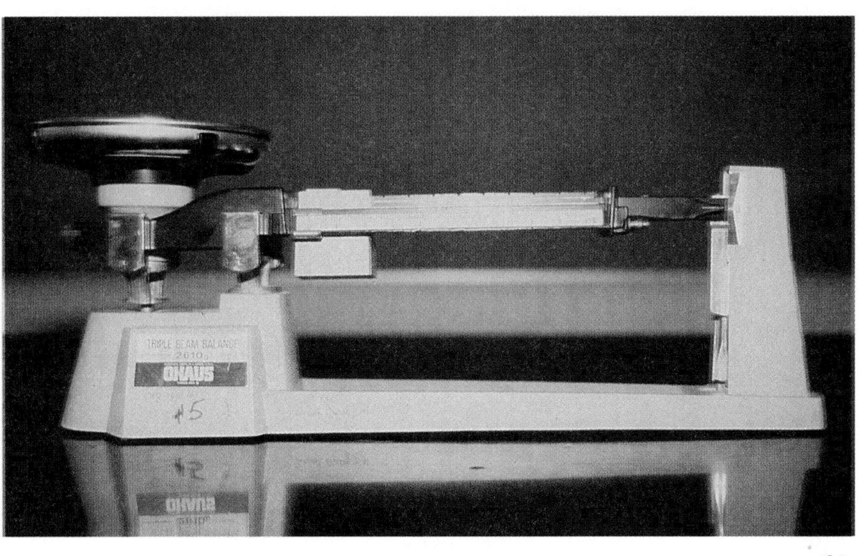

(b)

FIGURE 2.15 ◀
The old pharmacist's balance, (a), is not as accurate as today's school balance. Scientists would put the substance they wanted to measure the mass of in one pan, then add standard, metal masses to the other until the indicator at the top lined up with a vertical line. The beam balance, (b), is accurate to 0.1 g. To make the pharmacist's balance this accurate, what size would the smallest standard mass have to be? Do you think such a small standard mass would be practical?

ACTIVITY 2D / MASS OF REACTANTS AND PRODUCTS

In this activity, you will measure and compare what happens to the total mass of reactants and products during a chemical reaction.

MATERIALS

safety goggles
apron
small test tube
sodium carbonate solution
calcium chloride solution
Erlenmeyer flask (250 mL)
rubber stopper for flask
balance

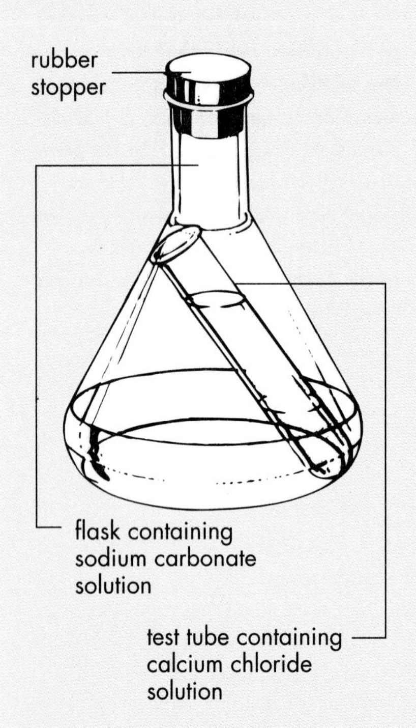

FIGURE 2.16
The test tube should be small enough to fit inside, but large enough that it cannot lie flat.

PROCEDURE

1. As a class, discuss and predict how a chemical reaction might affect the total mass of reactants and products. If you can, give examples of chemical changes to support your predictions.
2. Put on your safety goggles and apron.
3. Check that your test tube is the correct size for your flask (Figure 2.16).
4. Carefully pour the sodium carbonate solution into the flask to a depth of about 1 cm.
5. Pour calcium chloride solution into the test tube until it is about three-quarters full. Carefully dry off the outside of the test tube and gently slide it into the flask by tapping the flask a bit.
6. Put the rubber stopper firmly into the flask. Check that the outside of the flask is dry.
7. In your notebook, record a description of both of the reactants.
8. In your notebook, make a data table like Table 2.4. Find the total mass of the reactants and their containers (the test tube and the flask) and record the mass in your data table.

IF A GAS IS PRODUCED, WILL THE PRODUCTS FLOAT AWAY?
WOULD A SOLID PRODUCT HAVE MORE MASS?
WHAT ABOUT THE ENERGY?

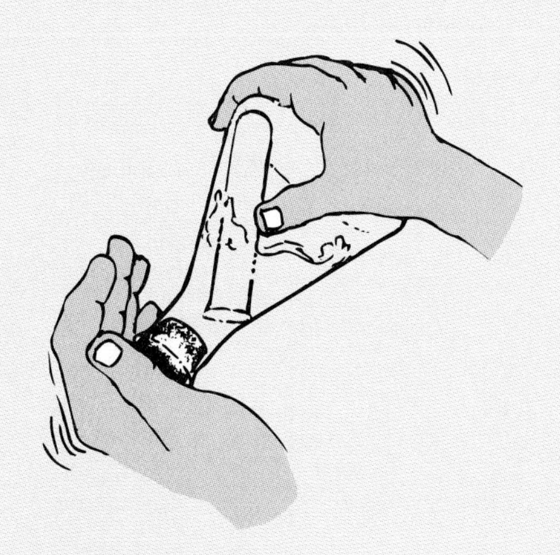

FIGURE 2.17
Step 9

EXTENSION

■ In Activity 2D, you investigated a chemical change in order to find out if the total mass changed during the reaction. Does mass change during a physical change? Design an experiment to compare the mass of an ice cube with the mass of the water that remains when it melts.

TABLE 2.4
Sample Data Table for Activity 2D

Mass of container and reactants	=	______	g
Mass of container and products	=	______	g
Difference	=	______	g

9. Carefully turn the flask upside down while you hold the stopper firmly (Figure 2.17).
10. In your notebook, describe the chemical reaction. Then use the balance to measure the mass of the flask and its contents. Record the mass in your data table.
11. Wash your hands thoroughly after you have completed this activity.

DISCUSSION

1. (a) Name the reactants in this chemical reaction.
   (b) What evidence is there that a chemical reaction has occurred?
2. (a) Calculate the difference between the mass of the reactants and the mass of the products.
   (b) Record your results, along with the results of other groups, in a class data table. How many groups found that the mass increased? How many groups found that the mass decreased?
   (c) Calculate the average amount that the mass appeared to change.
   (d) How important do you think errors in measurement might be in your results and in the average results for your class?
3. The products of this chemical reaction are calcium carbonate and sodium chloride. Write a word equation for this reaction.
4. There was another substance present in both the flask and the test tube, but it did not take part in the chemical reaction. Suggest what substance this was.
5. Which of the products do you think you could see in the flask after the reaction? Suggest a reason. ❖

MISSING MASS?

In some chemical reactions, matter may appear to be destroyed. For example, if you trust your senses, you might think that burning destroys matter, reducing a pile of wood to a mere handful of ashes. Following a forest fire, for instance, a large tree can become a small pile of ashes. How can you explain this?

ACTIVITY 2E / MASS AND GAS

In this activity, you will observe how mass is affected in a chemical reaction in which a gas is produced.

MATERIALS

safety goggles
apron
sodium hydrogen carbonate
dilute hydrochloric acid
beaker (250 mL)
test tube
2 mL measuring scoop
balance

■ **Dilute hydrochloric acid is very corrosive and can cause damage to your eyes and skin. If you get any in your eyes or on your skin, immediately rinse with cold water for 15 to 20 minutes and inform your teacher.**

■ **The reaction of the hydrochloric acid and the sodium hydrogen carbonate may cause splattering. Do not place your hand directly over the beaker when adding the acid to the sodium hydrogen carbonate. Keep your face away from the beaker also.**

PROCEDURE

1. Put on your safety goggles and apron.

2. Be sure that the test tube and beaker are clean and dry. Pour dilute hydrochloric acid into the test tube to a depth of about 5 cm.

3. Use the balance to measure 3 g of sodium hydrogen carbonate. Put the sodium hydrogen carbonate into the beaker.

4. Place the test tube in the beaker and measure the total mass of the beaker and test tube and the substances in them (Figure 2.18). Record the total mass in your notebook.

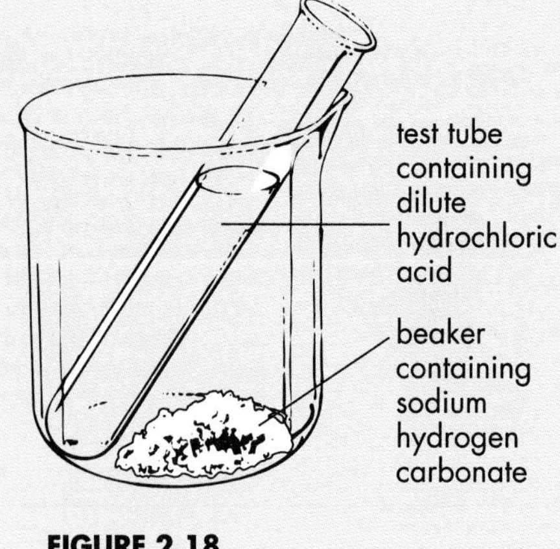

FIGURE 2.18
Step 4

5. Remove the beaker from the balance. Slowly pour the acid into the beaker. Observe and record what happens.

6. Put the test tube back into the beaker, and again measure the total mass. Record the total mass in your notebook.

7. Wash your hands thoroughly after you have completed this activity.

DISCUSSION

1. What evidence is there that a chemical reaction occurred?

2. (a) Name the reactants in this chemical reaction.
(b) Did the total mass increase, decrease, or stay the same? If there was a difference, how large was it?

3. (a) What do you think might cause the difference in mass?
(b) If you were to make a fair comparison between this activity and Activity 2D, how would you change the procedure? (That is, what is the main difference between the set-up in Activity 2D and the set-up in this activity?)
(c) How could this change be dangerous?
(d) What property of one of the products causes the possible danger?

4. What do you think the products of this chemical reaction are?

5. Suggest a reason why you were told to remove the beaker from the balance before mixing the two reactants.

6. In her science class, Ms. Cardinal showed her students a chemical reaction. She carefully measured the mass of a piece of magnesium ribbon. Then she burned it, being careful to collect all the pieces of white ash (Figure 2.19). Finally, she measured the mass of the ash. Look at her results, shown in Figure 2.20, and try to explain them. (Hint: First write a word equation for the reaction. There are two reactants and one product.) ❖

FIGURE 2.19
When magnesium burns, it produces a brilliant light that can harm your eyes.

FIGURE 2.20 ▶
The results of Ms. Cardinal's investigation

CONSERVATION OF MASS

When you are measuring the mass of reactants and products, you must consider the mass of any gases as well as the mass of the solids and liquids. For more than 200 years, scientists tried to find methods to trap the gases that are used or produced in chemical reactions, and to measure their masses. After years of experimenting, in which the masses of all the reactants and all the products were determined, scientists came to agree that mass is neither gained nor lost in any chemical reaction. It does not matter whether the reaction is slow or fast, spectacular or dull; matter is neither gained nor lost. This conclusion is stated as a scientific law. A **scientific law** is a general statement that sums up the conclusions of many experiments. This law is known as the **law of conservation of mass:** *In a chemical reaction, the total mass of the reactants is always equal to the total mass of the products.*

In Activity 2D (Figure 2.16, page 34) the stoppered flask prevented any materials from being added to or taken away from the chemical reaction. In Activity 2E, however, matter could freely enter or leave the beaker. When matter can enter or leave as a reaction takes place, it may be difficult to determine the mass of all the reactants and all products. To demonstrate the law of conservation of mass, you need to include the mass of all the reactants and all products.

In calculating the final mass of a chemical reaction, it is easy to forget about gases that may be reactants or products. The mass of any gases cannot be ignored. Thus, when a piece of iron rusts, the mass of the rust that is produced will be greater than the mass of the piece of iron, because oxygen has combined with the iron. In contrast, when sodium hydrogen carbonate reacts with hydrochloric acid, as in Activity 2E, carbon dioxide gas escapes into the air. As a result, the mass of the products will seem to be less than the mass of the starting materials.

DID YOU KNOW

■ The law of conservation of mass was first described by the French chemist Antoine Lavoisier. His research provided information that is essential to our modern understanding of chemistry. Unfortunately, Lavoisier had many enemies, and he was guillotined during the French Revolution. Another famous scientist at the time, Joseph Louis Lagrange, exclaimed, "A moment was all that was necessary to strike off his head, and probably a hundred years will not be sufficient to produce another like it!"

In this chapter, you have seen several chemical reactions that can be carried out in a science laboratory. There are also chemical reactions occurring all around and within you. For example, chemical reactions are used to produce many common products that you can buy in any supermarket (Figure 2.21). Can you think of other products that you use every day that are the result of chemical reactions?

FIGURE 2.21 ▶
Chemical reactions are used to produce this plastic film.

▶ REVIEW 2.3

1. When you heat powdered copper metal over a Bunsen burner, you see that a black substance forms (Figure 2.22). This black substance is called copper oxide. Which do you think would have more mass, the powdered copper or the black copper oxide? Explain your answer.

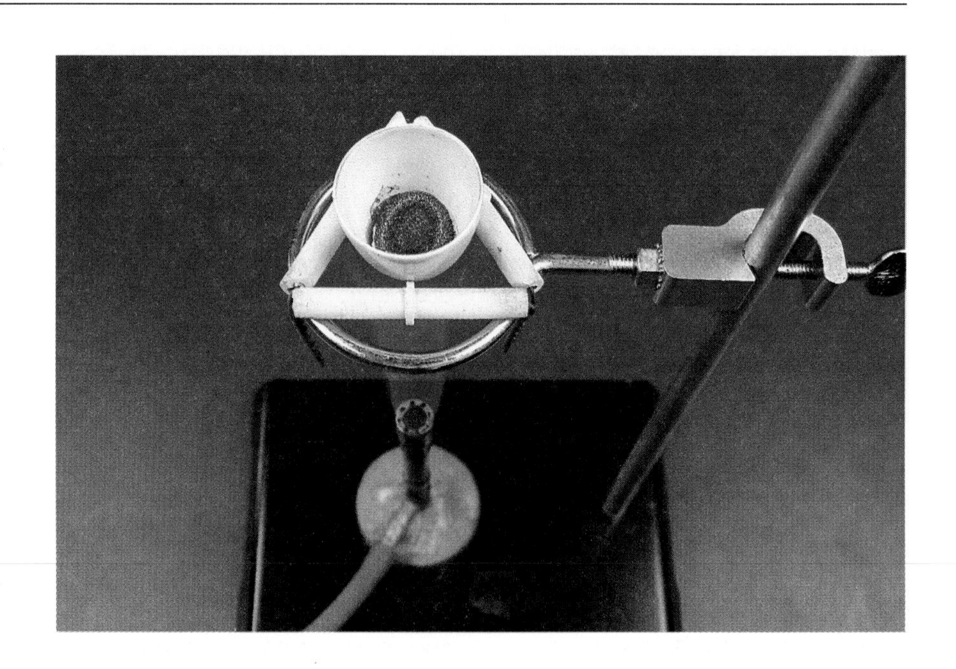

FIGURE 2.22 ▶
When heated strongly, reddish-brown copper metal forms a black compound.

FIGURE 2.23
Magnesium reacts with dilute hydrochloric acid.

2. (a) Write a word equation for the reaction of magnesium and dilute hydrochloric acid in which magnesium chloride and hydrogen gas are produced (Figure 2.23). (You may have seen this reaction in Activity 2B, page 24.)
(b) How do you think the mass of the test tube and contents would change as the reaction continues for 30 min?

3. (a) Why do you think it would be difficult to demonstrate the law of conservation of mass using the reactions that occur as wood burns?
(b) For the burning of wood, name as many of the reactants and products of the reactions as you can.

4. Ace Transport Company in Los Angeles receives a container of expensive wool from New Zealand. The container, still in perfect condition, with its seal unbroken, is labelled "105 kg" and "Contains mothballs to prevent insect damage." The Ace Transport agent finds that the package has a mass of only 102 kg. How could you explain the "missing" mass? Try to suggest two or three explanations.

CHAPTER REVIEW

KEY IDEAS

- Matter can be classified into two categories based on its properties: pure substances and mixtures.
- Pure substances can be classified as elements or compounds.
- Mixtures include solutions, suspensions, and mechanical mixtures.
- In a physical change, no new substance is produced.
- In a chemical change, one or more new substances are produced. Each of these new substances has properties that are different from the properties of the starting materials.
- A chemical reaction can be written as a word equation showing the reactant(s) and product(s).
- In chemical reactions, the total mass of the products equals the total mass of the reactants. (This rule is called the law of conservation of mass.)

VOCABULARY

matter
property
mixture
pure substance
mechanical mixture
suspension
solution
element
compound
physical change
chemical change
chemical reaction
reactant
product
word equation
corrosion
reactive
combustion
scientific law
law of conservation of mass

V1. Use all the words in the vocabulary list, and any other words or ideas you need, to create a mind map about changes in matter. Remember that you may connect any two parts of your map with lines and words describing how they are related.

V2. Some of the words in the vocabulary list have other meanings besides their meanings in chemistry. For "property," "suspension," "product," and "law," give the meaning of the word in everyday life.

CONNECTIONS

C1. Give two examples of elements and two examples of compounds.

C2. A black material is heated. A gas is released and a white substance remains. Was the original material an element or a compound? Explain your answer.

C3. For each situation shown in Figure 2.24, decide whether the change is physical or chemical. Give your reasons in each case.

C4. (a) What do all chemical reactions have in common?
(b) Give four examples of differences between chemical reactions.

C5. Look around your home and choose three situations where chemical reactions are occurring. For each, explain how you know there is a chemical reaction.

C6. The following word equations describe three different chemical reactions, X, Y, and Z.

Reaction X
calcium carbonate + hydrochloric acid → calcium chloride + carbon dioxide + water

Reaction Y
lead nitrate + sodium iodide → lead iodide + sodium nitrate

Reaction Z
potassium chlorate → potassium chloride + oxygen

(a) Which of these chemical reactions has (have) only one reactant?
(b) Which has (have) a product that is a gas?

C7. Identify the reactants and products in the following chemical reactions. Then write the word equation for each.
(a) Hydrochloric acid reacts with sodium hydroxide to produce sodium chloride and water.
(b) Zinc sulfide is produced when zinc reacts with sulfur.
(c) Hydrogen peroxide gradually breaks down, forming water and oxygen.
(d) Carbon dioxide and water are produced in our bodies. These are the end products of a series of chemical reactions, starting with sugar in the food we eat and oxygen in the air.

(a)

(b)

(c)

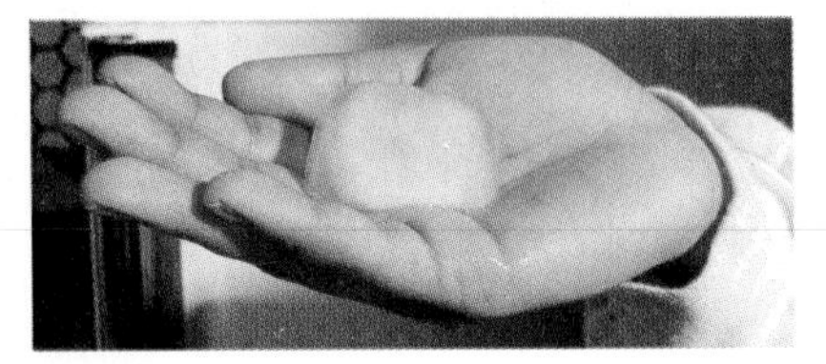

(d)

(e)

FIGURE 2.24
(a) A space shuttle blasts off.
(b) Wood rots.
(c) Jelly changes from a liquid to a solid.
(d) An ice cube melts.
(e) Paint dries.

C8. When limestone is heated, it breaks down into calcium oxide (a solid) and carbon dioxide (a gas).

(a) Write a word equation for this chemical reaction.

(b) If you heated some limestone, would the mass of the limestone that you started with be greater or less than the mass of calcium oxide? Explain your answer.

EXPLORATIONS

E1. Hydrogen and helium have some similar properties and some different ones. Use a reference book to find out how these two gases are used (Figure 2.25).

FIGURE 2.25
In 1937, 36 passengers and crew members died when the airship Hindenburg caught fire and exploded. The Hindenburg was filled with hydrogen. What property of hydrogen caused the explosion? Modern airships could not explode in this way. Why not?

E2. When a welding torch is used, the following chemical reaction takes place.

acetylene + oxygen ⟶ carbon dioxide + water

(a) What forms of energy are produced in this chemical reaction?

(b) Name the reactants and products in this chemical reaction.

(c) Classify the pure substances in your word equation as either elements or compounds.

(d) If a welding torch is improperly adjusted, it can produce a smoky yellow flame. If the flame touches a metal object, it leaves black marks on the metal. What do you think the black material is? Explain your answer.

E3. The care of a swimming pool involves a lot of chemistry.

(a) Why is the water in a swimming pool chlorinated?

(b) To chlorinate a pool, you do not use chlorine gas directly. Use a reference to find out about some of the properties of chlorine that make it dangerous.

(c) From a pool supervisor or a store that sells pool supplies, find out what compounds are added to pools in order to keep the quantity of chlorine at a safe level.

E4. In strong sunlight, fair-skinned people may find that their skin becomes sunburned quickly. Sunscreens, purchased in a pharmacy and applied to the skin, slow down or prevent sunburn.

(a) Find out what causes the chemical change called sunburn.

(b) How do sunscreens prevent sunburn?

(c) What other characteristics, besides the ability to prevent sunburn, would you want a sunscreen to have?

REFLECTIONS

R1. In your learning journal, read the list of changes that your group thought of in Activity 2A.

(a) Does your list include any changes that are not changes in matter? If it does, write them in a separate list. If it does not, list one or two of the examples on page 20.

(b) With your group, decide which of the changes in matter are physical changes and which are chemical changes. (Remember that you should consider several clues in each case.) In your learning journal, mark each item in the list with a P or a C to indicate whether it is a physical change or a chemical change.

(c) Write word equations for three of the chemical changes in your list.

CHAPTER 3

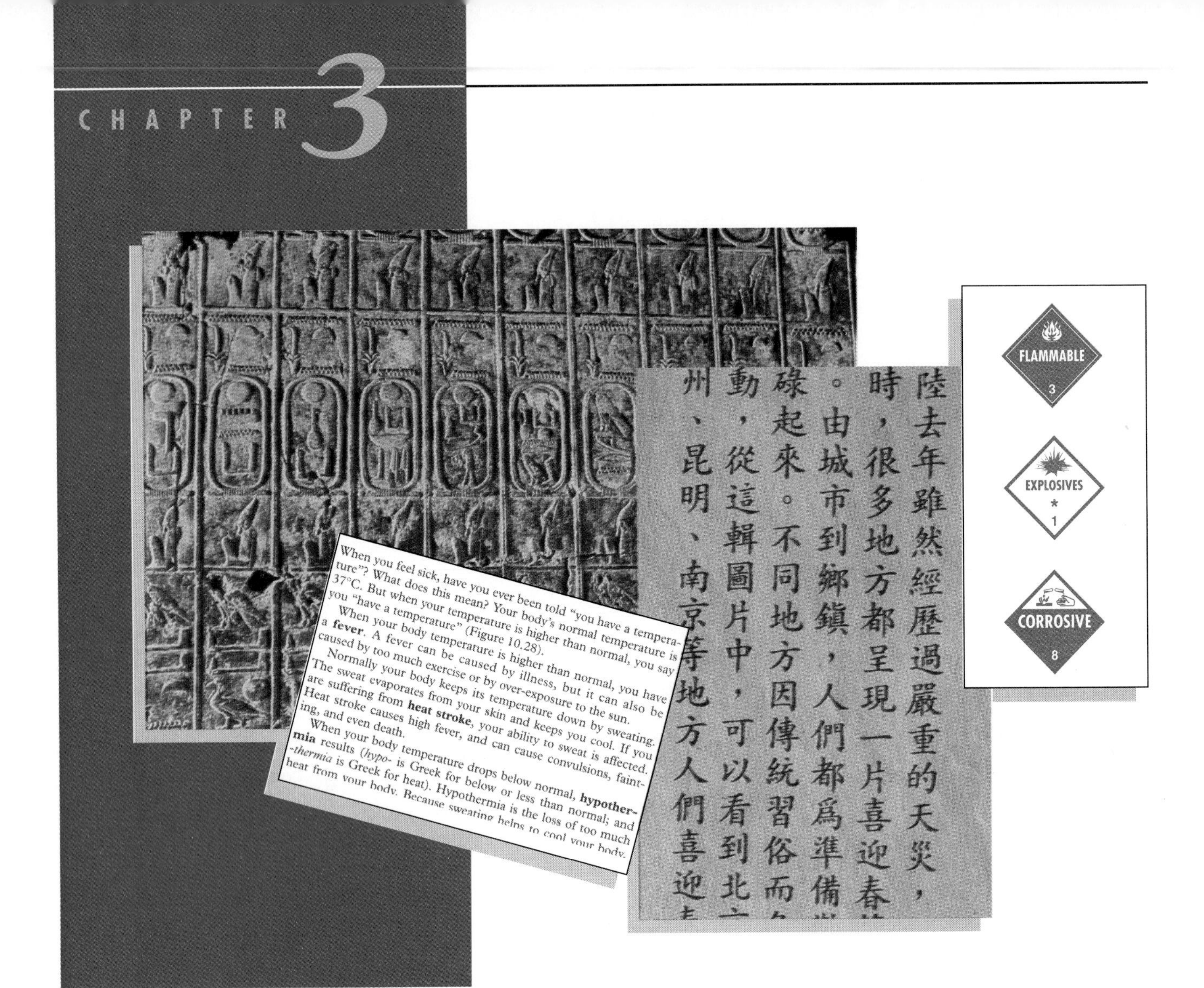

SYMBOLS AND FORMULAS

The early chemists, working in several countries and speaking different languages, wanted to tell each other about their new discoveries. They wanted to find out what other scientists were thinking and what kinds of experiments they were doing. They needed a way of writing about elements, compounds, and chemical reactions that scientists throughout the world could understand.

Some of the symbols shown on this page and page 43 can be understood by anyone who reads English. Others can be understood by people who read other languages using the same alphabet. Still others can be understood by people who use other alphabets and other ways of writing. And some, such as the skull and crossbones, are understood almost everywhere in the world. Chemists wanted a truly international system of symbols, to be used the world over. In this chapter, you will find out about that system.

ACTIVITY 3A / SYMBOLS

1. With your group, think of ways in which symbols could be used in chemistry. How could they be used to tell you about elements and compounds and how they react?
2. Examine the symbols in Table 3.1 and suggest how those symbols were chosen. Repeat for Tables 3.2 and 3.3. Your group could brainstorm possible explanations for the symbols in Table 3.3.
3. The three tables list symbols for 21 elements. There are over 100 elements in total. There are millions of different compounds made up of these elements; how do you think you could use these same symbols to represent compounds? ❖

Element	Symbol
Carbon	C
Hydrogen	H
Iodine	I
Nitrogen	N
Oxygen	O
Phosphorus	P
Uranium	U

TABLE 3.1

Element	Symbol
Aluminum	Al
Arsenic	As
Calcium	Ca
Cobalt	Co
Magnesium	Mg
Manganese	Mn
Platinum	Pt
Zinc	Zn

TABLE 3.2

Element	Symbol
Copper	Cu
Gold	Au
Iron	Fe
Lead	Pb
Mercury	Hg
Silver	Ag

TABLE 3.3

■ 3.1 ELEMENTS AND THEIR SYMBOLS

Elements

hydrogen nitrogen carbon

oxygen phosphorus sulfur

I Z L

iron zinc lead

C ch T

copper chlorine tin

Compounds

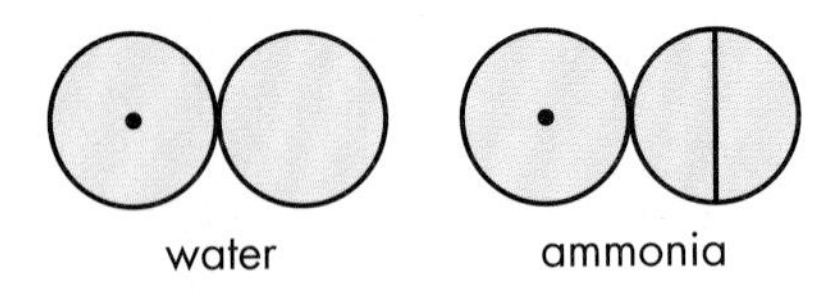

FIGURE 3.1
Some of Dalton's symbols for elements and compounds. Scientists found that Dalton's system was difficult to use.

Around 400 B.C., a Greek scholar, Democritus, suggested that matter is made up of tiny particles. We call these particles **atoms**. More than 2000 years later, in 1808, John Dalton (1766–1844), an English scientist, suggested that each element has its own kind of atom. He also suggested that compounds are made up of combinations of different kinds of atoms. When explaining his model, Dalton represented elements by symbols and compounds by combinations of symbols (Figure 3.1).

Dalton's model has been useful in explaining many properties of matter. Over the years, the model has been modified as scientists have gained new knowledge, but much of it has remained the same. However, Dalton's system of symbols was never widely used.

Simpler symbols based on Dalton's model were suggested by Jons Jakob Berzelius (1779–1848). He suggested that all elements be represented by the first letter of their name in Latin, or by the first letter and another letter from the name. The first letter of the symbol would always be capitalized, and the second letter would be lower case. Berzelius's suggestion seemed so convenient to other scientists that his system became widely accepted and is still used today. These same symbols are now used in all languages, even those that use a different alphabet. In this way, the symbols are truly international.

THE ORIGIN OF SYMBOLS FOR THE ELEMENTS

You may wonder why the symbols for the elements are based on their Latin names. First, some of the elements were known to the Latin-speaking Romans as early as two thousand years ago and were therefore named by the Romans. A few elements occur in nature in almost pure form. Thus, they are easy to separate and identify. An example of such an element is gold (*aurum* in Latin), whose symbol is Au.

Second, for more than a thousand years, Latin was used throughout Europe and was considered to be the language of learning. Long after the Roman Empire had collapsed, scholars in Europe continued to use Latin for their writings. As alchemists and scientists discovered new elements, it was natural for them to give the elements Latin names. To keep the system consistent, even elements discovered very recently are given Latinized names. Examples are uranium, einsteinium, and californium.

The English names of certain elements are the same as or very similar to their names in Latin, so it is easy to recognize these symbols. Examples are carbon (from *carbonem* in Latin, meaning charcoal), whose symbol is C, and sulfur (*sulfur* in Latin), whose symbol is S. Other elements, however, have English names that are very different from the Latin names.

DID YOU KNOW

■ Latin was the language of the Roman Empire. It was spoken for many centuries, from about 300 B.C. to A.D. 500. Modern languages such as Italian, Spanish, and Portuguese developed from Latin. Mexico and countries in South America and Central America are known as Latin American, because most of their citizens speak Spanish or Portuguese.

Thus, their symbols are less easy to recognize. An example is silver (*argentum* in Latin), whose symbol is Ag.

Table 3.4 lists both the English and the Latin names of some elements and their symbols.

English name	Symbol	Latin name
Antimony	Sb	Stibium
Arsenic	As	Arsenicum
Bismuth	Bi	Bismutum
Carbon	C	Carbonem
Copper	Cu	Cuprum
Gold	Au	Aurum
Iron	Fe	Ferrum
Lead	Pb	Plumbum
Mercury	Hg	Hydrargyrum
Silver	Ag	Argentum
Sulfur	S	Sulfur
Tin	Sn	Stannum

TABLE 3.4 ***English and Latin Names of Some Elements***

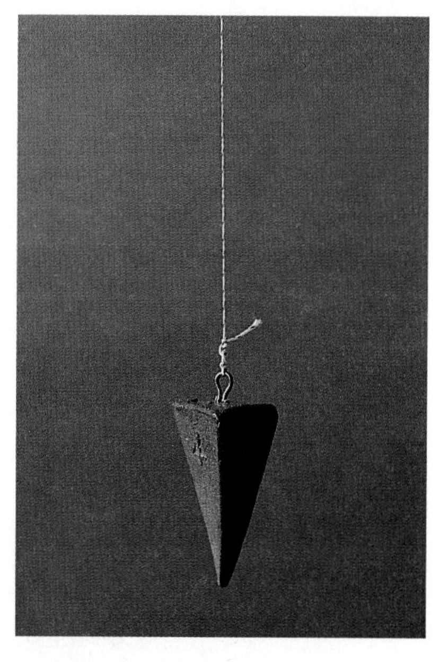

FIGURE 3.2
A heavy weight on a string is called a plumb bob. At one time, plumb bobs were often made of lead. Where do you think the name plumb bob came from?

Some English words are related to the Latin names for elements. For example, the Romans used lead pipes in their water systems. The Latin word for lead is *plumbum*, so people who work with pipes and water systems today are called plumbers. A plumb bob is a heavy weight hung from a string. It is used for finding vertical lines (Figure 3.2). Since lead has a high density, it makes a good weight, and so lead, or *plumbum*, gave plumb bobs their name.

GROUPS OF ELEMENTS

There are over 100 elements. Although each element is different from every other, there are groups of elements that have similar properties. When elements are grouped according to their properties, naming the compounds they form becomes easier.

ACTIVITY 3B / CLASSIFYING ELEMENTS

In this activity, you will classify some elements based on their physical properties.

MATERIALS

safety goggles apron

samples from the following list of elements:

aluminum	chromium	iron	nickel	tin
bismuth	cobalt	lead	nitrogen	tungsten
calcium	copper	magnesium	silicon	zinc
carbon	iodine	mercury	sulfur	➡

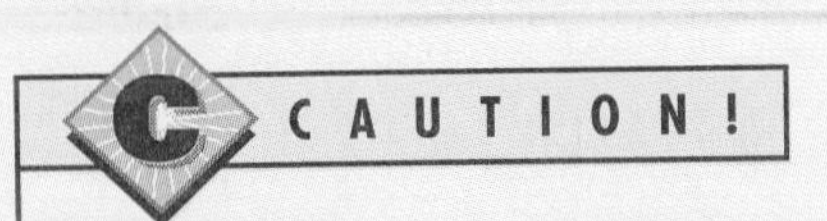

■ **Some elements are poisonous and dangerous. Follow your teacher's instructions very carefully for observing the elements.**

PROCEDURE

1. Draw a data table in your notebook, using the following headings: Element, Symbol, Properties. (Note: The column labelled "Properties" in your table will actually contain only a few of the many properties of each element.)
2. Put on your safety goggles and apron.
3. Carefully observe each of the elements that your teacher has made available. Record the name of the element and its symbol (Figure 3.3). (You can find the symbols of all the elements in the alphabetical list of elements in Appendix C.) Then fill in the column labeled "Properties." Use the following questions as a guide in your investigation:
   (a) What color is the element? (If the element is not visible, say so.)
   (b) What is the state of matter (solid, liquid, or gas) of the element?
   (c) Is the element shiny or dull?
   (d) Is the element heavy or light for its size compared with the others? (That is, how would you rate its density?)
   (e) Has the element any other special property that you notice?
4. Wash your hands thoroughly after you have completed this activity.

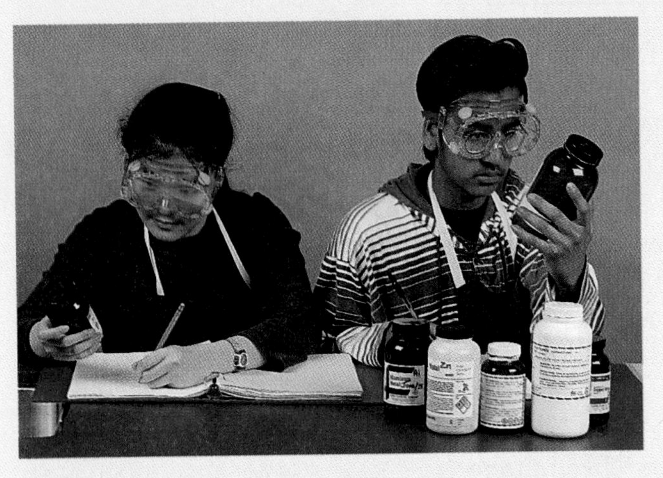

FIGURE 3.3
Observe each element carefully and record the name and the symbol of the element in your notebook.

DISCUSSION

1. One way to classify elements is according to whether they appear to be metals or not. In a table in your notebook, list the metals in one column and the non-metals in another column. If there are any elements that do not seem to fit into either category, list them in a third column.
2. Think of two other ways that you could classify the elements you studied in this activity. Discuss your ways of classifying the elements with other members of your class.
3. How did you decide whether or not an element was a metal? ❖

METALS AND NON-METALS

Based on experiments on many elements over many years, elements with certain characteristics are considered to be **metals**. Metals, such as the gold shown in Figure 3.4, typically have the following properties:

- They are shiny.
- They are **ductile**; that is, they can be pulled (stretched) into wires.
- They are **malleable**; that is, they can be beaten into sheets.
- They are good **conductors**; that is, they are able to conduct (transfer) electricity and heat.

Non-metals are elements that do not have these properties. If you look at Appendix B, The Periodic Table of the Elements, you will see that all the metals are grouped together on the left-hand side of the table, and all the non-metals, except hydrogen, are grouped together to the right of the metals.

As usual in nature, however, things are not so simple. Some elements do not fit very well into either group. Other elements have properties of both groups. Hydrogen is a gas, but when it forms compounds, it behaves as a metal would. Aluminum has the properties of a typical metal, but it sometimes acts like a non-metallic element. Silicon and germanium have both metallic and non-metallic properties.

DID YOU KNOW

■ Silicon and germanium conduct electricity, but not nearly as well as metals like copper and iron. For this reason, these two elements are called "semiconductors." Semiconductors are used in electronic devices such as radios, television sets, CD players, computers, and calculators. Complex electronic circuits are set into the surface of tiny chips made of silicon. As the technology of these microcircuits and semiconductors has become more advanced, electronic devices have become smaller and more convenient for people to use.

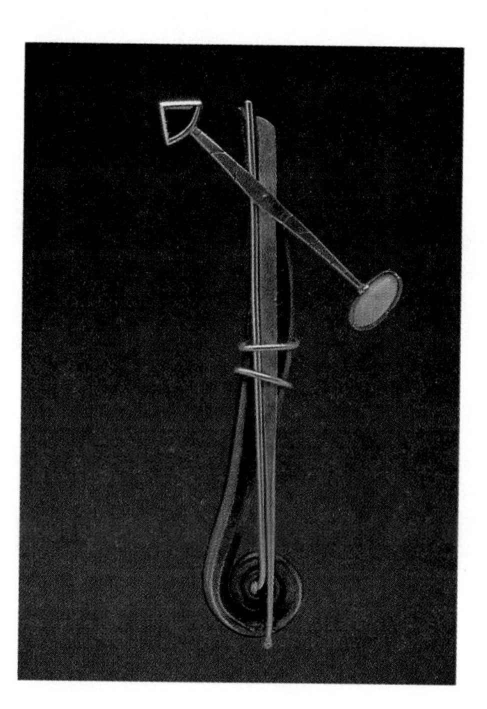

FIGURE 3.4 ◄
This brooch called Mood, was created out of sterling silver and gold by artist Andy Cooperman. What properties of gold make it easy for artist and jewellers to work with?

► REVIEW 3.1

1. Mercury is considered to be a metal, even though it is a liquid at room temperature. What properties do you think mercury has that make it a metal?
2. Summarize the rules for writing symbols for elements.
3. Imagine that you are a scientist who has just discovered a new element. You decide to name it either Montaium (after Montana) or after yourself. (For example, if your last name is Tang, the new element would be tangium.) Suggest two possible symbols for each name. To check that you do not use any existing symbols, refer to the alphabetical list of names and symbols in Appendix C.
4. Use Appendix C to explain why rechargeable batteries are sometimes called "NiCad" batteries.

ANDREW COOPERMAN: ARTIST AND METALSMITH

Andrew Cooperman is a jewelry maker and metalsmith who works in Seattle, Washington. He makes traditional jewelry to order for the retail jewelry store market and for private clients. Cooperman also creates innovative and unusual works that are exhibited and sold at art galleries. These works are nontraditional and more closely resemble small sculptures.

Cooperman is attracted to metalworking because of the challenge of taking a metal such as gold or silver and cutting, filing, and manipulating it so that it doesn't look like metal. Cooperman says when he first began working with metal, he felt he was not very good at it, so he enjoys the process of continually improving his technical mastery.

Cooperman works primarily with gold, silver, and bronze. In gold he works mostly with 18K. Pure gold is 24K; therefore, 18K is 75 percent pure gold. He uses three different colors of gold: white gold (75 percent gold, 25 percet nickel or palladium); yellow gold (75 percent gold, 25 percent copper); and green gold (75 percent gold, 25 percent silver). The raw gold with which he works is bought in thin sheets or pebbled, approximately 1/8 inch diameter clumps. Cooperman then hammers it into shape or heats it until it is red and then applies texture marks to it. Some of his pieces involve forging the gold or silver into a hollow form. He seldom works from a presketched idea, preferring to just start in and react spontaneously to whatever occurs.

Cooperman has a very strong sense of continuing the long line of metalsmiths through the centuries, starting at the beginning of the Bronze Age. Early peoples pounded tools and implements from metals mined from the ground. The early Egyptians were among the most talented goldsmiths history has known. The Inca of South America were also accomplished goldsmiths. During the Renaissance, goldsmithing was considered the highest art form. Cooperman says he finds metalsmithing particularly satisfying because it allows him to stretch his mind by using his hands.

In addition to his career as a jewelry maker and metalsmith, Andrew Cooperman is also an educator. He teaches his craft to students of all ages at the Pratt Fine Arts Center in Seattle. He also conducts workshops for advanced metalsmiths. Cooperman's own education consists of a degree in Fine Arts and Literature from the State University College at Oneonta, New York.

3.2 CHEMICAL FORMULAS

Just as symbols represent elements, combinations of symbols represent compounds. A **chemical formula** is the combination of symbols representing a particular compound. For example, water has the chemical formula H_2O. This formula tells you that the compound contains hydrogen and oxygen. It also tells you the relative numbers of atoms of hydrogen and oxygen in the compound. The small number 2, called a subscript and written below the line, refers to the symbol to its left, H. The number 1 is not usually written in chemical formulas. Thus, the formula H_2O tells you that in water there are two atoms of hydrogen for every one of oxygen.

Another familiar compound is table salt, or sodium chloride. It has the chemical formula NaCl. This formula shows that table salt is made up of the elements sodium (Na) and chlorine (Cl). It also tells you that there is one atom of sodium for every atom of chlorine.

The formulas for other compounds are similar. Calcium chloride, for example, has one atom of calcium for every two atoms of chlorine. Thus, its chemical formula is $CaCl_2$. Aluminum chloride has one atom of aluminum for every three atoms of chlorine, so its chemical formula is $AlCl_3$.

This system of chemical formulas is used for all compounds, including those that have more than two elements. Magnesium sulfate, for example, which contains atoms of magnesium, sulfur, and oxygen, has the formula $MgSO_4$. Baking soda (sodium hydrogen carbonate) has the formula $NaHCO_3$. Acetylsalicylic acid, commonly known as "ASA" or "aspirin," has the formula $C_9H_8O_4$.

Table 3.5 lists the common names of some compounds, their chemical formulas, and explanations of what the formulas mean.

Name of compound	Chemical formula	Meaning of formula
Table salt (sodium chloride)	NaCl	For every one atom of sodium, there is one atom of chlorine.
Chalk (calcium carbonate)	$CaCO_3$	For every one atom of calcium, there are one atom of carbon and three atoms of oxygen.
Copper sulfate	$CuSO_4$	For every one atom of copper, there are one atom of sulfur and four atoms of oxygen.
Baking soda (sodium hydrogen carbonate)	$NaHCO_3$	For every one atom of sodium, there are one atom of hydrogen, one atom of carbon, and three atoms of oxygen.

TABLE 3.5 ***Explanation of Some Chemical Formulas***

■ It sounds so simple — telling what kinds of atoms there are in a compound and how many of each there are. In fact, working out the chemical formula of a compound can require many investigations in many laboratories, and even then the formula may not be known for certain. Scientists are still discovering new compounds, many of them made by plants. Even in common foods, such as broccoli and carrots, there are compounds that are still unknown. In rain-forest plants, many of which are in danger of becoming extinct, scientists are finding many potentially valuable drugs.

Watch the news and read reports on the environment to find out about new and valuable materials that are being discovered. For one of these, make a poster showing where it is produced, how it is treated, and how it may be used.

MOLECULES

When atoms join together, they form a larger particle. This larger particle is called a **molecule**. A molecule of water, for example, is made up of two atoms of hydrogen and one atom of oxygen. Another compound, hydrogen peroxide, also consists of hydrogen and oxygen, but the relative numbers of the atoms are different. As a result, water and hydrogen peroxide have very different properties. Examine the molecules and formulas of these two compounds, shown in Figure 3.5. Figure 3.6 shows two other common compounds, both of which are made up of carbon and oxygen.

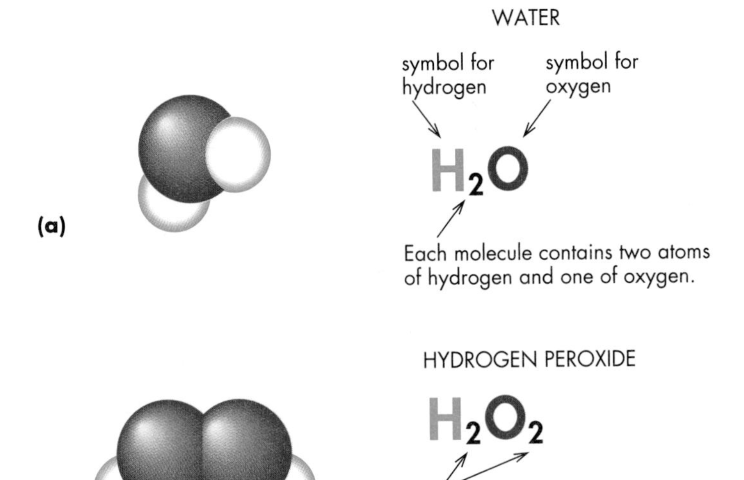
WATER
symbol for hydrogen
symbol for oxygen
H_2O
(a)
Each molecule contains two atoms of hydrogen and one of oxygen.
HYDROGEN PEROXIDE
H_2O_2
(b)
Each molecule contains two atoms of hydrogen and two atoms of oxygen.

FIGURE 3.5 ▶
(a) In water molecules, there are two hydrogen atoms for every oxygen atom.
(b) In hydrogen peroxide, which is used to lighten hair color and as a disinfectant for minor cuts, the number of hydrogen and oxygen atoms is equal.

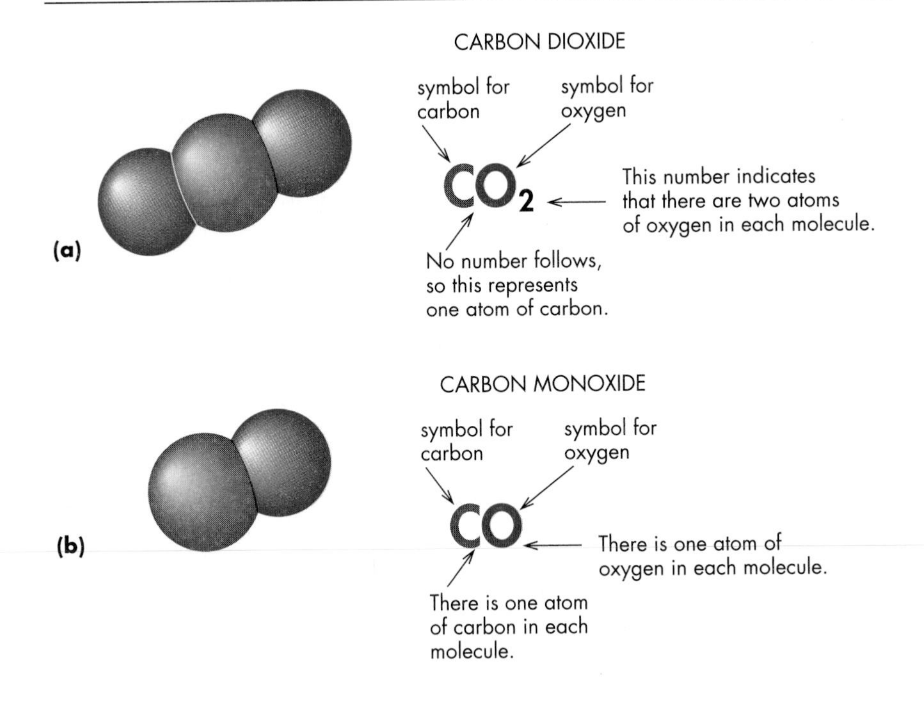
CARBON DIOXIDE
symbol for carbon
symbol for oxygen
CO_2
This number indicates that there are two atoms of oxygen in each molecule.
(a)
No number follows, so this represents one atom of carbon.
CARBON MONOXIDE
symbol for carbon
symbol for oxygen
CO
(b)
There is one atom of oxygen in each molecule.
There is one atom of carbon in each molecule.

FIGURE 3.6 ▶
Both carbon dioxide and carbon monoxide contain only atoms of carbon and oxygen, but these compounds have very different properties. Carbon monoxide (one of the gases in automobile exhaust) is extremely poisonous. Carbon dioxide is used by plants to make food.

MOLECULES OF ELEMENTS

The atoms of some elements join together with atoms like themselves. That is, the atoms of an element may form a molecule of the element. The oxygen in the air consists of oxygen molecules. The formula for these molecules is O_2 (Figure 3.7). Nitrogen, another gas in air, also consists of molecules. The formula for these molecules is N_2. A few other elements also consist of molecules; for example, sulfur molecules have eight atoms and are shown by the formula S_8 (Figure 3.8).

■ Atoms of oxygen form molecules that have two atoms each. These molecules are represented by the formula O_2. This form of oxygen makes up about 20 percent of the air. Under certain conditions — for example, in intense ultraviolet light — oxygen atoms can form a larger molecule that has three atoms. This form of oxygen is called ozone and has the formula O_3. Find out about the importance of the ozone layer of our atmosphere and why people are concerned about it.

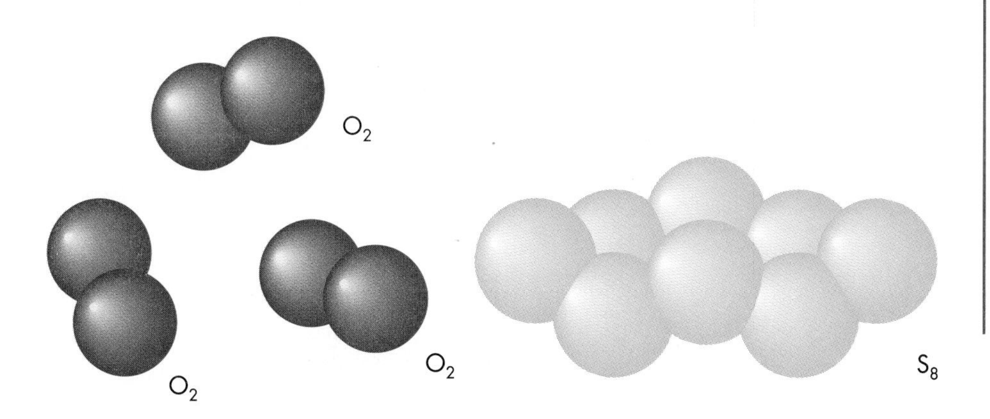

FIGURE 3.7
Each molecule of oxygen is made up of two atoms.

FIGURE 3.8
A molecule of sulfur is made up of eight atoms.

► REVIEW 3.2

1. (a) What is a molecule?
   (b) How is a molecule of a compound different from a molecule of an element?
2. Explain the meaning of the following chemical formulas. (If necessary, refer to the table in Appendix C for the symbols of the elements.)
   (a) H_2O (water)
   (b) CO_2 (carbon dioxide)
   (c) $MgSO_4$ (magnesium sulfate)
   (d) Fe_2O_3 (iron oxide)
3. Write a chemical formula for each of the following.
   (a) A molecule of hydrogen that is made up of two atoms of hydrogen.
   (b) A molecule of propane that is made up of three atoms of carbon and eight atoms of hydrogen.

■ 3.3 NAMING COMPOUNDS

There are thousands of chemical compounds, each with a different chemical name. The name of each compound tells you what elements are in it. Compounds may contain many elements or just a few elements (Figure 3.9). There are many common compounds that contain only two elements.

FIGURE 3.9
Many food colorings are compounds with complex molecules. For example, one red food coloring has the formula $Na_3C_{18}N_2S_3O_{11}$. A common blue coloring has the formula $Na_2C_{16}N_2S_2O_8$.

ACTIVITY 3C / LOOKING AT COMPOUNDS

In this activity, you will learn about compounds that are made up of only two elements. You will also discover how these compounds are named.

MATERIALS

sealed vials containing samples of five or more of the compounds listed in Table 3.6.

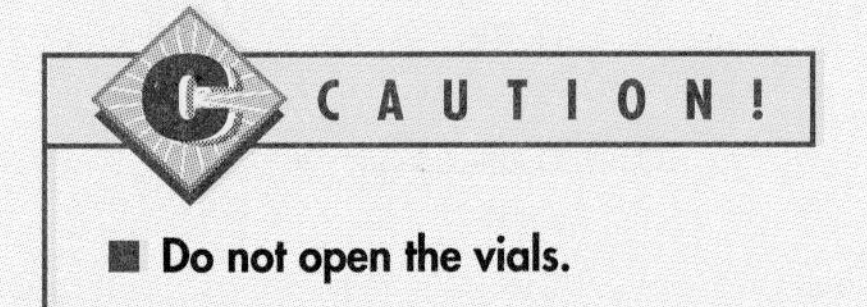

- **Do not open the vials.**

PROCEDURE

Examine the vials containing the compounds. In your notebook, list the names of the compounds. Write a description of each compound beside its name.

DISCUSSION

1. What properties do these compounds have in common?
2. What type of element is given first in the name of a compound?

TABLE 3.6
Some Common Compounds

Chemical name	Common name	Formula	Use
Boron oxide	——	B_2O_3	In heat-resistant glass; as a herbicide
Calcium carbonate	Chalk or limestone	$CaCo_3$	In toothpaste and white paint; limestone is used in building construction
Calcium chloride	——	$CaCl_2$	In bleaching powder; for melting ice on roads
Calcium oxide	Quicklime or lime	CaO	In mortar and plaster for construction
Copper chloride	——	CuCl	To make red-colored glass
Potassium iodide	——	KI	In table salt, to make "iodized" salt
Silver chloride	——	AgCl	In photography
Sodium chloride	Salt	NaCl	In food; for melting ice on roads

3. (a) What type of element has its name changed?
   (b) How is the name changed?

4. Use your answer to questions 2 and 3 to complete the following word equations.
   (a) iron + fluorine →
   (b) magnesium + oxygen →
   (c) nickel + iodine →

5. Suggest general rules for naming compounds, based on what you have learned in this activity. ❖

▶ REVIEW 3.3

1. (a) What information is given by the chemical name of a compound?
   (b) What additional information is given by the chemical formula?

2. Divide a page in your notebook into two columns. Label one column "Metals" and the other column "Non-metals." List the following elements in the appropriate column. (Recall that in the periodic table in Appendix B, metals appear to the left of the staircase line and non-metals appear to the right.)
   sodium
   nitrogen
   carbon
   calcium
   lead
   chlorine
   helium
   zinc

FIGURE 3.10
Sea salt is "harvested." Can you tell from this picture how the process works?

3. Complete the following word equations:
   (a) calcium + oxygen →
   (b) sodium + sulfur →
   (c) aluminum + chlorine →

4. Sea salt (Figure 3.10) contains a number of compounds that are formed by the reaction of metallic elements with non-metallic elements. In a reference, find out what compounds are found in sea water, and how these compounds can be used.

■ 3.4 WRITING CHEMICAL FORMULAS

The name of a compound gives some useful information, but it does not show the composition of the compound. For example, from the name "calcium chloride," it is not evident that the composition of this compound is two chlorine atoms for each calcium atom. The chemical formula $CaCl_2$ tells you this. How do scientists know that a compound like calcium chloride should have the formula $CaCl_2$, rather than $CaCl$, $CaCl_3$, or some other formula?

COMBINING CAPACITY

After performing many experiments, scientists have learned how elements combine with one another to form compounds. By analyzing many compounds, scientists have found out how many atoms of each of the different elements are in a molecule of each of these compounds. From all these experiments, scientists have inferred that different elements have different abilities to combine with other elements. **Combining capacity** is sometimes used to describe the ability of an element to combine with other elements.

Scientists have given a numerical value to the combining capacity of each element. Sodium has been assigned a combining capacity of 1. In the compound sodium chloride ($NaCl$), there is one atom of chlorine for each atom of sodium. This means that chlorine also has a combining capacity of 1. In the compound calcium chloride ($CaCl_2$), there is one atom of calcium for every two chlorine atoms. Each of the two chlorine atoms has a combining capacity of 1, so calcium must have a combining capacity of 2. In aluminum chloride ($AlCl_3$), for each aluminum atom there are three chlorine atoms. Each chlorine atom has a combining capacity of 1, so the combining capacity of aluminum must be 3.

Tables 3.7 and 3.8 list the combining capacities of some metals and some non-metals. If you know the combining capacities of the elements, you can predict the chemical formulas of compounds that contain only two elements.

Element	Symbol	Combining capacity
Aluminum	Al	3
Barium	Ba	2
Calcium	Ca	2
Magnesium	Mg	2
Potassium	K	1
Silver	Ag	1
Sodium	Na	1
Zinc	Zn	2

TABLE 3.7
Combining Capacities of Some Metals

TABLE 3.8 ▶
Combining Capacities of Some Non-metals

Element	Symbol	Combining capacity	Combined name
Bromine	Br	1	Bromide
Chlorine	Cl	1	Chloride
Fluorine	F	1	Fluoride
Iodine	I	1	Iodide
Oxygen	O	2	Oxide
Sulfur	S	2	Sulfide

It is simple to predict the chemical formulas of compounds in which the two elements have the same combining capacity. For example, the compound sodium bromide is made up of sodium and bromine. Each of these elements has a combining capacity of 1. This means that for each atom of sodium in the compound, there is one atom of bromine. Thus, the chemical formula is $NaBr$.

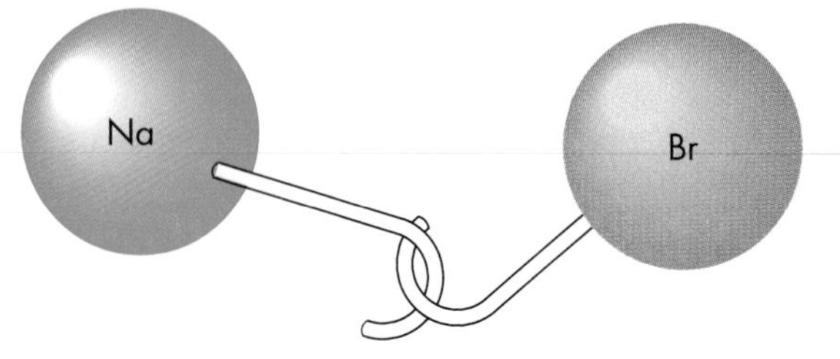

The compound calcium oxide is made up of calcium and oxygen. For both of these elements, the combining capacity is 2. Thus, one atom of calcium can combine with one atom of oxygen. The chemical formula of calcium oxide is CaO.

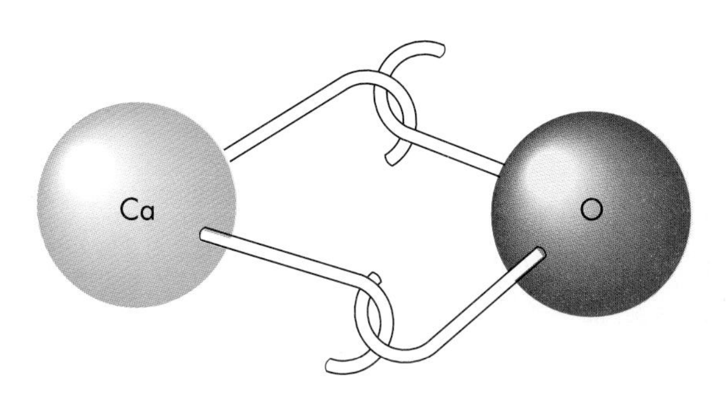

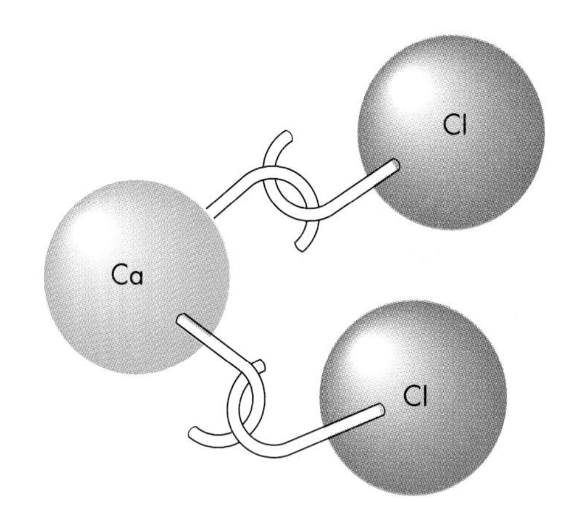

If the combining capacities of the two elements in a compound are different, then the numbers of atoms are also different. For example, calcium has a combining capacity of 2, and chlorine has a combining capacity of 1. Therefore, in the compound calcium choride, *one* atom of calcium combines with *two* atoms of chlorine. The chemical formula of calcium chloride is $CaCl_2$.

Aluminum has a combining capacity of 3. Oxygen has a combining capacity of 2. Therefore, in the compound aluminum oxide, *two* aluminum atoms must combine with *three* oxygen atoms. The chemical formula of aluminum oxide is Al_2O_3.

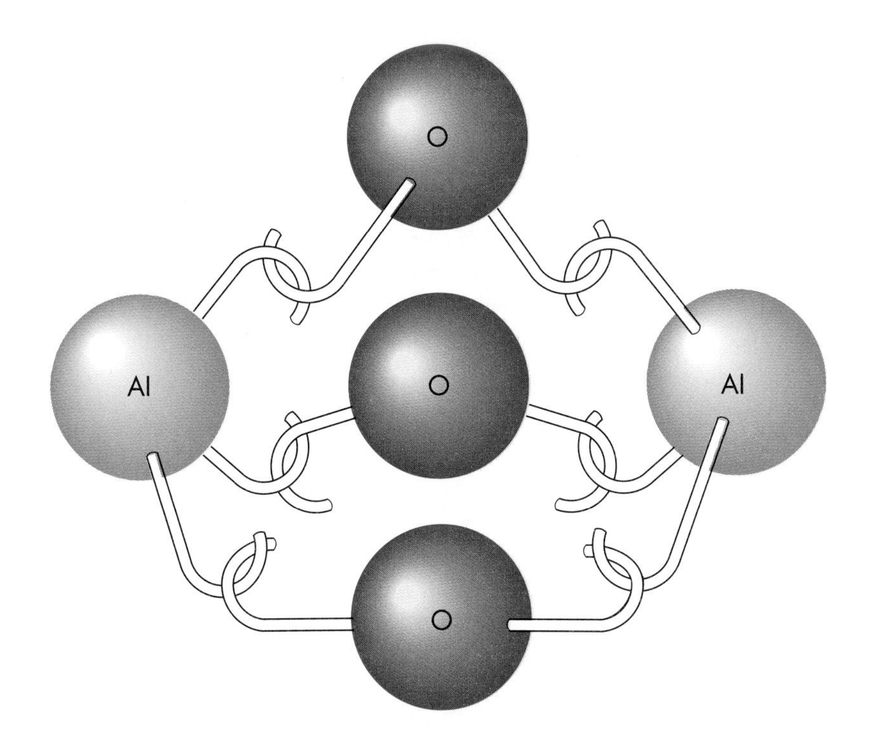

Some metals have different combining capacities in different compounds. You can see from Table 3.9 that copper, iron, lead, and nickel all have more than one possible combining capacity. In naming compounds of these metals, their combining capacity is shown in Roman numerals following the name of the metal. For example, there is a compound named iron(II) oxide ("iron-two-oxide") and another compound named iron(III) oxide ("iron-three-oxide"). Similarly, there are two compounds of copper and oxygen, shown in Figure 3.11.

Element	Symbol	Combining capacity
Copper	Cu	1, 2
Iron	Fe	2, 3
Lead	Pb	2, 4
Nickel	Ni	2, 4

TABLE 3.9
Some Elements That Have More Than One Combining Capacity

FIGURE 3.11 ▶
Copper(I) oxide is red and copper(II) oxide is black. The names are read as "copper-one-oxide" and "copper-two-oxide."

DID YOU KNOW

■ In a chemical change, an atom may combine with another atom or atoms to form a molecule. The combining atoms gain, lose, or share one or more electrons from their outer shell. The number of electrons of an atom that are gained, lost, or shared is called the atom's combining capacity, or valence.

▶ INSTANT PRACTICE

1. Here are some examples of chemical compounds. Use Tables 3.7 and 3.8 and the above method to determine their chemical formulas.
   (a) sodium fluoride
   (b) magnesium fluoride
   (c) potassium iodide
   (d) potassium chloride
   (e) zinc oxide
   (f) silver oxide
   (g) aluminum fluoride
   (h) aluminum sulfide
2. Refer to Table 3.6 on page 52. The table gives the names and chemical formulas of compounds containing copper and iron. For each of these compounds, write the formula in your notebook. Beside the formula, write the chemical name, including the combining capacity.

GROUPS OF ATOMS

Scientists have found that there are groups of atoms that sometimes act together as if they were a single atom. Several of these, shown in Table 3.10, act like non-metallic elements. These groups can combine with metals, forming compounds. Some examples of compounds that you may have already used or read about are sodium hydrogen carbonate ($NaHCO_3$), calcium hydroxide ($Ca(OH)_2$), and potassium nitrate (KNO_3).

■ John Polanyi, a chemist, won the Nobel prize in chemistry in 1986. When he tells of his studies of molecular motion, he describes a chemical reaction as a dance. "At the beginning of the dance—the chemical reaction—two atoms are holding hands. At the end, two others are holding hands. What happened?"

Name of group	Formula	Combining capacity
Carbonate	CO_3	2
Hydrogen carbonate	HCO_3	1
Hydroxide	OH	1
Nitrate	NO_3	1
Phosphate	PO_4	3
Sulfate	SO_4	2

TABLE 3.10 ***Some Common Groups of Atoms That Act Like Non-metals***

You are now able to understand the name of a familiar compound, copper sulfate, in which copper and the sulfate group are combined. In this compound, the combining capacity of copper is 2, so the complete name for the compound is copper(II) sulfate. The formula is $CuSO_4$.

■ A group of atoms that acts as a single unit with an overall charge is also called a radical.

CHEMICAL FORMULAS AND THE LAW OF CONSERVATION OF MASS

In Chapter 2, you learned that in a chemical reaction, the total mass of the products is always equal to the total mass of the reactants. Now that you are familiar with atoms, molecules, chemical symbols, and chemical formulas, you can understand why this is so. In a chemical reaction, the atoms are rearranged, but the total number of atoms is the same before and after the reaction. Also, the number of atoms of each element on the products side of the equation is the same as the number of atoms on the reactants side. Atoms are not created or destroyed during the reaction. For example, Figure 3.12 shows what happens when hydrogen burns in air. Hydrogen molecules and oxygen molecules are broken apart, and water molecules are formed.

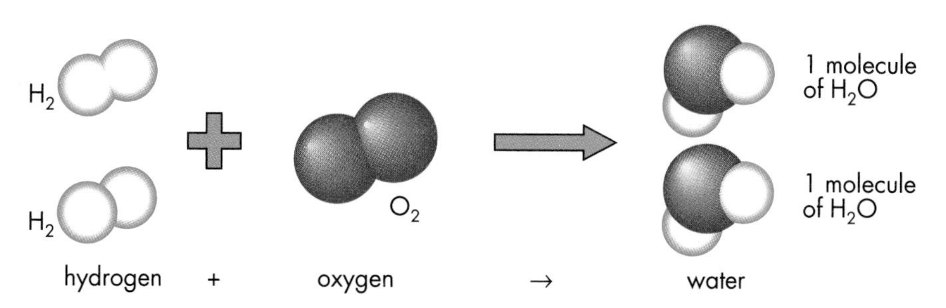

FIGURE 3.12 ◀
The combustion of hydrogen

▶ REVIEW 3.4

1. (a) What is a subscript?
   (b) In a chemical formula, what does a subscript tell you? Give two examples.
2. Name the elements present in the following compounds. Also tell the numbers of atoms of each element represented in the formula. (If necessary, refer to Appendix C to recall the symbols for the elements.)
   (a) ZnS (d) $FeCl_3$
   (b) $CuCl_2$ (e) Al_2O_3
   (c) $FeCl_2$ (f) Mg_3N_2
3. Name the following compounds. (If necessary, refer to Appendix C.)
   (a) $NaBr$ (c) ZnS
   (b) MgO (d) Al_2O_3
4. Write a chemical formula for each of the two compounds whose molecules are shown in Figure 3.13.

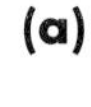

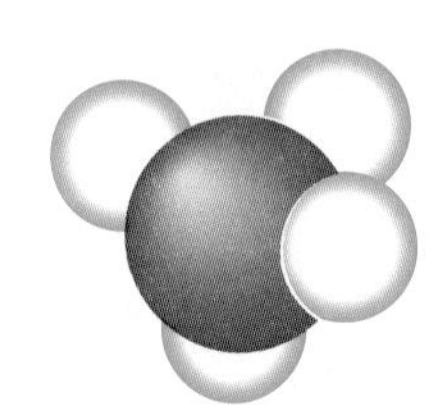

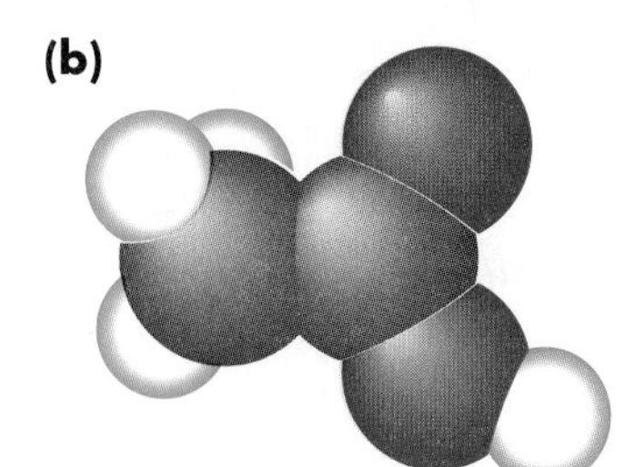

FIGURE 3.13
Methane, (a), is the main component of natural gas. It is used in home heating and for most Bunsen burners. Each molecule of methane contains one atom of carbon and four atoms of hydrogen.

Acetic acid, (b), which in diluted form is called vinegar, has molecules made up of two atoms of carbon, two atoms of oxygen, and four atoms of hydrogen.

5. Write a chemical formula for each of the following. For each, state whether it is an element or a compound.
   (a) one molecule of nitrogen that is made up of two atoms of nitrogen
   (b) sodium nitrate
   (c) one molecule of octane (a compound found in gasoline) that is made up of 8 atoms of carbon and 18 atoms of hydrogen

CHAPTER REVIEW

KEY IDEAS

■ Every element can be represented by a symbol that consists of one or two letters. This system of symbols is understood throughout the world.

■ Elements can be classified as metals or non-metals.

■ The names of many compounds indicate what elements are in them.

■ Chemical formulas tell what elements are present in a compound and the relative numbers of the different kinds of atoms.

■ Molecules consist of two or more atoms joined together to make a larger particle.

■ The combining capacity of an element can be used to predict how that element will join with the atoms of other elements to form compounds.

■ In a chemical reaction, atoms are rearranged but are not created or destroyed.

VOCABULARY

atom
metal
ductile
malleable
conductor
non-metal
chemical formula
molecule
combining capacity

V1. In a mind map based on the word "symbol," show how all the terms in the list above can be linked.

V2. Using as many words as possible from the vocabulary lists in this chapter and in Chapter 2, write a story about the discovery of a new element.

CONNECTIONS

C1. Identify the elements in the following compounds, and tell the relative numbers of atoms of the elements.

(a) K_2CO_3 (c) Ag_2S
(c) $CuBr_2$

C2. Which of the following formulas are of elements and which are of compounds?

(a) P_4 (c) $PbSO_4$
(b) C_3H_8 (d) Br_2

C3. Write names for the following compounds:

(a) $CdCl_2$ (d) FeI_2
(b) CaI_2 (e) PbS
(c) CuF_2

C4. (a) What does "molecule" mean when it refers to an element?
(b) How is the meaning different when "molecule" refers to a compound?

C5. Write chemical formulas for the following compounds:

(a) magnesium chloride
(b) silver iodide
(c) zinc oxide
(d) aluminum sulfide

C6. (a) State the law of conservation of mass.
(b) State the law in another way, using "atoms" in your statement.

C7. Using what you know about the combining capacity of elements and the combining capacity of groups of atoms, explain the meaning of the following chemical formulas.

(a) $Cu(NO_3)_2$
(b) $Al(OH)_3$
(c) $Ca_3(PO_4)_2$

EXPLORATIONS

E1. Helium, a very unreactive element, belongs to a "family" of elements called the noble gases. Do some research on this family to find out the names of the other members and where they are found on the periodic table of the elements (Appendix C).

E2. Mercury is an element that is both useful and very poisonous. Do some research and write a report on the uses of mercury. Include information about mercury poisoning in your report.

E3. Find out how carbon dioxide is used in fire extinguishers. In your notebook, draw a diagram showing what is inside the extinguisher.

REFLECTIONS

R1. Examine Table 3.3 on page 43. Explain why the ancient Romans were familiar with the elements listed in this table but were not familiar with many others known to us today.

R2. Compared with a science student in China or the Middle East, what advantage do you have in learning the chemical symbols?

R3. What advantages can you think of for having a set of chemical symbols that are accepted everywhere in the world?

CONTROLLING CHEMICAL REACTIONS

In a campfire, chemical reactions release energy that is given off to the surroundings. As the wood burns, light and heat are produced. By adding more wood to the fire, you can keep the reaction going for hours, producing more heat.

Other chemical reactions absorb energy rather than releasing it. When a cake bakes in an oven, for example, the chemical changes happen only while the cake is being heated. That is, they happen only while the cake absorbs, or takes in, heat from its surroundings.

If you want to bake a cake in a hurry, you could try to cook it more quickly by turning up the heat in the oven. If you do, you might find that different reactions occur. The cake might turn out to be hard and blackened. If you try to cook it more slowly in a cooler oven, the reactions might not occur at all — the cake might remain uncooked all day.

As you cook or as you burn a campfire, you are making use of chemical reactions. You are also making them happen more quickly or more slowly to suit your purpose. In this chapter, you will learn more about how energy may be involved in chemical reactions and how you can control the speed of reactions.

ACTIVITY 4A / KEEPING WARM WITHOUT BURNING UP

When you are camping on a cold night, a fire keeps you warm. It heats your soup and takes the chill out of your bones. But a fire that is out of control can be dangerous. You need to know how to keep a fire just the right size. On a hot July evening, you might want just enough heat to cook your dinner, but for a Halloween party in October, you might want a big bonfire.

Divide a page in your learning journal into two columns. In one column, describe several methods you could use to slow down a fire while still allowing it to burn. In the second column, list several ways in which you could speed up the fire. ❖

■ 4.1 ENERGY IN CHEMICAL REACTIONS

Chemical energy is energy that is stored in chemical compounds. The compounds that you eat, such as carbohydrates, proteins, and fats, have chemical energy stored in them. As these substances take part in reactions in your body, energy is gradually released. You use this energy to carry on your daily activities.

EXOTHERMIC AND ENDOTHERMIC REACTIONS

A chemical reaction that releases energy to its surroundings is known as an **exothermic reaction**. For example, when a flashlight is turned on, an exothermic reaction occurs in the batteries. Chemical energy, stored in the chemical compounds in the batteries, is changed to electrical energy and then to light. Burning is an exothermic reaction as well. During burning, energy is released in the form of heat and light. The exothermic reaction shown in Figure 4.1 also produces sound and mechanical energy.

FIGURE 4.1 ◀
An explosion is an exothermic reaction that releases a lot of energy in a very short time.

EXTENSION

■ You can easily observe some endothermic reactions that use light as a source of energy. Obtain a dark-colored sheet of construction paper. Place objects such as scissors, a coin, and an eraser on the paper and leave them in bright sunlight for at least one day. Where on the paper do the endothermic reactions occur?

Rapid exothermic reactions can be extremely dangerous, particularly in an enclosed space like an underground mine. For example, in coal mines, methane gas often seeps into the air from the coal. Methane can collect in small spaces, then react explosively with oxygen if a small amount of energy is supplied to get the reaction started. In the days before electricity, when miners worked by candlelight, the heat from a candle could be enough to set off an explosion. Even with modern equipment and safety rules, mines are dangerous places. In 1992, an explosion in a coal mine killed 26 miners.

For some chemical reactions, energy is needed to make the reaction occur. Such a reaction absorbs energy from its surroundings and is called an **endothermic reaction**. The most familiar endothermic reactions are those that occur during cooking. As you fry an egg, for example, the white part of the egg begins to become solid. This change happens only while the egg is being heated. If you were to remove the egg from the heat, the chemical reaction would stop. The egg would remain half cooked until you heated it again. Another familiar endothermic reaction is the one that occurs when you take a picture with a camera. Light that reaches the film causes chemical reactions, producing a pattern of colors on the film. More chemical reactions occur when the film is developed and the color photographs are made. The reaction that occurs as you recharge a battery is also an endothermic reaction (Figure 4.2).

Many of the reactions carried out in industry are endothermic reactions (Figure 4.3). Industries using endothermic reactions require large amounts of energy. Oregon is a good site for such industries since hydroelectric energy is plentiful and relatively inexpensive (Figure 4.4).

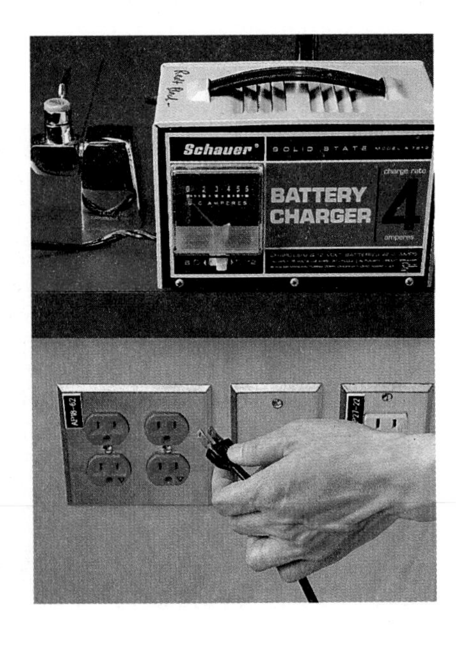

FIGURE 4.2 ▶
Plugging in a battery charger causes an endothermic chemical reaction in the batteries. The reaction, which is caused by electrical energy, results in chemical energy being stored in the batteries. The chemical energy is released later when the batteries are used.

FIGURE 4.3
This aluminum smelter uses large amounts of energy to produce aluminum metal from its ore, bauxite.

FIGURE 4.4
Without the electrical energy carried from the dam, the aluminum smelter (Figure 4.3) could not operate.

STARTING A CHEMICAL REACTION

If burning is an exothermic reaction, why do you need to add heat to start something burning? Why do you need to use a match? Most chemical reactions involve at least two steps. First, the atoms of the reactants are broken apart from one another. Then, these atoms recombine in new ways to form the products. For example, to start a match burning, you heat one end by striking it (Figure 4.5). When a small part of the match is burning, it produces enough heat to start more of the match burning, and so on.

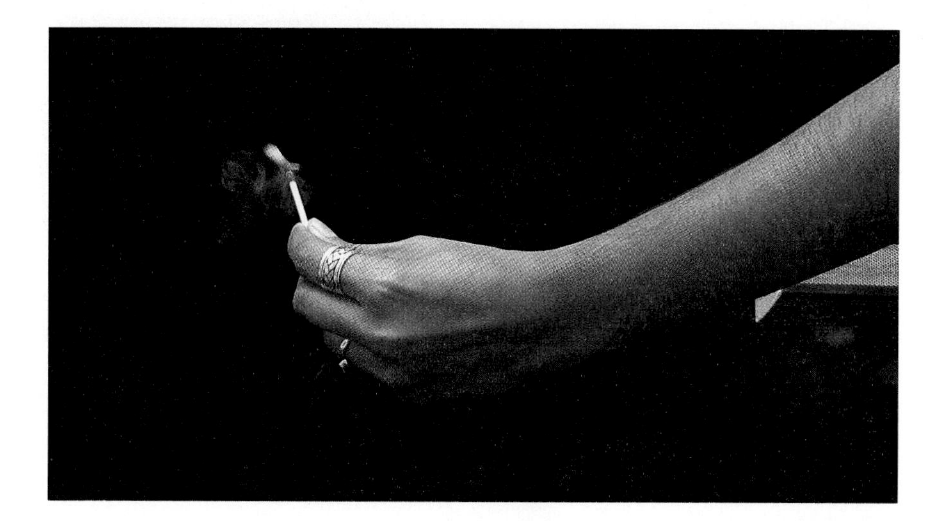

FIGURE 4.5 ◀
Although a little energy must be used to start the reaction, burning is an exothermic reaction.

ENERGY FROM FOOD

Light energy can be absorbed by green plants and then stored as chemical energy in the plants through a series of chemical reactions called "photosynthesis." The chemical reactions involved in photosynthesis are endothermic. Some of the energy may be stored in the plants in their roots, stems, or fruits. When you eat these parts, the energy is released in your body in a series of exothermic reactions.

ACTIVITY 4B / ENERGY STORED IN FOOD

Have you ever wondered how much energy is stored in food? How much energy is there, for example, in a bag of peanuts? In this activity, you will burn a peanut and use the energy given off to heat water. From your measurements of the mass of the water and the change in temperature, you will calculate the amount of energy that is given off. You will then be able to estimate the amount of energy provided to you when you eat peanuts.

MATERIALS

safety goggles
apron
one fresh peanut
balance
paper clip
100 mL beaker
thermometer
clamp
rubber stopper
100 mL graduated cylinder
aluminum foil
wire
ring stand
match or Bunsen burner

PROCEDURE

1. In your notebook, prepare a data table like Table 4.1.
2. Put on your safety goggles and apron.
3. Measure and record the mass of a peanut. Follow your teacher's instructions for measuring mass accurately in this activity. Then set the peanut on a stand made from a bent paper clip (Figure 4.6).
4. Set up apparatus similar to that shown in Figure 4.7. Arrange the aluminum foil so that it keeps the heat near the beaker as you burn the peanut.

Volume of water	=	______	mL
Mass of water	=	______	g
Final temperature of water	=	______	°C
Starting temperature of water	=	______	°C
Change in temperature	=	______	°C
Mass of peanut before burning	=	______	g
Mass of peanut remaining	=	______	g
Mass of peanut that burned	=	______	g

TABLE 4.1 ***Sample Data Table for Activity 4B***

5. Place 50 mL of water in the beaker. Record the starting temperature of the water.
6. Ignite the peanut with a match or Bunsen burner. Quickly move the lighted peanut under the beaker and fold the aluminum foil to keep the heat in.
7. As soon as the peanut has stopped burning, record the final temperature of the water.
8. Measure the mass of any of the peanut that did not burn.

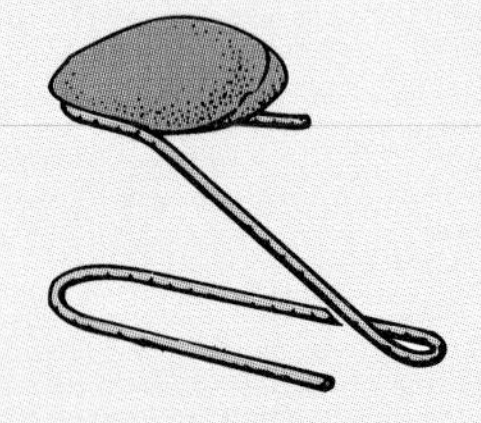

FIGURE 4.6 ◀
Bend the paper clip to make a stand that holds the peanut off the desktop.

EXTENSION

■ How do you think you could compare the energy per gram in other types of nuts? Devise a method, ask your teacher's permission to carry out your experiment, and try it. How do the results compare with the results for peanuts? Based on your results, which type of nut supplies the most energy per gram? Check these results by comparing them with the energy values in a reference book.

Type of exercise	Energy used (kJ/min)
Walking	21
Bicycling	36
Running	84

TABLE 4.2
Energy Used During Exercise

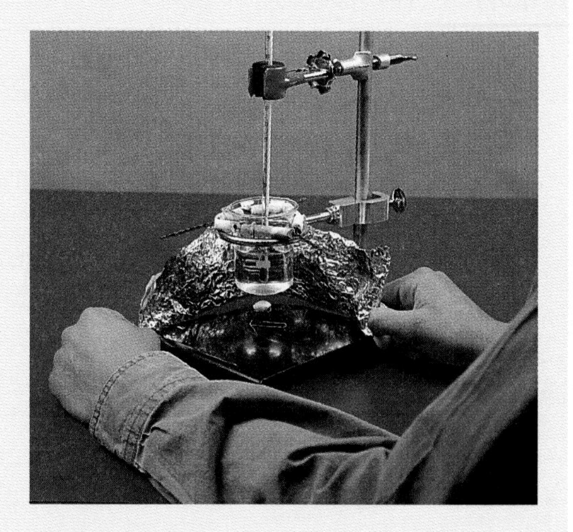

FIGURE 4.7 ▶
Cover the top and sides of the beaker with aluminum foil and make a "chimney" around the burning peanut. What does the foil do? Do you think it's best to have the foil's shiny side toward the peanut, or the outside air?

DISCUSSION

1. Why do you think you were told to set the peanut on a stand before burning it?
2. Calculate the mass of the water you used. Remember that 1 mL of water has a mass of 1 g.
3. Calculate the change in temperature of the water used in this activity.
4. Calculate the mass of the peanut that burned.
5. Energy is measured in joules (J). (To change the temperature of 1 g of water by 1°C, 4.2 J are required.) Calculate the amount of energy given off in your experiment, using this equation:

$$\text{Energy} = (\text{mass of water}) \times (\text{change in temperature}) \times 4.2 = \_\_\_\_\_\_ \text{ J}$$

6. Calculate the amount of energy given off per gram of peanut, using this equation:

$$\text{Energy} = \frac{(\text{energy given off})}{(\text{mass of peanut that burned})} = \_\_\_\_\_\_ \text{ J for each gram}$$

7. (a) When you eat peanuts, chemical reactions in your body release the energy that was stored in them. Calculate how much energy you obtain when you eat a 500 g bag of peanuts. (Convert your answer to kilojoules: 1000 J = 1 kJ.)
   (b) Refer to Table 4.2 and calculate how many minutes you could walk, bicycle, or run using the energy from a 500 g bag of peanuts.
8. You probably found a black solid on the bottom of the beaker after this investigation.
   (a) What do you think this black material is?
   (b) Do you think that the production of the material has any effect on your calculation of the energy released? Explain.
9. Is a burning peanut an example of an exothermic reaction or an endothermic reaction? Explain. ❖

ENERGY AND THE ENVIRONMENT

The earth is constantly bathed in light from the sun. Plants absorb some of this energy and store it as chemical energy through photosynthesis. This energy is then passed on to animals when they eat the plants. In this way, the light from the sun is the basis for all the food chains and food webs on our planet. All the forests, farms, and gardens in the world use energy from the sun.

Scientists and technologists have learned how to use some of the energy from the sun—for example, in the solar cells that are used in calculators and cameras. But humans have not yet learned to use the energy as efficiently as plants do in photosynthesis. For the near future, our lives will continue to depend on food produced by plants in a healthy environment.

▶ REVIEW 4.1

1. (a) What is the difference between exothermic and endothermic reactions?
   (b) What are the similarities?
2. Give examples of two exothermic reactions that you use in your daily life.
3. Give two examples of endothermic reactions that you use in your daily life.
4. What endothermic reaction is taking place in Figure 4.8?
5. Imagine an endothermic chemical reaction in which two reactants, *A* and *B*, produce two products, *C* and *D*. This chemical reaction can be written this way:

   $A + B \rightarrow C + D$

   (a) What would you expect to observe during the reaction?
   (b) Which part of this chemical reaction contains more stored energy, *A* and *B* together, or *C* and *D* together?
   (c) On which side of this chemical equation would you place the word energy? Explain your reasoning.

FIGURE 4.8
Leaves and light are needed for a very important chemical reaction.

■ 4.2 FACTORS AFFECTING REACTION RATE

You can cook an egg in only three minutes. It takes longer to cook a hamburger, and even longer to cook a whole chicken or a turkey. Other reactions also take a lot of time. You may have painted a room using paint that becomes "dry to the touch" in 2 hours. But the instructions tell you not to put on a second coat for at least 24 hours. The extra time is required for further chemical reactions that make the paint drier and harder.

Reaction rate is the speed of a chemical reaction. Some chemical reactions have a naturally fast rate; others are naturally slow. Sometimes people want to slow down the rate of reactions. For example, how do you slow down the rate at which food spoils? How can you slow down or stop the rusting of iron? Sometimes people want to speed up the reaction rate. When you cook hamburgers, what can you do to make them cook more quickly? If you were developing a new headache tablet, you would want it to work as fast as possible. But you would need to make sure that speeding up the effect was not harmful to people. When products are made in factories, chemists work to find just the right conditions so that certain reactions will occur more quickly and others will be slowed or stopped.

In this section, you will investigate how four factors affect the rate of chemical reactions: temperature, surface area of solids, concentration of solutions, and the presence of a catalyst.

TEMPERATURE

Temperature affects the rate of all chemical reactions. For some reactions, you might not be able to notice or measure the effect. In other reactions, the effect may be easy for you to see. For example, with just the right temperature, a hamburger cooks to perfection. But if you try to hurry it up with too much heat, other reactions occur as well. You could end up with a blackened lump!

ACTIVITY 4C / HOT OR COLD?

In this activity, you will see how temperature affects reaction rate.

MATERIALS

safety goggles
apron
four large beakers
(250 mL or larger)
ice
cool water
hot water
very hot water
(e.g., from a kettle)
thermometer
four Alka Seltzer® tablets
watch or timer

■ **Use boiling water only in heat-resistant glassware (e.g., Pyrex). With ordinary glass, use hot, not boiling, water.**

PROCEDURE

1. Prepare a data table in which you can record your results for each of the four glasses. Your table should have two columns, headed "Temperature" and "Time."
2. Put on your safety goggles and apron.
3. Fill one beaker with ice and water. Fill the others with cool, hot, and very hot water.
4. Use the thermometer to measure the temperature of the water in each of the four beakers, and record these temperatures in your table.
5. Before the hot water has a chance to cool, drop one Alka Seltzer® tablet into each of the four beakers at the same time. Record how long it takes each tablet to completely dissolve.
6. Wash your hands thoroughly after you have completed this activity.

DISCUSSION

1. What happened to the Alka Seltzer® tablets when you put them in the water?
2. What evidence do you have that a chemical reaction occurred?
3. (a) In which beaker was the chemical reaction the slowest? In which beaker was it fastest?
   (b) How did temperature affect the rate of the chemical reaction?
   (c) Draw a graph of time and temperature for these results. Label the x-axis and the y-axis. Give your graph an appropriate title. Plot your four points on the graph and join them with a smooth curve.
   (d) Use your graph to predict the amount of time that would be necessary for this reaction to occur at 40°C and at 60°C.
4. Suppose you obtained data for this reaction at temperatures of 0°C, 20°C, 50°C, and 95°C. Suppose that your friend obtained data for 0°C, 30°C, 70°C, and 80°C.
   (a) How do you think your two graphs would be different?
   (b) How would you expect them to be the same?
5. According to the **kinetic molecular theory,** particles in matter are constantly moving, and they move more quickly at higher temperatures. Use this theory to explain your results in this investigation. ❖

SURFACE AREA OF A SOLID

For two substances to react, they must come into close contact. The closer the contact, the more likely they are to react. When one of the reactants is a solid and the other is a liquid or a gas, the chemical reaction occurs on the surface of the solid. The amount of surface area affects the rate of the chemical reaction. Which do you think has a larger surface area: a large chunk of a solid, or the same mass of solid broken into tiny pieces?

ACTIVITY 4D / COMPARING SURFACE AREA

In this activity, you will compare the surface area of a large block and the surface area of the same block cut into smaller blocks.

PROCEDURE

1. Examine Figure 4.9. In both part (a) and part (b), the volume of the blocks is 8 cm^3.

2. Calculate the total surface area of the large block.

3. Suppose the large block were cut into eight smaller blocks of 1 cm^3 each. What is the total surface area?

FIGURE 4.9
Which has more surface area, the block in (a) or the blocks in (b)?

DISCUSSION

1. When a large piece of a solid is broken into smaller pieces, how is the total surface area affected?

2. Predict how the size of pieces might affect the rate of a chemical reaction. Use your result from question 1 to explain your prediction.

3. Which burns more quickly, one large log or several small sticks? Explain this in relation to the surface area of the wood.

4. Why do you think that laundry soap is sold as a powder instead of in cubes? ❖

CONCENTRATION OF A SOLUTION

As you learned in your earlier study of solutions, **concentration** refers to the amount of solute that is dissolved in a certain amount of solution. In Activity 2B, you used dilute hydrochloric acid, which contains relatively little acid in water. (Concentrated hydrochloric acid contains much more acid in the same volume of water and is too dangerous to use in a classroom.) How do you think the rate of a chemical reaction would be affected if the acid were more concentrated or more diluted?

ACTIVITY 4E / TESTING PREDICTIONS

In this activity, you will make and test predictions about the reaction between calcium carbonate and hydrochloric acid (Figure 4.10). The word equation for this reaction is:
calcium carbonate + hydrochloric acid →
carbon dioxide + water + calcium chloride

You will investigate either how the surface area of the solid affects the rate of reaction or how the concentration of a solution affects the rate of the reaction.

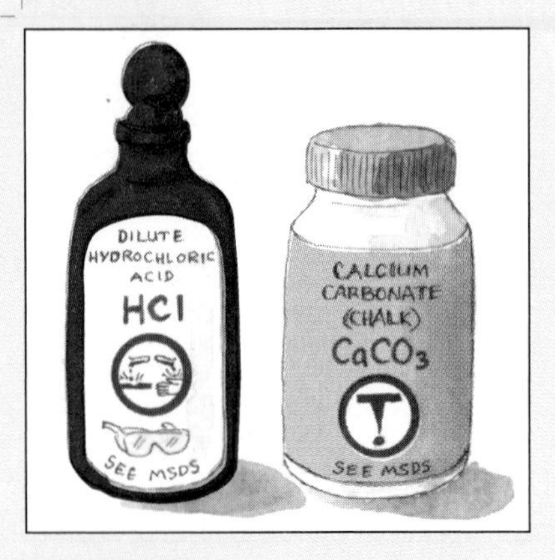

FIGURE 4.10
These two substances react together.

■ **Hydrochloric acid is very corrosive and can cause damage to your eyes and skin. If you get any in your eyes or on your skin, immediately rinse with cold water for 15 or 20 minutes and inform your teacher.**

■ **To mix, make gentle circular motions. Do not shake the tube up and down. Point the test tube opening away from yourself and others.**

■ **Be sure to wear your safety goggles and apron at all times.**

PROCEDURE

1. Decide which of the two variables your group will investigate. Design a controlled experiment.
2. Write a step-by-step procedure, including a list of the materials you will need. (An easy way to make small pieces of chalk from a larger piece is to use a mortar and pestle, as shown in Figure 4.11.) Obtain your teacher's permission, and carry out your investigation.
3. Wash your hands thoroughly after you have completed your investigation.

DISCUSSION

1. (a) Describe the reactants you used in this reaction.
   (b) Describe the products of the reaction.
2. What evidence do you have that the same reaction occurred in each test tube?
3. Did any of your observations suggest that the reaction was endothermic or exothermic? Explain your answer.
4. (a) How did the variable you investigated affect the rate of the reaction between calcium carbonate and hydrochloric acid?
   (b) Explain the effect.
5. Choose one member of your group to present your results to the rest of your class so they may compare everybody's results.
6. How do you think the rate of this reaction would be affected if the reactants were heated? ❖

FIGURE 4.11
If a mortar and pestle are available, use them to make powdered chalk from larger pieces.

CATALYSTS

A **catalyst** is a substance that speeds up a chemical reaction without being used up in the reaction. Catalysts may have this effect because they lower the amount of energy required for the reaction to occur. This means that the reaction can occur at a lower temperature.

ACTIVITY 4F / INVESTIGATING CATALYSTS

Hydrogen peroxide is a compound that you can buy at a drugstore. Hydrogen peroxide decomposes, or breaks down, fairly easily, producing a gas. In this activity, you will identify the gas. You will also investigate the effect of two substances on the rate of decomposition.

MATERIALS

Part I

safety goggles
apron
3% hydrogen peroxide solution (H_2O_2)
manganese dioxide powder (MnO_2)
two 2 cm × 15 cm test tubes
bent glass tubing with single-hole rubber stopper at one end
400 mL beaker
ring stand
ring and wire gauze
two clamps

Part II

safety goggles
apron
two test tubes
test tube rack
10 mL graduated cylinder
3% hydrogen peroxide solution (H_2O_2)
raw liver (small cube, about 0.5 cm^3)
cooked liver (small cube, about 0.5 cm^3)

PART I Teacher Demonstration

PROCEDURE

1. Put on your safety goggles and apron.
2. Your teacher will set up the apparatus shown in Figure 4.12, then add 0.5 g manganese dioxide powder to the hydrogen peroxide and immediately replace the rubber stopper. Record your observations in your notebook.
3. Your teacher will collect the gas given off in this chemical reaction and then remove the test tube while covering its mouth.
4. Your teacher will test for oxygen by holding a glowing splint to the gas. Record your observations in your notebook.

DISCUSSION

1. What was the effect of the manganese dioxide on the hydrogen peroxide?
2. As far as you can tell, was the manganese dioxide changed during this reaction? Explain your answer.
3. After many experiments like this one, scientists have concluded that manganese dioxide acts as a catalyst in this chemical reaction. Describe what experimental procedure you could follow to test this.
4. Write a word equation for the decomposition of hydrogen peroxide. (Hint: There are two products. To help you discover the name of the second product, refer to the model of the hydrogen peroxide molecule in Figure 3.5 on page 50.)
5. The gas produced in this reaction was collected by a technique called the displacement of water (Figure 4.12). Explain this technique for collecting a gas.

PART II Student Experiment

PROCEDURE

1. Put on your safety goggles and apron.
2. Carefully pour 10 mL of hydrogen peroxide solution into each of two test tubes. Observe the solution carefully: if the bottle was freshly opened, you may see bubbles form very slowly as the hydrogen peroxide decomposes.

EXTENSION

■ Your teacher may choose to demonstrate how a cube of sugar burns and how it burns if it is first sprinkled with ashes. How do the ashes affect the reaction?

3. To one test tube, add the piece of raw liver; to the other, add the piece of cooked liver. Record your observations in your notebook.
4. Wash your hands thoroughly after you have completed this activity.

DISCUSSION

1. (a) What was the effect of the raw liver on the hydrogen peroxide solution? (b) What was the effect of the cooked liver?
2. Liver contains a catalyst that affects how quickly the hydrogen peroxide decomposes. What can you infer from this activity about the effect of cooking on this catalyst? ❖

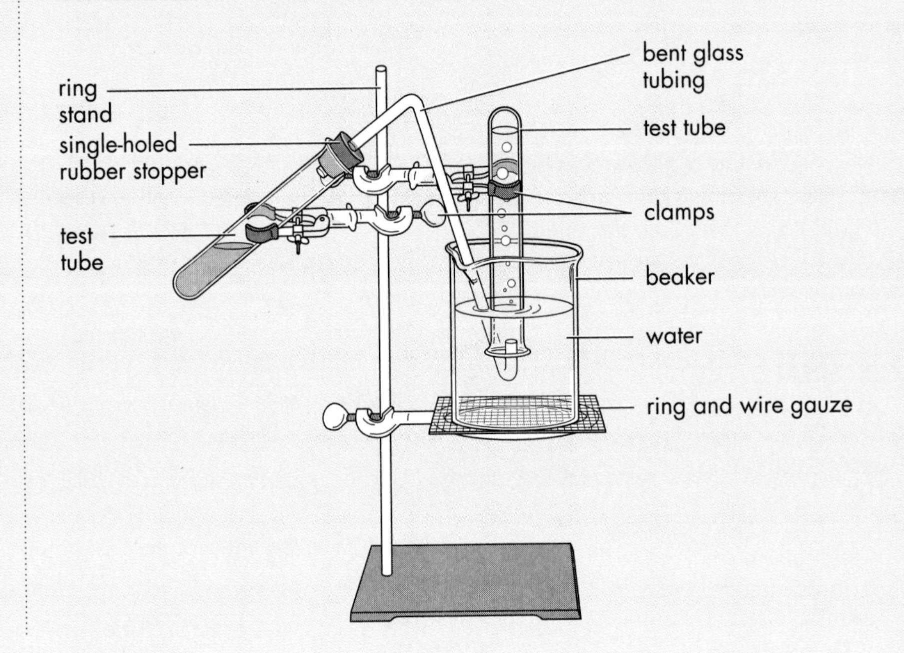

FIGURE 4.12 ▶
When the apparatus is set up, the inverted test tube is full of water. The gas produced in the reaction displaces the water and bubbles up to the closed end of the inverted test tube.

DID YOU KNOW

■ All mammals have a liver. The liver produces essential chemicals for the body, stores supplies of other chemicals, helps to digest food, and removes toxic materials from the blood. Beef liver, pork liver, and chicken liver are commonly sold in the meat department of grocery stores.

USING CATALYSTS

Catalysts are used in automobiles. A car engine produces small amounts of dangerous gases. In the past, these gases were released to the air through the car's tailpipe. To reduce the amounts of these gases, laws require that all new cars have a mixture of catalysts in the exhaust system. These catalysts are contained in a device called a catalytic converter (Figure 4.13). This device causes exhaust gases to react more completely. As a result, even smaller amounts of dangerous gases are emitted.

In your body, because of the action of catalysts, chemical reactions can release the energy stored in the food you eat. **Enzymes** are catalysts that are found in living organisms. Enzymes are essential to life. They affect the chemical reactions that go on both inside cells and throughout the body. For example, in Activity 4B, you released the energy stored in a peanut by burning it. The energy could not be

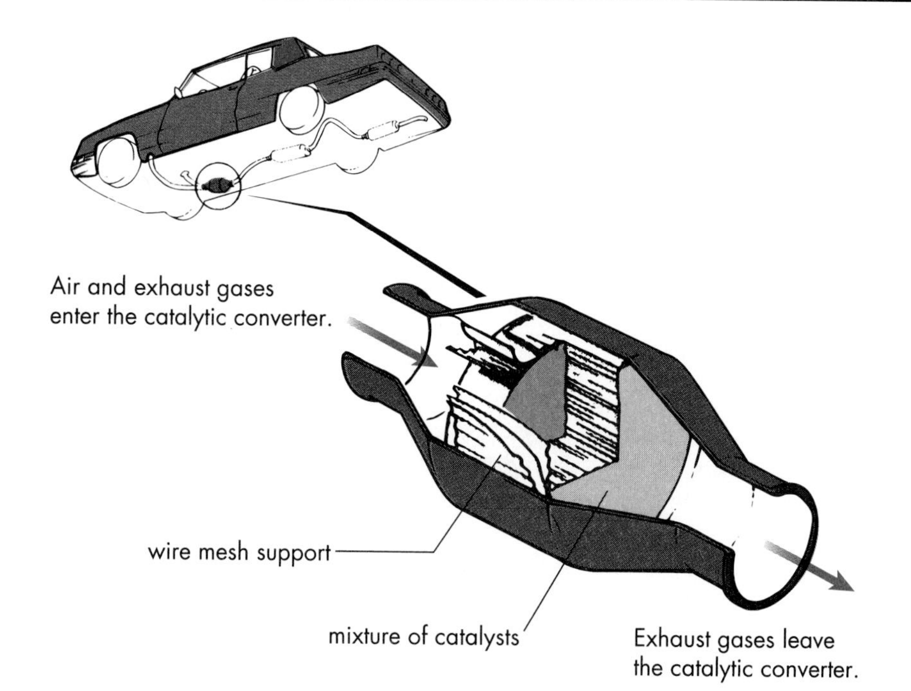

Air and exhaust gases
enter the catalytic converter.
wire mesh support
mixture of catalysts
Exhaust gases leave
the catalytic converter.

FIGURE 4.13 ◀
The insides of a catalytic converter are heavily folded. Why do you think a catalytic converter is designed this way? (Hint: see Activity 4D.)

released in the same way inside your body, because the high temperature would damage your body. Because of enzymes, the energy is released gradually in many small steps.

Enzymes are also essential for other kinds of reactions in your body. There are many different enzymes, each catalyzing a different small step in a series of reactions. Some poisons act by destroying enzymes. For example, the deadly poison cyanide blocks the action of the enzyme that allows cells to use oxygen. Oxygen is necessary for the cells to release energy from food, so when this enzyme is blocked, cells die.

Enzymes can also be destroyed by too much heat. In Activity 4F, you saw that the enzymes in the cooked liver had been destroyed by cooking, so they no longer acted as catalysts in the breakdown of hydrogen peroxide. In the raw liver, the enzymes were still active.

DID YOU KNOW

■ The reaction between wood and oxygen in woodstoves is usually not complete. The incomplete burning of wood produces a black, oily material called creosote, which is deposited inside the chimney. The creosote itself may start to burn if the fire gets too hot, causing a fire in the chimney. Some woodstoves have built-in catalysts that cause the wood to burn more completely, thus preventing creosote from forming.

▶ REVIEW 4.2

1. (a) What is meant by "reaction rate"?
(b) Name four variables that affect reaction rate.

2. List three chemical reactions in your everyday life whose reaction rate you might want to control.

3. What is a catalyst?

4. How does a catalyst change the rate of a reaction?

5. What are enzymes and why are they essential to life?

6. Why do you think a small amount of catalyst or enzyme can have such a large effect on a chemical reaction?

7. Give an example of a catalyst that is not an enzyme.

8. Enzymes occur in the walls of our intestines to digest food. The walls have many folds in them. What purpose could the folds serve?

COMPUTER APPLICATION

HOW COMPUTERS ARE USED IN CARS

Have you ever seen a picture of a 1958 Cadillac? It had enormous metal fins swooping up on either side at the back.

Since then, cars have certainly changed, and not just in appearance. Today's cars also operate very differently. Much of this difference has to do with computers.

In most modern cars, a computer manages the engine. It controls the timing of the ignition spark, the mix of air and fuel, and the idling speed. The computer can make adjustments depending on how cold the engine is, the temperature of the outside air, or how fast the car is driven.

This engine management computer plays an important role in how well the car's catalytic converter works. A catalytic converter is a box that sits in front of the muffler underneath the car. The box contains the rare elements platinum and rhodium. They act as catalysts, encouraging exhaust gases that flow from the car's engine through the converter to burn up more completely. This added combustion greatly reduces the amount of polluting gases, such as unburned hydrocarbons, that spew from the car's tailpipe.

Several chemical reactions can occur within a converter. For instance, unburned hydrocarbons react with oxygen to form carbon dioxide and water. Carbon monoxide reacts with oxygen to form carbon dioxide. Since oxygen is a reactant in these chemical reactions, it is important that the exhaust gases going into the converter contain enough oxygen. An oxygen sensor located just in front of the converter sends information about oxygen levels to the car's computer. It adjusts the amount of air in the engine's fuel-air mix so that the converter works properly.

In some new cars, computers do much more than fine-tune the engine. They automatically check fuel and oil levels, and turn on headlights and windshield wipers.

Computers control the brakes and prevent the wheels from locking up. They can switch a car from two-wheel drive to four-wheel drive when the slip of the primary wheels exceeds a certain point.

Car manufacturers are testing a new cruise control for cars. You'll be able to set how fast you want your vehicle to go. The computer will readjust that setting and slow down the car if traffic ahead is moving too slowly.

Some cars have computerized navigation systems to help drivers find their way. Drivers punch in information that tells where they are when they start, and the screen displays their whereabouts on a map as they drive.

Today's computer-managed car may not look as flashy as the '58 Caddie, but it has features that drivers back then could only dream of.

What computer-controlled features would you like to see in a car?

This on-board computer, called TravTek, can provide drivers with video map displays, audio and visual driving directions, and information about traffic conditions.

4.3 METALS AND CORROSION

Various metals corrode in different reactions and at different rates. The corrosion of iron, called "rusting," can eventually eat up all the metal. Gold, however, does not corrode. Aluminum and copper do corrode, but only at the surface of the metal; the product of the corrosion forms a thin layer on the surface that protects the metal underneath. Some metal **alloys**, which are solid solutions of metals, do resist corrosion, but they are too expensive to use in large-scale construction. Brass, which is a solution of copper and zinc, is an example of an alloy.

The Sun/Tronic House, located in Greenwich, Connecticut, is covered with sheets of copper (Figure 4.14). Over a period of years, the red-brown color of copper is gradually replaced with a green layer as the surface of the copper reacts with gases in the atmosphere. The green layer is a mixture of several compounds of copper, including copper(II) carbonate and copper(II) sulfate.

Silver becomes covered with a black coating of tarnish if it comes into contact with sulfur-containing foods such as eggs or mustard. The black coating is silver sulfide. Silver tarnishes slowly if left out in air — the more sulfur-containing pollutants in the air, the more quickly it tarnishes. The black layer can be removed by polishing the silver.

FIGURE 4.14
The Sun/Tronic House has a copper roof. What other buildings have copper roofs? Why do you think the architects chose copper?

Iron and steel are used extensively in construction. For both of these substances, corrosion is a serious problem. The rust that forms is a crumbly, flaky compound that does not protect the metal underneath from further corrosion. In fact, further corrosion is speeded up by the presence of rust.

ACTIVITY 4G / RATE OF RUSTING

In this activity, you will investigate variables that affect the rate of rusting.

MATERIALS

safety goggles
apron
rusted objects (or photographs of rusted objects)
small iron objects
250 mL beakers
plastic wrap
cotton balls or paper towels
various substances, depending on factors to be investigated (e.g., salt, dilute hydrochloric acid, copper, zinc)

■ **Hydrochloric acid is very corrosive and can cause damage to your eyes and skin. If you get any in your eyes or on your skin, immediately rinse with cold water for 15 or 20 minutes and inform your teacher. Do not use your sense of taste in this activity.**

Dry conditions	versus	damp conditions
In water with no air	versus	in water with air above
Damp conditions	versus	damp conditions with acid
Damp conditions	versus	damp conditions with salt
Damp conditions	versus	damp conditions, touching copper
Damp conditions	versus	damp conditions, touching zinc
Damp conditions in freezer	versus	damp conditions at room temperature

TABLE 4.3 ***Suggestions of Variables to Investigate***

PROCEDURE

1. Put on your safety goggles and apron.
2. Observe several rusted objects or several photographs of rusted objects. Some should be only a little rusty and some should be almost completely rusted away. As a class, decide on a "rating scale" for rusting. For example, an object with a very small amount of rust might be rated 1, whereas an object that is almost completely rusted might be rated 10.
3. Decide what variable(s) you will investigate. Table 4.3 lists some possibilities.
4. Plan your procedure, obtain your teacher's approval, and set up the equipment needed (Figure 4.15). Be sure to wash your hands after handling the materials used for the set-up and at the end of your investigation.
5. Carry out your investigation for at least one week and record your results in a data table.

DISCUSSION

1. (a) List conditions that favor the formation of rust.
   (b) List any conditions in which the formation of rust is slowed or prevented.
2. Rate the amount of rusting of all the objects used by members of your class according to the scale agreed upon in step 2.
3. Think back to Activities 4C and 4E, in which you investigated reaction rates. Use those results in making the following predictions.
   (a) Which do you think would rust away more quickly, steel wool or steel nails? Why?

FIGURE 4.15 ◀
Compare the rate of rusting under different conditions. To keep the contents damp, put moist cotton balls or paper towels in the beaker. Why must the cover be moisture-proof?

(b) Which might become more badly rusted, a bicycle left outside all year in Death Valley or one left outside all year in Virginia? Explain your answer.

4. How does the presence of salt affect the rate of corrosion of iron?

5. If you or your classmates used copper or zinc in this investigation, you may be able to make some inferences about these two metals. Which metal—copper, zinc, or iron—do you think reacts more readily with other substances? Explain what observations led you to make this inference. ❖

▶ REVIEW 4.3

1. (a) What would be the advantage in using brass or stainless steel for the body of a car?
(b) Why do you think cars are not made of these materials?

2. The following types of care could have slowed the rusting of the car shown on the left in Figure 4.16. Use the ideas that you have learned in this section to explain the effect of each type of care.
(a) parking the car outside rather than in a heated garage during the winter
(b) washing the car frequently, especially in winter
(c) putting rubber mats over the carpet in the car
(d) repairing chipped paint promptly

3. Imagine a metal with properties that make it perfect for every use. Write a paragraph describing this ideal material. Write a second paragraph telling how you would use it to improve the quality of your life.

4. Suppose you are a scientist and you discover a new material that acts as a catalyst in dozens of common reactions. This could be a blessing to humankind, but in the wrong hands, it could be dangerous!
Write a short story describing what reactions are affected by the catalyst and how you could use the new chemical.

FIGURE 4.16
Rusting can be slowed.

CHAPTER REVIEW

KEY IDEAS

- In an exothermic reaction, energy is released to the surroundings.
- An endothermic reaction absorbs energy from its surroundings.
- The rate of many useful reactions can be controlled.
- Reaction rate may be affected by temperature, surface area of a solid, concentration of a liquid, and the presence of a catalyst.
- The rate of corrosion of metals can be controlled in the same way that other chemical reactions are controlled.

VOCABULARY

chemical energy
exothermic reaction
endothermic reaction
reaction rate
concentration
catalyst
enzyme
alloy

V1. Construct a mind map for this chapter. Include all the words from the vocabulary list as well as words from your personal experience.

V2. Write a short story about selling or buying a used car. The appearance of the car, including the amount of corrosion, should be part of your story.

FIGURE 4.17
An endothermic reaction

CONNECTIONS

C1. In Figure 4.17, an endothermic reaction is about to take place. Where will this reaction occur, and what form of energy will cause it?

C2. Suppose you have just finished a woodworking project at school. You want to varnish your handiwork, but you may not have the time needed for putting on the required number of coats during class time. Varnish hardens by reacting with oxygen in the air. Suggest ways of speeding the drying rate of each coat of varnish.

C3. On the containers of some medicines there is an instruction that says, "Store in the refrigerator for up to 14 days."
(a) Why do you think the medicine is to be kept in the refrigerator? (Hint: In your answer, use the idea of reaction rate.)
(b) What do you think would happen if you did not follow this instruction?

C4. (a) Explain how the surface area of a solid reactant affects the rate of a chemical reaction.
(b) Explain how the concentration of a solution affects the rate of a chemical reaction.

C5. (a) When you strike a match against a rough surface, heat is produced. Why is this heat necessary to start the match burning?
(b) Is the burning of the match an exothermic or endothermic reaction? Explain.

C6. Hydrogen peroxide solution must be fairly fresh to be effective. How could you store hydrogen peroxide to keep it fresh?

C7. (a) Name the process by which plants convert energy to food.
(b) State whether this reaction is exothermic or endothermic and describe how you could show this.

EXPLORATIONS

E1. The heat from a candle could be enough to set off an explosion in a coal mine. But if the candle is enclosed in a screen made from copper wire, the temperature outside the screen does not rise enough to cause an explosion (Figure 4.19). In a reference book, find out how this device works.

FIGURE 4.18
Grain is unloaded from elevators into ships.

FIGURE 4.19
Before the development of electric lighting, the Davy safety lamp was used to reduce the risk of explosions in coal mines.

E2. Dust explosions occasionally occur in grain elevators, where grain is stored (Figure 4.18). A small flame or spark begins the explosion, and tremendous damage may result. Use what you have learned about chemical reactions and how to control them to explain why you think these explosions occur. What do you think could be done to prevent them?

E3. Before vegetables are stored in a freezer, they are usually cooked slightly. Why do you think this is necessary?

REFLECTIONS

R1. In Activity 4A, you listed ways of speeding up or slowing down a combustion reaction. Look back at your list and try to relate each point to one of the control methods discussed in this chapter.

R2. Food "spoils" when mold or bacteria grow on it, or when chemical reactions occur in the food itself.

(a) List at least three methods by which you can slow the spoiling of food.

(b) Explain how each method works, using the idea of reaction rate.

R3. Suppose you are testing a newly developed headache tablet to find out how quickly it acts. What other factors, besides the rate of the reaction, would be important to test before the new tablet is sold to the public?

CHAPTER 5

HOUSEHOLD CHEMICALS

You and your family use household chemicals to do many jobs. For example, you may use different chemical cleaners for the stove, the drains, and the bathroom sink. Although these chemicals are useful, they can be dangerous if handled improperly or stored carelessly.

The labels on household chemicals tell you how they are to be used and why they are dangerous. If they weren't there, you wouldn't know which product you were spraying on your window. However, labels seldom explain why a chemical is useful for a particular purpose. You need a basic knowledge of chemicals to understand how to use certain products effectively. This chapter provides this information and gives you the opportunity to consider some common chemical hazards.

ACTIVITY 5A / NEW MATERIALS

Look around you in your classroom. What do you see that you could not have seen in a classroom 100 years ago? Plastics, of course, some of the fabrics in students' clothing, construction materials, and many other substances have been developed by chemists and chemical engineers.

PROCEDURE

1. With your group, list as many new materials as you can, including products from both home and school.
2. From your general knowledge, try to list these in order from the oldest inventions to the most recent.
3. Think about what people used before these new materials were developed. Beside as many items in your list as you can, describe what was used before the new material was available. ❖

■ 5.1 CHEMICALS IN THE HOME

Most people use household chemicals. In your home, you may use paints, varnishes, deodorizers, construction materials, pesticides, fertilizers, and many types of cleaners. Do you really need all these products? What did people use 50 or 100 years ago to do the same jobs? Do you think we will still be using these products 50 years from now?

For many of these products, there are alternatives (Figure 5.1). To make sensible choices about which products to use, some knowledge of chemistry is helpful.

One very common household chemical is soap or detergent. Water alone can wash away dirt, but it cannot remove compounds such as oils, fats, and greases. Soap or detergent must be used along with water to remove these compounds.

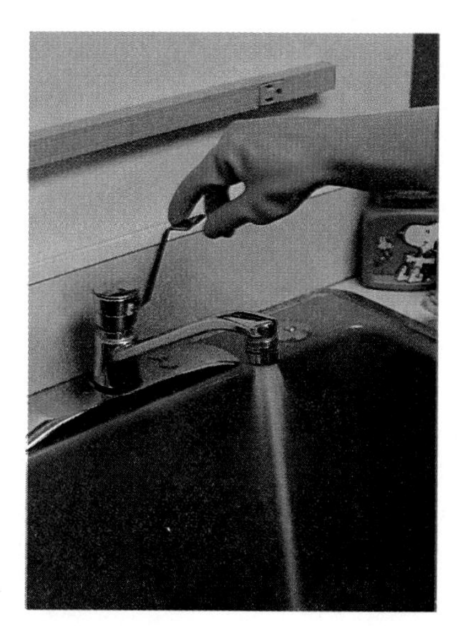

FIGURE 5.1 ◀
Is there another way to clean the sink?

Soaps and detergents contain long molecules that are responsible for the cleaning action. One end of each molecule is attracted to water, and the other end is attracted to the other compounds.

When detergent and water are used to clean soiled objects, one end of each detergent molecule is attracted to the oily or greasy material and helps to break it up into smaller drops. Then these smaller grease drops are surrounded by the water and float away. Thus, the oily or greasy dirt is removed.

ACTIVITY 5B / THE WHITEST WHITES

In this activity, you will design a method for comparing various laundry detergents.

MATERIALS

safety goggles
apron
rubber gloves
several laundry detergents
15 cm × 15 cm squares of dirty white cloth
other equipment, if necessary

PROCEDURE

1. In your group, hold a brainstorming session. Think of all the ways you could test the laundry detergents. What factors could you vary in your experiments? What factors should you keep the same? Out of all the tests that you think of, select one that your group will do.
2. Design a procedure for the test that your group selected. (Do not forget that you will need to include one step in which you use no detergent, just tap water, as a control.) Decide exactly how you will rate the effectiveness of the detergent and how you will record your results.
3. Discuss the method you have designed with your teacher. Modify your design, if necessary, then carry out the test.
4. Record your observations.
5. Be sure to rinse the detergent off your hands at the end of this activity.

DISCUSSION

1. (a) Did one detergent work better than the others? How much better did it work?
   (b) How much more effective were detergents than tap water alone?
   (c) Why did you need to include a control in your comparison?
2. (a) How did your group's results compare with those of other students in your class?
   (b) Are such comparisons fair if the groups used different procedures? Explain your answer.
3. (a) Find the prices of the detergents you used.
   (b) What relationship was there between price and how well the detergent worked?
4. If you are planning an experiment to compare different products, you could follow the instructions given on each package. Or you could carry out the same tests on each product. Which do you think is a fairer way of comparing the products? ❖

■ Some household cleaners use enzymes to remove dirt and stains from clothing. Obtain a cleaner of this type and design a procedure to test its ability to remove stains. Does the enzyme cleaner work if you heat it before you use it? Does the presence of other chemicals affect how well it cleans? With your teacher's permission, carry out the procedure. Be sure to wear your safety goggles and apron.

► REVIEW 5.1

1. Some students plan to test three brands of bathtub cleaner, Brand A, Brand B, and Brand C, to find out which one cleans best. To do this, they mark off a bathtub into four equally soiled sections. They use the same method to clean one section with water, a second with Brand A, a third with Brand B, and the last with Brand C. Then they compare the results.
   (a) Is this a controlled experiment? If so, what is the control?
   (b) Why is a control often necessary in an experiment?
2. (a) List six ways in which you use chemicals in your home.
   (b) If you did not have each of these chemicals, what could you use instead?
3 (a) From your list for question 2, list any uses that you think are not absolutely necessary.
   (b) How would your life be different if you did not have those chemicals?
4. The compound perchlorethylene is often used as the liquid cleaner in dry cleaning. Perchlorethylene does not dissolve in water.
   (a) What sorts of dirt do you think perchlorethylene would remove?
   (b) What kinds of dirt do you think it would not be able to remove?

■ 5.2 ACIDS AND BASES

"Basic" and "acidic" are common words, used in many everyday situations (Figure 5.2). You are probably also familiar with many substances that are acids or bases. You have used many at home, and you have used some in the science classroom.

FIGURE 5.2
The words "acid" and "base" are used in many ways, but only one of the examples shown here has a chemical meaning. Which one is it? For the other examples, give a synonym or an explanation for the words related to "acid" or "base."

The teacher, still angry, managed to produce an acid smile.
37
"The next tryout will be the acid test."
"Oooooh. . .I've got acid indigestion!"
"Let's make this our base."
"The basic problem is that . . ."
He slid into third base.
2

An acid is a compound that has certain properties. **Acids** taste sour, are soluble in water, and take part in chemical reactions in a similar way. A base is also a compound with certain properties. **Bases** taste bitter, are soluble in water, feel slippery, and react with acids. Substances that are neither acids nor bases, such as pure water, are said to be **neutral**.

How can you tell, without tasting, whether a substance is an acid or a base? You can use an acid-base indicator. An **acid-base indicator** is a substance that is a different color in an acid than it is in a base. One of the most useful indicators is litmus (Figure 5.3), but many other common substances also change color in acidic or basic solutions.

FIGURE 5.3 ◀
Litmus is a dye produced from lichen, like the lichen growing on this rock. Litmus paper has been soaked in a solution of litmus and then dried.

ACTIVITY 5C / ACID-BASE INDICATORS

In this activity, you will try out some materials to see if they are acid-base indicators.

■ **Dilute hydrochloric acid and dilute sodium hydroxide are corrosive and can cause damage to your eyes and skin. Wash spills and splashes off your skin and clothing immediately, using plenty of water. If you get any in your eyes, immediately rinse them for 15 or 20 minutes and inform your teacher.**

MATERIALS

safety goggles
apron
red cabbage
water
tea bag
dilute hydrochloric acid
dilute sodium hydroxide (a base)
balance
two 250 mL beakers
one 100 mL beaker
sharp knife
Bunsen burner
ring stand
wire gauze
red and blue litmus paper
test tube stand
three test tubes
tape for labels
other sources of possible indicators, such as blueberries, beets, cranberries, leaves of plants, purple grape juice

PROCEDURE

1. Put on your safety goggles and apron.
2. Shred about 40 g of red cabbage leaves and put them in a 250 mL beaker. Add enough water to cover the leaves, then boil for 5 min as shown in Figure 5.4. Allow the cabbage water to cool for several minutes, then pour off the water into the second beaker. Discard the cabbage as directed by your teacher.
3. Place the tea bag in a 100 mL beaker. Add very hot water and stir. After 5 min, remove and discard the tea bag as directed by your teacher.
4. Prepare any other solutions as directed by your teacher.

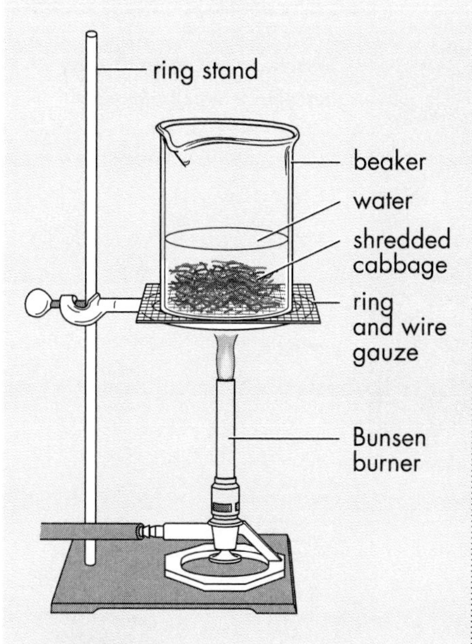

FIGURE 5.4
Preparing an indicator from red cabbage

5. Put labels on the three test tubes, marking them "acid," "base," and "control." In your notebook, draw a table like Table 5.1 for your observations.
6. Slowly pour the cabbage solution into each of the labelled test tubes to a depth of about 2 cm. Add an equal amount of acid, base, or water to the appropriate tube, and record your observations.
7. Discard the solutions as directed by your teacher. Rinse the test tubes thoroughly. Repeat Step 6 for all the other indicators. (When testing the litmus paper, place it in the test tube, then add acid, base, or water.)
8. Wash your hands thoroughly after you have completed this activity.

DISCUSSION

1. Which test tube was the control in this experiment?
2. Which of the substances tested are effective acid-base indicators?
3. Which indicators do you think would be easiest to use? Explain. ❖

Indicator	**Color**		
	In acid	In base	In water
Red cabbage solution			
Tea			
Red litmus paper			
Blue litmus paper			

TABLE 5.1 ***Sample Data Table for Activity 5C***

COMMON ACIDS AND BASES

Acids and bases are common in fruits and vegetables as well as in many other foods and drinks. They are also found in household products such as cleaners, shampoos, drugs, and batteries.

ACTIVITY 5D / HOUSEHOLD ACIDS AND BASES

In this activity, you will test some common substances with acid-base indicators.

TABLE 5.2 ▶
Sample Data Table for Activity 5D (Procedure)

Indicator	**Color in acid**	**Color in base**
Litmus	Red	Blue
Phenolphthalein		
Bromothymol blue		

■ Dilute hydrochloric acid and dilute sodium hydroxide are corrosive and can cause damage to your eyes and skin. Wash spills and splashes off your skin and clothing immediately, using plenty of water. If you get any in your eyes, immediately rinse them for 15 or 20 minutes and inform your teacher.

■ Some products, especially cleaners, may be hazardous. Be aware of the safety labels on the packaging and handle with care. Pour slowly, using small amounts. If any are spilled on the skin, wash immediately with cold water. Never mix cleaners, since they may react together.

FIGURE 5.5 ◀ ***Marking your test tubes carefully will help you keep track of your results.***

MATERIALS

safety goggles
apron
water
dilute hydrochloric acid
dilute sodium hydroxide solution
test tube rack
test tubes
indicators, such as red and blue litmus paper, phenolphthalein, bromothymol blue, grape juice, tea, etc.
samples of materials to be tested, such as salt water, swimming-pool water, lake or river water, cola drink, other soft drinks, vinegar, various fruit juices, oven cleaner, drain cleaner, household ammonia, baking soda solution, cream of tartar solution, various detergents, shampoo, milk, sour milk, solution of aspirin tablets or antacid tablets, bleach, egg white, water from boiled vegetables

PROCEDURE

1. Put on your safety goggles and apron.
2. In your notebook, draw a table like Table 5.2.
3. Add a sample of each indicator to water, acid, and base. Record the colors in your table.
4. Test each of your samples with the indicators as directed by your teacher. Keep a careful record of your results (Figure 5.5).
5. Follow your teacher's instructions for disposing of the materials you used in the activity.
6. Wash your hands thoroughly after you have completed this activity.

DISCUSSION

1. List any materials that were not definitely acidic or basic.
2. Make a table like Table 5.3. List the materials that were definitely acidic or basic in one of the four columns.

TABLE 5.3 ▶
Sample Data Table for Activity 5D (Discussion)

Foods		Non-foods	
Acids	Bases	Acids	Bases
Lemon juice			
etc.			

3. Examine the four lists in your table. What conclusions can you make? For example, do acidic foods have any other properties in common? Are cleaning materials generally acids or bases?

4. Suppose your teacher forgot to order red litmus paper for your class. The supplies cupboard is overstocked with blue litmus paper, but the students need both kinds to do an activity. What would you advise your teacher to do?

5. Blueberries in muffins that are baked with yeast remain blue. But blueberries in muffins that are baked with baking soda may turn slightly green. Suggest an explanation for this difference. ❖

HOW ACIDIC? HOW BASIC?

Some acids are called strong acids and are more acidic than others. Strong acids can cause serious skin damage and can produce fumes that can harm your lungs (Figure 5.6). Strong bases, such as the bases used in household drain cleaners, can also be dangerous, for the same reasons. Very weak acids and bases are not as harmful (Figure 5.7).

POISON

6

FIGURE 5.6
Products that contain strong acids and bases are marked with a special label to warn of their danger.

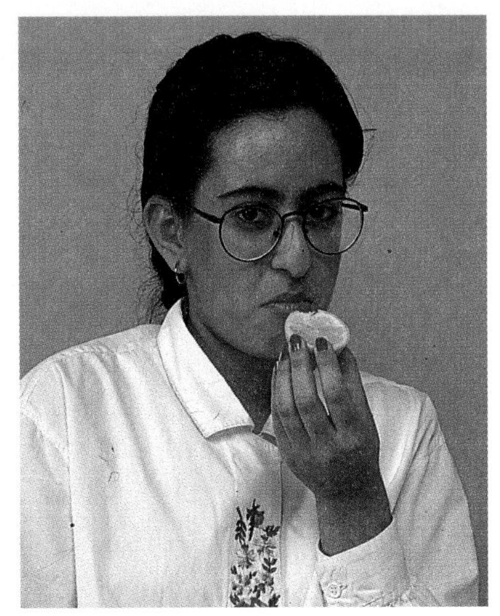

FIGURE 5.7
Strong acids and bases, such as the base used to clean drains, are dangerous. But a very weak acid, such as lemon juice, can be pleasant to eat.

■ Many fish oils are bases that have a distinctive odor. Lemon juice neutralizes the bases and eliminates the "fishy" smell.

MIXING ACIDS AND BASES

When an acid and a base are mixed together, there is a chemical reaction called **neutralization**. This reaction produces water and a type of compound called a **salt**. For example, when hydrochloric acid reacts with sodium hydroxide, sodium chloride (table salt) and water are produced:

hydrochloric acid + sodium hydroxide →
sodium chloride + water

Weak acids and bases are often used to neutralize spills of stronger, more dangerous bases or acids. For example, vinegar can be used to neutralize spills that involve bases. Baking soda, a base, can neutralize acid spills (Figure 5.8).

FIGURE 5.8 ▶
This student is using baking soda to neutralize an acid spill.

▶ REVIEW 5.2

1. What is an acid-base indicator?
2. (a) What are the characteristic properties of an acid?
   (b) Give three examples of common acids.
3. (a) What are the characteristic properties of a base?
   (b) Give three examples of common bases.
4. (a) Name some color indicators that show if a chemical is an acid or a base.
   (b) How do their colors change?
5. What effect do an acid and a base have on each other?

■ 5.3 HAZARDOUS CHEMICALS IN HOUSEHOLD PRODUCTS

It can be difficult to find information about chemical household products. Often the label on a product does not fully describe the contents. When it does, the proper chemical names may not be used. In addition, the amount of information on the label of a product depends on the type of product it is and what it is used for. How can you find out whether or not a product is safe?

ACTIVITY 5E / EXAMINING HOUSEHOLD PRODUCTS

In this activity, you will examine the packaging of several common household products.

■ **Do not attempt to open these products or to use them. Always put the products back in a safe, secure storage spot.**

■ Do research to find out more about one of the products you observed in Activity 5E. Write a report about this product that includes the chemical formula of the product, how the product is used, when and how it was discovered, companies that manufacture it, the chemical reactions involved in manufacturing the product, and the chemical reactions involved in using it.

PROCEDURE

1. Your teacher will give you cleaned, empty containers of six common household products, some of which you may have used. Observe each carefully.
2. Read the following questions, then design a table for your observations.
   (a) How is the product used?
   (b) What chemicals are in the product? (If you cannot find this information on the label, leave a blank in your table.)
   (c) What warning labels are on the product?
   (d) What special safety features are there in the packaging?
   (e) Describe any safety-related instructions on the package.
   (f) If the household product is poisonous, describe the instructions that are printed on the label.

DISCUSSION

1. What kinds of labels are used to indicate that a product might be dangerous?
2. (a) What kinds of special packaging are used to prevent small children from opening containers of household chemicals?
   (b) Do you think that this packaging is effective? Explain.
3. Some of the products you observed in this activity should be stored in a special way. Describe how you would store each product in your home. Explain why the special storage is necessary.
4. List 10 chemical products found around your home that were not used in this activity. If the contents of the product are given on the label, list them beside the name of the product. ❖

REACTIVE CHEMICALS

Certain products are useful in the home because the chemicals they contain do not react easily with other materials (Figure 5.9). For example, you would not want plastic bowls to react with the food stored in them, leaving a taste or odor on the food. Nor would you want clothes made from fabrics that fell apart during normal use. The less reactive some household products are, the better.

FIGURE 5.9
The glass on the left was filled with a reactive chemical. What would happen if all household chemicals were this reactive?

FIGURE 5.10
What is the polish reacting with?

Other household products, however, are useful because they *do* react with other materials (Figure 5.10). **Reactive chemicals** are ones that react readily with other chemicals. Reactive products may help with cleaning, polishing, painting, and many other activities. Glues, solvents, polishes, and paints are examples of such products. These products work because they contain reactive chemicals.

Products that contain reactive chemicals are much more dangerous than those that do not. You need to take special care in storing and using these products. They must be kept safely away from children and pets. Often they must be kept away from moisture and extremely high or low temperatures. Your eyes and skin can be seriously damaged by some of these products. The chemicals contained in some common products that your family might use are listed in Table 5.4.

Chemical	Found In	Chemical Formula	Hazard
Caustic soda (sodium hydroxide)	Oven cleaners	$NaOH$	Poisonous; burns skin; corrosive
Caustic potash (potassium hydroxide)	Drain cleaners	KOH	Poisonous; burns skin; corrosive
Muriatic acid (hydrochloric acid)	Iron cleaners; concrete etching fluid	HCl	Poisonous; burns skin; corrosive
Toluene	Contact cement	C_7H_8	Poisonous; flammable
Acetone	Some types of glues; nail polish remover	C_3H_6O	Poisonous; flammable
Ketones	Some types of glues	Various compounds	Poisonous; flammable
Petro products*	Spray paint; furniture polish; paint solvents; thinners	Mixtures of compounds called hydrocarbons; C_5H_{12}, C_8H_{14}, and C_7H_{16} are examples.	Poisonous; flammable
Pentachlorophenol	Wood preservatives	C_6HCl_5O	Poisonous
Methanol (wood alcohol)	Paint solvent	CH_3OH	Poisonous; flammable
Calcium hypochlorite	Tablets for treating water in swimming pools	$Ca(ClO)_2$	Acidic; corrosive

*These are compounds related to gasoline and natural gas.

TABLE 5.4 ***Common Hazardous Chemicals in Household Products***

The chemicals listed in Table 5.4 are named on product labels because they are dangerous, either to human health or to the environment. Such chemicals are called **hazardous chemicals**. Products that contain hazardous chemicals always have some warning on the container. The container might also describe a first-aid treatment. Before you use a household product, be sure you are aware of any hazardous chemicals it contains. Learn the proper first-aid treatment for these chemicals.

Chemicals may be hazardous for several reasons. The warning symbols used on hazardous materials identify four kinds of products—poisonous, flammable, explosive, and corrosive. These symbols are shown on page xii, as part of the Safety Rules for this book.

Some hazardous chemicals work quickly. For example, the effect of a concentrated strong acid, such as sulfuric acid, is instantly obvious—it burns the skin, puts holes in clothes, and eats into metals. For this reason, great care must be used when handling car batteries that contain sulfuric acid. There are also some bases, such as those used in drain cleaners, whose concentrated solutions are highly corrosive.

■ Find out how reactive chemical products are stored in your home. If there are small children in your home, describe any special storage precautions taken to protect them from these products.

■ In a library, find a reference book about family medicine. Read about what you should do if a small child eats or drinks something that you know is a poison. Create a poster for your class. On the poster, show the local phone number to call for poisoning emergencies. Illustrate what you should do until you can get the victim to a doctor.

Other hazardous chemicals take a longer time to show their effects. For example, leaded gasoline contains a lead compound that causes changes in red blood cells, general weakness, and paralysis. But it can be many years before these effects appear. Leaded gasoline is no longer available in the U.S.

DISPOSAL OF HAZARDOUS CHEMICALS

Waste products—garbage, sewage, or waste water from factories — may contain hazardous materials. But these wastes are only part of the problem. Since they come in large quantities from identified locations, factory wastes are often easier to deal with than the wastes produced by individual people. Although the amount of hazardous waste from any one home may be very small, the amount from all the homes in a city is a serious problem. In many communities, there are special collection depots where people can take their cans of leftover paint, garden chemicals, and unused cleaners for safe disposal.

OUR USE OF CHEMICALS

You are a member of a society that uses many chemicals. Chemicals can be very helpful if they are understood and used properly. Used improperly, however, they can be dangerous to humans and to the environment. During your lifetime, you will have to make many decisions about chemical use in your home, in your community, and in the businesses in your community. To make these decisions, you will need to apply your knowledge of science to everyday life.

▶ REVIEW 5.3

1. List three examples of unreactive household products that you use in your home.
2. Sketch three labels that warn you of how a household product might be dangerous.
3. Choose two types of hazardous household chemicals. Describe how each of the products should be stored in your home.
4. Design a container that cannot be opened by young children. Make a labelled drawing of the container.
5. (a) Describe two types of effects that hazardous chemicals can have on people.
   (b) Give an example of each type of hazardous chemical.
6. (a) List three hazardous chemicals that you have found in household products.
   (b) Explain why each one is hazardous.

DISPOSING OF HOUSEHOLD HAZARDOUS WASTE

Let's go searching for hazardous chemicals in your home. Look for words such as "Danger," "Poison," or "Caution"—good indications that the contents are hazardous.

You could start under the kitchen sink. Perhaps you'll find oven cleaner, drain cleaner, or bleach. In the bathroom, you might discover powdered cleansers, hair sprays, or nail polish. If you have a garage, you might find used oil, car batteries, or barbecue starter there. You might also find oil-based paints, paint thinners, and glues somewhere in your home.

We often think that chemical industries are the big polluters that release hazardous wastes into the environment. But we all use products containing hazardous chemicals, and we contribute to pollution if we fail to dispose of these products safely. One household may not produce a large volume of hazardous waste, but with millions of households in the United States, the overall effect is enormous.

Much of this waste is put out with the garbage or poured down the drain. Hazardous wastes dumped at a landfill site can contaminate the soil and groundwater. Besides, such dumping is illegal. Pouring toxic chemicals down the drain pollutes drinking water, since sewage disposal systems do not remove these chemicals. Burning hazardous wastes contributes to air pollution. Burning can also be dangerous since some of these chemicals are highly explosive.

Some communities have hazardous waste collection depots that are open occasionally during the year. If you take in your hazardous chemicals, experienced workers will sort them, package and seal them, and send them to a toxic waste incinerator. Some materials, such as used motor oil, paints, and car batteries, can be recycled.

Another way to tackle the problem of hazardous household waste is to use fewer harmful products in the first place. When buying a hazardous product, buy only as much as you'll need. Try the non-toxic alternatives to many of these products. For instance, baking soda and vinegar can be used to clean surfaces in your home. Water-based latex paints can be used instead of oil-based paints. Use compost instead of fertilizer in your garden. You can get rid of garden pests without using chemicals, too.

Find out how hazardous household products can be disposed of in your community. You can phone your local recycling center or local government office.

CHAPTER REVIEW

KEY IDEAS

- Most homes contain chemical household products such as paints, varnishes, and many types of cleaners.
- Some common household chemicals can be easily tested to find out how well they work.
- Many common household chemicals are acids or bases.
- When an acid and a base react, they form water and a salt. This reaction is called neutralization.
- Some household products contain hazardous chemicals. The labels on the containers of these products indicate the type of hazard and how dangerous it is.
- The effects of a hazardous chemical may appear immediately, or they may not appear for a long time.
- Wastes containing hazardous chemicals can be dangerous to humans and to the environment. All members of society must make responsible decisions about how to use and dispose of hazardous chemicals.

VOCABULARY

acid
base
neutral
acid-base indicator
neutralization
salt
reactive chemical
hazardous chemical

V1. Create a mind map around the phrase "household chemicals." Use the words from the vocabulary list and any other words that you think are appropriate.

V2. Dr. Rajiv Joshi was asked to give a talk on hazardous chemicals to a group of students. Write a short script for Dr. Joshi's speech.

CONNECTIONS

C1. Name a hazardous chemical that you have used. Explain how you should discard any of it that you do not need.

C2. Why is it easier to control wastes from a factory than wastes from households? Suggest one or more reasons.

C3. In your notebook, complete the comparison of acids and bases as in Table 5.5.

C4. Sodium hydroxide is a strong base used in oven cleaning. When it reacts with fat, soap and water are produced.

(a) Write a word equation for this reaction.

(b) After using an oven cleaner, some people rinse the oven with a dilute solution of vinegar in water. What is the purpose of the vinegar rinse?

C5. (a) What type of chemical reaction takes place when an acid reacts with a base?

(b) Write a word equation to show an example of this type of reaction.

	Acids	Bases
Properties	sour-tasting	
Examples		baking soda

TABLE 5.5 ***Comparison of Acids and Bases***

FIGURE 5.11
How do you think any hazardous chemicals that were dumped here would affect this area after the landfill site is closed?

EXPLORATIONS

E1. Disposing of chemical products, whether hazardous or not, can cause problems. In a reference book, find meanings for the following terms. Then explain how they relate to the disposal of chemical wastes.
(a) non-biodegradable
(b) landfill
(c) runoff
(d) incinerator

E2. Contact your local government or a company in your area that disposes of solid waste or sewage. Find out what laws your community has for disposing of hazardous household chemicals (Figure 5.11).

E3. (a) Design a method to test the effect of different chemical fertilizers on a patch of lawn. As a control, you would need to leave some of the lawn unfertilized.
(b) Consider the length of time that you would need to assess the fertilizers. (For example, you might find that the lawn improves for several years but that later the soil becomes less productive.) What period of time would you consider to be long enough to test the various lawn treatments?
(c) Describe exactly what effects you will consider in your test. For example, will you include any effects on wildlife in the area, such as earthworms, insects, or fish in a nearby stream?

REFLECTIONS

R1. Refer to Activity 5A in your learning journal, in which you wrote a list of household products developed by chemists over the last 100 years.
(a) From that list, pick one product that you think has provided a large number of benefits for humanity. List three benefits in your learning journal.
(b) For the product you chose in part (a), list any disadvantages you think it has.

R2. (a) In Activity 5A, you listed materials invented in the last 100 years, but even if you consider materials only developed in the last 25 years, you would still have a long list. Discuss with an adult how common household products were different 25 years ago. List some of these differences in your learning journal.
(b) What new products do you think will be invented in the next 25 years? Write a short story illustrating how these new products could affect your life in the future.

R3. (a) How do you think the environment might be affected by household chemical products? Make a list of at least eight common products. Then rate each one on a scale from 0 to 10, with 0 indicating "low hazard to the environment" and 10 indicating "highly hazardous."
(b) How necessary are common household chemicals? Rate the products you listed in part (a) on another scale. On this scale, 0 should indicate "completely unnecessary" and 10 should indicate "essential."
(c) Compare the ratings on your two lists. Are there any chemical products that you might do without?

TREASURE

UNDERSTANDING YOUR LIVING BODY

As a young adult, you are making many choices for yourself. Some of your choices will affect your health for the rest of your life. To make the best choices, you need to understand how your body works, what care it needs, and how your choices affect your body.

Your body is like a living machine, with many connected parts and systems that must all be in working order if your body is to run smoothly. The food you eat supplies energy for your body as well as providing many other substances you need for good health.

In this unit, you will learn how some of your most important body systems work. Like all the systems of your body, your digestive, respiratory, circulatory, and excretory systems help to keep you alive. You will also learn about nutrition and exercise—two of the most important tools you can use to keep those systems "tuned up" and working well. Most important, you will gain the information you need to make personal choices that will affect your health.

You can think of this understanding as an investment in your own future. Good health is an investment that will pay off now and throughout your life.

UNIT II

CHAPTER 6 *Nutrition*

CHAPTER 7 *Digestion*

CHAPTER 8 *Respiration*

CHAPTER 9 *Circulation and Excretion*

CHAPTER 10 *Fitness and Health: A Way of Life*

CHAPTER 6

NUTRITION

Whether you are at home or in a restaurant, you must often make choices about what to eat. Imagine that you, like the people in the photograph, are reading a menu. How would you decide what to eat? Would you think about the taste of the food or its price? Would you think about how each food might affect your health? Whatever food you choose, it would probably keep you from being hungry. Would it, however, provide all the things your body must get from the food you eat? Do you know what these things are?

Often, poor food choices do not affect your health right away. The effects of poor choices may show up much later in your life. This chapter will provide you with some of the information you need to make wise choices.

ACTIVITY 6A / CHOOSING A MEAL

The Student Café

Soups

beef noodle
cream of mushroom
vegetarian vegetable

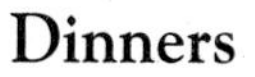

Dinners

The slice–
pepperoni pizza

The fowl–
fried chicken breast with peas and carrots

The fish–
baked salmon with baked potato, tossed salad, and Italian bread

The burger–
hamburger on a bun with french fries or potato chips

The salad–
large fresh garden salad with lettuce, tomato, radishes, peppers, and alfalfa sprouts

Desserts

chocolate éclair
apple pie with ice cream
canned peaches

Beverages

milk (whole)
orange juice
cola drink
tomato juice
chocolate milkshake

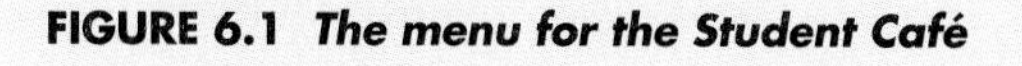

FIGURE 6.1 ***The menu for the Student Café***

PROCEDURE

Figure 6.1 shows the menu of an imaginary restaurant. From the choices available, select what you would enjoy for dinner: a soup, a main dish, a dessert, and a beverage. Record your selections in your learning journal in the form of a list entitled, "Most Enjoyable Dinner." Next, choose a meal that will be the most nutritious—one that will give your body what it needs for good health. Record your choice of soup, main dish, dessert, and beverage and entitle your list "Most Nutritious Dinner." At the end of this chapter, you will look at these lists again.

DISCUSSION

1. Were the choices you made for "Most Enjoyable Dinner" different from your choices for "Most Nutritious Dinner"? If so, why?

2. Why do you think some foods are not as nutritious as others?

3. (a) A nutritious dinner is often described as a "well-balanced" meal. What do you think this term means?
(b) Do you think your "Most Nutritious Dinner" was well balanced? Why or why not?
(c) Do you think your "Most Enjoyable Dinner" was well balanced? Why or why not?

4. (a) Imagine that you have just eaten a meal or snack. You enjoyed it very much, but you think that it might not have been good for your health. Do you think that this poor choice could cause you any health problems? Explain why or why not.
(b) If your answer to part (a) was yes, what do you think you could do now to overcome or prevent these problems? ❖

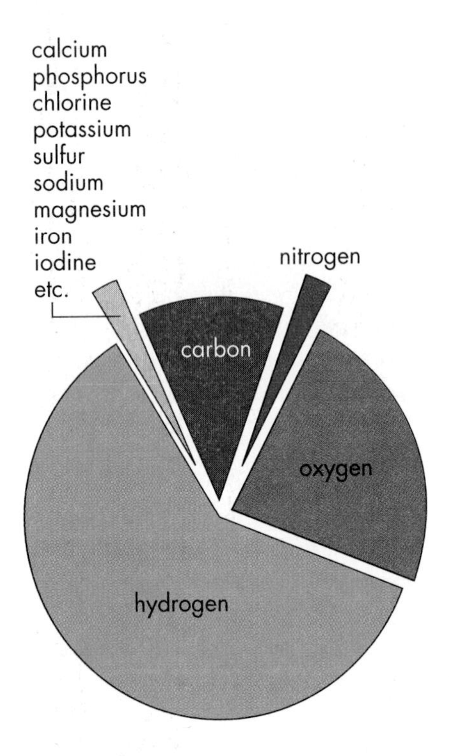

FIGURE 6.2
In this diagram, the circle represents the human body. Each segment represents the fraction of the body that is made up of the element or elements listed. Which three elements are most common in the human body?

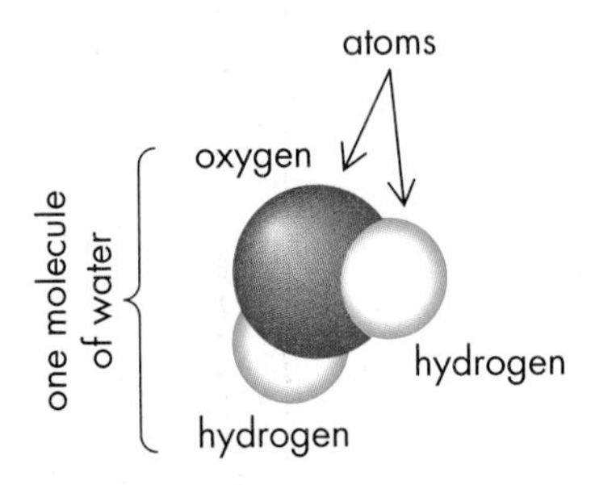

FIGURE 6.3
Water is a compound that contains two elements. A single molecule of water is made of one atom of oxygen joined to two atoms of hydrogen.

6.1 WHY YOUR BODY NEEDS FOOD

Your body is made up of skin, muscle, bone, blood, and many other materials. Where do these materials come from? How does your body repair broken bones, or replace blood when you bleed? As you grow, your body makes new skin and muscle, as well as bone and other types of material. Where does this new material come from? Just as a car needs fuel to run, your body needs energy to power all its activities. Where does your body get its energy?

The answer to all of these questions is, "from the food you eat." (Food includes the things you drink, like milk, juice, and soft drinks.) Your food contains many different substances. Some of these substances provide you with energy as well as with the materials your body needs to grow and repair itself. In some very real ways, you are what you eat.

YOUR BODY — WHAT IS IT MADE OF?

Just like your desk or your textbook, your body is made of matter. The smallest particles of all matter are atoms. Two or more atoms joined together form larger particles called molecules.

There are many different kinds of atoms. An **element** is a substance that contains only one type of atom. Although there are over 100 different elements, only some of them are found in your body. In fact, the human body is made up mostly of only four elements (Figure 6.2). Hydrogen, oxygen, carbon, and nitrogen make up about 96 per cent of your total mass.

Other elements make up the remaining portion of a human's mass. For example, there are several grams of phosphorus and more than 1000 g of calcium in the body of a typical adult. Elements such as iron and iodine occur in even smaller amounts. Although these and other elements are present in very small amounts, they are still important. Without them, your body cannot work properly.

Most of the elements found in the human body are not present in pure form, but are joined together with other elements to form molecules called compounds. For example, most of the oxygen atoms and hydrogen atoms in your body are joined together to form molecules of water (Figure 6.3). Water is a compound made up of these two elements.

Some of the elements and compounds found in your body are used to build or repair your body cells. Other elements and compounds supply energy to the cells in your body. Cells are the tiny subunits that make up all living things. Inside your cells, many chemical reactions occur. It is these chemical reactions that keep your body alive and active. For the reactions to occur, your cells need energy. This energy, as well as the substances your cells need for growth and repair, comes from your food.

NUTRITION

Nutrients in the food you eat contain the many different elements and compounds that your body needs. A nutrient is any material that can be taken into your body cells and that is useful to your body. The nutrients in food supply the energy and the materials your body needs to grow and survive.

Nutrition is the study of the nutrients in food and the effect of these nutrients on your health. Sometimes people use "nutrition" and related words to refer to healthful foods or good eating habits. For example, a meal that contains a good balance of the nutrients your body needs is often called "nutritious." If a person practices good nutrition, she or he selects nutritious foods and eats just enough of them to supply the body's needs.

FIGURE 6.4
Different people need different amounts of energy.

Good nutrition is especially important for you as a teenager because of the many changes taking place in your body. Adolescence is a time of rapid growth and development. Many of your bones grow longer, causing your height and overall size to increase. Your muscles also grow larger, increasing your strength. Your body may produce more fat. Such body growth requires a lot of energy and large amounts of certain elements and compounds. Teenagers need different amounts of nutrients than either children or adults. In addition, no two teenagers have exactly the same nutritional needs. Your own needs depend on many factors, including your mass, age, and level of activity (Figure 6.4).

There is also a difference between the nutritional needs of girls and boys. At the age of 10 or 11, girls usually begin a growth spurt, a time of rapid growth. Their growth spurt is often greatest at age 12 and ends by age 15. In boys, this spurt does not usually begin until 12 or 13 years of age. It is greatest at about age 14 and is usually over by age 19. A rapidly growing, active boy of 14 needs to supply his body with more nutrients than a girl of the same age whose growth spurt has already stopped.

Your body needs many different nutrients. In the following sections you will learn about six groups of nutrients. The nutrients in each group are important since they help your body grow, repair itself, or work properly.

REVIEW 6.1

1. Why do you need food?
2. (a) List, from the largest amount to the smallest amount, the four most common elements in your body.
   (b) List five elements that are found in small amounts in your body.
3. What is a nutrient?
4. (a) Explain why a 12-year-old girl might require more nutrients per day than a 12-year-old boy.
   (b) Why might the situation change when they are both 15 years old?

ACTIVITY 6B / THE NUTRIENT GROUPS AT A GLANCE

You may find it helpful to design a table to record information as you read through the next three sections. Your table should allow you to record the following information:

- the name of the nutrient group
- the function (job) or functions of the nutrient group
- examples of foods that contain one or more of the nutrients in the group
- other information that is important to you ❖

■ In some books and pamphlets, energy is measured in units called calories (cal), rather than in joules. When nutrition is being discussed, the units used for food energy are sometimes called Calories (note the capital "C"). One Calorie is equal to 1000 calories, which is equal to approximately 4.2 kJ.

■ 6.2 NUTRIENTS FOR ENERGY

Your body needs about 50 different nutrients. Most of these nutrients can be classified into one of the following six groups: carbohydrates, fats, proteins, minerals, vitamins, and water. In this section you will learn about carbohydrates and fats. The nutrients in these two groups supply your body cells with most of the energy they need to function.

Energy is the ability to do work. **Food energy** is the energy your cells obtain from the food you eat. Food energy allows your cells to do their work—work that makes it possible for you to think, move, breathe, and perform all sorts of other activities. **Joules** (J) are the units used to measure energy. Because the amount of energy needed by your body is so great, food energy is usually measured in kilojoules (1 kJ = 1000 J.) Different types of nutrients supply your body with different amounts of energy.

CARBOHYDRATES

Carbohydrates are the group of nutrients that are the major source of energy for all your body needs. About 55 percent of the energy needed by your body should come from carbohydrates.

Carbohydrates are substances that contain carbon, hydrogen, and oxygen atoms arranged in certain ways. Different combinations of these three elements form different types of carbohydrates. There are two main types: simple carbohydrates and complex carbohydrates.

Sugars, also known as simple carbohydrates, are made up of relatively small, simple molecules. You can easily recognize the name of a sugar because it ends with "ose." For example, fructose is a type of sugar that is found in many fruits, such as apples and peaches. Lactose is found in milk. The sugar you might sprinkle on grapefruit or cereal, or put into coffee and tea, is called sucrose. It comes from two types of plants: sugar beets and sugar cane. The most common type of sugar is glucose. It is found in maple syrup, honey, and many foods that contain carbohydrates. Glucose is the major source of energy for all your body cells—as well as for the cells of many other organisms.

Complex carbohydrates consist of larger molecules. These large molecules are actually made of many sugar molecules joined together. Think of the small sugar molecules as building blocks; a complex carbohydrate molecule is "built" by joining many of these small sugar "blocks" together. A complex carbohydrate that you often eat, which is made up of many glucose molecules, is **starch**. Starch comes from plants. When you eat fruit, vegetables, or bread, for instance, one of the nutrients you obtain is starch. Different plant foods contain different amounts of starch. Another complex carbohydrate in food is **glycogen**, which comes from meat and fish (Figure 6.5). Glycogen, like starch, is made up of glucose molecules.

FIGURE 6.5
Which ingredients of pizza contain glycogen? Which parts contain starch?

Complex carbohydrates cannot immediately supply energy to your body cells. First your body must break down these carbohydrates into the small sugar molecules that they are made up of. In your body **digestion**, the breaking down of larger molecules into smaller ones, takes several steps. (You will learn more about digestion in Chapter 7.) Once a molecule of starch or glycogen has been digested (broken down), many smaller glucose molecules become available. These can then be taken into your cells, where they are used to provide energy.

When you eat a meal, some of the glucose you obtain from simple and complex carbohydrates will be used right away for energy. Often, however, there are some glucose molecules left over. Some of these are joined together and stored for future use as glycogen molecules. Later, your body can break the glycogen down into glucose molecules if your cells need energy. All animals (not just humans) store glucose molecules as glycogen. Glycogen is stored mainly in the muscles and liver. You obtain complex carbohydrates in the form of glycogen if you eat meat or fish (Figure 6.6).

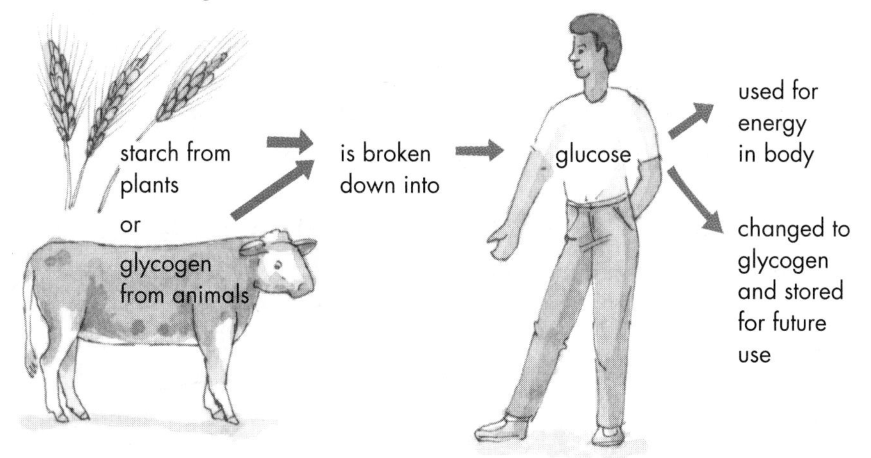
starch from plants
or
glycogen from animals
is broken down into
glucose
used for energy in body
changed to glycogen and stored for future use

FIGURE 6.6 ◀
Glucose to power the body comes from both glycogen and starch.

FATS

Fats are another important source of energy for your body. Most of the fats you might eat come from meats, margarine, eggs, and dairy products (such as milk, butter, and cheese). Nuts and cooking oils are also sources of fats. Although you should get most of your energy from carbohydrates, fats can also help supply your body with energy. A gram of fat gives your cells twice as much energy as a gram of carbohydrate.

Like carbohydrate molecules, fat molecules are made up of carbon, hydrogen, and oxygen atoms. The difference is that fat molecules contain many more hydrogen atoms and far fewer oxygen atoms than carbohydrate molecules contain. However, by adding hydrogen atoms, and making other changes, your body can convert glucose molecules into fat molecules. Eating too many carbohydrates can, therefore, increase the number of fat molecules in your body.

Fatty acids are one of two building blocks for fats. There are many different kinds of fatty acids, just as there are different kinds of sugars. The various kinds of fatty acids are the building blocks that make up different types of fats. (Some other molecules are also present in fats.) The flavor, color, and other features of a fat depend on which fatty acids are joined together to make up the fat molecule.

FIGURE 6.7 ▶
If you eat more fatty acids than your body needs right away, they are stored as fat. Why do you think the body stores extra fatty acids?

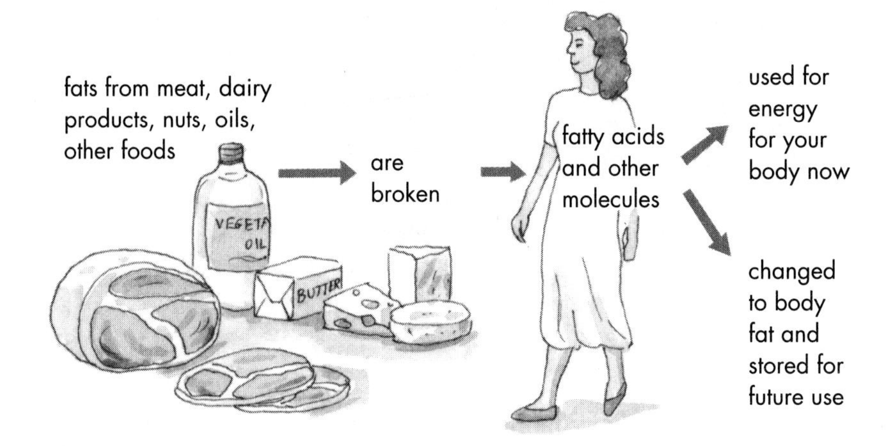
fats from meat, dairy products, nuts, oils, other foods
are broken
fatty acids and other molecules
used for energy for your body now
changed to body fat and stored for future use
BUTTER

Your body cannot use fats in the form that you eat them. Like carbohydrates, fats must be digested in order to release their energy. When a fat is digested, its fatty acid molecules are separated from each other. Sometimes your body uses the energy from these molecules right away. If the energy is not needed immediately, your blood carries the fatty acids to storage areas throughout your body. There, the fatty acids are joined together again to form body fat (Figure 6.7).

Your body stores much of its energy in your body fat. This stored energy can then be used to meet your body's energy needs at some time in the future. Your body fat also helps keep you warm and protects your body organs from knocks and bumps.

You may have heard that you should not eat too many **saturated fats**. These are fats that contain as many hydrogen atoms as they possibly can. In other words, the molecules that make up these kinds of fats have no spaces left into which hydrogen atoms could be added. (The word "saturated" means full.) Saturated fats are found in animal products like meat, milk, cheese, eggs and butter.

Unsaturated fat molecules have space left in them in which one or more hydrogen atoms could be added. Unsaturated fats come from plant products such as corn, peanut oil, and certain types of margarine (Figure 6.8).

FIGURE 6.8 ◀
Which of these foods contain saturated fats? Which contain unsaturated fats? How can you find out?

DID YOU KNOW

■ Unsaturated fats, such as those contained in olive oil and corn oil, are usually liquid at room temperature. Saturated fats, like those in butter or lard, are usually solid at room temperature. Food manufacturers sometimes add hydrogen atoms to unsaturated fats so that they will be more solid at room temperature. For example, this is done with some types of peanut butter and margarine. These hardened fats are said to be hydrogenated. Do you think that hydrogenating fats would make them better or worse for your health?

Many doctors and nutritionists (people who study nutrition) suggest that most Americans need to reduce the amount of fat they eat. No more than 30 percent of your food energy should come from fats. Nutritionists also recommend that no more than 10 percent of your food energy should be from *saturated* fats. There is evidence that, for some people, higher amounts of saturated fats increase the amount of **cholesterol** in the blood. Cholesterol is a fat-like compound found in animal products such as butter and egg yolks. Cholesterol is also found in all your body cells. Your liver makes all the cholesterol that your body requires, so eating foods that are high in saturated fats or foods that contain cholesterol can result in your body having too much cholesterol. In some people's bodies, too much cholesterol can lead to heart disease and other serious health problems.

ACTIVITY 6C / A SIMPLE TEST FOR FATS

In this activity you will test various types of foods to see if they contain fat.

PROCEDURE

1. Put on your safety goggles and apron.
2. Use a pencil to divide a sheet of paper in such a way that you leave a square about 3 cm x 3 cm for each sample that you want to test. You will also need one square without a sample as a control. ➡

MATERIALS

safety goggles
apron
starch suspension
glucose solution
paper towel or brown paper bag (unglazed paper)
fat sample (such as margarine or salad oil)
raw potato, bread, lard, boiled egg white, and at least two other foods that you would like to test for the presence of fat

■ **Do not taste any materials, including foods, in the laboratory.**

3. Label each square before you add each sample. Place a couple of drops of the starch suspension, the glucose solution, and the fat sample (if liquid) on separate squares (Figure 6.9). If you use a solid fat sample, carefully rub a very small portion of it on a separate square for a few seconds.
4. Allow the paper to dry. Hold the paper up to a light. Compare the square containing the fat sample with the squares containing the glucose and starch samples and with your control square. Record your observations. (By observing how the fat sample affects the paper you have a way to test whether fat is present in the food samples.)
5. Repeat Steps 3 and 4, using the food samples listed. Record your observations in a table of your own design.
6. Wash your hands thoroughly. Follow your teacher's instructions for discarding waste.

DISCUSSION

1. (a) What result indicates the presence of fat? (This is called a positive test for fat.)
   (b) Why did you test the glucose and starch samples for the presence of fat?
   (c) Why did you have one square as a control?
2. Of the foods you tested, which contained fat?
3. (a) Look at your test results. Try to rank the food samples in order from the most amount of fat to the least amount.
   (b) Why did you rank the samples in this order? Can you be sure that this ranking is correct? Explain your answer. ❖

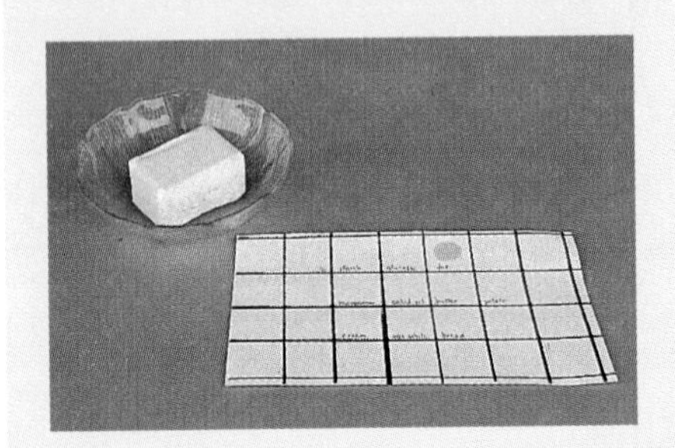

FIGURE 6.9 ◀
What might happen if your sample goes outside of its square?

▶ REVIEW 6.2

1. Why do you need to eat carbohydrates?
2. (a) List several foods containing carbohydrates that you would enjoy for breakfast. Make another list of carbohydrate-rich foods you would enjoy for lunch and a third list of similar foods you might eat at dinner.
   (b) Do all these foods contain the same type of carbohydrates? Explain your answer.
3. You need to eat some fats. List the functions of fats in your body.
4. (a) Name some foods that you regularly eat that contain fats.
   (b) Does each of these foods contain mostly saturated fats or mostly unsaturated fats?
   (c) Why might it be important to know whether a food contains unsaturated or saturated fat?

5. Imagine that for breakfast you had an egg, two strips of bacon, and a piece of toast with jam on it.

(a) What kinds of carbohydrates are contained in this meal?

(b) Are there any fats in these foods? If so, are they mostly saturated or unsaturated fats?

(c) What does your body do with these foods before it can use the energy in the food?

(d) What does your body do with this energy?

■ 6.3 PROTEINS: NUTRIENTS FOR GROWTH AND REPAIR

Proteins form a very important nutrient group with several functions. The main jobs of proteins are to help your body cells work properly and to help in building and repairing these cells. Many different foods supply protein, including meat, eggs, fish, dairy products, dried beans, rice, and corn (Figure 6.10).

FIGURE 6.10 ◀
Protein is available from many kinds of foods. Suggest other foods that are high in protein.

Like carbohydrates and fats, protein molecules are made up of chains of smaller molecules joined together. These smaller molecules—the building blocks of proteins—are **amino acids**. Every amino acid molecule is composed of nitrogen, hydrogen, oxygen, and carbon atoms. There are many different amino acids, each with a different combination of these four elements. A molecule of any protein is composed of many of these amino acid molecules joined together. There are many different types of proteins, depending on the number and order of the amino acids. Each type of protein has a different job in your body.

■ For infants and young children, there are *nine* essential amino acids. Their bodies cannot yet make enough of the ninth essential amino acid to meet their needs. Therefore, they must get it from the food they eat.

When you eat food containing protein, the protein molecules are broken down into their amino acid "building blocks." Your blood carries the amino acid molecules to places in your body where they are needed. There, they are used as the building blocks for new proteins needed by your body.

You require 20 different amino acids to make all the proteins your body needs. Your body can make 12 of these amino acids from atoms supplied by nutrients in the food you eat. These 12 are called **non-essential amino acids** because you do not have to eat foods containing them.

There are eight **essential amino acids**, those that your body cannot make at all or cannot make in sufficient quantity to meet your needs. You must eat foods containing these amino acids. If you lack even one of these amino acids, your body cannot make any of the proteins that contain it. And almost all of the proteins your body makes must be present if your body is to work properly.

Some of the proteins you eat are called **complete proteins**. They contain all eight essential amino acids in adequate amounts. **Incomplete proteins** are some of the other proteins you eat that are missing one or more of the essential amino acids. You can get complete proteins from animal products, such as meat, eggs, fish, and dairy products. Proteins from plant products (for example, from nuts, cereals, peas, and beans) are usually incomplete. However, it is not necessary to eat animal products for good nutrition. It is possible to obtain all eight essential amino acids from plant proteins if you eat certain combinations of foods (Figure 6.11).

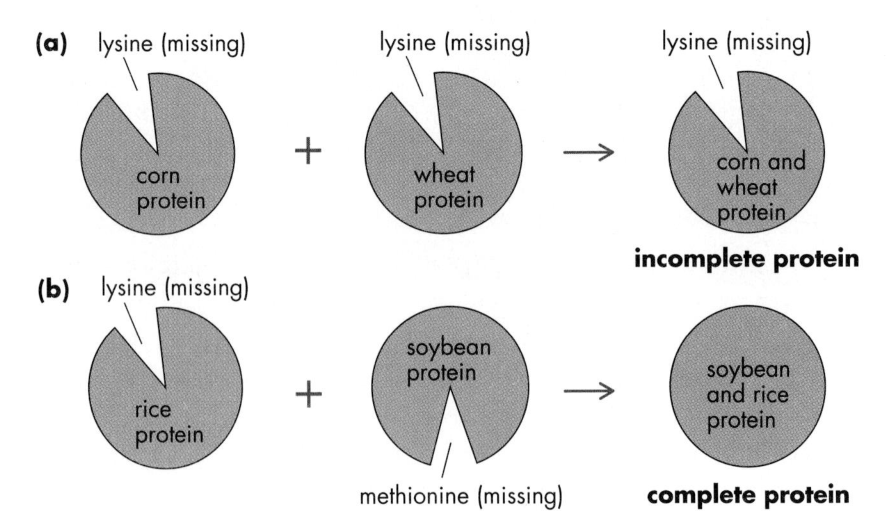

FIGURE 6.11 ▶
Lysine and methionine are essential amino acids. What do you think might happen to people who tried to get all of their protein from corn and wheat (a)? Would the same things happen to people who ate only rice and soybeans (b)?

HOW YOUR BODY USES PROTEINS

Proteins are needed in every cell of your body. The many different types of proteins do various important jobs in your body. Some types of proteins form a major part of your bones, hair, skin, and nails. Your sense

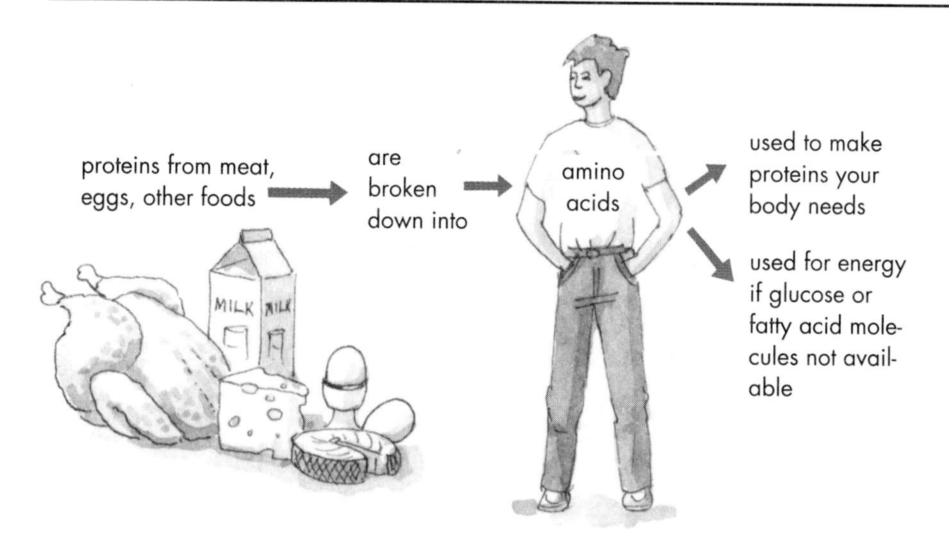
proteins from meat,
eggs, other foods
are
broken
down into
amino
acids
used to make
proteins your
body needs
used for energy
if glucose or
fatty acid mole-
cules not avail-
able
MILK

FIGURE 6.12 ◀
Animal and plant proteins are broken down to amino acids in the body, and then put together again into human proteins. In emergencies, amino acids can be used to supply energy to your body. Why do you think the body draws on stored glucose and fat for energy before starting to use amino acids?

organs, nerves, and brain would not work without the help of other types of proteins. When you cut yourself, certain proteins help your blood to clot (thicken and stop flowing out of your body). Still other types of proteins help you fight illness.

Some of the most important proteins are the **enzymes**. Enzymes are proteins that speed up the rates of chemical reactions within your body. All living things need enzymes. Without them, most chemical reactions would take place too slowly and organisms would die. Many different enzymes are needed to keep your body working properly. (In Chapter 7, you will learn more about how enzymes help digestion.)

After being broken down into their amino acid molecules, proteins can also be used to supply your body cells with energy. For energy, your body first uses any glucose or fatty acid molecules that are available. If none of these molecules are in your food, or stored in your body, then amino acids must be used. Using too many amino acids for energy is not usually good for your body. Amino acid molecules must be available for building the many proteins that your body needs (Figure 6.12).

Your body cannot store amino acids in a reserve supply, the way it can store glucose and fatty acids. If you eat too many proteins—and thus too many amino acids—the oxygen, hydrogen, and carbon atoms they contain are converted into fat. However, because the nitrogen atoms in the protein are removed when this is done, your body cannot remake amino acids from fat. The only place amino acids *are* held in the body is in your body parts—in the proteins that make up your muscles, for example. If you do not eat enough protein, your body starts to break down these body parts to obtain the amino acids it needs. Using amino acids from body parts can lead to illness and even death (Figure 6.13). You must eat enough carbohydrates and fats to supply your energy needs. Then the proteins you eat can be used to build and repair your growing body.

FIGURE 6.13
People become ill if they do not eat enough protein. This illness is common in parts of the world where people do not have enough nutritious foods. People with this illness are weak, and their skin becomes red and swollen. It can be corrected with a proper diet.

MAUREEN ARDELL

It began when Maureen Ardell entered Grade 9. "I started to really hate myself," she remembers. "I wanted to change everything about me—lose some weight, reshape my body, and improve my marks." What started as a sensible diet turned into an illness called anorexia nervosa.

Ardell refused to eat foods she loved. She kept track of every bit of food she ate, and even measured out each tiny scoop of margarine. "Somehow I felt that the less I ate, the purer I became," she says. Yet as her weight dropped she became more depressed, anxious, and tired. Her long blond hair started falling out. She withdrew from activities she used to enjoy with her friends, and hid in her room studying.

At 28 kg, Ardell was admitted to the psychiatric ward of a local hospital. Her doctor told her, "Being anorexic doesn't show strength. Recovery—that's what takes strength."

He was right. As Ardell wrote in her diary at the time, "I keep hearing the voice of common sense telling me to keep eating. But I'm afraid to listen to it because another voice says, *You are weakening, you are giving in.*" She had to force herself to eat each chocolate bar.

Three months later, with a mass of 43 kg, Ardell was released from hospital. There were still many difficult times ahead. Today, nine years later, she feels it is finally all over. "I am having the kind of life I thought I would never have. I don't think about my weight anymore. I don't even own a scale."

Ardell is very fortunate; many others with anorexia never recover fully. They often end up in the hospital again. Their hair and teeth may fall out. Many become very depressed. Some die. Ardell still suffers from occasional depression, and she has had to learn to keep herself mentally healthy. "Happiness is not something I sit around and wait for," she says. "It's something I work at every day of my life."

Today Ardell regrets that she missed out on a normal adolescence. "Sometimes I feel almost overcome with anger that those years are gone and I can never have them back," she says. "Teenagers should just know how humiliating anorexia is—it's nothing to be proud of. Your whole life, your dignity, your everything—it just disappears."

Anorexia most often affects teenage girls. The reasons for this eating disorder are complex, but one factor that seems to contribute is society's obsession with being thin. Look carefully at the magazine models who represent "beauty" today. Many are unnaturally and unhealthily thin. What do you think would be a better "ideal" that teenagers might aspire to?

REVIEW 6.3

1. Why do you need to eat proteins?
2. List some foods containing proteins that you eat regularly.
3. (a) What are essential amino acids?
   (b) Why do we need to be concerned about them?
4. How can a person who eats no meat or dairy products obtain enough essential amino acids? Explain your answer using the terms "incomplete proteins" and "complete proteins."

6.4 MINERALS AND VITAMINS

Your body needs large quantities of carbohydrates, fats, and proteins. The amounts of these nutrients that you should eat daily are measured in grams. Other nutrients, although necessary for your good health, are required in only very small amounts. Your daily requirements of these other nutrients are measured in micrograms. (One microgram is one one-millionth of a gram; 1 µg = 0.000 001 g.) This section describes the two types of nutrients that you need in such small quantities: minerals and vitamins.

Before reading further, stop and think about what you already know about minerals and vitamins. Can you name some? What foods contain them? Why does your body need them?

EXTENSION

■ Is fluoride added to the drinking water in your area? Find out why fluoride is added to some water supplies. Are there any possible risks to people when fluoride is added to their drinking water? How much fluoride do you need every day? Officials at your municipal hall or public health unit may be able to help you to determine why fluoride is or is not added to your drinking water.

MINERALS

Minerals are elements that your body needs for many important jobs. You need larger amounts of the minerals calcium, chlorine, magnesium, phosphorus, potassium, sodium, and sulfur. You need smaller quantities of other minerals, such as iron, copper, iodine, and zinc. Table 6.1 (see page 112) lists some of the minerals you need and the reasons you need them. Two minerals that are especially important for teenagers are calcium and iron.

CALCIUM: THE MOST IMPORTANT MINERAL

About 2 percent of your body's total mass is made up of calcium, most of it contained in your bones and teeth. A small amount of calcium can also be found in your muscles as well as your blood and other body fluids. The main source of calcium for most Americans is milk. Other dairy products, such as cheese and yogurt, are also good sources of calcium. If you eat the small, soft bones contained in some types of canned fish (for example, salmon), you are eating calcium (Figure 6.14).

Calcium is not easily absorbed by your body cells. In other words, it is difficult for your cells to pick calcium up from food as it is digested and take this nutrient into the cells. If a nutrient cannot get into the cells where it is needed, it cannot do its job.

FIGURE 6.14
These foods are good sources of calcium.

Mineral	Function	Sources
Calcium	Forms teeth and bones; helps blood clotting and muscle function	Milk, cheese, yogurt, dried beans
Phosphorus	Combines with calcium to help build bones and teeth	Meats, grains, fish, dairy products
Iron	Helps red blood cells carry oxygen	Meats, liver, nuts green leafy vegetables
Sodium	Regulates water levels in cells and blood; necessary for sending nerve impulses	Table salt, most foods
Iodine	Used by the thyroid gland to make hormones	Iodized salt, seafood
Potassium	Helps keep kidneys, heart, and muscles healthy	Meats, milk, fruits, green vegetables
Magnesium	Necessary for release of energy from sugars; required for muscle and nerve function	Nuts, potatoes, meats, fish

TABLE 6.1 ***Some Minerals That Are Important for Your Health***

■ Many people must reduce or eliminate eating dairy products because they have a condition called lactose intolerance. Find out the causes of lactose intolerance. How can people with lactose intolerance obtain enough calcium if they restrict or eliminate milk products from the set of foods they eat?

What helps calcium get into cells? Calcium is absorbed more easily when another nutrient, vitamin D, is present. In many countries, including the U.S., vitamin D is added to milk. In this way, vitamin D will be present along with the calcium to help a person's body cells absorb more of the calcium.

Your body uses calcium mainly for the growth and repair of your bones. Even when you have stopped growing, you need calcium to keep your bones healthy. In addition, calcium is needed in your nerve cells; it helps nerve impulses travel along and between these cells. This mineral also helps your blood to clot where skin is cut. You also need calcium for your muscles to work properly, including the most important muscle in your body— your heart.

YOUR NEED FOR IRON

Have you ever had your blood tested to see what might be making you feel unwell? Your doctor might have been checking to see if you have enough of the mineral iron in your blood. You might not have enough iron if you are not eating foods that contain iron. The best source of iron is meat. People who eat little or no meat must make sure that they obtain enough iron from other types of food (Figure 6.15).

You need iron in your blood cells to help those cells pick up oxygen, carry it throughout your body, and deliver it to other cells. All cells need oxygen to survive. Iron is also present in muscle cells, where it

FIGURE 6.15 ◀
These foods are good sources of iron.

helps these cells store oxygen. Muscle cells often need extra oxygen supplies because they are so active.

People who do not eat enough foods containing iron may have iron-deficiency anemia. ("Deficiency" means not enough of something.) The body cells of people with iron-deficiency anemia receive less oxygen than they need. These individuals may feel tired and weak.

Iron-deficiency anemia is a health problem for a large number of Americans. Teenagers especially need to watch for this condition for two reasons. First, they have a greater need for iron in their diets than do most men or children. When menstruation begins, teenage girls lose blood (and the iron it contains) and need to replace it every month. Teenage boys often have a rapid increase in their muscle mass, and iron is an important part of muscle cells. Second, many teenagers are at risk for iron-deficiency anemia because some of the foods they regularly eat contain little iron. For example, milk products and many common snacks (such as potato chips and chocolate bars) contain little or no iron. People who do not get enough iron from their food may need to take an iron supplement for a time. An iron supplement might be taken as a pill to provide extra iron to the body. However, too much iron may be harmful for your body. As for any nutrient, you should ask your doctor whether you need an iron supplement.

VITAMINS

Hundreds of years ago, explorers travelled the world's oceans in wooden ships. At that time, during long voyages, many sailors died of a mysterious disease called scurvy. Eventually the sailors learned that they could avoid getting scurvy by eating citrus fruits, such as oranges, grapefruits, and lemons. However, it was a very long time before anyone discovered the reason for this. Today we know that citrus fruits contain large quantities of a nutrient called vitamin C, which helps prevent scurvy. Other vitamins that are important to your good health have also been discovered.

Vitamins, which your body needs in very small amounts, are nutrients that usually act as "assistants" to enzymes. They help enzymes speed up chemical reactions. Nearly every enzyme needs the help of a specific vitamin in order to do its job. Table 6.2 outlines the functions of many important vitamins and lists the best food sources of each.

■ There are a few vitamins that the human body can produce itself. Vitamin D, for example, is produced when the sun shines on your skin. Vitamin K and some B vitamins are produced by helpful bacteria that live in human intestines. You also obtain some of each of these vitamins from the food you eat. You must get the other vitamins that you need (for example, C and E) from food.

Vitamin	Function	Sources
A	Promotes good night vision; helps body resist infections; helps keep skin and hair healthy	Milk, liver, egg yolk, butter, margarine, green and yellow vegetables
B1 (Thiamine)	Helps release energy from carbohydrates; promotes appetite	Meats, liver, nuts, enriched and whole-grain cereals and breads, cheese, eggs, legumes, green leafy vegetables
B2 (Riboflavin)	Helps release energy from carbohydrates, fats, and proteins	Liver, meats, milk, enriched and whole-grain cereals and breads, cheese, eggs, legumes, green leafy vegetables
B3 (Niacin)	Helps release energy from carbohydrates, fats, and proteins	Enriched and whole-grain cereals and breads, fish, meat, poultry, liver, legumes
B12	Helps nervous system function; needed for red blood formation	Meat, fish, eggs, dairy products
C	Acts as an antioxidant in the blood; maintains teeth, gums, ligaments, tendons, and other supportive tissue	Citrus fruits, potatoes, tomatoes, berries, green leafy vegetables
D	Helps the body absorb calcium and phosphorus for bones and teeth	Milk, fish, liver
E	Helps keep muscles healthy; acts as an antioxidant in the blood; needed in reproduction	Vegetable oils, margarine, enriched and whole-grain cereals
K	Promotes normal blood clotting	Green leafy vegetables, peas, grains

TABLE 6.2 ***Some Vitamins That Are Important for Your Health***

Certain vitamins are needed for your body to use other nutrients. For example, you read about how vitamin D helps your cells absorb calcium better. Vitamin C helps your body cells absorb iron. Since teenagers should be concerned about getting enough iron to their cells, their meals should include foods containing both vitamin C and iron.

CLASSIFYING VITAMINS

Vitamins can be divided into two groups, depending on whether they dissolve in water or in fat. Vitamin C and the B vitamins are water-soluble vitamins because they dissolve in water. The other vitamins listed in Table 6.2 are fat-soluble. These vitamins cannot be absorbed by your body unless you eat some fats along with these vitamins. They are often found in foods that contain fat, such as oils, meat, and dairy products.

If you eat more of a fat-soluble vitamin than you need, the extra amount is stored in your body fat. This is why people should not take vitamin pills that they do not need; their bodies may store toxic (poisonous) amounts of fat-soluble vitamins. Taking in too much of certain vitamins can cause serious problems. Toxic amounts of vitamin A, for example, can cause loss of hair, blurred vision, dry skin, nausea, and diarrhea. Most healthy Americans get enough fat-soluble vitamins from their food.

Because vitamins B and C are water-soluble, any extra amounts of these nutrients dissolve in your body fluids. They leave the body in your urine rather than being stored. You need to eat enough of these vitamins every day. Since these vitamins are water-soluble, they can dissolve in water when they are cooked. Eating vegetables raw, or steaming them instead of boiling them, will help the vitamins stay in the food.

▶ REVIEW 6.4

1. (a) List three minerals that your body needs, and state why each is necessary for good health.
(b) What foods supply those minerals?

2. Why does your body need vitamins?

3. (a) How are vitamin C and the B vitamins different from vitamins A, D, E, and K?
(b) When you plan your meals for a day, what difference does it make if a vitamin is fat-soluble or water-soluble?
(c) List one food source for each of the fat-soluble and water-soluble vitamins.

4. Why might some fruits or vegetables give your body fewer vitamins than expected?

5. Broccoli and spinach contain iron. Why would sprinkling lemon juice on a salad made with these vegetables be a nutritious practice?

6.5 WATER: THE MOST IMPORTANT NUTRIENT

DID YOU KNOW

■ Water makes up 55 to 65 percent of the human body. If you lose even one-tenth of this water, you can have serious health problems. A person cannot survive if more than one-fifth of the body's water is lost.

EXTENSION

■ Write the procedure for an experiment to determine how much of an apple's mass is water. Discuss the procedure with your teacher to obtain approval, and then carry out the procedure.

Water may not seem like a nutrient at all. Water does not supply your body with energy. It does not usually need to be purchased or prepared in any way. It often has little taste. Water is so much a part of our lives that we seldom think about its importance— until we have to go without it.

Water is the most important nutrient for all animals, including humans. You could survive without food for weeks, but without water you would die in a few days. Water has several important functions in your body.

Water is essential for cells—the basic unit of all living things. Water carries nutrients and other needed materials into your body cells. It also carries waste products away from your cells and out of your body. Your cells need water for many of the chemical reactions that break down large nutrient molecules such as carbohydrates, fats, and proteins. In fact, water is the main component of all the cells in your body. Without it, your cells could shrink and eventually collapse—and so would you.

Water also helps your body in other ways. When you sweat, the water that appears on your skin helps to cool your body. In this way, water helps you to maintain a safe body temperature even on very hot days or when you are very active. Water is also a lubricant, just as grease or oil is a lubricant for a hinge or for the gears of a bicycle. For example, water lubricates your joints—the places where your bones are joined together—so that the bones move more easily. The saliva in your mouth is mainly water; it lubricates food so that you can swallow it more easily.

Where do you think your body gets the water it needs? Most of the solid foods we eat, especially fruits and vegetables, contain a great deal of water (Figure 6.16). In addition, many nutritionists recommend that we drink six to eight glasses of water or other fluids a day.

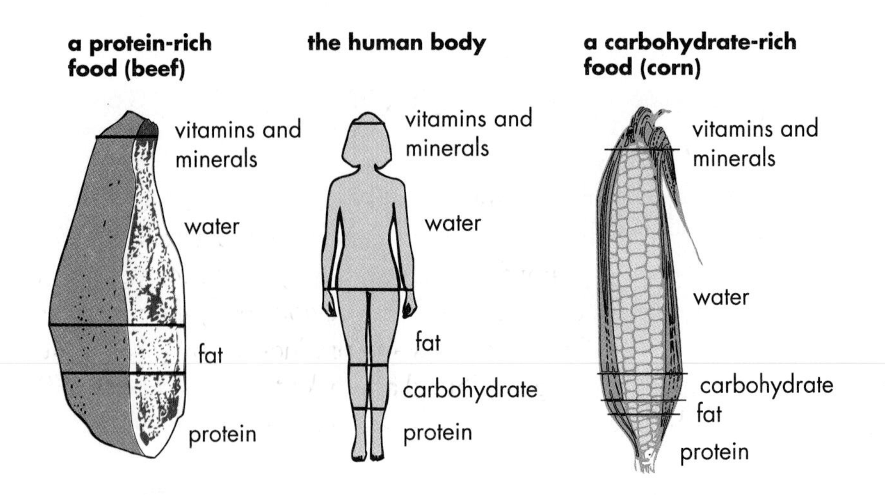

FIGURE 6.16 ▶
Which nutrient is most common in both of these foods and also in the human body?

Although the easiest place to get water may be from the kitchen tap, milk and juice are nutritious sources of water. Many Americans drink tea, coffee, alcoholic beverages, or soft drinks. Some of these drinks contain substances such as caffeine, which may be harmful if consumed in large quantities. Soft drinks and alcoholic beverages provide many kilojoules of energy from sugar, but few other useful nutrients. Drinking large amounts of these beverages may lead to an unwanted increase in body mass. As well, a person who drinks these beverages regularly may not drink more nutritious liquids, such as milk or fruit juice.

REVIEW 6.5

1. Why is water considered a nutrient?
2. List five functions of water in your body.
3. Why might you need to drink more water
   (a) on a hot day than on a cool day?
   (b) if you spent a day hiking than if you spent the day in school?
4. (a) How do you fill your own water needs—what beverages do you regularly drink?
   (b) Do you think you should drink more plain water every day? Why or why not?

ACTIVITY 6D / SELL THAT NUTRIENT

GROUP ACTIVITY · CROSS-CURRICULAR

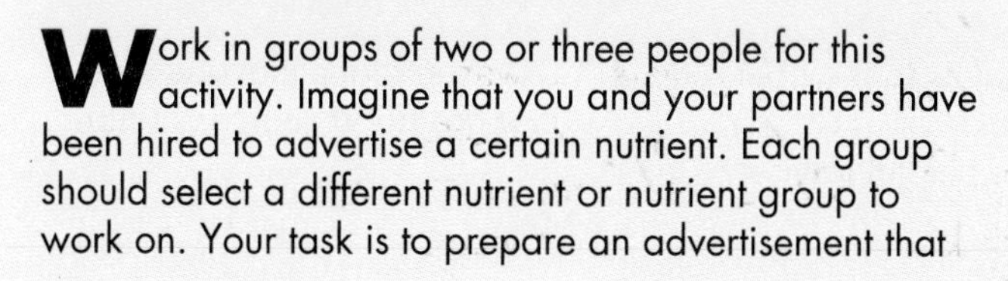
Work in groups of two or three people for this activity. Imagine that you and your partners have been hired to advertise a certain nutrient. Each group should select a different nutrient or nutrient group to work on. Your task is to prepare an advertisement that will "sell" people on the need to obtain enough of your selected nutrient. Will you choose to advertise on the radio or television, on a billboard, or in a newspaper or magazine? Be creative, and be sure your advertisement will catch people's attention! ❖

6.6 FIBER: NOT A NUTRIENT, BUT STILL IMPORTANT

You have now studied the six nutrient groups. All nutrients are digested in your body and must enter your cells if they are to be useful. **Dietary fiber** is part of many foods, and it is important for your good health. However, it is not a nutrient because your body cannot digest it. The best sources of fiber are cereals and products made with whole-grain flour, such as some types of bread and baked goods (Figure 6.17). Fiber can also be found in fruits and vegetables. There are various types of dietary fiber.

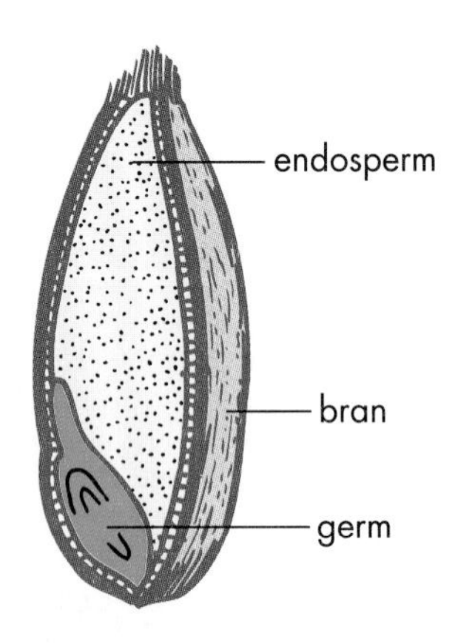

FIGURE 6.17
A kernel of grain. The endosperm and the germ contain starch, a source of energy. The bran is made up of carbohydrates that humans cannot digest. Bran is removed when making white bread, but is left in the flour for whole wheat and some other breads. How do you think such breads would help in digestion?

Dietary fiber is made up of complex carbohydrate molecules. However, unlike starch and glycogen, these complex carbohydrates cannot be broken down by your body into smaller glucose molecules. As a result, they cannot enter your cells and cannot be used to provide energy.

Because fiber cannot be broken down and absorbed by your body cells, it remains in and moves through your body after you eat it. Eventually fiber leaves your body with the feces—your body's solid waste. Some types of fiber hold on to water. This helps keep the feces moist, so they move more easily through and out of the body. In this way, fiber prevents or relieves constipation.

Many North Americans eat very little fiber compared with people in other parts of the world. Many more North Americans suffer from various diseases of the intestines, including cancer, than do people in countries where a lot of fiber is eaten. Some people think that eating foods that are high in fiber may reduce the chance of getting these diseases.

There are still many unanswered questions about dietary fiber. Most nutritionists agree that fiber is very important and that many Americans should eat more fiber. You will probably get enough fiber if you eat complex carbohydrates for a large part of your energy needs. This is because carbohydrates from plant foods contain fiber as well as starch.

REVIEW 6.6

1. Why is dietary fiber not considered a nutrient?
2. (a) What happens to the fiber that you eat?
   (b) Why is this helpful to your body?
3. List three foods that you enjoy that are good sources of fiber.

EXTENSION

■ Like humans, cows cannot break large molecules of fiber into smaller molecules of simple carbohydrate. Even though cows are not able to digest the fiber themselves, they do use it as a nutrient, obtaining energy from it. Find out how this is possible. What is special about the digestive system of cows?

■ 6.7 FOOD PROCESSING

Selecting foods that provide you with a balance of fiber and nutrients is an important part of healthy eating. Did you know that the packaging, storage, and preparation of food can change the amount of nutrients it contains? In some cases, treating food in these ways increases the number of nutrients it contains; or it may change the nutrients so that your body can use them more easily. In other cases, the treatment reduces the number of nutrients. To make decisions about good nutrition, you sometimes need to know something about the "history" of the food items you are eating.

Food processing refers to anything that is done to plant or animal foods before they are eaten. Cooking, packaging, freezing, and drying are some methods of food processing. Although much food processing is done in factories, people also do a lot of food processing at home (Figure 6.18).

The most important reason for processing food is to **preserve** it, or keep it from spoiling ("going bad"), so that it will be safe for you to eat. Processing may also improve the flavor, color, texture, or nutrient content of food.

FIGURE 6.18
Three ways to process foods: drying chilis, smoking salmon, and pickling. Can you think of others?

PRESERVING FOOD

Some foods, such as potatoes, can be stored for months without spoiling if they are kept in a cool, dry place. Grain will remain good to eat for years if stored properly. Most food, however, will begin to spoil in a few days or even hours.

Often, it is microorganisms that cause food to spoil. Microorganisms are organisms so small that they can only be seen with a microscope. Some, such as bacteria and fungi, live in all types of food eaten by humans. Microorganisms can damage or destroy a lot of food. A major goal of many types of food processing is to prevent the growth of microorganisms.

People can become ill after eating food that contains some types of microorganisms. Illness caused by microorganisms in food is often called **food poisoning**. Food poisoning can produce headaches, nausea, vomiting, diarrhea, and occasionally death.

Some people process various types of food at home by canning, freezing, drying, pickling, or smoking it. These types of processing all help to preserve food—to keep food from spoiling or to slow down the rate at which it spoils. Most preserving is done in factories that prepare food for sale in stores (Figure 6.19). For example, milk is treated (pasteurized) to kill microorganisms and is put into containers before it is sold. Many people work in factories where fish, vegetables, and fruit are put into cans. No matter who does the preserving, or what type of food is preserved, the goal of this type of food processing is to keep food from spoiling.

FIGURE 6.19
What advantage does processed food have over fresh food?

ACTIVITY 6E / METHODS OF FOOD PROCESSING

Work individually or in small groups to find out about different methods of food processing. You might choose a method used at home or in a factory such as a dairy, fish cannery, or fruit-processing plant. You might also consider methods used by stores to keep food safe and fresh.

Do research in the library or interview people in your community, such as homemakers, elders, or workers who process food.

Share your research with the rest of the class in the form of a pamphlet, poster, or display. Include information such as the following:

- the purpose of this type of food processing
- a description of this method of food processing
- the advantages and disadvantages of this type of food processing
- any other interesting information you have gathered ❖

FOOD ADDITIVES

A **food additive** is any substance added to any food during processing. Thousands of substances with many different purposes can be considered additives. Table 6.3 lists some of the food additives used in the United States. As this table shows, some additives are used to give foods a particular color, odor, or texture. Other additives are used to keep food moist or to keep it dry. Many additives are used to preserve food. These substances, called **preservatives**, help to keep food from spoiling.

■ Some people have allergies to certain food additives. After eating food containing these additives, these people might have an allergic reaction. They might have difficulty breathing, get a headache, or feel ill. Sulfites are additives that were once used in restaurants to keep fruits and vegetables fresh. Because some people have trouble breathing when they eat sulfites, it is no longer legal to use them in U.S. restaurants. Monosodium glutamate (MSG) is an additive that brings out flavors in foods. Do you know anyone who is allergic to MSG?

What does the additive do?	Some examples of additives	Some examples of foods in which they are used
Keeps powders dry, prevents clumping	Calcium aluminum silicate, calcium silicate	Dessert mixes, drink mixes, mixes for coating foods (for example, chicken, fish), salad dressing mixes
Adds color	Caramel, cochineal, chlorophyll, fast green, iron oxide, silver	Margarine, cheese, candies, jams and jellies, pudding mixes
Brings out certain flavors	Monosodium glutamate (MSG)	Powdered soups, smoked meats
Preserves foods by preventing or delaying spoilage	Acetic acid, ascorbic acid, wood smoke, calcium sorbate, potassium bisulfite, sulfur dioxide, sodium nitrate, BHT, BHA	Potato chips, vegetable shortening, canned and frozen fruits, breads, cookies, cheese, fruit juices
Maintains a consistent feel (texture) to foods	Agar, carrageenan, gelatin, methylcellulose, pectin polysorbate, sodium alginate	Candies, sauces, jams and jellies, cake mixes, chocolate, bread, pudding mixes, ice cream, cream cheese, baked goods
Promotes desirable chemical reactions in foods	Amylase, catalase, cellulase, lactase, lipase, milk-coagulating enzyme, papain, pepsin, rennet	Cheese, cream cheese

TABLE 6.3 ***Some Food Additives***

Additives may be nutrients, such as vitamins and minerals. Sometimes nutrients are added to foods in which they are not normally present. For instance, the vitamin D added to milk and the iron added to some breakfast cereals are additives. These foods are then described as being **fortified** with these nutrients. In other cases, an additive increases the amount of a nutrient that was already present in a particular food. Often these added nutrients replace nutrients that were removed during processing. Such foods are said to be **enriched**. Many types of breads and flours are enriched with B vitamins, for example.

It is important to think about both the advantages and disadvantages of using food additives. For example, chemicals called nitrites are added to meats such as bacon, ham, and hot dogs as preservatives. On the one hand, nitrites prevent spoiling. On the other hand, some researchers think that nitrites themselves may help cause cancer. In the U.S., only a small amount of nitrites may be added to meats. Today, these preservatives are needed because no other chemicals are known to protect these meats from spoiling the way nitrites do. The use of nitrites is a good example of weighing the risk of using an additive against its benefits. It is something you should think about when you choose what foods to eat.

ACTIVITY 6F / HOW LONG WILL BREAD LAST?

In this activity, you will design your own controlled experiment to determine which bread will remain unspoiled for the longest time. Fungi (sometimes called molds) are usually the first types of microorganisms to start growing in bread. Fungi prefer to live in environments that are damp and warm and have dim light. Design an experiment to determine the effectiveness of using preservatives to reduce or prevent the growth of fungi in bread. To do this, you need to compare bread that contains preservatives with bread that does not. You might also want to see if different kinds of bread (for example, rye, white, or whole-wheat) spoil at different rates. Or you might want to see if one type of preservative works better than another. Different students or groups might study different aspects of this question and then share their results with the rest of the class.

When designing your experiment, remember to include an experimental control. Also make sure you have included proper safety procedures. Predict which of your bread samples will spoil first and which will stay fresh the longest.

When you have prepared your procedure, show it to your teacher for approval. Carry out your experiment using bread samples brought in by your class. Label each type of bread used. Write a report on the results and your conclusions. Include a description of how you would improve on your procedure if you were to repeat this experiment. ❖

ADVANTAGES AND DISADVANTAGES OF FOOD PROCESSING

Food processing offers many advantages. Preserving food allows it to be stored longer. Imagine if you had to shop daily because you could not keep food fresh for any longer than 24 hours.

Some types of food processing add to convenience in other ways. For example, it is much easier to buy pre-cut french fries than it is to cut your own, and it is easier to mix up frozen orange juice than it is to squeeze oranges (Figure 6.20). Other types of food processing are helpful or absolutely necessary. For example, when you cook meat, you are breaking down many of the large protein molecules it contains. This breaking down makes it easier for your body to digest the meat and obtain useful amino acid molecules from it. Some types of grain, such as wheat, cannot be eaten until they are ground or cooked.

Sometimes the goal of food processing is to add nutrients to food. As you have read, vitamin D is added to milk, and various vitamins and minerals are added to some breakfast cereals.

Food processing can also have disadvantages. In many cases, processed food (such as applesauce) costs more than the same amount and type of fresh food (such as apples). Some types of processing can remove vitamins or minerals. For example, boiling vegetables rather than steaming them can remove some water-soluble vitamins. Also, it is better to peel or trim fruits and vegetables as little as possible because most of the nutrients are contained in the outer leaves or the outer layers of the skin.

In some food processing, substances that are added to food might be harmful to some people. For example, salt and sugar are sometimes added to food to prevent spoiling. Although these substances can be nutrients, too much salt or sugar can be harmful to some people's health. In general, people should avoid eating a great many processed foods that contain large amounts of salt and sugar. There may be risks with other additives too.

FIGURE 6.20
There's more than one way to drink an orange. What are the advantages of fresh, reconstituted, and frozen orange juice?

FOOD LABELING

Do you ever read the labels on packages of food? If so, you may have noticed that one of these labels lists all the ingredients, or items, in that food. This label will tell you what additives the food contains, as well as what animal or plant products are in it. Knowing what is contained in the foods you buy is a very important part of choosing nutritious foods. In the U.S., the law states that every package of food must have a label that lists all the ingredients. The ingredients must be listed in order, starting with the ingredient that makes up the largest part of the product, and ending with the one that makes up the smallest part. Food labels can help you avoid foods containing a lot of sugar or other ingredients that you might not wish to eat (Figure 6.21).

The labels on some foods also tell you about the amounts of the nutrients in a serving (Figure 6.22). Sometimes the label lists how many kilojoules of energy you obtain from a serving. Some products also have a label providing information about the freshness of the product. You may see this as a label saying "best before" a certain date. Being an informed consumer is an important part of selecting nutritious meals (Figure 6.23).

INGREDIENTS

SUGAR, CORN FLOUR, WHITE FLOUR, WHOLE OAT FLOUR, HYDROGENATED COCONUT OIL, SALT, COLOR, NATURAL FRUIT FLAVORING, CITRIC ACID, VITAMINS (THIAMIN HYDROCHLORIDE, NIACINAMIDE, PYRIDOXINE HYDROCHLORIDE, FOLIC ACID, d-CALCIUM PANTOTHENATE), MINERALS (REDUCED IRON, ZINC OXIDE). BHT ADDED TO PACKAGE MATERIALS TO MAINTAIN PRODUCT FRESHNESS.

FIGURE 6.21
All the ingredients in this food product are listed on the label. How many of them can you recognize? Can you guess what this product is?

EXTENSION

■ You may have seen unsaturated fats labeled as either "monounsaturated" or "polyunsaturated." What do these terms mean? Is one of these two types of unsaturated fats better for your health than the other?

The ingredients are listed here in order from the largest to the smallest amount.

INGREDIENTS: RICE, SUGAR AND/OR GLUCOSE-FRUCTOSE, SALT, MALT FLAVORING, THIAMIN HYDROCHLORIDE, NIACINAMIDE, PYRIDOXINE HYDROCHLORIDE, FOLIC ACID, d-CALCIUM PANTOTHENATE, REDUCED IRON, BHT.

NUTRITION INFORMATION

		PER 30 g SERVING CEREAL (250 mL, 1 CUP)	PER 30 g SERVING CEREAL + 125 mL, P.S. MILK†
ENERGY	Cal	110	175
	kJ	470	740
PROTEIN	g	2.2	6.5
FAT	g	0.1	2.6
CARBOHYDRATE	g	25	32
SUGARS	g	2.9	9.1
STARCH	g	22	22
DIETARY FIBER	g	0.3	0.3
SODIUM	mg	310	375
POTASSIUM	mg	35	235
PERCENTAGE OF RECOMMENDED DAILY INTAKE			
VITAMIN A	%	*	7
VITAMIN D	%	*	23
VITAMIN B_1	%	46	50
VITAMIN B_2	%	*	14
NIACIN	%	8	13
VITAMIN B_6	%	10	13
FOLACIN	%	8	11
VITAMIN B_{12}	%	*	25
PANTOTHENATE	%	7	13
CALCIUM	%	*	15
PHOSPHORUS	%	4	15
MAGNESIUM	%	4	12
IRON	%	28	19
ZINC	%	5	11

†P.S. Milk = Partly Skimmed Milk
*Less Than One Percent

Serving size

This list provides all the cereal contents *plus* the nutrients you will get if you *add* 125 mL of milk.

The nutrients are grouped for convenience — protein, fat, vitamins, and minerals. These are not in the order of quantity supplied.

This recommended intake may not be the same as your body's actual requirements at certain times.

Nutrient value listed for 30 g of cereal alone. This is useful for comparison with other similar products.

FIGURE 6.22 ***This is the label from a breakfast cereal.***

FIGURE 6.23
How can reading food labels help you decide which products to select?

ACTIVITY 6G / COMPARING FOOD LABELS

This activity will give you an opportunity to look closely at the information on food labels and to compare the nutrient value of similar types of food.

MATERIALS

food labels from at least three different kinds of breakfast cereal

PROCEDURE

1. Examine the food labels of the cereals you have chosen. Design a table in which you can record all the ingredients in the three cereals. Your table should allow you to record the amount of each of these ingredients.
2. Most dry cereals list ingredients for a serving of 50 g. Check whether your three cereals list amounts for the same serving size. If not, adjust the amounts so that they are based on the same serving size, or choose another cereal.
3. Complete your table by filling in the list of ingredients and the amounts in each cereal. If a particular ingredient is not present in one or two of the cereals, indicate this by leaving a space or putting a dash in the appropriate place.

DISCUSSION

1. Compare the amounts of some of the nutrients that your cereals contain. Are there any differences? If so, what are these differences?
2. (a) Based on what you have learned in this chapter, which of these cereals do you think is the most nutritious? Why do you think this?

(b) If you were to have this cereal for breakfast, would you need to eat anything else to obtain a good balance of nutrients for this meal? If so, what other foods might you eat?

3. Are there any ingredients listed that you have not heard of before? (Look back at Table 6.3, page 121, to help you identify some additives.) If so, list three questions that you have about one or more of these ingredients. ❖

▶ REVIEW 6.7

1. What is food processing?
2. List three advantages and three disadvantages of food processing.
3. Think about what you had for breakfast. Was anything that you ate processed? If so, describe the ways in which the food was processed and explain why you think this type of processing was done.
4. (a) What is a food additive?
   (b) Give three examples of helpful food additives.
   (c) List three food additives that may be harmful for some people.
5. Imagine that you are having some type of packaged food for lunch. What information might you find on the labels of this food product?
6. What kinds of experiments might a food processing company do in order to determine the "best before" date?

■ 6.8 NUTRITION AND DIET

Your **diet** consists of all the food that you regularly eat and drink. You may not have the same diet as your classmates. Your diet is affected by many things, such as the foods available where you live or your family's customs and religious beliefs. Your age, likes and dislikes, and health may also affect your diet. People with certain diseases or health problems, for example, may have to eat certain foods and avoid others. Individuals who are trying to reduce their body mass often say they are on "a diet." Their diets may consist of foods that contain fewer kilojoules of energy than usual. These diets should still, however, provide a proper balance of all of the nutrients necessary for good health.

Your diet may provide you with enough of the nutrients and fiber that you need—or it may not. This depends on what you choose to eat. As you have read, your food choices have a very important effect on your health. Your diet does not consist of just one meal, but of all the foods you regularly eat. Occasionally, therefore, you can decide to eat what might be a rather unhealthful meal or snack—as long as you usually eat nutritious foods and avoid eating foods with too many kilojoules of energy. To make decisions about what to eat, you need to know what nutrients you need and how to obtain them. In other words, you must know what makes up a healthy diet.

■ Because your body goes without food all night, it needs to be "recharged" with a meal in the morning to carry you through your daily activities. Studies have shown that, on average, students who eat breakfast do better on tests, are able to learn more, and are less likely to get tired during the day than those who skip this meal. Do you know why our morning meal is called "breakfast"?

A BALANCED DIET

To help people plan a healthy diet, the U.S. Department of Agriculture has prepared *The Food Guide Pyramid*, similar to the one shown in Figure 6.24. The guide divides all the foods we eat into five major groups: 1) Grain Group (bread, cereal, rice, and pasta); 2) Vegetable Group; 3) Fruit Group; 4) Milk Group (milk, yogurt, and cheese); and 5) Meat Group. The "Others" group at the top of the food pyramid consists of fats, oils, and sweets. The guide suggests the number of food servings people should eat every day from these categories. The guide gives information for different groups, such as children, teenagers, and pregnant women. The amount of each type of food suggested will allow you to obtain the nutrients you need without eating too many kilojoules of energy. This information will allow you to plan a balanced diet: one that allows you to obtain just the right amounts of all types of nutrients.

Selecting items from each of the five major food groups is only your first step in planning a nutritious diet; the groups are very broad, and each contains many different foods. The guide advises you to include a variety of foods from each group in your diet. In this way you will eat all the nutrients your body needs. The guide also suggests that Americans should choose foods with lower amounts of fats and added sugars more often.

The daily servings suggested by *The Food Guide Pyramid* are suitable for the average healthy person. However, every person is different. Your needs will depend upon your health, your level of physical activity, your age, and perhaps even on your personal goals.

■ Start a class bulletin board for newspaper and magazine articles related to human population growth, the lack of food in some parts of the world, and problems in agriculture and the fishing industry. Hold a class discussion on how all these subjects might be related.

NUTRITION AND WORLD HEALTH

Humans have existed for about 3 million years. During most of this time, the number of people on Earth was relatively small. They survived by eating wild plants and animals. About 10,000 years ago, people began to grow crops and raise domestic animals, such as goats and cows. With more food available, the human population grew to 10 to 20 million people. Earth's resources could still easily support that number.

During the last 150 years, scientists and doctors have discovered how to treat and prevent many of the diseases that had previously killed many children. With better health care, more young people than before grew up and had children of their own. As a result, the number of people on Earth began to increase rapidly. Since the 1950s, the human population has been "exploding," or growing *very* quickly. In 1987, the number of people on Earth reached 5 billion. Now, more than 90 million babies are born every year. The human population is predicted to reach 6 billion by about the year 2000.

DAILY FOOD GUIDE PYRAMID

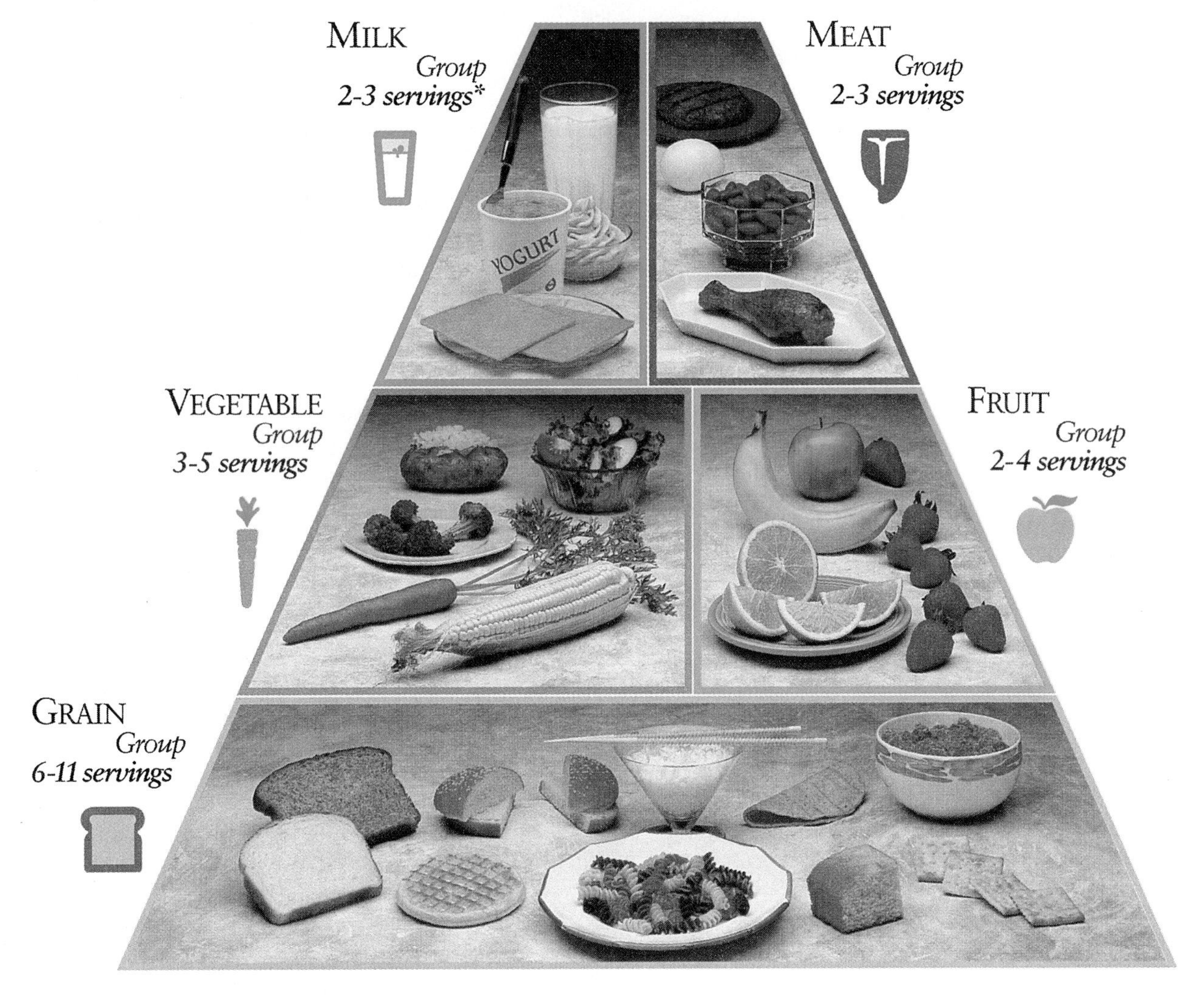

*Preteens, teens, and young adults (age 11 to 24) and pregnant and lactating women need 4 servings from the Milk Group to meet their increased calcium needs.

FIGURE 6.24
Daily Food Guide Pyramid

For a long time, better methods of farming, fishing, and food processing allowed food production to keep up with the growing human population. Recently, the world has not been able to produce enough food to feed everyone. Pollution, soil erosion, natural disasters, overfishing, overgrazing, and many other factors have all resulted in poorer quality and a lower quantity of many types of food (Figure 6.25). An additional problem is that the world's food is not spread out among the world's people. In some parts of the world, people have more food than they need. In other places, people do not have enough to eat, and many die.

FIGURE 6.25 ▶
The Sahel, in Africa, is slowly turning into desert. The Sahara Desert to the north of the Sahel is spreading south. What kind of food could people living in the Sahel grow?

EXTENSION

■ The diets of people in different parts of the world are often quite different. Study the traditional diet of a group of people, choosing a diet that is different from your own. In what ways is it different from yours? Why do you think these differences exist? You might choose to study the diet of one of the following peoples: Szechuan Chinese, Korean, Inuit, Kalahari Bushmen, Masai, Caribbean, Indonesian. Or you could select another traditional diet. Your research might take you to the library or to restaurants. You might also ask one or more of your friends or classmates.

In countries like the U.S., the main nutritional problem is overnourishment: eating more food than the body needs. Overnourishment results in obesity—too much body fat. This is particularly a problem for American adults.

In many other countries, millions of people are undernourished. They do not have enough to eat, so they cannot obtain all of the nutrients they need. Many of these people die as a result of starvation. Even in the U.S., many people are undernourished. In some cases, this happens simply because people try to be too thin. Concern for their appearance causes these people to eat an improper diet. Because they do not eat enough, they do not obtain the nutrients their bodies need. Serious health problems, and even death, can result.

Overnourishment and undernourishment may seem to have little in common. Yet both are forms of **malnutrition**. ("Mal" means "bad.") In most cases, malnutrition is caused by a lack of one or more important nutrients in a person's diet (Figure 6.26). One of the biggest challenges facing world leaders today is how to manage the world's resources so that all people can obtain enough food.

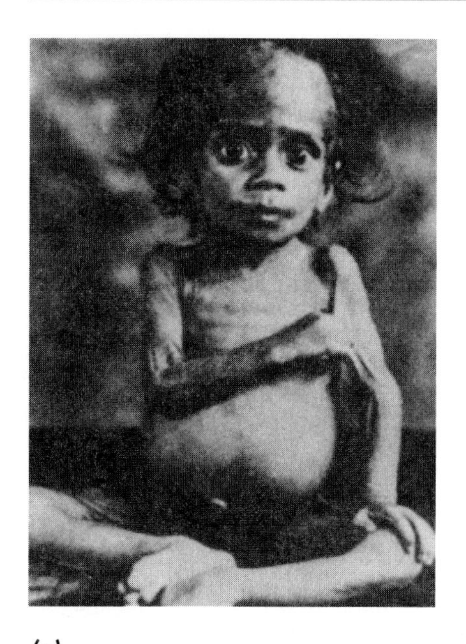

(a)

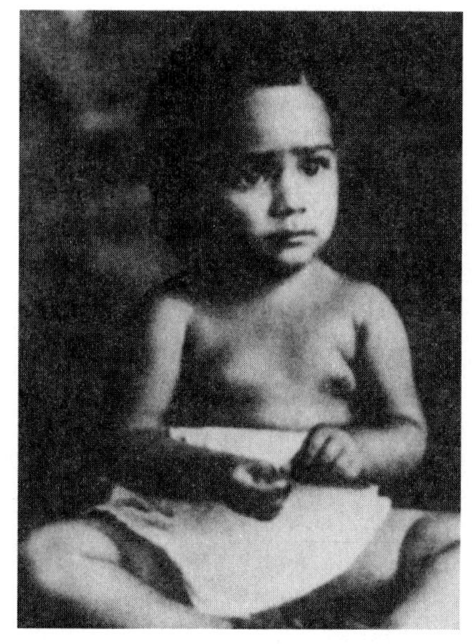

(b)

FIGURE 6.26
(a) This photo shows a child suffering from a type of malnutrition. The child's diet did not include enough carbohydrates, fats, and proteins.
(b) When fed a more balanced diet, the child once again became healthy.

REVIEW 6.8

1. What is meant by "diet"?
2. What factors affect the foods in your diet? If you cannot or do not eat certain foods, explain why.
3. (a) What are the five food groups listed in *The Food Guide Pyramid*?
   (b) Every day you should eat a variety of foods (not just the same food) from each of the four food groups. Why?
4. Why do some people suffer from malnutrition?
5. How do you think Americans might help solve some of the problems that cause malnutrition either in the U.S. or in other countries?

CHAPTER REVIEW

KEY IDEAS

■ Food supplies your body with (1) the elements and compounds it needs in order to build and repair cells and (2) the energy it needs.

■ A nutrient is any substance that is taken into your body cells and that serves a useful purpose there.

■ The human body needs about 50 nutrients. These can be classified into six major groups: carbohydrates, fats, proteins, minerals, vitamins, and water.

■ Carbohydrates and fats supply most of the energy your cells require.

■ Proteins are used for building and repairing body cells. They can also be used to supply your body with energy.

■ Vitamins and minerals serve a variety of important purposes in your body, although you need only small amounts of these nutrients each day.

■ Water is sometimes called the most important nutrient. You can live only a few days without water.

■ Although fiber is not a nutrient, it is still important to your health. Fiber helps digestion and perhaps helps to reduce the risk of heart disease and some types of cancer.

■ Food processing has many purposes, including helping to keep food from spoiling. Substances called additives may be added to food during processing. Some additives help to preserve food, while others add nutrients or improve the taste, texture, or appearance of the food.

■ To have a balanced diet, you need to know what nutrients your body needs and what foods contain these nutrients. *The Food Guide Pyramid* and food product labels can help you plan nutritious meals and select foods wisely.

VOCABULARY

nutrient
nutrition
food energy
carbohydrate
simple carbohydrate
sugar
complex carbohydrate
digestion
fat
fatty acid
saturated fat
unsaturated fat
cholesterol
protein
amino acid
non-essential amino acid
essential amino acid
complete protein
incomplete protein
enzyme
mineral
vitamin
water
dietary fiber
food processing
preserve
food poisoning
food additive
preservative
fortified
enriched
diet
malnutrition

V1. Create a mind map using some of the words in this vocabulary list. Place the word "nutrition" at the center of your map.

V2. Select six words from the vocabulary list. For each of these words, write a sentence or two that explains what you have learned about the word from this chapter.

CONNECTIONS

C1. (a) What are the six major nutrient groups?
(b) Describe at least one important function performed by the nutrients in each group.

C2. List at least three good food sources of some of the nutrients in each of the six major nutrient groups.

C3. What are the smaller, "building-block" molecules that combine to form
(a) carbohydrates?
(b) fats?
(c) proteins?

C4. (a) Give four reasons why it is important for you to include some fat in your diet.
(b) Why is it important for you to avoid eating too much fat?

C5. What happens to carbohydrate molecules, fat molecules, and protein molecules between the time you eat them and the time they can be used by your cells?

C6. In an experiment, some rats were fed on brown bread and water; these rats thrived. Other rats were given the same amount of white bread and water; these rats became ill. Give a possible explanation of these results.

C7. (a) Why do you think *The Food Guide Pyramid* divides all of the foods we eat into five groups?
(b) Why do you think *The Food Guide Pyramid* recommends that you eat foods from all five groups every day?

C8. Some foods may contain one or more of the nutrients that you need. However, it may not be wise for you to eat large quantities of these foods. Suggest two reasons why this might be so.

C9. (a) List three ways in which food processing is helpful.
(b) List three possible disadvantages of food processing.

C10. (a) What is malnutrition?
(b) Give two types of problems with a diet that would cause a person to suffer from malnutrition.
(c) How might each of the problems you mentioned in (b) affect a person's health?

EXPLORATIONS

E1. Having too much or too little of certain nutrients in the body can cause various illnesses. Select one of the following health problems to study: diabetes, rickets, scurvy, hypertension, osteoporosis, pellagra, beriberi, goiter, pernicious anemia, xerophthalmia, osteomalacia, kwashiorkor. Find out the following:

- What causes the illness?
- What type of person is most likely to have this health problem?
- What effects does it have on a person's body?
- How can it be treated or prevented?

Present the information you discover in the form of a pamphlet like one you might find in a doctor's office or public health unit.

E2. Many meals today consist of "fast food"—food that is prepared quickly, either in a restaurant or at home. Select several fast-food items and do research to discover what nutrients and how much of each nutrient is in these foods. Your research might include reading food labels or library books, or obtaining nutritional information from the restaurant that serves the food.

Find out the number of milligrams of fat, protein, salt, vitamins, and other nutrients provided per serving or per gram of each food. Also find out the number of kilojoules of energy provided by each food item. Looking at the information you have collected, do you think that fast food is nutritious? Explain your conclusion. Present your information to the class.

E3. Some athletes eat certain types of food and avoid other types, especially just before they compete. For example, cyclists (Figure 6.27), long-distance runners, sprinters (people who run very fast over a short distance), football players, and swimmers often follow special diets at certain times. Choose one of these sports or another one for which

FIGURE 6.27
Sprint cyclists need huge amounts of energy, fast. What kind of foods should they eat before they compete?

competitors follow a special diet. What does the special diet include? Do you think this special diet can help a person's athletic ability? Are there any dangers for a person following this diet? What do you think an athlete should do before deciding to follow a special diet?

E4. What are some ways people can reduce the amount of saturated fats in their diets? To discover how this might be done, do some research—which might include visiting a grocery store and reading food labels. Share your suggestions with the class in the form of a poster or a pamphlet.

REFLECTIONS

R1. Look back at your answers to the discussion questions in Activity 6A. Are your answers to these questions different now that you have read the chapter? If so, how? Do you have any new questions about nutrition or about your diet that you would like to have answered? If so, record them in your learning journal. You may find the answers in the other chapters in this unit.

R2. Think about your diet. (You might want to record, in your learning journal, everything you eat for several days before doing this.) Compare what you eat with the recommendations made in *The Food Guide Pyramid.* Are you eating enough from each food group? Are you eating too many foods from one or more groups? If so, how could you correct this problem? Imagine that your body mass is increasing too much. Do you think it would be better for you to increase your level of activity—to use up additional energy—or to change your diet? (In Chapter 10, you will learn more about balancing what you eat with what you do.)

CHAPTER 7

DIGESTION

You may have heard someone say, "Just the smell of that pizza makes my stomach growl." Or, "My stomach is about to burst. I couldn't eat another bite."

It is amazing how much time people spend talking about their stomachs. Then again, maybe it is not so surprising. Your stomach, along with the rest of your digestive system, is responsible for keeping your body "fueled" and ready for action. Since you are an active, growing teenager, your digestive system is having the busiest time of your entire life.

When you, like the people in the photograph, choose something to eat, you are setting in motion a series of events in your body. These events help your body to obtain every possible nutrient from your food. In this chapter, you will learn about how food travels through your digestive system and how this system helps you get the most from the food you eat.

ACTIVITY 7A / YOUR DIGESTIVE SYSTEM

What do you already know about your digestive system? What parts of your body act on the food you eat?

Your teacher will provide you with an outline drawing of the human body. Inside this outline, draw the parts of the digestive system that you know about. Label all the parts you have drawn. Under each label, state what you think is the function, or job, of each part of the digestive system. Check and revise your drawing as you work through this chapter. ❖

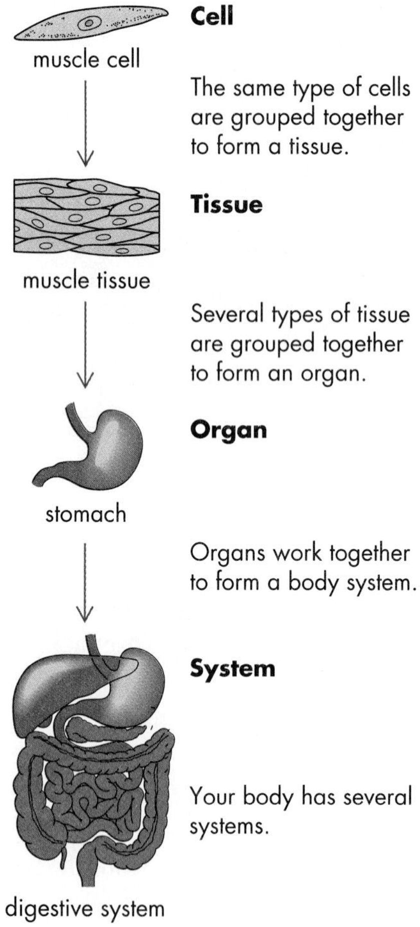

muscle cell

Cell

The same type of cells are grouped together to form a tissue.

muscle tissue

Tissue

Several types of tissue are grouped together to form an organ.

stomach

Organ

Organs work together to form a body system.

digestive system

System

Your body has several systems.

FIGURE 7.1
How your body is organized

7.1 MAKING USE OF THE FOOD YOU EAT

Chapter 6 explained that nutrients must enter your cells before they can be used by your body. The cells in your body use the nutrients in your food for the many activities that keep them—and you—alive. Although all the cells in your body are similar in many ways, there are several types of cells. Nerve cells, muscle cells, and skin cells are three different types. Each type of cell has special features that make it good at doing specific tasks, or functions. Nerve cells, for example, carry messages (nerve impulses) from one part of your body to another. Figure 7.1 shows how your body is organized—from tiny cells to your whole living body.

When cells with similar tasks are grouped together, they form a **tissue**. There are different types of tissues. Muscle tissue, for example, is made up of muscle cells. Nervous tissue is made up of nerve cells.

When several types of tissue are grouped together for a specific task, this group forms an **organ**. Your stomach is an organ. It is made up of several tissues, including muscle tissue and nervous tissue. Your heart is also an organ, made up of different tissues working together.

Organs that work together to do a specific task make up a **system**. You have several body systems. For example, your **digestive system** includes organs such as your stomach and intestines. The function of your digestive system is to break down food into substances small enough to enter your cells. In other chapters in this unit, you will learn about three other body systems—your respiratory system, your circulatory system, and your excretory system.

Why do you need a digestive system to break down your food? The nutrients from the food you eat are used *inside* the cells of your body. So all the food you eat must be changed or broken down into substances small enough to pass into your cells (Figure 7.2). **Digestion** is the process of changing food into these smaller substances. Each part of your digestive system has a specific job, starting from the moment you place food in your mouth.

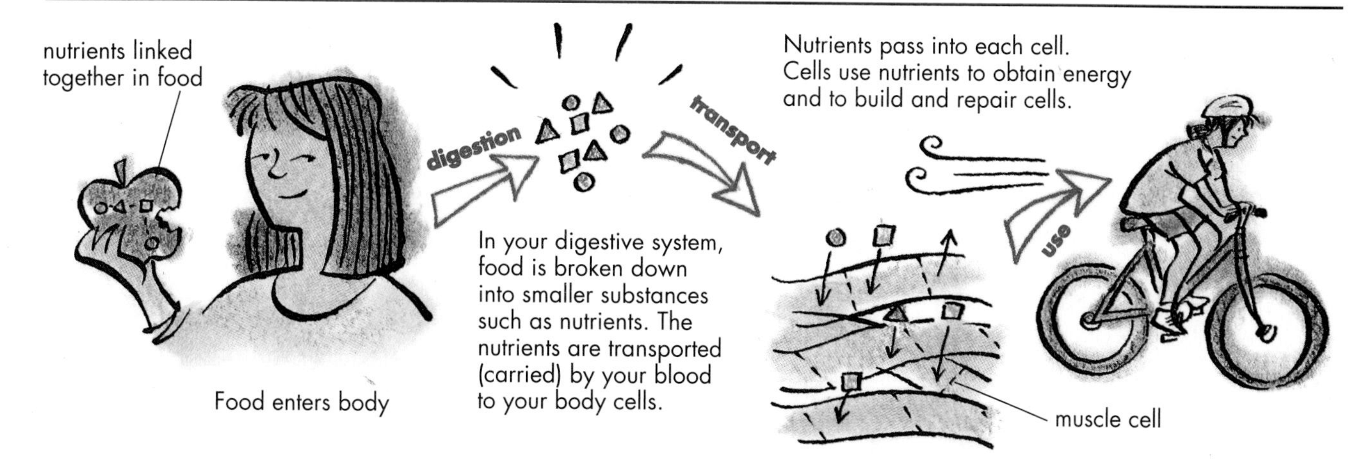

nutrients linked together in food
digestion
transport
Nutrients pass into each cell. Cells use nutrients to obtain energy and to build and repair cells.
use
In your digestive system, food is broken down into smaller substances such as nutrients. The nutrients are transported (carried) by your blood to your body cells.
Food enters body
muscle cell

FIGURE 7.2
Why is digestion necessary before your body can use the nutrients from your food?

GETTING STARTED: YOUR MOUTH

Several parts of your mouth help to break down the food you eat. Teeth are able to tear apart, cut, crush, and grind food (Figure 7.3). At the same time, the food is mixed with **saliva**, the watery fluid in your mouth. Saliva moistens your food, making it easier to swallow. Your lips hold in the food and saliva. Your tongue moves the food about. It pushes food onto the grinding teeth and puts the chewed food into the right position for swallowing.

Two kinds of digestion happen in your mouth. **Mechanical digestion** is the physical breaking down of food into smaller pieces. Your teeth do this job. **Chemical digestion** involves breaking apart the chemical bonds that hold the molecules of food together. To help break these bonds, your digestive system produces special enzymes, called **digestive enzymes**. Enzymes in your saliva begin the process of chemical digestion (Figure 7.4).

■ Although mostly muscle, your tongue also contains cells that let you taste food. These taste cells are located along the edge of your tongue and in certain areas on top of your tongue. What are the basic tastes these cells can sense?

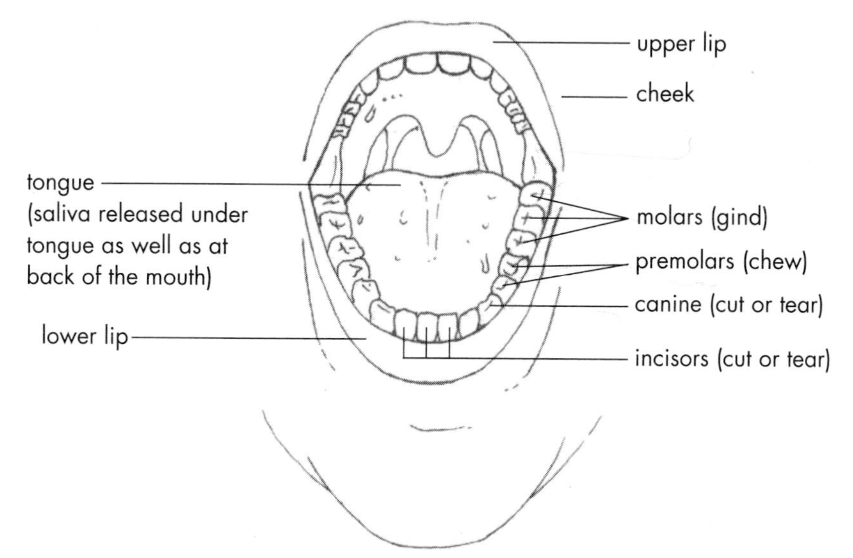

FIGURE 7.3
What roles do the lips, tongue, and teeth play in the first stage of digestion?

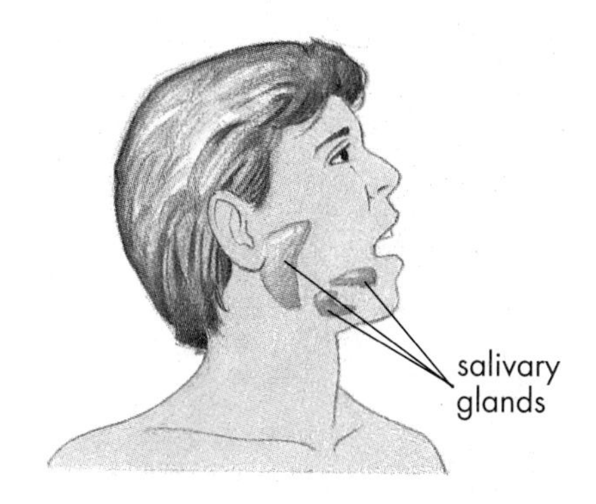

FIGURE 7.4
Three pairs of salivary glands produce saliva. Saliva contains digestive enzymes and water. Why do you think saliva is mostly water?

■ What happens when you swallow? Chew a piece of bread and swallow slowly, trying to feel what is happening in your mouth and throat. Hold your hand on your throat and describe what happens. Try to tell when you have no further control over swallowing and a reflex action starts to move the bread along automatically.

The mechanical digestion of food by your teeth is important for two reasons. First, only small pieces of food can fit into the tube that carries food down to your stomach. Second, the digestive enzymes can mix with your food more easily if the food is in small pieces. The digestive enzymes in saliva start to break down large carbohydrate molecules into smaller sugar molecules. For example, starch starts to break down into glucose molecules.

Think about what happens when your tongue pushes a piece of food to the back of your mouth. You automatically swallow. Swallowing moves your food into your throat. The small ball of moist food is then ready for the next part of your digestive system.

THE ESOPHAGUS

At the back of your throat, there are two separate tubes. The **esophagus** is a narrow tube that carries food to your stomach (Figure 7.5). The other tube, the trachea, carries air to your lungs. It is not part of your digestive system. Figure 7.6 shows how your body keeps food or liquid from entering the trachea. When you swallow, a fleshy flap of tissue called the **epiglottis** automatically covers the opening of the trachea. Food then moves down the esophagus to your stomach.

FIGURE 7.5 ▶
Your esophagus runs from your throat to your stomach. After you swallow, you don't have to think about moving food down your esophagus and into and beyond the stomach. Why do you think control of the digestive system is mostly automatic?

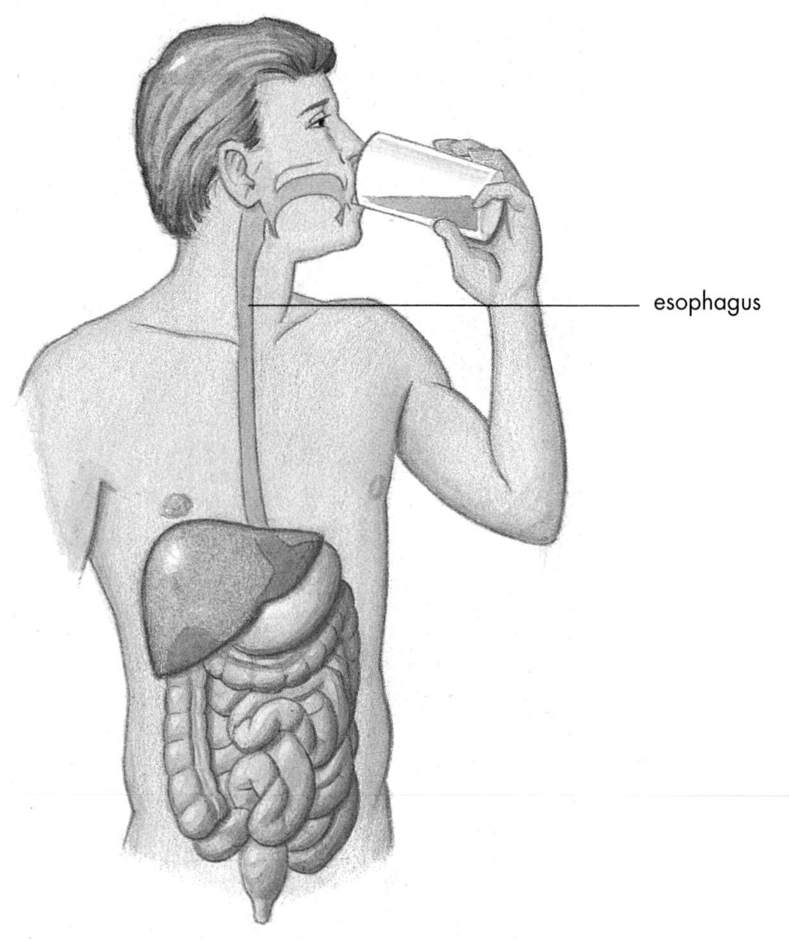

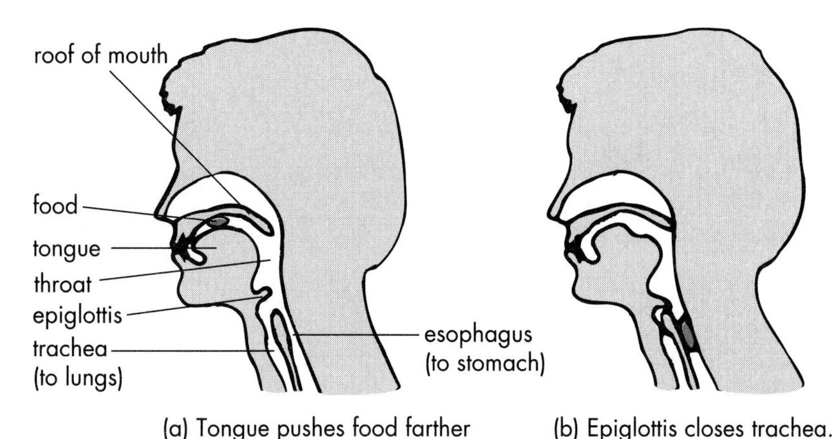

(a) Tongue pushes food farther back.

(b) Epiglottis closes trachea. Food moves into esophagus only. (Notice how the roof of the mouth presses up to close off the tube leading into the nostrils.)

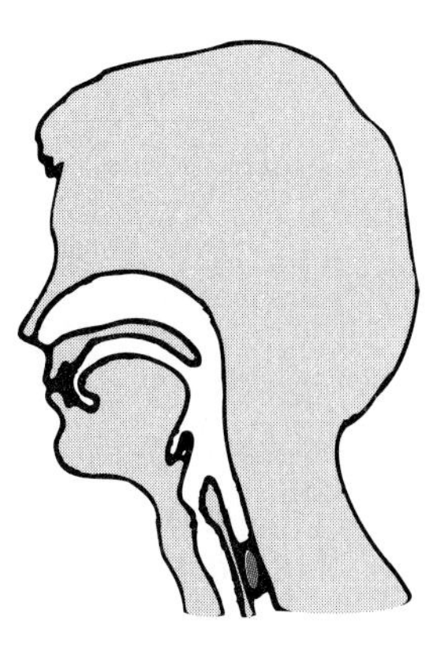

(c) Epiglottis opens. Food continues through esophagus to stomach.

FIGURE 7.6
Swallowing. After your tongue has moved a piece of food toward the back of your mouth, the rest of the swallowing process is automatic — a reflex action. What stops food or liquid from entering the trachea (the tube to the lungs)?

The walls of the esophagus are lined with muscles. These muscles help move each ball of food down the esophagus to your stomach. The esophagus lining also produces mucus, a slimy liquid that helps the food to move through the tube more easily.

Food moves through your esophagus when the muscles that run along and around the esophagus tighten, or contract. As these muscles squeeze from the top to the bottom, they push food toward your stomach. The series of muscle contractions (squeezings) is called **peristalsis** (Figure 7.7). To help you picture peristalsis, think about how you could push a glass marble through a garden hose by squeezing the hose just behind the marble. As well as moving pieces of food down the esophagus, peristalsis moves material along other tubes in your digestive system.

■ Peristalsis usually moves material toward your stomach, but sometimes the movement is reversed. An irritation in your digestive system can cause vomiting. Your epiglottis closes over the trachea, and reverse peristalsis moves the stomach contents back to your mouth. The powerful muscles of your stomach and lower abdomen help to force out the irritating substance. Vomiting helps to quickly remove some poisons from your digestive system.

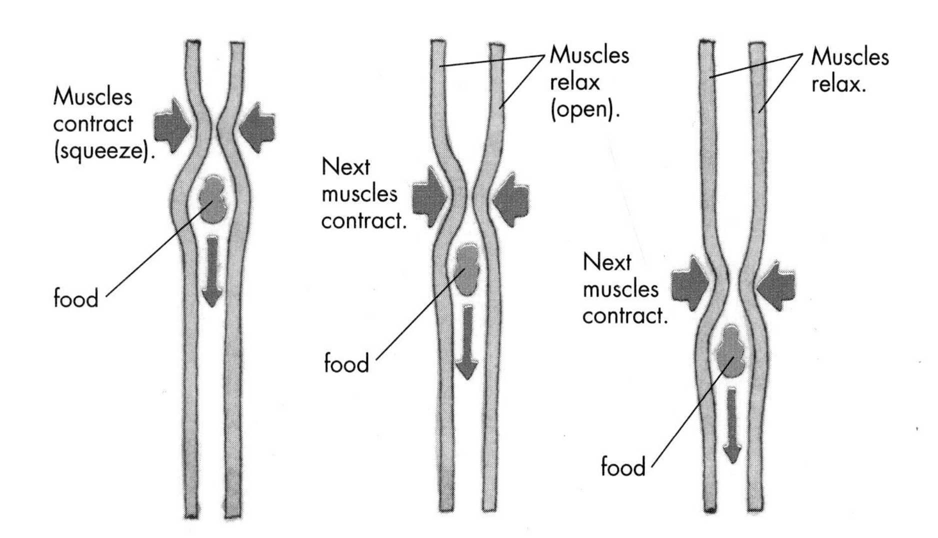

FIGURE 7.7 ◀
Peristalsis is the series of muscle contractions that push food along the esophagus and other tubes of your digestive system. What do you think happens if a piece of food is too large?

▶ REVIEW 7.1

1. How is your body organized, from individual cells to systems? Give one example of each.
2. What is the main function of your digestive system?
3. What is the difference between mechanical and chemical digestion? Give one example of each.
4. Describe the job each of the following parts of the digestive system does in digestion.
   (a) teeth
   (b) tongue
   (c) salivary glands
   (d) epiglottis
   (e) esophagus

Type of body part	Example	Function (job) in body
Cell	Nerve cell	Carries messages
Tissue	Muscle	______
Organ	Esophagus	______
Organ	Tongue	______
System	Digestive	______

TABLE 7.1 ***Sample Data Table***

5. What is peristalsis and where is one place it occurs? (You may wish to use diagrams to help answer this question.)
6. Copy Table 7.1 into your notebook. Complete the table using what you have learned so far about how your body digests food.

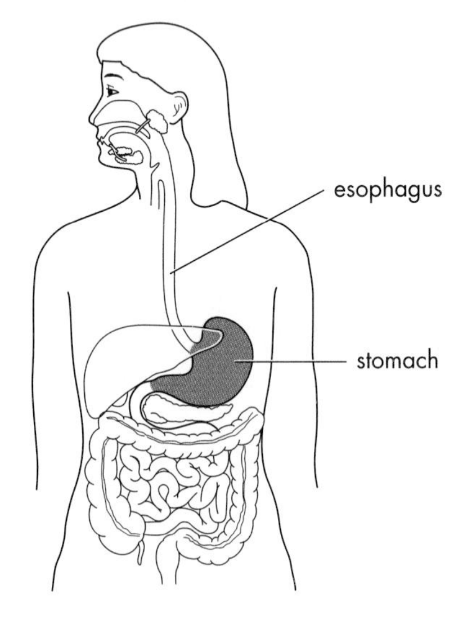

FIGURE 7.8
Your stomach receives food from your esophagus. Why do you think it is important that the stomach is able to expand? (In this drawing, the organs of the digestive system are shown farther apart than they actually are.)

■ 7.2 THE STOMACH: DIGESTION AND STORAGE

All the food you eat travels down your esophagus to your stomach (Figure 7.8). Your stomach is like a thick balloon made of muscle. It can stretch to hold as much as 2 L of food and it shrinks when it is empty. Put your hand where you think your stomach lies. You may be surprised to know that your stomach is tucked under the ribs on the left side of your body.

DIGESTION IN THE STOMACH

Like your mouth, your stomach breaks down food in two ways: mechanically and chemically. The mechanical digestion in the stomach is shown in Figure 7.9. The wall of your stomach has powerful muscles that contract, squeezing the food in the stomach much as a baker kneads bread dough. This churning helps to mix your food with a fluid produced by your stomach. This fluid is responsible for further chemical digestion of the food in your stomach.

Your stomach produces different substances to moisten food and help chemical digestion. The fluid produced by your stomach is made up of digestive enzymes, an acid, and mucus. The function of each of these substances is listed below.

- Pepsin is one of the digestive enzymes. It starts to break down protein molecules into smaller molecules.
- Hydrochloric acid is a powerful, corrosive acid. Pepsin works best at breaking down proteins when there is acid present.
- Mucus is a protective material that coats the walls of your stomach. Without this protective lining, the acid and pepsin would digest the walls of the stomach. Mucus from the lining mixes in with the partially digested food.

Anything that causes the stomach to become too full, such as overeating, can push the acidic contents of your stomach upward into the lower part of your esophagus. This causes a burning pain that you feel in the middle of your chest, near the heart. This pain is called "heartburn."

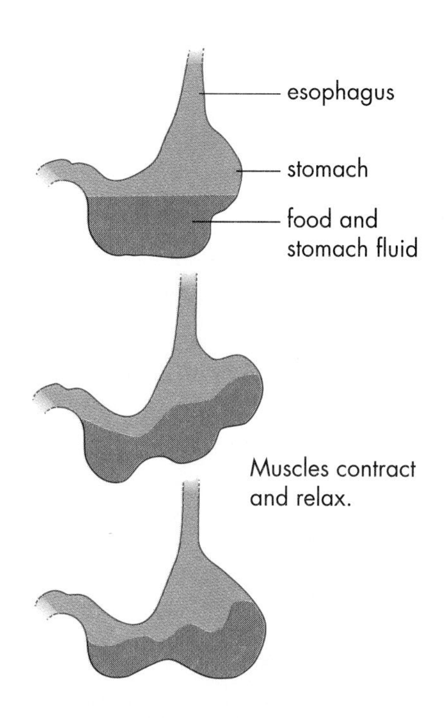

FIGURE 7.9
The muscles in the wall of the stomach help mix the food and fluid. What do you think causes the growling or gurgling sound of a stomach with no food in it?

THE STORAGE FUNCTION OF THE STOMACH

Another major function of your stomach is to store partially digested food so that it does not enter the next part of your digestive system too quickly. A circle of muscle, the sphincter muscle, is able to open or close the bottom end of your stomach (Figure 7.10). This opening from your stomach usually stays closed, keeping food in the stomach. While food is in your stomach, peristalsis causes the food to be churned about until it becomes a creamy mixture, made up of stomach fluid and partially digested food. Only small amounts of this mixture are pushed through the circle of muscles at any one time. All the food from a meal has usually left the stomach within two to four hours.

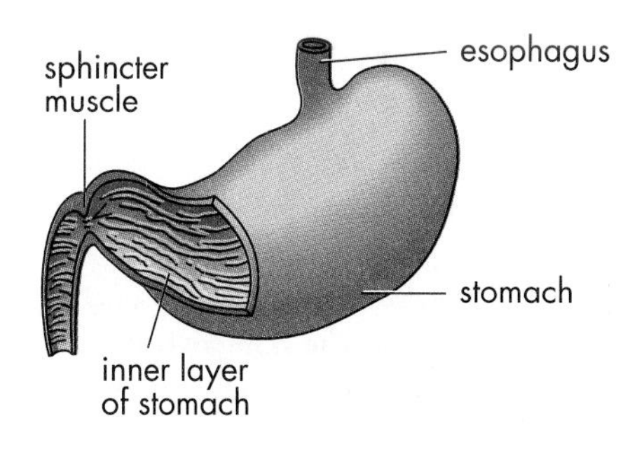

FIGURE 7.10
A circle of muscle, the sphincter muscle, relaxes to allow food to leave the stomach.

REVIEW 7.2

1. What happens to food from the time it arrives in your stomach until it leaves the stomach?
2. What is the function of the powerful muscles in the walls of your stomach?
3. What are the two main functions of the stomach? For each, describe the parts of the stomach that help the stomach carry out that job.
4. Make a two-column chart in your notebook. In the left-hand column, list the three substances produced by the stomach. In the right-hand column, describe the function of each.
5. There are nerve cells located in the wall of your stomach. These nerve cells signal the brain when the stomach muscles are stretched. Why do you think these nerve cells are important?

7.3 THE INTESTINES

After partially digested food is squeezed from your stomach through the opening at the bottom of the stomach, it enters the longest section of your digestive system. The **small intestine** is a tube about 2.5 cm in diameter and 4 to 6 m long. Stretched out, an adult's small intestine would be longer than a full-sized car. The last section of the digestive system is another tube, the **large intestine**. At 6 cm in diameter and 1.5 m long, the large intestine is wider than the small intestine but much shorter. (The two intestines are sometimes called the small bowel and the large bowel.)

THE SMALL INTESTINE AND DIGESTION

When the creamy mixture enters the small intestine, it is very acidic since it contains the hydrochloric acid from the stomach fluid. For protection, the small intestine produces large amounts of mucus to coat its inside wall. Peristalsis moves the creamy mixture through the small intestine. This usually takes three to six hours. As the food mixture is moved along, more digestion takes place in the first 25 cm of the small intestine.

The small intestine is very important for completing the process of digestion (Figure 7.11). If your stomach were removed, you could still digest well-chewed food. Without your small intestine, however, your food could not be broken down well enough. The breaking down of carbohydrates begins in your mouth, and protein molecules start to be broken down in your stomach. However, the resulting molecules are still too large for your cells to absorb and use. Proteins, fats, and carbohydrates

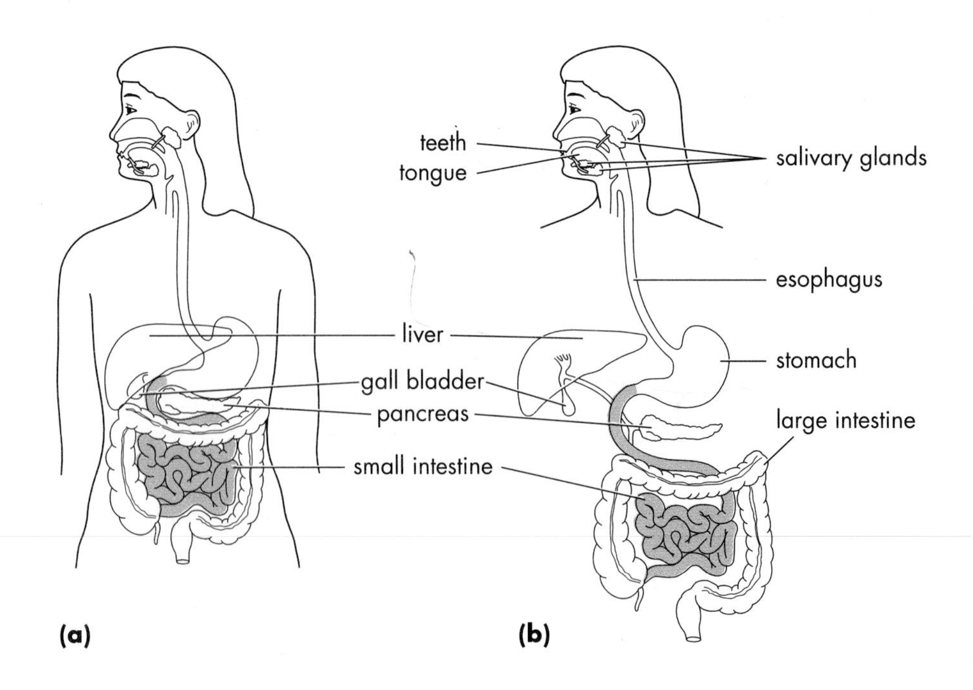

FIGURE 7.11 ▶
(a) The small intestine, pancreas, liver, and gall bladder.
(b) This diagram shows these and other organs of the digestive system separated for a clearer view.

must be further chemically digested in your small intestine with the help of more digestive enzymes. For example, proteins are broken down into their basic "building blocks," amino acids. Carbohydrates are broken down into smaller molecules of sugar. Vitamins and minerals are released from food, ready to be used by your body. By the time food has passed through the small intestine, most of the nutrients have been broken down into molecules small enough to enter your cells.

The cells lining the wall of the small intestine produce some digestive enzymes. Several other digestive enzymes are added to the small intestine from the **pancreas**, an organ located near your stomach. The pancreas also produces a substance to help to neutralize the hydrochloric acid. The pancreas is considered part of your digestive system, although it has other jobs besides digestion.

The **liver** is another organ that is part of your digestive system. This large organ lies above the stomach and fills most of the space in the upper part of your abdomen. The liver helps in the digestion of fats.

To help the digestion of fats, the liver produces a green fluid called **bile** that is stored in the **gall bladder**. Bile is not a digestive enzyme, but it is necessary for the digestion of the fat in your food. Bile helps to break apart large drops of fat into small droplets, much the way soap breaks apart grease in dishwater. This process exposes a greater surface area of fat for the digestive enzymes to work on as they chemically break down fat molecules.

Figure 7.12 shows the role of the small intestine, pancreas, and liver in digestion. The liver has many functions, helping other systems besides the digestive system. For example, the liver controls the storage of useful substances and releases them into the blood when your body needs them. The liver also breaks down substances that your body cannot use so they can be removed from the body.

■ Gallstones may be formed in the gall bladder if there is too much cholesterol in the bile. If this happens, the cholesterol may form crystals that stick together. Gallstones can be a serious and painful medical problem if they block the opening that leads out of the gall bladder. Find out what medical treatments are used for people with gallstones.

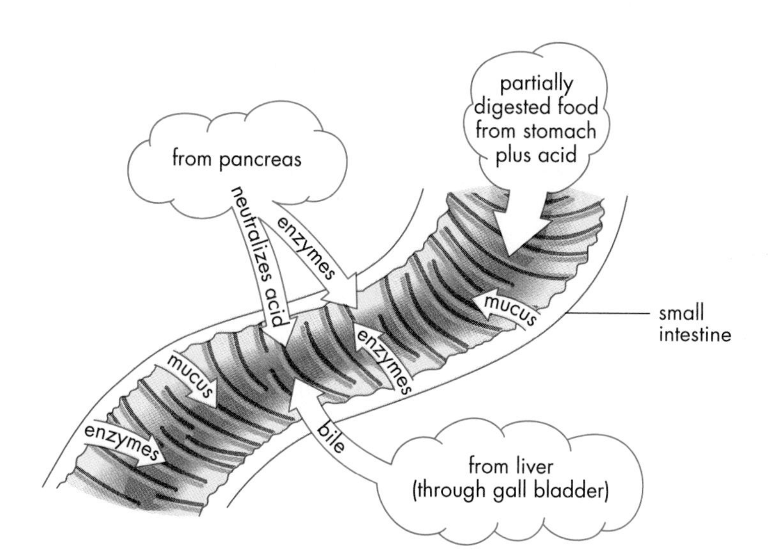

FIGURE 7.12 ◀
The small intestine receives "shipments" of partially digested food from the stomach. Digestive enzymes help to release nutrients from the food. Notice how the different organs work together to allow your body to digest the remaining food.

INSULIN

Seventeen-year-old Daniel David is a competitive diver, among the top in the world in her age group. She trains three hours each day and keeps her grades up in school. She also injects herself with insulin every day.

Insulin is a lifesaving medication for Daniel David and the thousands of other people who have Type 1 diabetes. Their bodies produce little or no insulin. In people without diabetes, this hormone is produced in the pancreas and released into the bloodstream. (With Type 2 diabetes, the most common type, the pancreas produces insulin but the body cannot use it effectively.)

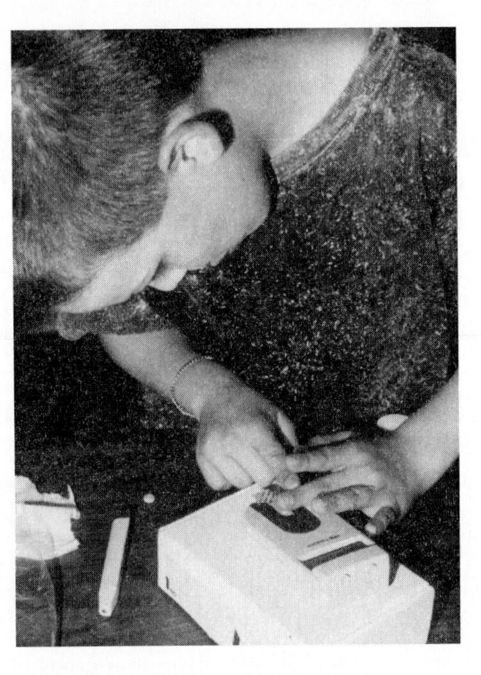

This boy is checking his blood glucose level with a portable monitor.

The body needs insulin so it can get energy from food. After food is digested, molecules of glucose are released into the blood. Insulin allows glucose to be absorbed from the blood into the cells. Without insulin, glucose cannot enter the cells to be a source of energy. As a result, the cells cannot carry out their many functions. The amount of glucose in the blood (the blood glucose level) rises to dangerous levels. After a person with diabetes takes an injection of insulin, the glucose can be absorbed into the cells.

Many people with diabetes check their blood glucose levels at home. They put a drop of blood from a finger onto a special test strip. The strip is inserted into a small, portable glucose meter to obtain a read-out of the blood glucose level. This information helps people with diabetes decide how well they have balanced their insulin, exercise, and food. New glucose monitors being tested are able to give glucose levels without needing a drop of blood. They use infrared light beamed through the finger instead. How often people check their blood glucose depends on the individual, but usual times are before meals and before exercise.

People with diabetes can often control their blood glucose levels with proper exercise, a loss of excess body mass, and proper meal planning. They need a nutritious diet, low in fat and sugar. Some also take insulin injections or pills.

Insulin was discovered over 70 years ago by Canadian researchers. It has saved many lives and helps people to live with this disease, but it is not a cure.

Researchers today are still looking for a cure as well as for better ways to help people manage everyday living with the disease. About 1 in 20 people will develop diabetes in their lifetime. Someday, people with diabetes may be able to have an artificial pancreas implanted in their bodies to produce insulin.

Daniel David says that coping with diabetes has made her a stronger person. Have you had physical disabilities or other challenges to overcome in your life? How have they made you stronger?

THE SMALL INTESTINE AND ABSORPTION

After passing through the first 25 cm of the small intestine, the food that started in your mouth has now been broken down into molecules small enough to enter your cells. These molecules must be removed from the creamy fluid inside the small intestine in order to be carried to cells throughout the body. **Absorption** is the process by which substances, such as nutrients, enter the cells of the small intestine. Absorption takes place in the cells that line the wall of the rest of the small intestine.

Many millions of cells are needed to absorb all the nutrients your body requires. To fit this many cells into the lining of your small intestine, the inside walls are folded again and again. This folding is shown in Figure 7.13. Each fold has many finger-like projections. As a result there is a surface area inside the small intestine hundreds of times larger than if the inside were smooth. You might compare the folded wall to a thick terry towel, which is made of much more thread than a smooth tea towel is.

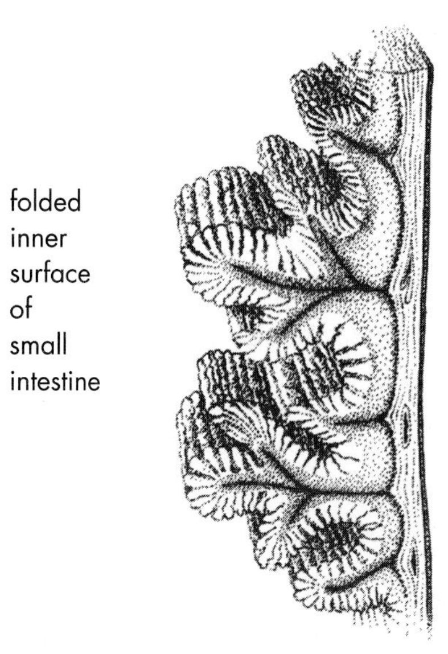

FIGURE 7.13
If you were to stretch out the inside surface of an adult's small intestine, it could cover the floor of a large room. Why do you think it is important for the small intestine to have so many folds?

THE LARGE INTESTINE AND ELIMINATION

The remains of the food mixture leave the small intestine and enter the last part of the digestive system, the large intestine (Figure 7.14). Like the esophagus, the wall of the large intestine produces mucus to help the remains of the food mixture move easily.

The mixture entering the large intestine consists mainly of water and waste materials that cannot be digested, such as fiber. The main function of the large intestine is **elimination**, the removal of wastes from your body. (You may have heard the term "bowel movement" when someone is talking about elimination.) If wastes from digestion were not eliminated, they would block the digestive system.

The mixture entering the large intestine is very watery. When the wastes leave, they are drier, more solid. The wastes you eliminate are called **feces**. Two processes in the large intestine produce feces.

First, the cells of the large intestine absorb about 1.5 L to 1.8 L of water and chemicals from the waste materials every day. This makes the feces drier. The water is reused by your body.

The second way that the large intestine produces feces is through the work of bacteria. Your large intestine is a perfect place for bacteria to grow and reproduce. It is warm and moist, and bacteria can use the feces as their nutrients. These bacteria help you. They collect important minerals from the waste materials for your body to use. They also manufacture vitamins that your body absorbs. The bacteria also consume some of the waste material, reducing the amount you have to eliminate.

Feces are about 75 percent water and 25 percent solid matter. The solid matter includes dead bacteria and wastes such as fiber. Fiber is an

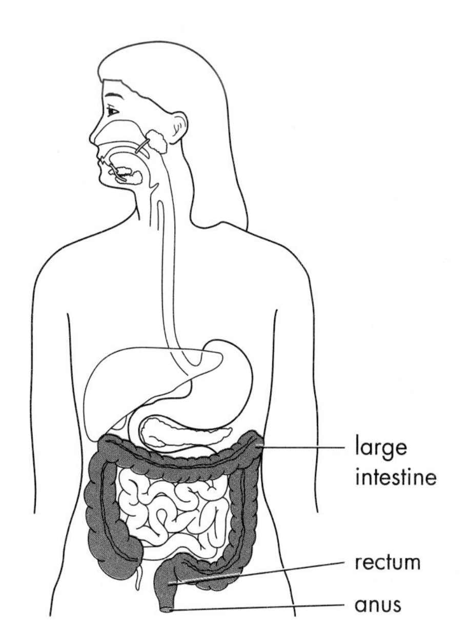

FIGURE 7.14
The large intestine prepares wastes for elimination. It also removes water from the wastes.

■ You might take antibiotics to kill bacteria that are making you ill. Sometimes antibiotics kill the normal, helpful bacteria in your large intestine as well. As a result, you may not receive the minerals and vitamins that are normally made available by the helpful bacteria. When taking antibiotics, you may also have more frequent bowel movements with more watery feces.

important part of your food. It helps the feces hold enough water to move easily through the large intestine.

It takes from 18 to 24 hours for material to move through the large intestine. Peristalsis moves the material along, as it does in other tubes in the digestive system. The feces then reach the **rectum**, a small section at the end of the large intestine. The rectum expands to hold and store the feces until they are eliminated.

When the rectum is full, your nervous system signals your brain that you need to push the feces out of your body. The feces leave through an opening called the **anus**. From about the age of 18 months, a person is able to control the act of elimination long enough to reach a toilet.

THE WHOLE SYSTEM

Take a moment to think about what has happened to the bite of food that you read about at the beginning of this chapter. It has passed through the digestive system, being broken down by mechanical digestion and chemical digestion. Valuable nutrients have been absorbed and waste material has been eliminated. This takes place in a system of connected tubes and body parts: mouth, esophagus, stomach, small intestine, and large intestine. Other organs also help the process of digestion (Figure 7.15).

■ Use household materials, such as old stockings, to make a three-dimensional model of the digestive system.

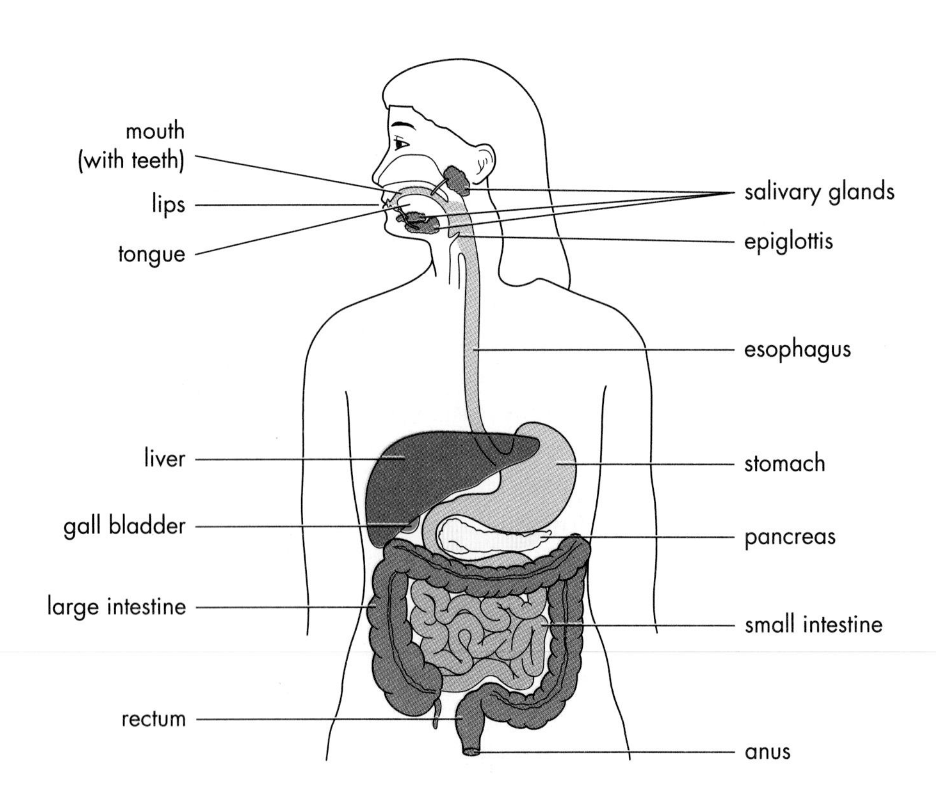

FIGURE 7.15 ▶
Follow the process of digestion through the various parts of the digestive system.

ACTIVITY 7B / CHECK IT OVER

Take a moment to look at your drawing from Activity 7A.

Compare your drawing with Figure 7.15. Add any new information to your drawing or make notes.

Use your finger to trace the path a bite of food takes through the complete digestive system shown in Figure 7.15. Use a colored pencil or pen to trace a similar path on your own drawing. ❖

► REVIEW 7.3

1. (a) What two other organs help the small intestine in its job of digesting food?
(b) What substance does each of these organs produce?
(c) Describe the function of each substance you listed in (b).

2. The first section of the small intestine produces large amounts of mucus. Explain why this is necessary.

3. (a) What is absorption?
(b) Why is it necessary?
(c) Where does most absorption of nutrients take place in your body?

4. If you were to compare a piece of your small intestine and a piece of your large intestine, you would see that the inside wall of the small intestine is very folded. What is the purpose of these folds?

5. (a) What is the main function of the large intestine?
(b) How does the large intestine perform this function?
(c) What other important things happen in the large intestine to help your body?

6. Imagine that two animals with diets similar to that of a human are born with problems with their digestive systems.
(a) One animal has no stomach. What problem will it have in digesting food?
(b) One animal has no small intestine. What problem will it have in digesting food?

7.4 TAKING CARE OF YOUR DIGESTIVE SYSTEM

Each of the people in Figure 7.16 shared the same meal. Everyone enjoyed the same food. A short while later, one person complained of heartburn. A teenager said he was still hungry, although most of the adults felt too full. Later that night, one adult had stomach cramps. The next morning, one person woke up "starving," while another still felt too full to eat breakfast. Do all these people have problems with their digestive systems?

The answer is that they probably do not have serious problems. People differ in how quickly or how well their digestive systems work. Sometimes, though, small pains and difficulties are warning signs (symptoms) of problems with your digestive system.

Most people have stomach pains or heartburn at some time. These problems do not usually last long, but in some cases they could be the first sign of an ulcer. An ulcer is a sore in the wall of the digestive system. It is often caused by the hydrochloric acid and pepsin produced

EXTENSION

■ What kind of problems or diseases affect the human digestive system? Find out about one of the following: gastritis, food poisoning, colitis, hiatus hernia, hepatitis, or stomach cancer. Use the information you find to write a newspaper article to explain the problem or disease to readers.

DID YOU KNOW

■ Scientists have found that over one dozen types of bacteria can live in the strong acid of the human stomach. These bacteria can attack the walls of the stomach and cause ulcers. New methods of detecting and treating this problem are being developed.

by the stomach. Normally your digestive system is protected from these substances by a layer of mucus. If this mucus layer is weakened, the acid and pepsin may attack the exposed area. An ulcer can occur in any lining that the acid and pepsin can reach: the lower esophagus, the stomach, or the first part of the small intestine. With a very bad ulcer, there may even be a hole in the wall of the organ. When this occurs, the contents can leak into the body and harm other organs.

FIGURE 7.16 ▶
The enjoyment and nutrition people obtain from food will depend on the health of their digestive systems.

Sometimes people have problems with elimination. Most people pass feces every day, but two or three times a week is normal for others. Sometimes a person gets **constipation**. Feces build up within the rectum and the rest of the large intestine for longer than usual. The large intestine becomes swollen. In Chapter 6, you learned of the important role of dietary fiber in preventing constipation.

Sometimes people have the opposite problem. Their feces are watery and they need to eliminate them frequently. This condition is called **diarrhea**. The cause can be an infection attacking the walls of the large intestine. Nervousness and stress can also cause diarrhea. If diarrhea is very bad or if it lasts a long time, the body may have a serious loss of water and nutrients. In many parts of the world, diarrhea is a major cause of death in babies and young children.

DID YOU KNOW

■ Indigestion, loss of appetite, or even mild diarrhea are some of your body's normal reactions to stress. But if these symptoms continue or if they happen several times, you should see your doctor.

■ The most important way to prevent disease throughout the world has been to provide people with clean drinking water. Diseases such as typhoid and cholera are spread when people drink water contaminated by feces.

Many minor problems of the digestive system can be reduced by a well-balanced diet. For example, constipation in children and teenagers is usually the result of a poor diet. Eating more fiber, drinking more fluids, and exercising regularly are the best ways to prevent constipation. Heartburn and an upset stomach can often be prevented by eating smaller amounts of food and eating in a slower, more relaxed way. Avoid foods that you know upset your stomach or give you heartburn. Over-the-counter drugs can relieve symptoms such as constipation and heartburn, but they do not treat the cause (Figure 7.17). If you rely on such medications, you might wait too long to see a doctor for a serious problem just because you feel better for a little while.

How and when you eat, as well as what you eat, affect digestion in your body. By eating regular meals, you can avoid getting very hungry and then eating too much and overloading your stomach. Take time to enjoy your meals. If you eat slowly, you will probably chew your food properly. When your stomach is full, your brain receives a message to let you know that you have had enough to eat, but this process takes about 20 minutes. If you eat very quickly, you can eat too much, not realizing that your stomach is already full. There is another advantage to being relaxed when you eat. Your digestive system works best when you are not in a rush or feeling stress.

Taking care of your digestive system is as simple as taking good care of your entire body. Eat a balanced diet of healthy foods, get enough rest, keep fit, and eat regular meals in a relaxed way. These practices will help you keep your digestive system in good working order.

DOSAGE FOR RELIEF OF DIARRHEA
Adults: 4 tablets initially then 4 tablets after each subsequent bowel movement. Not to exceed 28 tablets in 24 hours. **Children 6 to 12 years:** 2 tablets initially then 2 tablets after each subsequent bowel movement. Not to exceed 14 tablets in 24 hours. **Children 3 to 6 years:** 1 tablet initially then 1 tablet after each subsequent bowel movement. Not to exceed 7 tablets in 24 hours

WARNING - Do not use for more than 2 days, or in the presence of high fever, or in infants or children under 3 years of age. If diarrhea persists, consult a doctor.

FIGURE 7.17 ◀ ***Over-the-counter drugs for digestive problems have warnings on their labels. These warnings tell you to see a doctor if the problems continue. Why is this good advice?***

▶ REVIEW 7.4

1. (a) Why is the mucus in your stomach important?
(b) What medical problem might result from a lack of mucus?

2. (a) Make a table with three columns similar to Table 7.2. Choose three foods, one with carbohydrates, one with protein, and one with fat. Complete the table.
(b) How do you think eating a great deal of any one of these foods could affect your digestive system?

3. Matthew often had stomach pains after eating a big meal. He found that chewing an antacid tablet helped. Lately, though, he seemed to need more and more tablets to make the pain go away. What advice would you give Matthew? Explain your reasoning.

Name of a food I enjoy	An important nutrient in this food	Where in my body this nutrient is digested

TABLE 7.2 ***Sample Data Table for Question 2(a)***

CHAPTER REVIEW

KEY IDEAS

■ Your digestive system is made up of several organs and tubes: mouth, esophagus, stomach, small intestine, large intestine, pancreas, liver, and gall bladder. These all work together to break down food into particles small enough to enter your cells.

■ As food is chewed in your mouth, it starts to be broken down by mechanical digestion and chemical digestion.

■ Your esophagus is a tube that carries food and liquid from the mouth to the stomach.

■ Food is pushed through the tubes of your digestive system by a series of muscle contractions known as peristalsis.

■ When food reaches your stomach, further mechanical and chemical digestion takes place. The stomach produces a fluid containing hydrochloric acid, mucus, and digestive enzymes.

■ The small intestine is the main organ that breaks down your food into small, usable nutrients. Digestive enzymes are produced by the pancreas and the small intestine. In addition, the liver produces bile to help you digest fat.

■ Nutrients from your digested food are absorbed into the cells lining your small intestine. This is called absorption. The inside wall of the small intestine has many folds; with its large surface area, the small intestine can absorb all the nutrients your body needs.

■ The large intestine converts the wastes that cannot be digested into feces. A large amount of water is absorbed for reuse by your body.

■ Many common problems of the digestive system can be prevented with a nutritious diet. Problems that do not go away quickly or that occur frequently are a warning that you should see your doctor.

VOCABULARY

tissue
organ
system
digestive system
digestion
saliva
mechanical digestion
chemical digestion
digestive enzyme
esophagus
epiglottis
peristalsis
small intestine
large intestine
pancreas
liver
bile
gall bladder
absorption
elimination
feces
rectum
anus
constipation
diarrhea

V1. Write "Digestive System" in the center of a page in your notebook. Use the terms from the vocabulary list to make a mind map that shows how the digestive system accomplishes its task. You may also wish to add other words.

V2. Organize the vocabulary words into a flow chart so that they describe what happens when a bite of food passes through your digestive system.

CONNECTIONS

C1. Explain the role of each of the following substances in digestion.
(a) mucus
(b) hydrochloric acid
(c) bile
(d) digestive enzymes
(e) saliva

C2. (a) What is peristalsis?
(b) Name three parts of the digestive system where peristalsis takes place.
(c) What does peristalsis do in each of these body parts?

C3. Describe two problems that can upset the normal functioning of the digestive system.

C4. Explain the role of the large intestine and the rectum in digestion.

C5. (a) What parts of your digestive system produce mucus?

(b) What is the function of the mucus in each of these parts?

C6. Compare the following parts of the digestive system. Give at least one similarity and one difference for each pair of body parts.

(a) esophagus/small intestine

(b) mouth/stomach

(c) large intestine/esophagus

C7. Copy the left-hand column of Table 7.3 into your notebook. Match each process with the body part involved (from the right-hand column). You may use a structure more than once.

Process (What happens?)	Body part (What structures are used?)
Chewing	Body cells
Swallowing	Large intestine
Contracting muscle	Teeth
Releasing nutrients from large molecules	Tongue
Using nutrients	Salivary glands
Elimination	Stomach
Digestion	Small intestine
Absorption	Pancreas
	Lips
	Esophagus

TABLE 7.3 ***Terms for Question C7***

EXPLORATIONS

E1. Find out more about the enzyme in your saliva. Put a cracker into your mouth but don't chew or swallow it. What happens after a few minutes?

E2. Contact a group that offers information about a disease or problem of the digestive system, such as diabetes or colitis. Find out the symptoms and how people deal with the problem. You may be able to arrange for a video or a speaker for your class.

E3. Examine a dentist's model of adult teeth, or look in the mirror at your own teeth. Prepare a chart to show the number, shape, and possible function of each kind of tooth you observe. Do the shape and function of each kind of tooth suggest a certain kitchen tool to you? If so, name the tool on your chart.

E4. Cow's milk contains many nutrients, including a carbohydrate called lactose. Some people do not have an enzyme in their small intestine to digest lactose. If these people drink regular milk, they may have painful cramps. Find out what lactose-free milk products are available for people who cannot digest lactose.

REFLECTIONS

R1. What is the most important thing you have learned about digestion in this chapter? Will you now make any changes in the way you treat your digestive system? Explain your answer.

R2. You may have tried some of the new products on the market that contain substitutes for sugar or fat. Now that you know more about your digestive system, what questions would you ask a food scientist about these products?

CHAPTER 8

RESPIRATION

You could live for several weeks without food and for nearly a week without water. Without oxygen, however, you could survive for no more than a few minutes. You obtain this important gas from the air you breathe. The astronaut in this photograph depends on the supply of oxygen within the space suit while working outside the spacecraft. The oxygen carried by the spacecraft must last for the entire journey away from Earth's atmosphere.

Who else carries a supply of oxygen? A scuba diver swimming under water needs to carry oxygen. So does a firefighter entering a smoke-filled building. A person having difficulty breathing may be given oxygen through a mask or tube.

You know that you obtain oxygen by breathing. You may also know that the oxygen in Earth's atmosphere is produced by green plants. Why is oxygen so important to you? How do you use this gas after it is in your body? In this chapter, you will investigate these and other questions about breathing and respiration.

ACTIVITY 8A / ALL IN A BREATH

Take a slow, deep breath. Besides oxygen, what enters your body when you breathe? Make a three-column table in your learning journal (Table 8.1). In the first column, list all the substances you can think of that might enter your body when you breathe. These might include gases that are in the air as well as other substances that might be in the air inside or outside the classroom.

In the second column, write down whether you think each substance is harmful, harmless, or helpful to your body. Put a question mark if you don't know.

In the third column, write down whether you have complete control, some control, or no control over the presence of each substance in the air you breathe. ❖

Substance in air	Effect on my body	Personal control
Nitrogen	?	No control
Dust	Harmful (allergies)	Some control

TABLE 8.1
Sample Data Table for Activity 8A

■ 8.1 THE NEED TO BREATHE

Air contains many different gases, including oxygen, nitrogen, carbon dioxide, and water vapor. Air may also contain solid particles, such as dust or pollen. Sometimes pollutants such as automobile exhaust fumes have been added to the air. When you breathe, however, the only substance your body is trying to obtain is oxygen.

A number of parts of your body help you obtain the oxygen you need. Just as your digestive system works to obtain nutrients from the food you eat, your body has a system that works to obtain oxygen from the air you breathe. Your **respiratory system** consists of the organs and tissues that move air in and out of your body (Figure 8.1). The process of breathing, **respiration**, is the moving of air in and out of your body. When you breathe in, you are inhaling. When you breathe out, you are exhaling.

Your body uses oxygen to release the energy stored in food molecules, such as glucose, a simple carbohydrate. Whether you are sleeping, sitting, or running, your cells use chemical energy (Figure 8.2). This energy is used to move muscles, send nerve impulses, and perform many other types of work in your body.

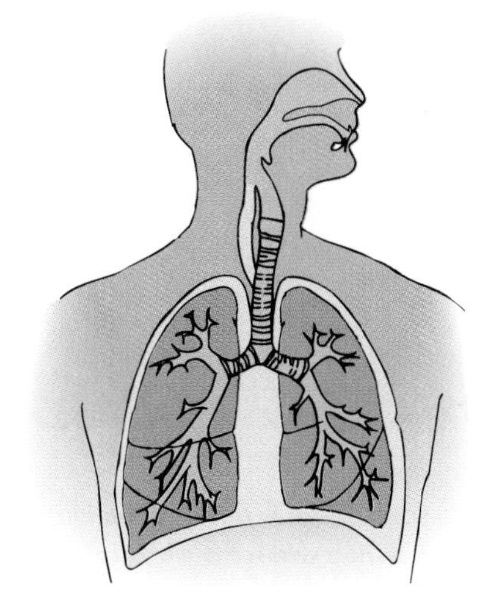

FIGURE 8.1
The human respiratory system is one of your body's most vital organ systems. Why do you think you can breathe through both your nose and your mouth?

FIGURE 8.2
Which of these people, the student doing school work or the hurdlers, needs more oxygen? To answer this question, think about how their bodies are using energy.

CELLULAR RESPIRATION

Your digestive system provides the cells of your body with nutrients. Some of these nutrients—carbohydrates, fats, and proteins—contain stored food energy. Your respiratory system provides your cells with oxygen from the air you breathe. **Cellular respiration** is the process by which your cells use this oxygen to release energy. Look at the word equation for cellular respiration:

$$\underbrace{\text{glucose (from food)} + \text{oxygen}}_{\textit{reactants}} \xrightarrow{\text{in living cells}} \underbrace{\text{chemical energy} + \text{carbon dioxide} + \text{water}}_{\textit{products}}$$

FIGURE 8.3
The loss of water through the lungs can be seen when the air is cold.

As you can see from the word equation, three products are formed during cellular respiration. Besides energy, water and carbon dioxide are produced. What happens to these substances?

Some of the water that is produced is used by your cells. The rest is removed through the process of **excretion**, your body's way of getting rid of waste. One place that excretion occurs is in your lungs. Water evaporates from your lungs into the air you breathe out. You can see this water vapor when you breathe out into cold winter air (Figure 8.3). Your skin and kidneys also help your body excrete excess water, as you will learn in Chapter 9.

The other product of cellular respiration is carbon dioxide. The carbon dioxide produced by your cells must be removed quickly, before it builds up to dangerous levels. Your blood carries the carbon dioxide to your lungs. The lungs remove the carbon dioxide from your blood, and you get rid of this gas when you breathe out (Figure 8.4). However, carbon dioxide has an important role in breathing while it is in your body.

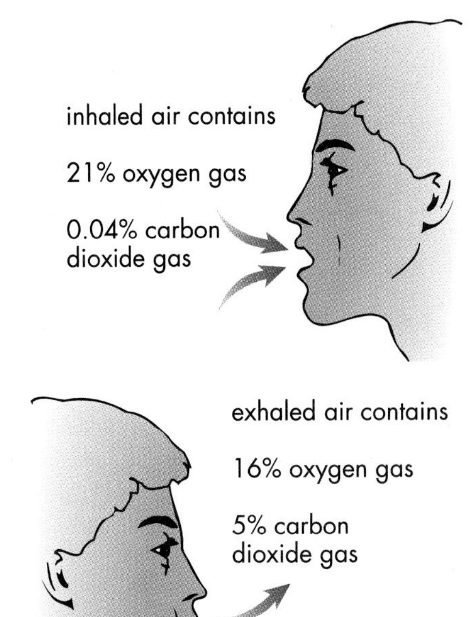

FIGURE 8.4
Why does the air you breathe out contain more carbon dioxide gas than the air you breath in?

BREATHING RATE AND CARBON DIOXIDE

Look at Figure 8.2 again. Which person would be taking more breaths per minute? When you are very active, your muscle cells use more oxygen to help release extra chemical energy. The cells also produce extra carbon dioxide, which enters your blood. The amount of carbon dioxide in your blood gives your body a way to measure how quickly you need to breathe. (The number of times you breathe in a minute is called your **breathing rate**.) Your nervous system constantly checks the amount of carbon dioxide in your blood. When there is a high amount of carbon dioxide, your nervous system responds by increasing your breathing rate.

As you breathe faster, carbon dioxide is removed through your lungs more quickly. At the same time, more oxygen is brought into your body. When the amount of carbon dioxide in your blood decreases to a normal level, your nervous system detects this change and returns your breathing rate to normal.

ACTIVITY 8B / CARBON DIOXIDE AND BREATHING RATE

How does carbon dioxide affect the rate at which you breathe? Your teacher will demonstrate this activity or have students work in small supervised groups.

MATERIALS

clean paper bag (lunch bag size)

■ **A volunteer must stop earlier if feeling any sign of dizziness or discomfort.**

PROCEDURE

1. For a demonstration, your teacher will select two or three volunteers in good health. For each volunteer, a student recorder will observe their breathing as follows. When each volunteer is sitting down and relaxed, the recorder will count the number of breaths the volunteer takes during 1 min. On a chart, this number is recorded as the volunteer's breathing rate "At Rest."
2. Each volunteer will remain seated and hold a paper bag over his or her nose and mouth. Under the teacher's supervision, the volunteers will breathe into and out of their bags for 30 s only and then remove the bag (Figure 8.5).
3. Immediately after the bag is removed, the recorder will count how many breaths the volunteer takes in 1 min. On the chart, this will be recorded as the volunteer's breathing rate "After Bag-Breathing."
4. Work out the average breathing rates before and after the volunteers breathed into the bag.

FIGURE 8.5
Breathe into and out of the bag for no more than 30 seconds.

5. For all students: Take a deep breath and hold it as long as is comfortable. Can you feel a tightness building? Stop as soon as you need to breathe.

DISCUSSION

1. After breathing into the bag, did the volunteers breathe more quickly or more slowly than their resting rate? Explain why you think this occurred.
2. (a) What do you think happened to the amount of oxygen in the bag as the volunteers breathed the same air over and over again? Explain your answer.
   (b) What do you think happened to the amount of carbon dioxide in the bag as the volunteers breathed the same air over and over again? Explain your answer.
3. (a) Which aspects of breathing seem to be under a person's own control (voluntary action)?
   (b) Which aspects do not seem to be under a person's own control (automatic action)?
   (c) Why do you think it is important to have both types of control? ❖

► REVIEW 8.1

1. (a) How does your body obtain oxygen?
   (b) Why do you need oxygen?
2. (a) Write the word equation for cellular respiration.
   (b) Where do the reactants come from?
   (c) What happens to each of the products of cellular respiration?
3. Compose sentences to show that you understand the importance of the following in your body: oxygen, carbon dioxide, water, energy, food.

■ 8.2 HOW YOU BREATHE

Imagine that you have been asked to design and build a robot that could breathe. What would you need? Breathing moves air both in and out of your body, over and over again. How does your body accomplish this task?

THE LUNGS AND CHEST CAVITY

To understand breathing, it helps to know more about the body parts you use to move air in and out. Your **lungs** are spongy organs that receive the air you inhale. The lungs are made up of clusters of tiny, hollow sacs called **alveoli** (singular: **alveolus**). Each alveolus is surrounded by blood vessels. The **chest cavity** is the large space in the upper part of your body where the lungs are located.

The chest cavity is sealed. It has airtight walls that keep your lungs (and heart) separate from the rest of your organs. The walls of your

■ The chest cavity is also called the thoracic cavity. The word "thoracic" is very similar to "thorax." You may have used the word thorax to refer to the middle body part of an insect (head, thorax, and abdomen). Why do you think scientists who compare the body parts of different animals would want to use similar terms for parts of a human and an insect?

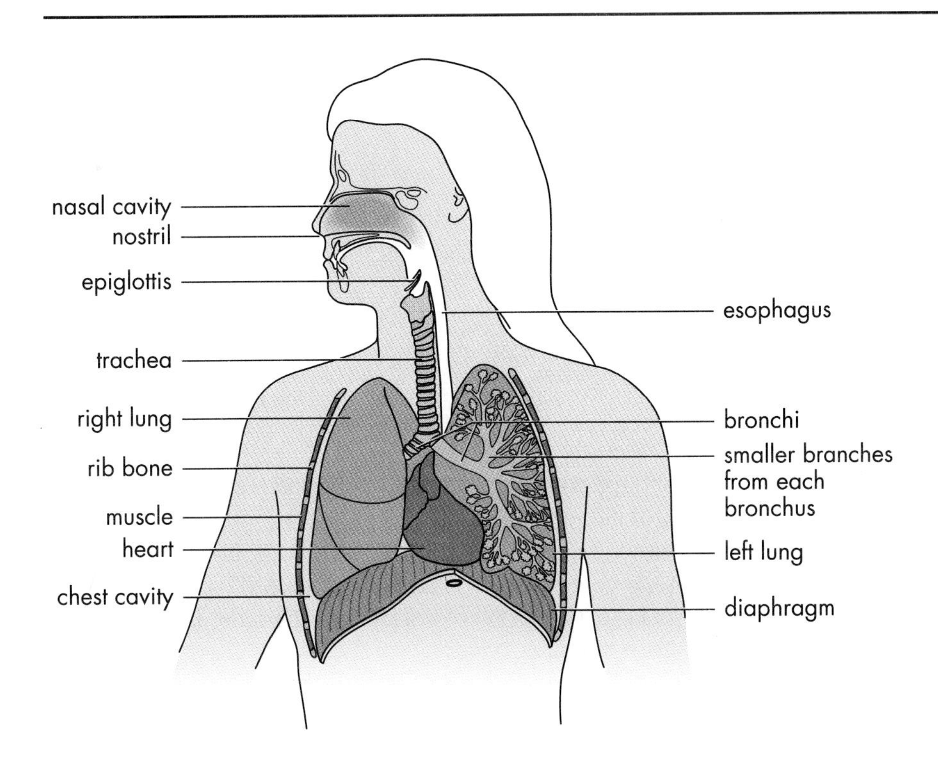

FIGURE 8.6 ◀
The structures of the respiratory system. (The heart is included so you can see its location, but it is not part of the respiratory system.) The rib cage and muscles lying in front of the lungs have been cut away. Notice that the lungs are well protected.

chest cavity contain rib bones and muscles. This is the **rib cage**. At the bottom of the chest cavity is a large sheet of muscle. This dome-shaped muscle is the **diaphragm** (Figure 8.6).

Think about how your chest is moving as you breathe. Your rib cage and diaphragm produce these movements.

ACTIVITY 8C / BUILDING A MODEL OF A LUNG

How does air move in and out of your body? In this activity, you will use a model of a lung to investigate this question.

MATERIALS

(Note: If your classroom already has a model of a lung, follow your teacher's directions.)

safety goggles
clear plastic cup with a hole in the bottom
clean straw
rubber bands
two balloons, one small and one large
modeling clay or putty

PROCEDURE

1. Put on your safety goggles. Construct a model of a lung using Figure 8.7 and the following instructions as a guide.
   (a) Put the straw into the neck of the small balloon. Keep it in place using a rubber band (Figure 8.7a). Test if you have an airtight seal by blowing air into the balloon through the straw.
   (b) Once you are sure your balloon and straw are airtight, push the open end of the straw through the hole in the plastic cup until the small balloon is inside the cup. Use modeling clay or putty to seal the hole around the straw (Figure 8.7b).
   (c) Cut the large balloon in half and stretch it to cover the top of the plastic cup. Hold it in place with a rubber band (Figure 8.7c).
2. Put your name on your model.
3. Draw your completed model in your notebook. On your drawing, label which parts of your model match these parts in the human lung: diaphragm (muscle), chest cavity, air tube (trachea) from lung to outside the body. ➡

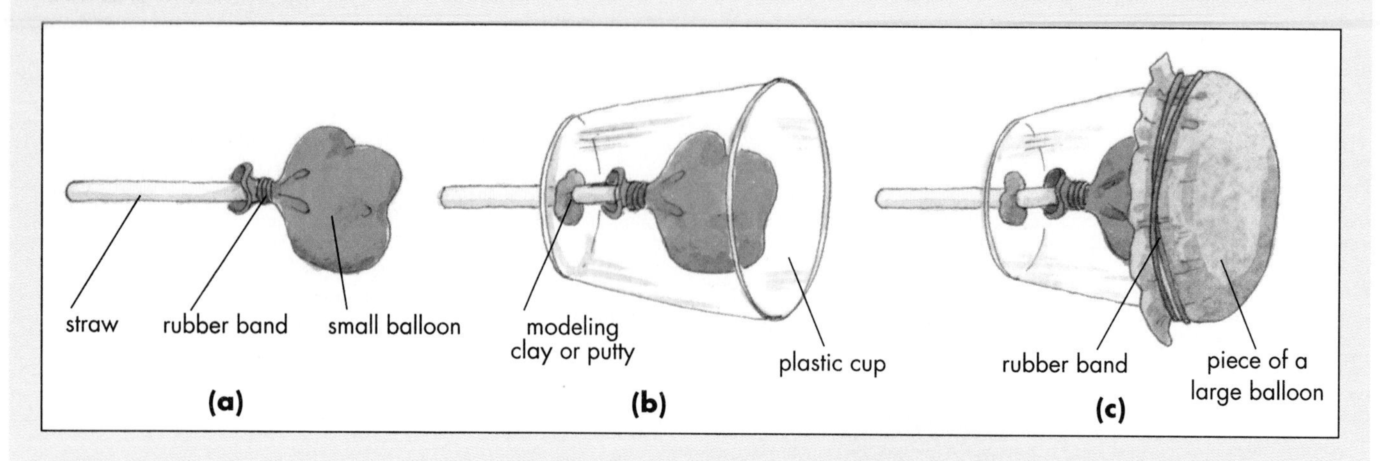

FIGURE 8.7
Preparing the model of a lung for Activity 8C

4. Perform the following experiments with your model and record the results:
   (a) Hold the sides of the plastic cup in one hand. Use your other hand to gently pull down on the outer balloon. What happens to the inner balloon when the outer balloon is pulled downward?
   (b) Let go of the outer balloon. What happens to the inner balloon when you release the outer balloon?

DISCUSSION

1. (a) Where did the air in the inner balloon come from?
   (b) How is this similar to what happens in your body?
2. What part of your body is sealed like the inside of the plastic cup? (Remember that the small balloon is open to the air outside the cup at all times.)
3. How is the movement in step 4 similar to the way your lungs work?
4. (a) Predict what would happen to the "lung" of your model if you pulled down the "diaphragm" of your model and then put a hole in it, puncturing the "diaphragm." Explain your reasoning.
   (b) If a person's chest is punctured, the lungs may collapse. Explain why this could happen.
5. Take your lung model home. Explain to a member of your family how your lung model works. Use the terms "chest cavity," "sealed," "air tube" (or "trachea"), "diaphragm," and "lungs." Ask this person to write a paragraph in your notebook that describes how air is moved in and out. Your teacher will mark the answer. ❖

HOW DO YOU BREATHE?

Press the palms of your hands tightly together. Now pull them apart quickly. You can feel air pushing into the space between your hands. A similar thing happens when you inhale. Figure 8.8 shows the diaphragm and the muscles of the rib cage that surround the lungs. When the diaphragm and the muscles of your rib cage tighten, or

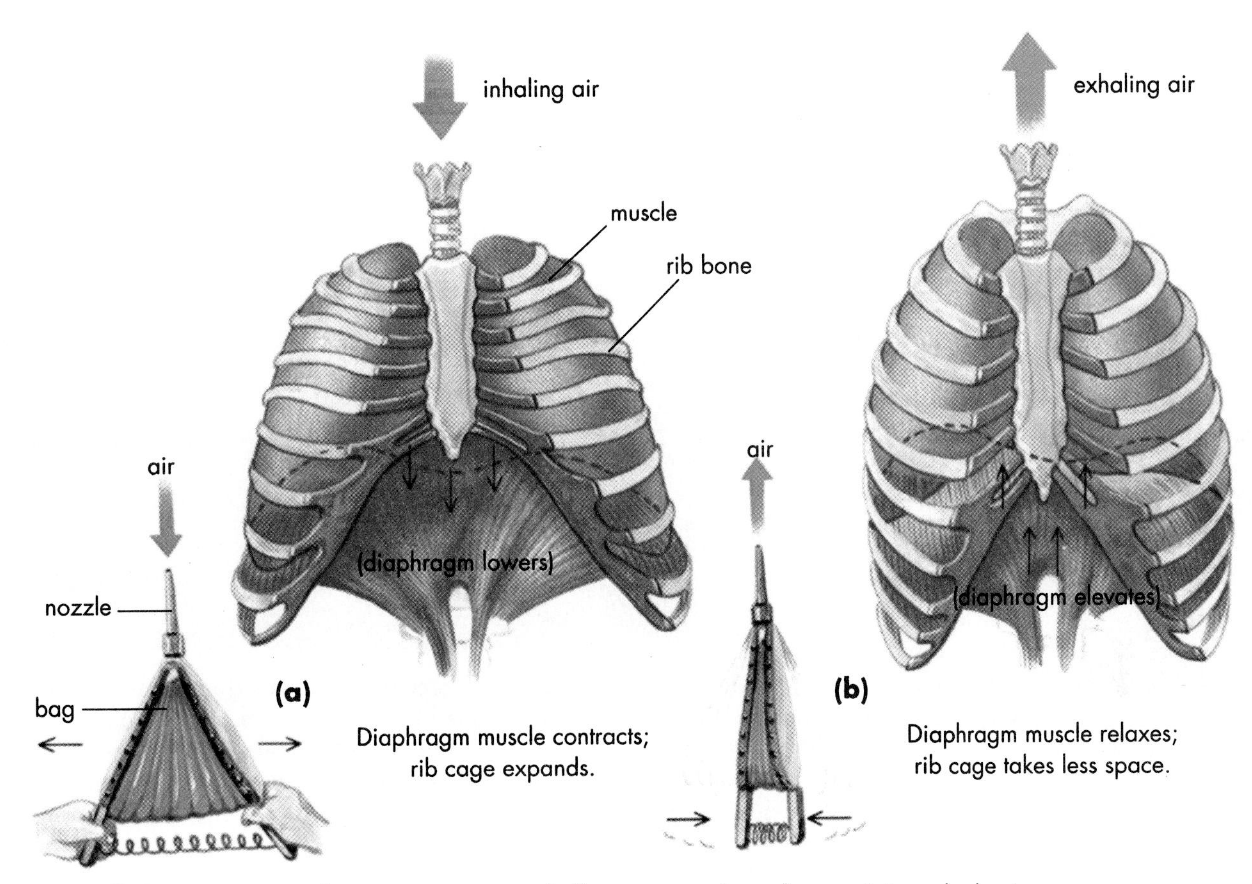

FIGURE 8.8
The muscles around the lungs act like bellows, drawing air in (a), and forcing it out (b). Which part of the system acts like the nozzle of the bellows? Which parts act like the handles? Which parts act like the bag of the bellows?

contract, they pull your lungs farther open (Figure 8.8a). Air from outside your body rushes in to fill the empty space.

What happens when you exhale? The same muscles of your rib cage and your diaphragm muscle now relax. Your rib bones move down and inward. The chest cavity and lungs decrease in volume. As a result, air is pushed out from your lungs into the tube that carries air into your mouth. All this happens automatically. If you want to speak or cough, for example, you can use the diaphragm and the muscles of the chest wall to force air out of your lungs.

If you did Activity 8C, you noticed that the inner balloon (which acts like a lung) collapsed when the air left it. Your lungs, however, do not collapse under normal circumstances. The outsides of your lungs are held tightly to the muscular walls of your chest cavity by a thin layer of moisture that coats the surface of the lungs. To understand how this moisture works in your body, think about how a person sometimes has difficulty separating wet dishes that are stacked on top of one another. The thin layer of water between the dishes provides a surprisingly strong "glue."

THE SCIENCE OF UNDERWATER DIVING

The human body is designed to live on land and breathe air. If you want to explore underwater, you need to use scuba equipment.

Scuba stands for Self-Contained Underwater Breathing Apparatus. This technology solves the problem of how to carry air to breathe. About 3 m^3 of air is compressed into a tank that divers wear on their backs. This air is under too high a pressure for a diver to breathe so a regulator reduces the air pressure before the diver breathes from the tank through a hose.

Another problem with diving is that human bodies are designed for the air pressure we feel on land. Water pressure is greater than air pressure. The deeper divers go, the greater the pressure on their bodies.

Imagine placing a closed plastic container 10 m underwater. Water pressure would squeeze the container to about half its original size. Bringing the container back up to the surface would cause it to expand and return to its original size.

As divers go deeper underwater, the pressure squeezes the alveoli, tiny sacs in the lungs. Scuba regulators keep the air from the tank at the right pressure for the depth so that the divers' lungs are not squeezed smaller and smaller.

If divers with full lungs 30 m underwater suddenly came up to the surface, their lungs would quickly expand, just as the plastic container expanded. The lungs, however, are elastic and would overexpand. To prevent that, divers rise to the surface slowly, breathing out all the way.

There is one more serious problem to overcome. As divers go deeper underwater, they breathe in more air to keep up the air pressure in their lungs. Since air is about 78 percent nitrogen gas, these divers take in more nitrogen than usual, which becomes dissolved in their blood.

To imagine what happens next, suppose you have a bottle of fizzy drink. While the cap is still on, the drink is under pressure and does not have any bubbles. When you take off the cap quickly, the pressure is released and many bubbles of gas form. If the cap is removed slowly, the gas is released in a more controlled way.

As divers rise to the surface, the pressure on their bodies becomes less. If they rise too quickly, the dissolved nitrogen in the blood forms bubbles. This painful condition is called decompression sickness, or "the bends." If the bubbles block an artery, they can cause a stroke or heart attack.

To prevent the bends, divers must rise to the surface slowly. Then the nitrogen from the blood is released as a gas slowly enough to be exhaled from the lungs. Divers also use tables to calculate how deep they can go and how much time they can safely spend at that depth.

Divers learn that "SAFE Dive" means "Slowly ascend from every dive." Why is this slogan helpful?

▶ REVIEW 8.2

1. Name two characteristics of the chest cavity that are important to the function of breathing. Explain your answer.
2. (a) How is air brought into your lungs?
   (b) How is air moved out of your lungs?
3. (a) A person with a wound in the chest has a collapsed lung—the lung deflates when it is empty of air. Explain why the lung collapsed.
   (b) To repair the damage, the surgeon first seals the wound and then removes as much air as possible between the collapsed lung and the walls of the chest cavity. The lung will partially inflate. When the patient next takes a breath, the lung inflates normally and does not collapse again. Explain why the lung was able to inflate (hold air) again after the wound was sealed.

■ 8.3 A SINGLE BREATH

Each breath of air enters your body through your nose or your mouth. If you pinched your nostrils closed or if your nose was blocked because of a bad cold, you could still breathe through your mouth. In fact, breathing through the mouth moves air in and out of your lungs more quickly.

Breathing in through your nose, however, has some important advantages. Your sense of smell is located within your nose (Figure 8.9). In addition, air from your nose is cleaned, warmed, and moistened before it reaches your lungs.

FIGURE 8.9 ▶
When you are near flowers, what's the best way to inhale?

ACTIVITY 8D / THROUGH THE RESPIRATORY SYSTEM

As you read the following section, use Figure 8.10 to help you follow the path that air takes through each part of the respiratory system. You might want to make a chart as you read; for each part, describe its structure and function. ❖

FROM YOUR NOSE TO YOUR LUNGS

Air enters your nose through your nostrils, the two openings that lead into the **nasal cavity**, the hollow space (Figure 8.10). Your nostrils are lined with tiny hairs. These hairs filter any large particles, such as dust, from the air you inhale.

Air passes from the nasal cavity into the throat, which divides into two tubes. In Chapter 7, you learned about one of these, the esophagus,

DID YOU KNOW

■ Your nose is so efficient at cleaning the air you breathe that it traps 99 percent of incoming dust and dirt.

■ A hiccup is a strong contraction of the diaphragm muscle. The sound is produced when the epiglottis slams shut over the trachea as air is drawn in quickly.

the tube that takes food to the stomach. The other tube, the **trachea**, carries air to your lungs. The entrance to the trachea is usually open. When you swallow food or drink, however, the flap of tissue called the epiglottis covers the opening of the trachea so that food cannot enter the trachea and go into the lungs.

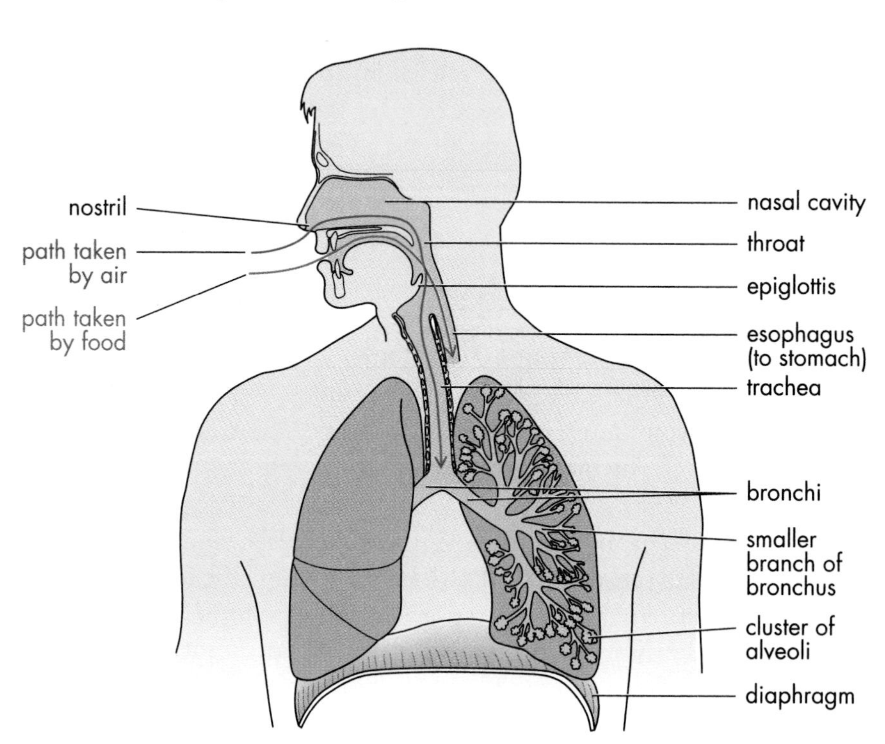

FIGURE 8.10 ▶
Food and air take different paths into your body. How else does the respiratory system prevent solid particles from reaching the delicate parts of your lungs?

Rub your finger along your trachea. You can feel a series of bumps. These are rings of a rigid material called cartilage. These tough rings hold the trachea open at all times so that you can breathe. At home, look at the exhaust pipe of a clothes dryer or the hose of a vacuum cleaner. Feel how the wire rings in the pipe or the hose hold it open. This is similar to your trachea.

As air travels through the nasal cavity and trachea, it is still being cleaned. Very small particles, such as pollen, are not trapped by the hairs in your nose and must be removed in a different way. Your air passages (nasal cavity, trachea, and the tubes leading into your lungs) are all coated with mucus. This sticky substance traps pollen, other particles, and bacteria.

These air passages are also lined with special cells that have tiny hair-like projections called **cilia** (Figure 8.11). The cilia wave back and forth, moving mucus away from your lungs. Any particles stuck to the mucus are moved toward the mouth and nose (Figure 8.12). The mucus and trapped particles are then coughed or sneezed out of the body, or swallowed into the digestive system.

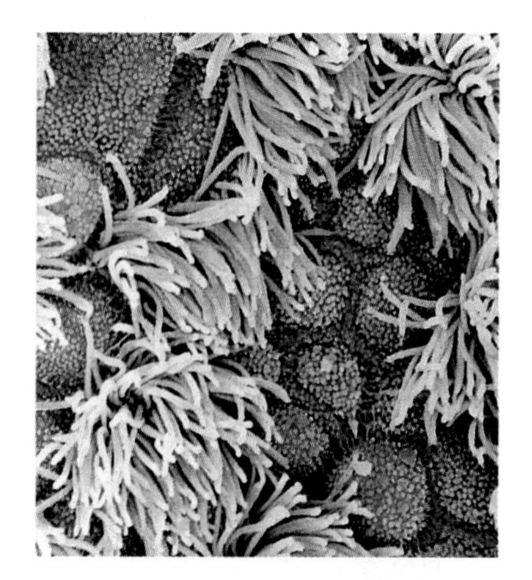

FIGURE 8.11
This photograph shows the cilia, tiny projections of the cells that line the air passages. (The round lumps between the cilia are parts of cells without cilia.)

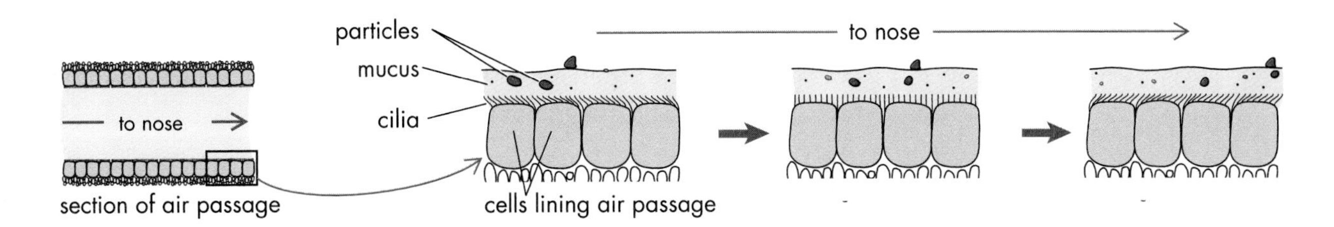

FIGURE 8.12
The cilia that line your air passages move mucus and trapped particles away from the lungs. If the cilia were injured, how would the injury affect the lungs?

Your lungs need moisture and warmth in order to survive. Your nasal cavity and trachea have many blood vessels full of warm blood. The air you breathe is warmed as it moves over these blood vessels. At the same time, the mucus that coats these passages adds moisture to incoming air.

The air you inhale continues down the trachea. After the trachea enters your chest cavity, it branches into two tubes called the **bronchi** (singular: **bronchus**). The bronchi carry air into the two lungs. In the lungs, air travels through smaller and smaller tubes, formed by the branching of each bronchus. Finally, the smallest branches bring air into the alveoli.

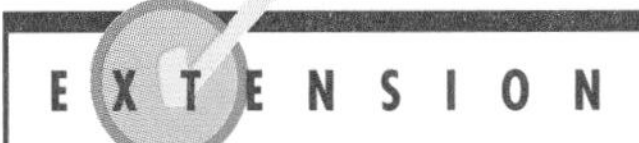

■ Find out how people make sounds such as speech. What structures in the trachea are used? How are volume and pitch of speech controlled?

EXCHANGE OF GASES IN THE ALVEOLI

Each alveolus is like a tiny balloon filled with air and surrounded by very small blood vessels (Figure 8.13a and 8.13b). The blood flowing into these blood vessels comes from all parts of your body. When blood arrives at the alveoli, it is low in oxygen and high in carbon dioxide because cellular respiration in your body's cells has used up oxygen and produced carbon dioxide.

When you breathe in, the air, which is rich in oxygen, fills the alveoli. By diffusion, oxygen moves from the air inside the alveoli into the blood. At the same time, carbon dioxide diffuses from your blood into the air inside the alveoli. Figure 8.13c shows this exchange of gases taking place. In this way, the blood that flows away from your lungs carries a fresh supply of oxygen to all your body cells. The air inside the alveoli now has the carbon dioxide it received from the blood. This exchange of gases takes place at all times, when you are active or at rest, and even while you hold your breath.

FIGURE 8.13
(a) Air entering the lungs ends up in the millions of tiny alveoli.
(b) Notice how blood vessels surround each alveolus.
(c) The exchange of gases takes place in the alveoli. Oxygen passes from inhaled air into the blood cells; carbon dioxide passes from the blood cells into the air.

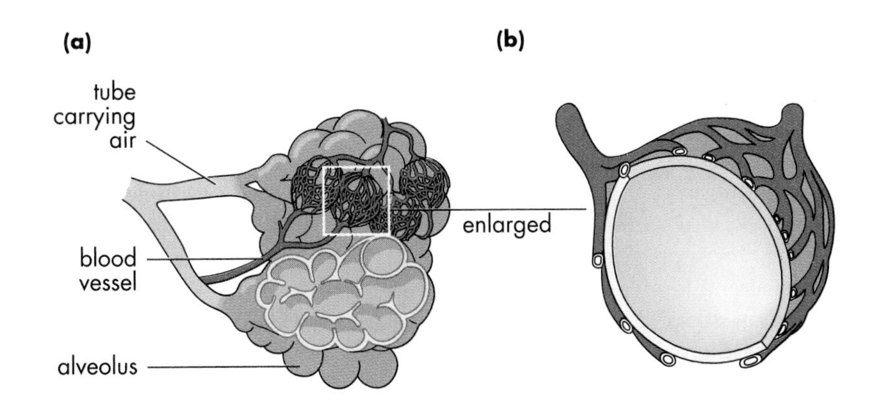

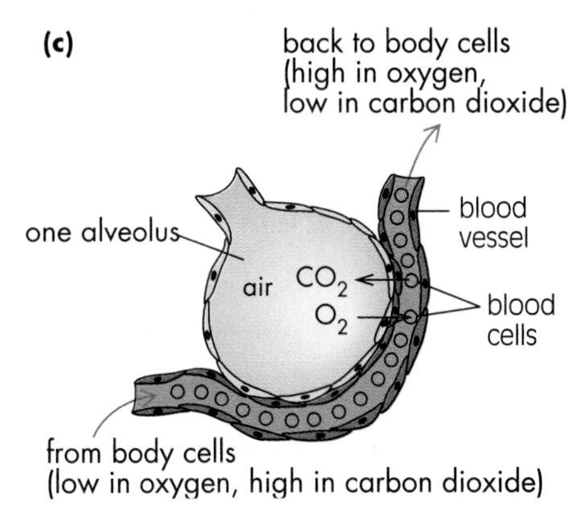

■ If a person has pneumonia, the alveoli become filled with fluid or blood cells or both. This makes it difficult for the exchange of gases to take place. Find out what causes pneumonia and how it can be treated.

Before you exhale, the air moves from the alveoli into larger air tubes in the lungs, then into the bronchi. The air travels along the trachea and leaves through your mouth or nostrils.

THE AMOUNT OF AIR IN A BREATH

Take a normal breath and hold it. Can you now breathe in more air without exhaling first? The amount of air you inhale in a breath changes with how deeply you inhale. A typical breath brings about 0.5 L of air into your body. Yet you could breathe in more than that. The greatest amount of air you can move in and out of your lungs in one breath is called your **vital capacity**. Your vital capacity is a measure of volume. Although different people have different vital capacities, an average adult has a vital capacity of about 4 L.

Vital capacity is a good way to measure how well your respiratory system can respond when you need more oxygen. You need more oxygen when you are active, so you breathe more deeply when you exercise. A fit person usually has a higher vital capacity than a less fit person because the muscles used in breathing become stronger with exercise. You will learn more about the connection between exercise and vital capacity in Chapter 10.

ACTIVITY 8E / VITAL CAPACITY

What factors affect a person's vital capacity?

MATERIALS

vital capacity bags from the Lung Association or the following materials:
safety goggles
apron
large rigid plastic bottle (4 L)
large, shallow container with an overflow spout (or sink)
75 cm of clean rubber tubing
clean straw
rubbing alcohol

PROCEDURE

1. With your teacher, discuss the factors that might affect a person's vital capacity. Working in small groups, determine the factor your group will investigate. For example, your group may decide to look at the effect of exercise on vital capacity. To do this, you would compare the vital capacity of classmates who exercise regularly with the vital capacity of classmates who do not. (You will need to define "regular exercise" first.) Or you might consider the effect of smoking on vital capacity. Form a hypothesis regarding the factor you select. You will test your hypothesis, and then share your hypothesis and results with other groups at the end of the activity.

2. Make a class table, using the factors chosen by the different groups as headings (Table 8.2). When each person's vital capacity is measured and recorded, a checkmark will be placed under every heading that applies to that student. For example, suppose the class table contains the headings: smoker, non-smoker, singer, non-singer, male, female. A non-smoking female singer would have a checkmark under three headings: non-smoking, singer, female. (Your class table may include many more headings than these.)

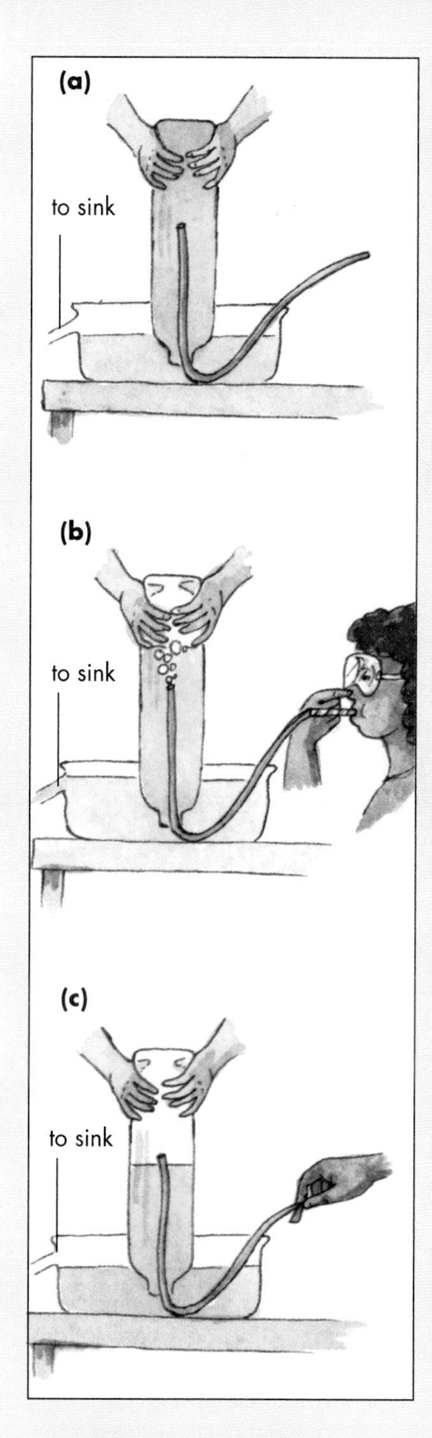

FIGURE 8.14
Apparatus for Activity 8E

Vital capacity	Factors that might affect vital capacity					
	Male	Female	Smoker	Non-smoker	Singer	Non-singer

TABLE 8.2 ***Sample Data Table for Activity 8E***

3. Measure your vital capacity. To do this, you might use vital capacity bags provided by the Lung Association or follow these steps using the apparatus shown in Figure 8.14 (or other apparatus prepared by your teacher). Be sure to put on your safety goggles and apron first.

(a) Fill the plastic bottle with water. As you do this, measure the volume that the bottle holds when full.

(b) Pour water into an overflow tray, or into a sink with a stopper, to a depth of approximately 5 cm. Turn the full bottle of water upside down so that the mouth is under water. Have a partner hold the bottle in the water. Put the end of the rubber tube into the mouth of the bottle. Your partner should continue to hold the bottle without pinching the tube (Figure 8.14a).

(c) Put a clean straw into the rubber tube.

(d) Practice taking a deep breath and exhaling. Now measure your vital capacity. Inhale as deeply as you can. Then put the straw into your mouth, pinch your nose closed, and exhale as much of your breath as you can into the straw (Figure 8.14b). IMMEDIATELY pinch the tube closed (Figure 8.14c).

(e) While you pinch the tube, have your partner keep the mouth of the bottle under water and remove the tube from the mouth. Have your partner cover the mouth of the bottle with his or her hand and remove the bottle from the tray and turn it upright, without changing the volume of water that remains in the bottle.

(f) Measure the volume of water that remains in the bottle. Subtract this volume from the total volume of the bottle (measured in step 3a). The resulting volume shows how much water was replaced by the air you exhaled.

(g) Record this volume in your notebook. This measures your vital capacity. Record it on the class table.

4. Discard your straw. Dip the end of the rubber tube into rubbing alcohol, rinse it with water, and dry it before allowing the next student to use the apparatus.

5. When all students have entered their data, calculate the average vital capacity for the class. ➡

6. Select the vital capacities you are interested in comparing in order to test your group's hypothesis. For example, you could calculate the average vital capacity for smokers and compare it with the average vital capacity for non-smokers. Graph the results.
7. Share your findings with the rest of the class.

DISCUSSION

1. Do the results support your hypothesis? Explain your answer.
2. For the students in your class, which factors do you think contribute to having a higher vital capacity?
3. Based on what you learned in this activity, suggest several ways you might increase your vital capacity. ❖

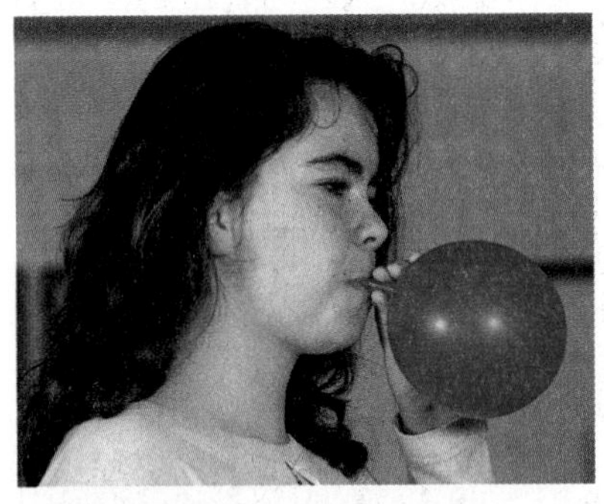

FIGURE 8.15
You normally breathe out without thinking about it, but you can control the movement of air out of your lungs when you want to.

EXHALING

It does not normally take any effort for you to exhale. Because your lungs, rib cage, and diaphragm are elastic, they automatically return to their former size and shape after being expanded. However, you can force your breath out, such as when you cough or blow up a balloon (Figure 8.15). Sneezing is an automatic, fast outward breath that helps remove something irritating from your nasal cavity.

Not all the air you inhale leaves your lungs immediately when you exhale. If it did, your alveoli would collapse like empty balloons. Even after you exhale as much as possible, about 1.5 L of air stays behind in the alveoli of your lungs. This "leftover" air is called **residual air**. Each breath replaces a bit of this residual air. It takes about four normal breaths before all of the residual air is replaced.

► REVIEW 8.3

1. (a) Is it better to breathe through your mouth or through your nose? Why?
   (b) When would the other way of breathing be better even though it has disadvantages? Explain your answer.
2. List two functions of the mucus in your air passages.
3. Make a three-column chart. In the first column, write "nasal cavity," "trachea," "alveoli." In the second column, describe the structure of each part. In the third column, explain how the structure of each part helps its function.
4. (a) What gases are exchanged in your lungs?
   (b) In what part of the lungs does this exchange of gases take place?
   (c) What is the role of your blood in this exchange of gases?
5. (a) What is vital capacity?
   (b) What is residual air?
   (c) Think about how much air you are breathing in and out as you read this sentence. Compare this amount of air with your vital capacity.
   (d) If you inhale one breath of air that contains a perfume, the air you exhale for the next few breaths will still contain some perfume. Explain how this happens.

8.4 TAKING CARE OF YOUR RESPIRATORY SYSTEM

Caring for your respiratory system has two parts: fitness and protection. Think of them as good habits that contribute to your overall good health.

Like any body part that contains muscle, your respiratory system benefits from exercise. When you are active, the muscles that you use in breathing are working harder too. Regular exercise helps make breathing easier and more efficient so that you can move more air in and out of your lungs with each breath. (You will learn more about the connection between fitness and your respiratory system in Chapter 10.)

Protecting your respiratory system is also important. You need to breathe, but sometimes the air you inhale contains substances that could harm your body. Your lungs do not have a tough coating of skin to protect them. The harmful substances you breathe in may damage your lungs. Sometimes these substances pass into your blood from your lungs and harm other parts of your body. The people in Figure 8.16 are taking care of their respiratory systems by preventing dangerous substances from entering their lungs.

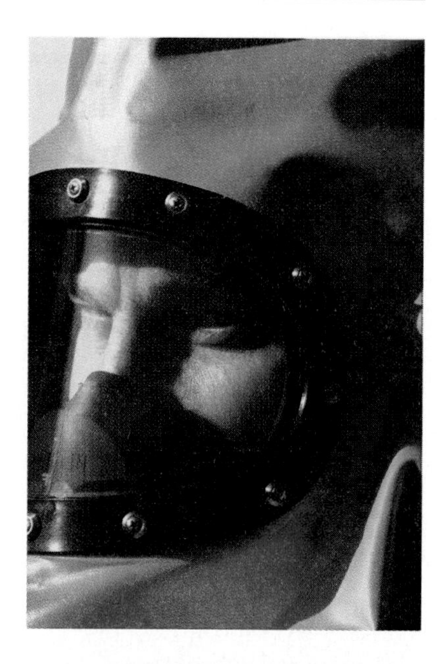

FIGURE 8.16
From what substances are these people protecting their respiratory systems?

How else can you take care of your respiratory system? It is important to pay attention to symptoms that might indicate problems. For example, a person with a cough that does not go away should visit a doctor. If necessary, a doctor will prescribe medications to help fight infections and make it easier for the person to breathe. Sometimes a doctor might suggest using a vaporizer to moisten the air.

If you keep fit and protect your respiratory system against harmful substances, you are taking care of yourself.

THE HABIT THAT KILLS

Think again about what substances might enter your lungs when you breathe (or refer to your answers in Activity 8A). Is cigarette smoke one of those substances? If cigarette smoke enters a person's lungs, harmful substances in the smoke enter the respiratory system.

Cigarette smoke does not just harm the minority of people who smoke. Cigarette smoke is a health risk to non-smokers as well. This is because a non-smoker can be exposed to sidestream smoke, or **second-hand smoke**. This unfiltered smoke enters the air directly from a burning cigarette and can be inhaled into the lungs of a person nearby. As well, the air exhaled by a smoker contains harmful substances.

Imagine that a person inhales a breath of cigarette smoke. The hot gases in the smoke enter the cooler areas of the mouth and air passages. There they condense into **tars**. Tars contain chemicals that stop the cilia from moving. The cilia stop their cleaning action and, over

■ Smokers often say, "I can quit anytime I want." Statistics do not agree. Addiction to nicotine usually occurs quite soon after a person takes up smoking. A child or teenager can become addicted to nicotine after smoking as few as four cigarettes. This young person then has a 90 percent chance of becoming a regular smoker as an adult.

■ Do research on the causes, symptoms, and treatment of asthma. Discuss the problems that a person with asthma might encounter. Find out what treatments are available for asthma.

time, are killed by these chemicals. Without the cleaning action of the cilia, sticky material becomes deposited in the air passages. As this material builds up, it damages the respiratory system. Over time, a smoker develops what is called a smoker's cough because he or she coughs frequently to try to remove the irritating substances.

As the hot gases of the cigarette smoke enter the lungs, they leave streaks of tar inside the smaller branches of the bronchi. Any gases that a person inhales finally reach the alveoli. Tar clogs their delicate, moist surfaces. Other substances in the smoke affect the exchange of gases. For example, carbon monoxide is a gas in cigarette smoke that will enter the blood instead of oxygen. The body's cells do not then receive all the oxygen they need.

Nicotine is another chemical from smoke that enters the blood. Within seven seconds, nicotine reaches the brain. This drug makes the heart work harder than it should and increases the risk of heart attack. Even when the person exhales the smoke, some smoke stays behind in the residual air of the lungs. Nicotine is the drug that people become addicted to if they smoke.

The harm done by continuing to smoke cigarettes builds up over the years. Cigarette smoke contains several chemicals that can cause cancer. Tar is one of these. Cancer in the lungs is hard to detect in early stages because the growing, cancerous tumors are not painful in the soft lung tissue. By the time the person has pain, it is usually too late to cure the cancer. In addition, the blood from the lungs can carry cancerous cells to other parts of the body. Researchers have estimated that a person who smokes just 10 cigarettes a day is much more likely to die from a smoking-related disease than from any kind of violence or accident.

Smokers—as well as non-smokers exposed to second-hand smoke—have more frequent chest infections, such as bronchitis, than people who are not exposed to this smoke. Bronchitis is an inflammation of the small air passages of the lungs caused by an infection. Anyone can get bronchitis, but smokers, and those near them, are more at risk. Substances in cigarette smoke damage the cilia; without the cilia to clean the air, more bacteria and viruses enter the lungs. In addition, a smoker's lungs can suffer when many of the smaller air passages become permanently blocked, leaving a smaller surface area for the exchange of gases (oxygen and carbon dioxide). This condition is called chronic (long-lasting) bronchitis.

The lung disease called emphysema can be more severe than bronchitis. Not only do the small air passages become blocked, but the walls of the alveoli break apart. A person with emphysema finds it hard to breathe. Over the years, the heart becomes overworked and the person may have a heart attack. Emphysema may be caused by breathing in harmful substances such as cigarette smoke, asbestos fibers, and mining dust. Unfortunately, emphysema is rapidly becoming a major cause of death among older people, as more long-time smokers reach their later years.

There is no doubt that smoking is a deadly habit. Why do people smoke? Part of the problem is that the harm done by smoking is hard to see and may take years to affect a person's health. Advertising can make smoking seem attractive. Ninety percent of smokers try to "kick the habit" by the time they are 30 years old. But by then, they are addicted to nicotine: their bodies depend on this drug, and they find it very hard to stop smoking. Most smokers wish they had not started smoking. Although most young people do not smoke at all, about 18 percent of high school students reported in a recent survey that they smoke at least once a week.

Your respiratory system brings in the oxygen that every cell of your body needs. How well this system works can affect all the activities you take part in and your enjoyment of life. Your respiratory system needs your help in dealing with pollutants in the air. Cigarette smoke is one pollutant over which you do have a lot of control.

ACTIVITY 8F / A QUESTION OF ATTITUDE

Over the years, cigarette manufacturers have used many kinds of advertising to sell tobacco to people. Often these advertisements make smoking seem attractive. They make it seem to be a part of everyone's lifestyle. You can work on the following activity individually or in small groups.

PROCEDURE

Examine several advertisements for cigarettes. Choose two or three that you find interesting. Think about why they catch your attention.

Design your own anti-smoking advertisement. See if you can use any of the interest-catching features of the cigarette advertisements. Display your advertisements in your classroom or the school hallway. Use this display to start a discussion with your classmates. What might make someone want to try smoking? What might help people choose not to smoke? ❖

▶ REVIEW 8.4

1. (a) How does your respiratory system protect your body against bacteria and harmful substances in the air you breathe?
(b) Explain why these defenses do not help against cigarette smoke.

2. (a) How does cigarette smoke affect the body right away?
(b) How can cigarette smoke affect the body over a longer period of time?

3. (a) What are the differences between bronchitis and emphysema?
(b) Why might a person with emphysema die?

4. Imagine that you get a job working in a restaurant with a non-smoking section and a smoking section. Does this arrangement protect you from cigarette smoke? What could you do about this situation? Explain your reasoning.

CHAPTER REVIEW

KEY IDEAS

■ Breathing (respiration) moves air into and out of the lungs. Breathing supplies oxygen to the body and removes carbon dioxide.

■ Oxygen is used by cells during cellular respiration to release energy from food. The carbon dioxide and some of the water that are produced are removed through the lungs.

■ When you inhale, your diaphragm and the muscles of the rib cage enlarge your chest cavity. Your lungs expand and air moves in.

■ When the diaphragm and the muscles between the ribs relax, you exhale. You can also force air out by tightening these muscles deliberately.

■ The air passages consist of the nose, nasal cavity, throat, trachea, and bronchi. These structures are lined with mucus and cilia. The mucus traps small particles and moistens the air; the cilia move the mucus away from the lungs.

■ The exchange of gases takes place in the alveoli. Oxygen moves into the blood, to be carried to the rest of the body. Carbon dioxide from the blood moves into the air in the alveoli, to be removed when you exhale.

■ The largest amount of air that you can breathe in and out is called your vital capacity. Normal breaths do not move all of the air in your lungs. Some residual air is always left.

■ Your respiratory system requires care that includes exercise and protection from harmful substances.

■ The chemicals in cigarette smoke damage the lungs and air passages, causing many health problems.

■ Cancer, bronchitis, and emphysema are some of the diseases that can affect lung tissues.

VOCABULARY

respiratory system
respiration
cellular respiration
excretion
breathing rate
lung
alveolus (plural: alveoli)
chest cavity
rib cage
diaphragm
nasal cavity
trachea
cilia
bronchus (plural: bronchi)
vital capacity
residual air
second-hand smoke
tar

V1. Make a three-column chart in your notebook. List the terms that refer to parts of your respiratory system in the first column. In the second column, describe the structure of each part (what it is made of or how it looks). In the third column, describe the function of each part.

V2. Write the word "air" in the center of a page in your notebook. Use the terms listed above, and any others you need, to make a mind map showing how air gets into your body and its importance.

CONNECTIONS

C1. Write the letters a to f in your notebook. Beside each, write the label that corresponds to that letter on the diagram of the respiratory system in Figure 8.17.

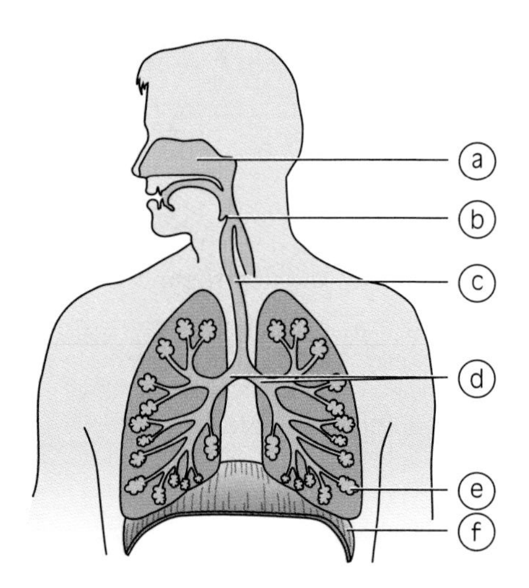

FIGURE 8.17
Use this diagram to answer question C1.

C2. How is the air you breathe in different from the air you breathe out? Suggest several ways.

C3. Explain why you may need to breathe more quickly in the following situations:
(a) during and after exercise
(b) in a crowded room without windows
(c) if some of your alveoli are damaged

C4. Draw a diagram to explain the exchange of gases in an alveolus.

C5. What benefits can someone expect from giving up smoking or not starting to smoke?

EXPLORATIONS

E1. Scientists are designing equipment to supply oxygen for either a space station or a human journey to the planet Mars. Find out about the breathing equipment now being used by astronauts in space. If astronauts are to take these longer journeys away from Earth, what problems in supplying oxygen and removing carbon dioxide must be solved? How would you solve them?

E2. Prepare a survey form to ask people's opinions about the issue of smoking in public areas. Check with your teacher before you conduct the survey.

E3. Work with a small group to produce a commercial that expresses your views on smoking. You might videotape your commercial or write a radio script.

E4. When green plants take in light energy and use it to produce food (glucose), they also use carbon dioxide and water. At the same time, oxygen is released by plants to the air.
(a) Write a word equation that describes the production of food by plants.
(b) Write the word equation for cellular respiration directly underneath your equation from (a).
(c) Look at the two equations. Draw lines to represent any possible links between food production by plants and the use of food (in cellular respiration) by your body. Discuss the importance of this relationship between humans and green plants.

E5. If you do the fish dissection in Appendix D, compare the respiratory system of the fish with that of a human. Do your comparison in the form of a chart or diagrams.

E6. Sometimes the words we use mean more than we think they do—at least to some people. For example, what do you think of when you hear or read about a "disease," or about someone who has a "disease"? Form a group of two or three other students and share your ideas. Each of you should write down your own definition of the word "disease." Next, look up the meaning of "disease" in several different dictionaries. (Try to include some very complete, or "unabridged," dictionaries.) Discuss how your definitions compared with those in the dictionaries.

In your group, discuss the following questions. Do you think using "disease" might cause problems for some people? Why? Are there any other words you could use instead? In what ways are these terms better or worse than the word "disease"? List some other words which, like "disease," might cause problems because they mean different things to different people.

REFLECTIONS

R1. Examine your list of substances from Activity 8A. Based on what you have learned in this chapter, make any changes you think necessary. Choose one substance you listed as harmful, other than cigarette smoke, and find out how it affects your respiratory system.

R2. What would you tell your own children about smoking? At what age do you think you would need to talk to them about it? Do you think adults can have much influence on whether or not young people smoke? Explain your answer.

CHAPTER 9

CIRCULATION AND EXCRETION

When you first stand up to give a presentation to your class or another group, how do you feel? It can be a bit frightening or stressful, even if you are used to public speaking. In this situation, you might notice changes in your body. Your heart might beat more quickly. Some people would blush or feel warm. Other people would have cold hands. These are all signs that your circulatory system, the body system that moves blood through your body, is reacting to your needs.

Your blood is one of the links that connect different body systems. For example, nutrients from your digestive system are carried to muscle cells by your blood. Wastes such as carbon dioxide or water are carried by blood from your cells to your lungs, skin, or other organs for removal from your body. In this chapter, you will learn about how blood moves through your body and how wastes are removed.

ACTIVITY 9A / THINKING ABOUT YOUR HEART AND BLOOD

In your learning journal, write several sentences stating what you already know about your heart and blood. To help you get started, consider questions such as the following:

- What does my blood do for me?
- What does my heart do for me?
- How can I take care of my heart?
- How are my heart and blood (my circulatory system) linked to my digestive system and my respiratory system?

Next, list some questions that you would like answered about your heart or your blood.

At the end of this chapter, review what you have written. You may find that your ideas have changed. You may have discovered the answers to some of your questions. You will probably have new questions. ❖

9.1 BLOOD: A VERY SPECIAL TISSUE

Each of the 60 trillion cells in your body depends on your blood. The movement of blood throughout your body provides a "transportation" system. This system carries oxygen and nutrients to each cell and takes away wastes such as carbon dioxide. Your blood is able to reach all your cells because blood travels through a network that carries it to within 0.01 cm of each cell.

Blood is a very unusual tissue because it moves through your body. Men have about 5 L of blood and women about 4 to 5 L. **Blood vessels** are tubes that carry the blood through your body. Your heart pumps the blood to keep it moving, or circulating. Your blood, heart, and blood vessels make up your **circulatory system** (Figure 9.1).

Every process in your body depends on the presence of blood. For example, wiggle your toes. The muscles of your feet need energy in order to move. Since that energy is released by cellular respiration, your feet need oxygen from the air and nutrients from food. These are carried by your blood. If the supply of blood to your feet were stopped for a period of time, you would not be able to wiggle your toes. If the supply of blood were cut off for too long, the cells making up your feet would die.

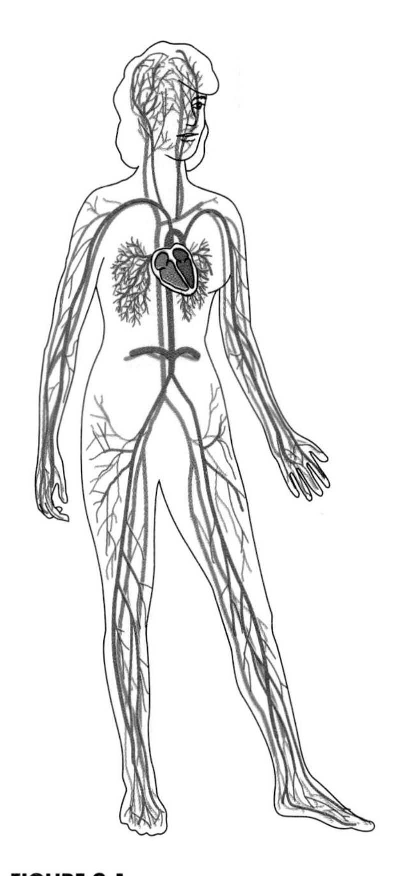

FIGURE 9.1
Your circulatory system includes your heart, blood vessels, and the blood that circulates through the heart and vessels. (Most of the small blood vessels have been left out of this drawing.)

WHAT IS BLOOD?

When you look at blood flowing from a cut, you see a red liquid. Blood can actually be separated into two parts: a liquid part and a solid part (Figure 9.2). The solid matter consists mostly of tiny blood cells. The liquid, **plasma**, is clear and yellowish. Plasma consists of about 92 percent water and 8 percent solids dissolved in the water. Table 9.1 lists some of these dissolved substances.

■ People with hemophilia do not produce certain clot-forming blood proteins. As a result, the blood of hemophiliacs does not clot properly. Before there was any treatment, hemophiliacs could bleed to death from even a minor accident. Now they are given clot-forming blood proteins from donated blood.

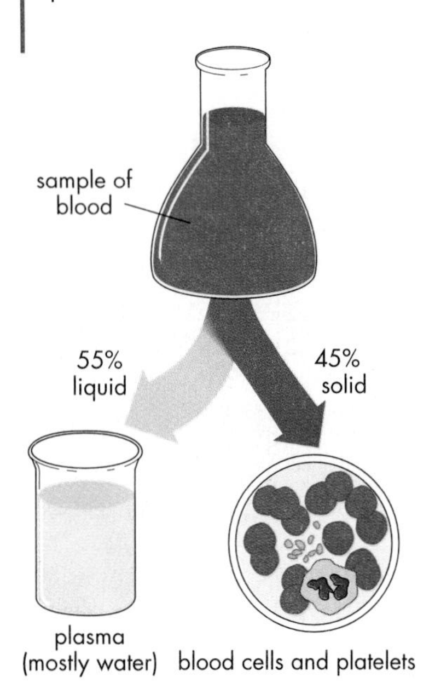

FIGURE 9.2
About 45 percent of your blood's volume consists of solids. The solids and the plasma can be separated by doctors. How would this be useful?

■ Prepare a report or poster on the causes and different types of anemia. What are the warning signs? Present your findings to the class. You might work on your own or in a small group.

Substance	Function in body
Glucose	Provides energy
Amino acids	Are used to build proteins
Minerals	Have several functions, including building bones and teeth
Vitamins	Have several functions, including helping enzymes speed up chemical reactions in cells
Carbon dioxide	Controls how quickly and deeply you breathe

TABLE 9.1 ***Some Dissolved Substances in Plasma***

Your plasma also contains **blood proteins**, which are proteins carried by the blood. You could not live without them. There are several types of blood proteins. Three are listed below.

- **Antibodies** "recognize" and try to destroy foreign matter such as bacteria. Antibodies are part of your body's **immune system**, the system responsible for fighting disease.
- **Hormones** are chemicals that direct and coordinate many processes in your body. A hormone may be produced in one part of your body but have an effect on another part. Adrenalin, for example, is produced by glands above your kidneys, but it acts on several body parts, including your heart and the muscles of your respiratory system. Your blood carries these hormones to the places where they are needed.
- **Clot-forming blood proteins** form a hard plug, called a clot, wherever a blood vessel is damaged. The clot helps to prevent the loss of blood.

BLOOD CELLS AND PLATELETS

The solid matter in your blood consists mostly of **red blood cells** (Figure 9.3a). The function of red blood cells is to carry oxygen. Red blood cells contain **hemoglobin**, a molecule that "grabs" oxygen when there is a high concentration of oxygen and "lets go" in areas of the body that are low in oxygen. The hemoglobin allows red blood cells to pick up oxygen in the alveoli of your lungs and release it near body cells that need oxygen. Your body cannot make hemoglobin without iron. Some iron comes from food. This source is especially important for people who are still growing and for females who are menstruating. In adults, most of the iron is "recycled" when red blood cells die.

Your body contains about 20 trillion red blood cells. Each red blood cell lives for about 120 days. New red blood cells are constantly being manufactured inside your bones. Your body replaces its red blood cells at a rate of 1 to 2 million cells per second. Old red blood cells are taken to your liver. The hemoglobin from these cells is recycled into bile, a substance that your digestive system needs in order to digest fats. The iron is split away from the hemoglobin and is recycled into new red blood cells.

White blood cells have a very different function from red blood cells (Figure 9.3b). White blood cells are part of your immune system. They guard against bacteria and other dangerous substances that enter your body. For example, certain kinds of white blood cells will consume bacteria. There are several types of white blood cells. Some types are made inside bones, whereas others are made in areas such as the tonsils and spleen.

Platelets look like tiny flat bags that float throughout the blood (Figure 9.3c). Platelets collect wherever the walls of blood vessels are damaged. They cling together and break open, releasing chemicals that cause the clot-forming blood proteins to form a clot.

■ A sample of your blood will contain more white blood cells than normal if your body is fighting off an infection. At other times, an abnormally high number of white blood cells could indicate leukemia. This type of cancer affects the tissue that produces white blood cells.

FIGURE 9.3
Blood cells and platelets make up the solid portion of the blood.

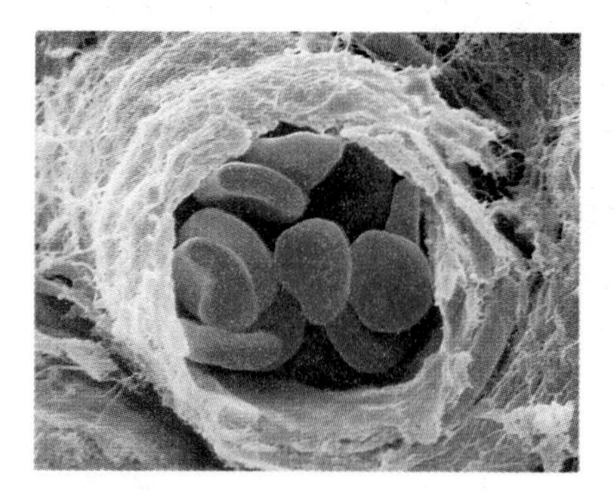

(a) Red Blood Cells
- ***disc-shaped; fit inside smallest blood vessels***
- ***contain hemoglobin, a molecule that can transport oxygen***
- ***most numerous blood cell***

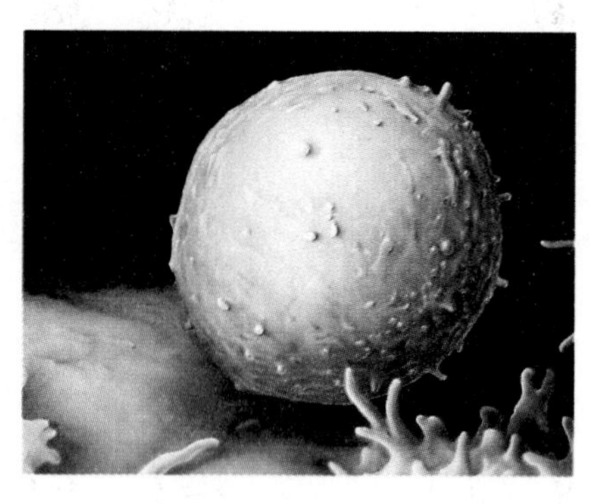

(b) White Blood Cells
- ***large, irregular shape; able to squeeze through the walls of blood vessels***
- ***seek out and consume foreign objects such as bacteria***
- ***much less numerous than red blood cells***

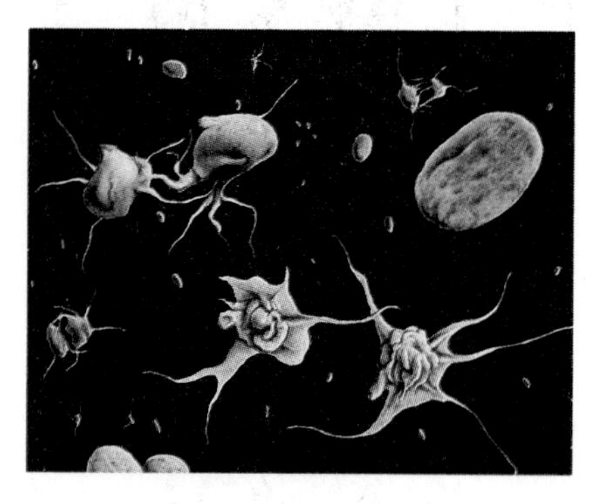

(c) Platelets
- ***much smaller than red blood cells***
- ***simple structure; "little flat bags"***
- ***contain additional clot-forming proteins that, when released, work with blood proteins to stop loss of blood.***

▶ REVIEW 9.1

1. Why do you need a circulatory system?
2. (a) What is plasma?
   (b) List five substances in plasma, and explain why each is important.
3. (a) What is the function of a red blood cell?
   (b) Why is a red blood cell able to perform its function so well?
4. (a) What is a blood clot?
   (b) What parts of your blood help to form blood clots?
   (c) Why is the ability of your blood to clot important to your survival?
5. What two parts of your blood help fight bacteria or other causes of disease?

■ 9.2 YOUR BLOOD VESSELS

Your blood circulates throughout your body in a system of blood vessels. There are three main types (Figure 9.4). The blood vessels carrying blood *away* from your heart are **arteries**. These have thick, muscular walls. The blood vessels returning blood *to* your heart are **veins**. Arteries and veins are connected to each other by **capillaries**, a system of very tiny blood vessels.

Think of your blood vessels as being highways and roads. Your blood is the traffic, and your body cells are the homes. If you were driving home in a car, you would have to leave the highway and drive along smaller streets until you reached your home. Arteries and veins act like highways in your circulatory system. Their function is to move large amounts of blood from place to place. Capillaries act like the smaller streets that take traffic directly to homes. The function of capillaries is to take blood within reach of all your body cells.

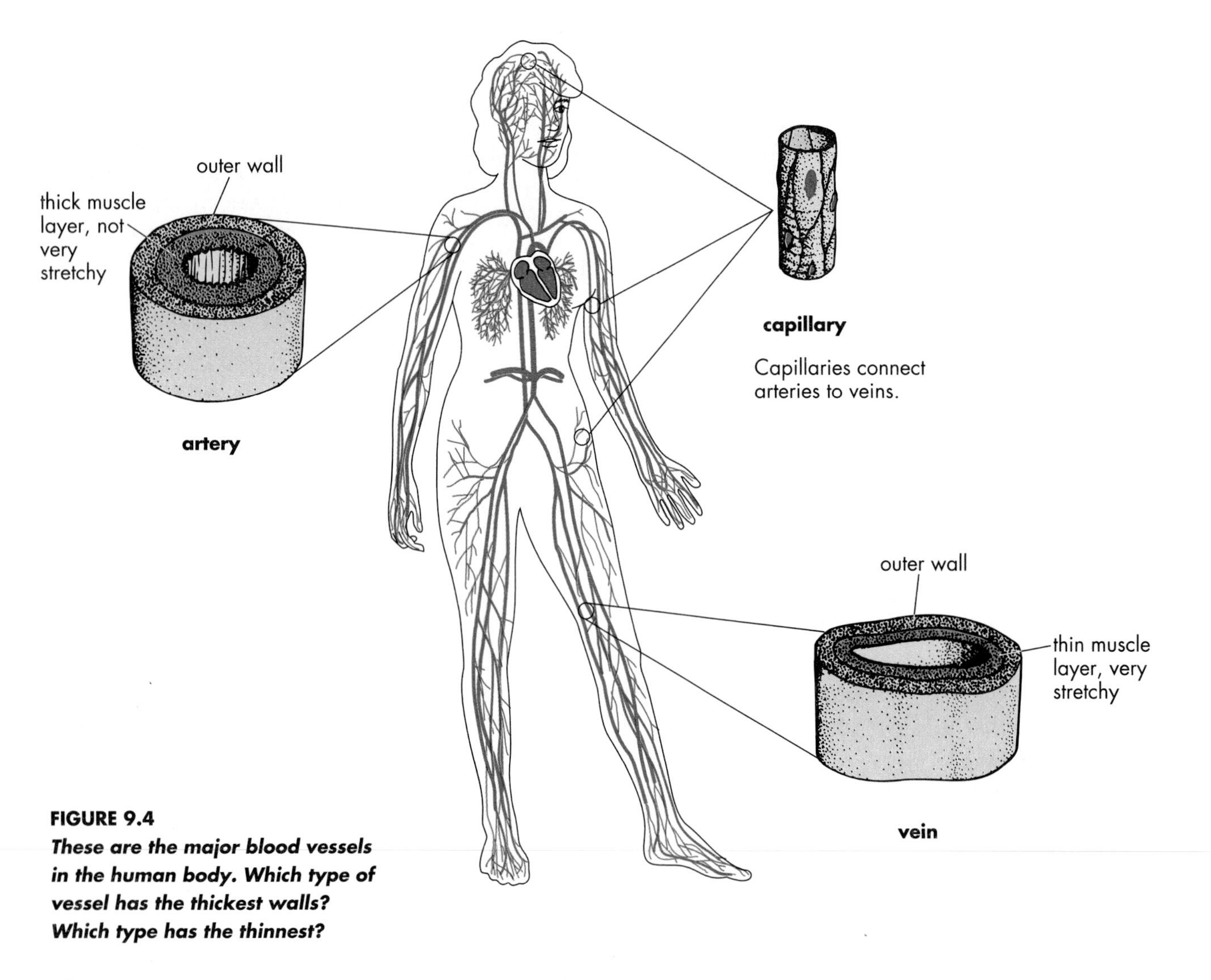

FIGURE 9.4
These are the major blood vessels in the human body. Which type of vessel has the thickest walls? Which type has the thinnest?

HOW BLOOD MOVES THROUGH YOUR BODY

Follow your blood on an imaginary journey through your body, starting from when it leaves your heart. As your heart contracts, it sends a burst of blood into your arteries. The blood in the arteries cannot flow back into the heart because of valves. **Valves** are flaps that open to allow flow in one direction and close to prevent flow in the opposite direction.

Your heart relaxes after each contraction. One cycle of contraction and relaxation is called a beat. Each contraction produces another surge of blood into the arteries. You can feel this rhythm as your **pulse** in any artery close to the surface of your body. Because your pulse is produced by the beating of the heart, it can be used to measure **heart rate**, the number of heartbeats per minute.

The muscular walls of the arteries are stretched when the contraction of the heart pushes blood into them. Then the artery walls squeeze to help move the blood along. The large arteries soon begin to branch into increasingly smaller ones. The artery walls are too thick for substances such as oxygen and nutrients to pass through. Before these substances can reach your cells, your blood must move from the arteries into the smaller capillaries. Capillary walls are very thin. Oxygen and nutrients in the blood easily pass through the capillary walls to reach nearby body cells. At the same time, wastes such as carbon dioxide move from these cells through the capillary walls into your blood (Figure 9.5).

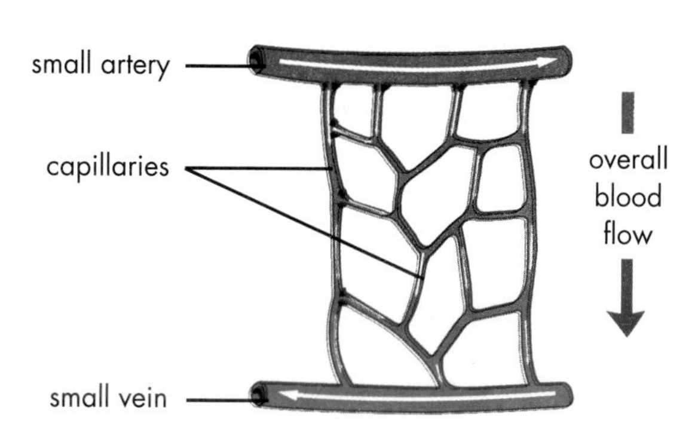

FIGURE 9.5 ◀
Between arteries and veins are tiny, thin capillaries. How do you think the thinness of capillary walls helps to transfer nutrients to cells?

As your blood passes through the maze of tiny capillaries, it loses most of the push it received from the beating of your heart and the squeezing of your artery walls. Blood from the capillaries begins the journey back to your heart by draining into small veins. The walls of your veins are elastic (Figure 9.4, inset). They stretch when they are full of blood. Unlike arteries, they are not lined with powerful muscle, so veins cannot squeeze to help push blood along. How then is blood returned through your veins to your heart? The return trip is accomplished by the large muscles of your body, such as those in your legs. When these muscles contract, they squeeze nearby veins, pushing the blood through the veins toward your heart. One-way valves inside the veins stop any blood from moving in the opposite direction (Figure 9.6).

■ The blood in the arteries travelling to your body cells is brighter and redder than the blood in the veins leaving your body cells. This is because hemoglobin is only bright red when it is saturated with oxygen (carrying as much oxygen as possible).

■ People who have jobs where they must remain still for long periods may get varicose veins. Find out the cause of this condition and the treatment available.

FIGURE 9.6 ▶
Any blood trying to move backward in a vein pushes the valve closed. This makes sure that blood travels only toward the heart. Where else do valves prevent blood from moving backward in blood vessels?

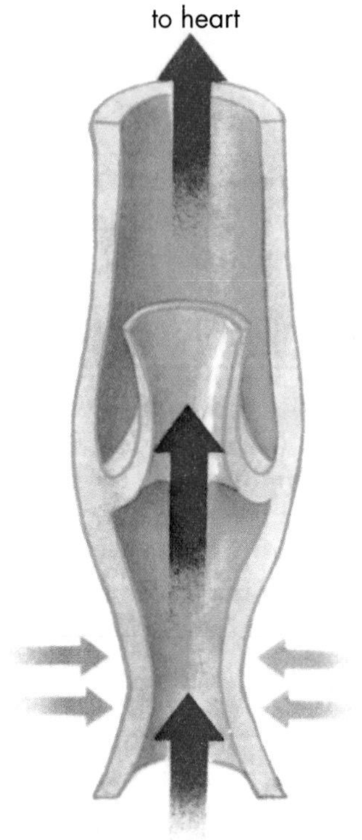

Blood being pushed toward heart when squeezed by body muscles.

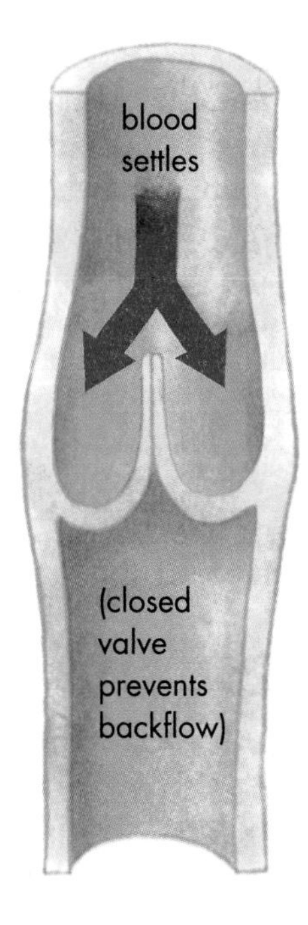

Blood cannot flow backward when body muscles relax.

ACTIVITY 9B / ARTERIES AND VEINS

The rhythmic squeezing of the arteries with each heart contraction is called your pulse. In Part I of this activity, you will feel your pulse at some of the locations shown in Figure 9.7. In Part II, you will learn about veins.

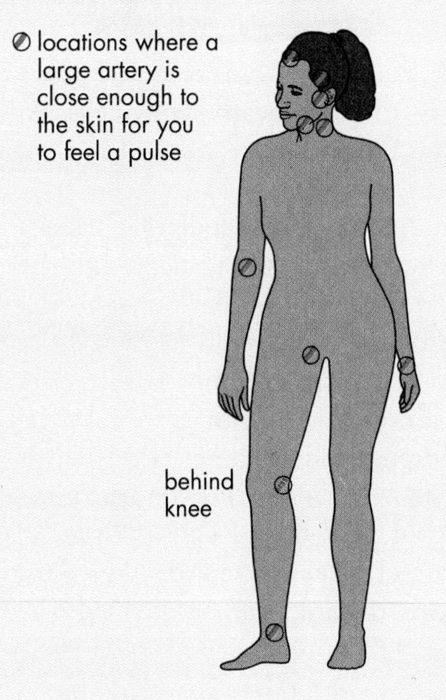

FIGURE 9.7 ▶
Points on the human body where the pulse can be taken. (Only one of each pair of points is shown here.)

PART I Finding Your Pulse

PROCEDURE

1. Find the pulse in your wrist, as shown in Figure 9.8. Use two or three fingers, not your thumb, to feel your pulse. Then find your pulse in the other wrist.
2. Check for a pulse at the side of your forehead, jaw, neck, elbow, and ankle, using Figure 9.7 as a guide. The pulse at your ankle may be found most easily by crossing one leg over the other and changing the angle of your foot.
3. Use one hand to find your ankle pulse. At the same time, use your other hand to locate the pulse in your neck.

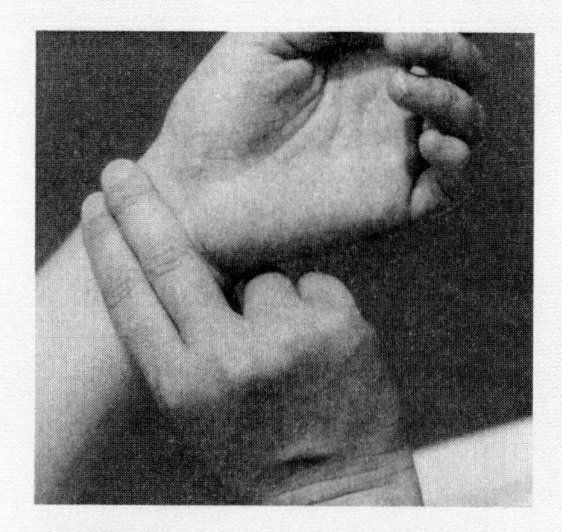

FIGURE 9.8
Locating the pulse in your wrist. Use the first two or three fingers on one hand to gently press the artery, as shown. If you have difficulty feeling a pulse, change the angle of your wrist slightly and try again.

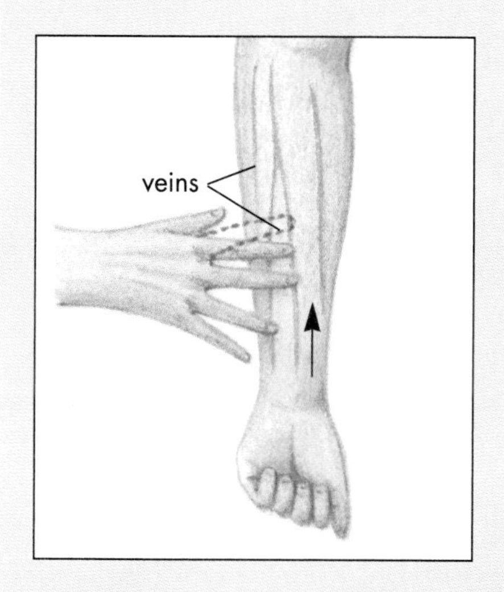

FIGURE 9.9
Select any vein on the underside of your arm that you can see clearly. If you have trouble locating a vein there, use one on the back of your hand.

Concentrate carefully and note when the pulse occurs in each location. Record which pulse occurs first and which appears to be stronger.

4. If a stethoscope is available, determine whether your pulse occurs at the same time as your heartbeat. Try this using your wrist pulse and then your ankle pulse.

DISCUSSION

1. (a) Which occurs first, the pulse in your ankle or in your neck?
   (b) Why do you think these pulses occur in this order?
2. Explain why your pulse can be used to measure how quickly your heart is beating.
3. Why do you think you can sometimes hear your pulse in your head?

PART II Your Veins

Note: If you are unable to observe the changes to your vein, make a fist, try another location, or observe another student.

PROCEDURE

1. Find a straight section of vein on the back of your hand or on your lower arm (palm facing you). You can identify a vein by its bluish color. Press one finger firmly over the vein.
2. Keeping your first finger where it is, push the blood along the vein toward your elbow with a second finger. Record what happens to the appearance of the vein between your two fingers (Figure 9.9).
3. Release your second finger, the one closest to your elbow. Note what happens.
4. Now release your remaining finger. Again, note what happens.

DISCUSSION

1. What did you discover about the direction of blood flow through your veins?
2. Why do you think the blood flows in only this direction?
3. What can you infer about the structure of veins?
4. If someone were bleeding from a vein, where could you apply pressure to slow or stop the flow of blood? Explain.
5. How could you decide if someone were bleeding from an artery or a vein? ❖

CARRYING OXYGEN AND WASTES

Think about a highway carrying traffic to and from a large city. The highway has two sets of lanes, those going toward the city and those leaving the city. The highway separates traffic and goods as they travel to or from the city.

How is this traffic system like your circulatory system? Blood also moves in a system that separates substances flowing to your cells from wastes leaving the cells (Figure 9.10). When you inhale, you bring oxygen into your body. In the alveoli of your lungs, this oxygen enters your blood. This blood, rich in oxygen, is called **oxygenated blood**. It travels to the left side of your heart where it is pumped into large arteries. The arteries carry the oxygenated blood throughout your body. When the blood enters your capillaries, it releases oxygen to the body cells and picks up carbon dioxide.

At this point, the blood has less oxygen and is called **deoxygenated blood**. (Oxygenated blood and deoxygenated blood are represented by different colors in the figures in this chapter. Blue is used for deoxygenated blood.) Deoxygenated blood travels to the right side of your heart through the veins. Your heart sends the deoxygenated blood to your lungs to begin the process again. In this way, oxygenated and deoxygenated blood are kept separate.

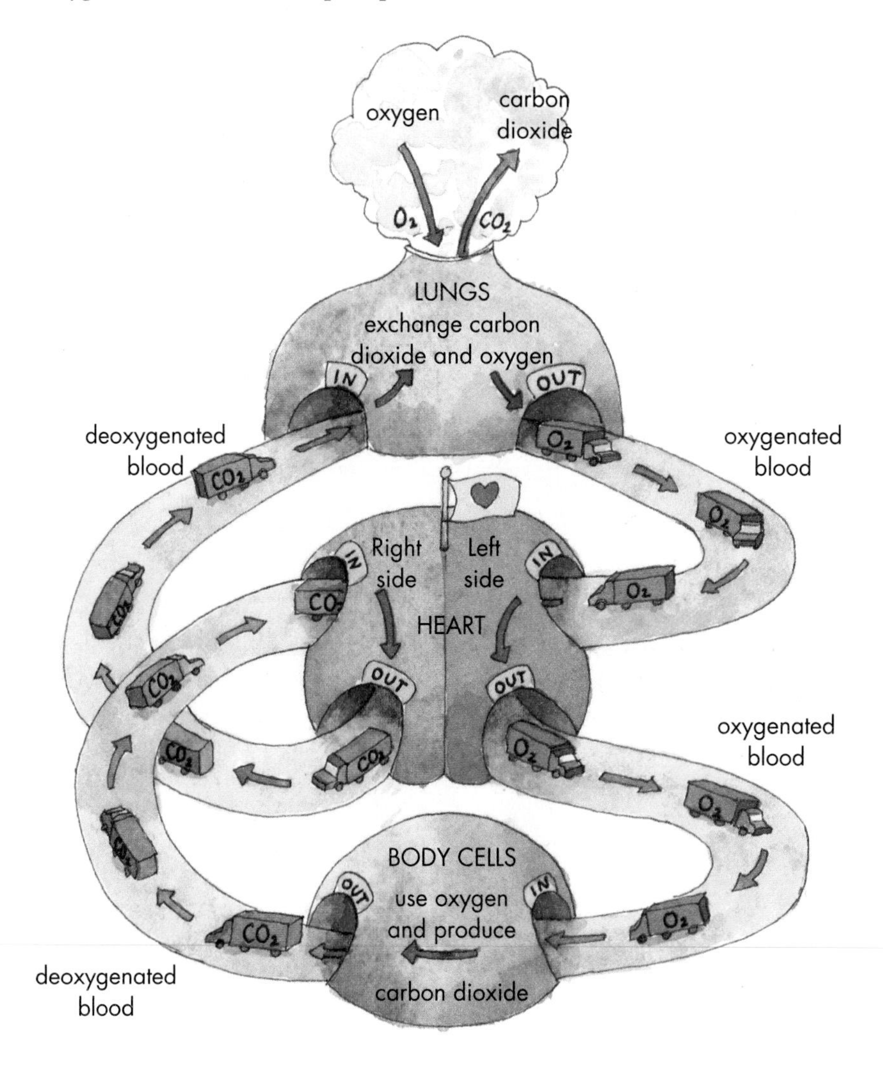

FIGURE 9.10 ▶
This diagram shows how oxygenated blood and nutrients flow to your body cells and how wastes are carried away. Trace these two routes with your finger. Where do they join together?

BLOOD AND BODY TEMPERATURE

As well as carrying substances such as oxygen and carbon dioxide, your circulatory system transports heat energy throughout your body. When you warm your hands over a fire, you might have noticed that your arms soon feel warmer. The blood in your hands has picked up heat energy through your skin. As the warm blood travels up your arm, some of this heat energy warms the cells of your arm.

Your body constantly produces its own heat energy. The chemical reactions (such as cellular respiration) that release chemical energy also release a tiny amount of heat energy. This heat energy passes into your blood and is carried throughout your body. The amount of heat that reaches any body part is controlled by the amount of blood travelling through its capillaries. For example, when you have a fever or when you exercise, the capillaries under your skin receive more blood than usual. This warms the skin and allows more body heat to be lost.

REVIEW 9.2

1. Make a three-column table. In the first column, list the three types of blood vessels in your circulatory system. In the second column, briefly describe the structure of each type. In the third column, explain the function of each type.
2. Your blood stays inside the vessels of your circulatory system. How is blood able to bring some substances to your cells and remove others?
3. (a) What are valves?
   (b) Why do you need valves in your veins?
   (c) Why do you need valves between your heart and your arteries?
4. Imagine holding a warm cup. Explain how the heat energy from the cup helps warm your whole body, not just your hands.

9.3 YOUR HEART

As you can see in Figure 9.11, a heart is mostly muscle. The job of this muscle is to keep the blood moving through the circulatory system. In order to do this, the heart muscle constantly beats by first squeezing (contracting) and then relaxing. Your heart normally beats about 65 to 75 times every minute. The muscle of the heart is called cardiac muscle. ("Cardiac" comes from a Greek word meaning heart.)

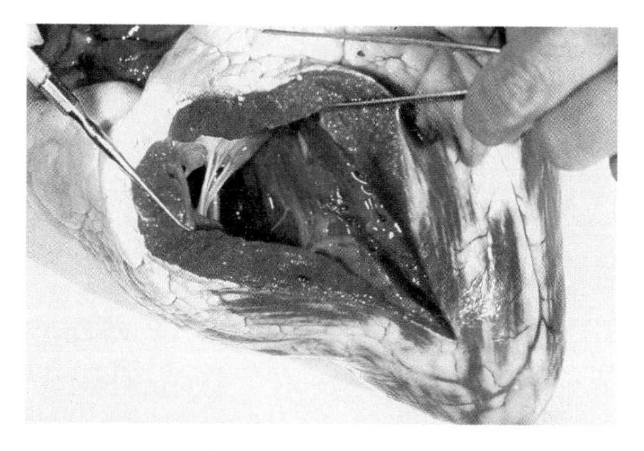

FIGURE 9.11
This beef heart is very similar to a human heart. Notice the thickness of the muscular walls of the heart. Why does the heart need so much muscle?

THE STRUCTURE OF YOUR HEART

Your heart contains four compartments, or chambers, that hold blood (Figure 9.12). The movement of blood in and out of each chamber is controlled by valves that act like doors opening in only one direction. The **atria** (singular: **atrium**) are the two chambers at the top of your heart. They are called the right atrium and the left atrium. The right and left **ventricles** are the two chambers at the bottom of your heart.

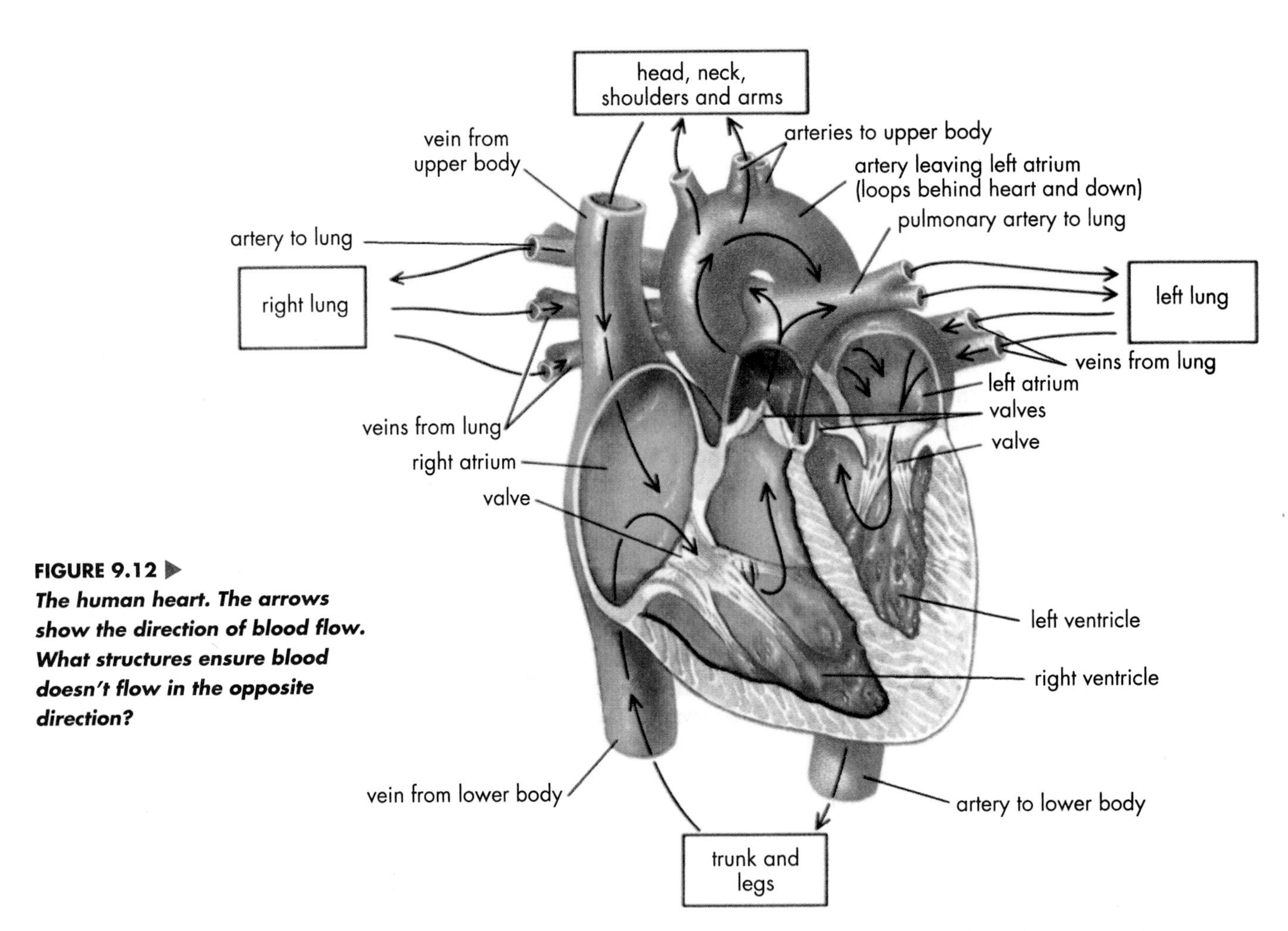

FIGURE 9.12 ▶
The human heart. The arrows show the direction of blood flow. What structures ensure blood doesn't flow in the opposite direction?

Figure 9.13 shows what happens during a heartbeat. The heart is relaxed in Figure 9.13a. When the muscles of the heart contract, they squeeze the blood inside the chambers of the heart. The direction in which the blood moves is controlled by valves. Here is how it works. The walls of the atria contract first during a heartbeat (Figure 9.13b). At the same time, valves open to allow blood to move from each atrium to the ventricle beneath it.

The ventricles contract after the atria (Figure 9.13c). The muscles surrounding the ventricles are more powerful than the muscles around the atria. The ventricles need this power to push the blood into and along the arteries. The valves between the atria and ventricles close to prevent blood from moving backward into the atria. The valves leading into the arteries are open so that blood can move into the arteries.

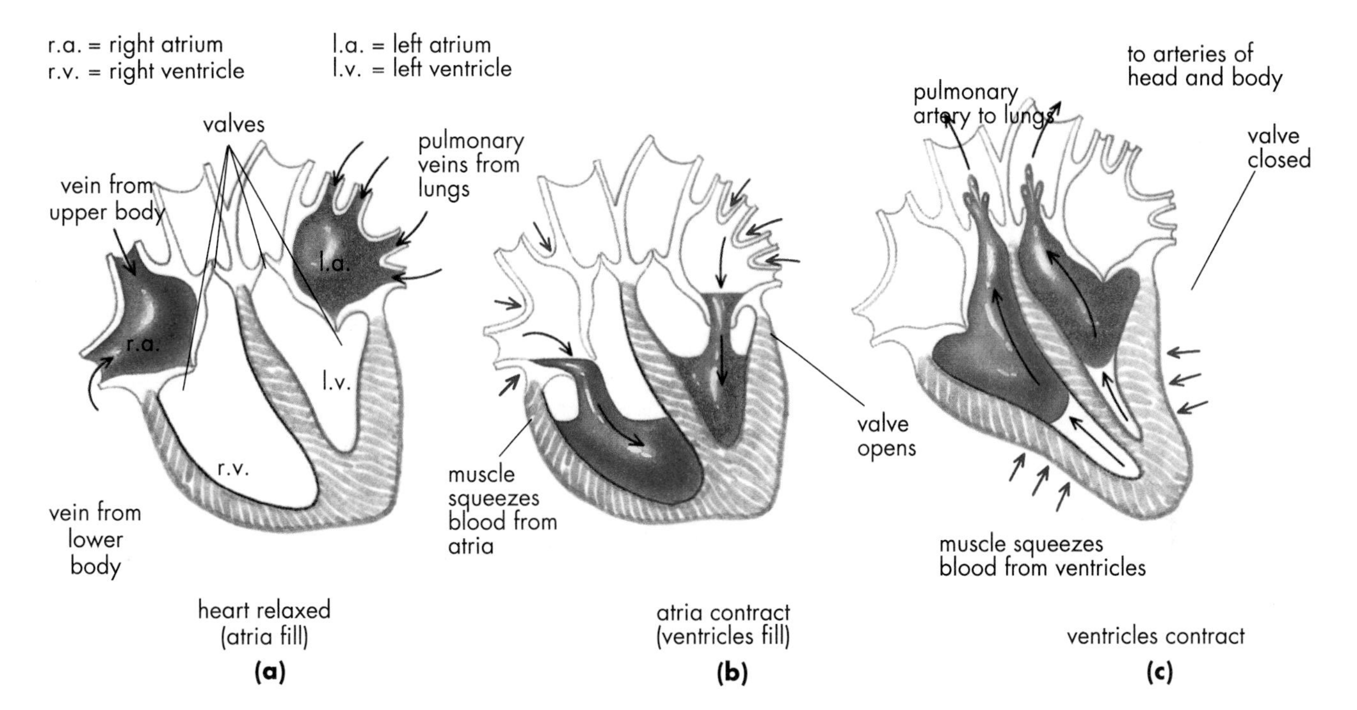

FIGURE 9.13
One heartbeat. The atria of the heart contract slightly before the ventricles. This moves blood from the atria into the ventricles. What happens when the ventricles contract?

THE MOVEMENT OF BLOOD

Figure 9.14 shows the path that blood takes as it enters and leaves your heart. The figure also shows the system of blood circulation throughout your body. The blood entering the right side of your heart is deoxygenated blood. The blood leaving the left side is oxygenated blood. As you read the following summary, trace the movement of blood through the diagram with your finger. The first five steps show how deoxygenated blood leaves your heart, goes to your lungs, and returns as oxygenated blood.

1. Veins from the upper and lower body bring deoxygenated blood into the right atrium of the heart.
2. When the atria contract, the deoxygenated blood in the right atrium is pushed into the right ventricle.
3. When the ventricles contract, the right ventricle pushes the deoxygenated blood to the lungs. Blood travels from the right ventricle to the lungs through two arteries called pulmonary arteries. (These are the only arteries in your body that carry deoxygenated blood.)
4. In the lungs, blood picks up oxygen and releases carbon dioxide.
5. Oxygenated blood from the lungs is carried by veins—the pulmonary veins—to the left atrium. (These are the only veins in your body that carry oxygenated blood.)

■ If you listen to your heart using a stethoscope, you will hear two distinct sounds making up each heartbeat. The first sound, a lower-pitched "lub," is heard when your ventricles start to contract. The second sound, a "dub," is heard at the end of their contraction.

DID YOU KNOW

■ The muscles of the heart are so powerful that in one day your heart could pump enough liquid to fill a 10,000 L tanker truck. What can you do to ensure that the muscles of your heart stay in excellent shape throughout your lifetime?

6. When the atria contract, the left atrium sends this oxygenated blood into the left ventricle.
7. When the ventricles contract, the left ventricle pushes oxygenated blood into the aorta, a very large artery that divides into branches leading to your upper and lower body.
8. In your body, oxygenated blood flows from arteries into capillaries. The blood releases oxygen to body cells and picks up carbon dioxide.
9. Veins from the upper and lower body bring deoxygenated blood into the right atrium of the heart. The cycle continues . . .

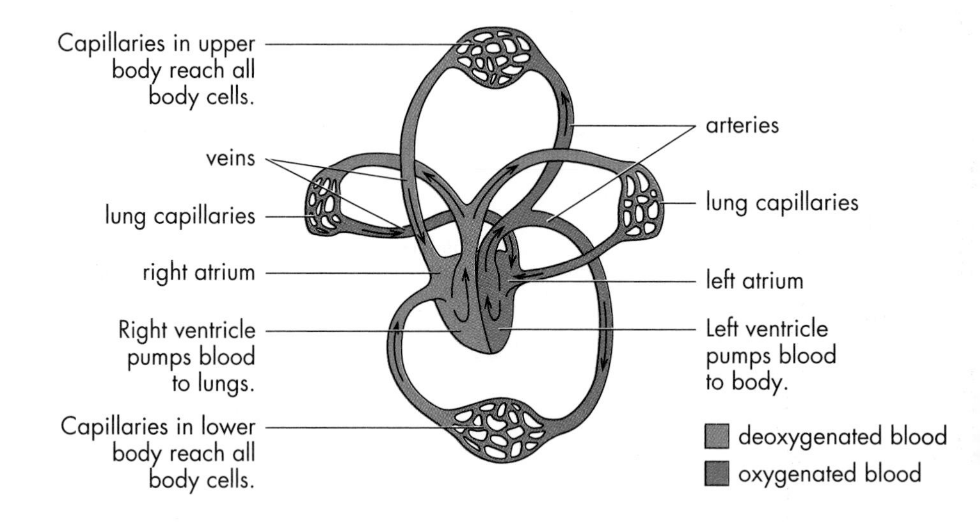

FIGURE 9.14 ▶
Reading about a round trip through your circulatory system takes far longer than the real thing. The average time for your blood to make one complete trip — from right atrium and back again—is only one minute.

THE HEART'S OWN BLOOD SUPPLY

You have learned how blood passes through the four chambers of your heart. The blood in these chambers cannot bring nutrients and oxygen to the cells of the heart muscle. These substances can only reach cells through the thin walls of capillaries. Therefore your heart has its own system of arteries, capillaries, and veins (Figure 9.15).

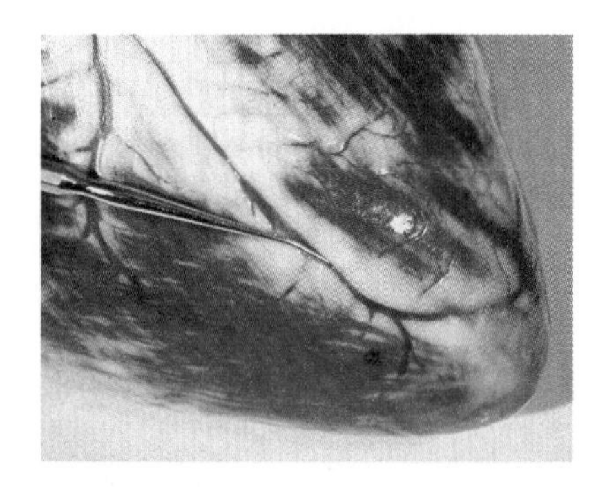

FIGURE 9.15
Find the blood vessels on this beef heart. Why does the heart need its own system of arteries, capillaries, and veins?

HEART RATE

Your heart rate does not stay exactly the same for very long. Your nervous system automatically adjusts your heart rate to meet the changing needs of your body. For example, when your muscles are working unusually hard, such as when you run, your heart rate increases. This increases the supply of oxygenated blood reaching your muscles, as well as speeding up the removal of deoxygenated blood. When your body's needs for oxygen decrease, such as when you sleep, your heart beats more slowly and gently—resting itself.

Just like any other muscle, your heart becomes stronger and healthier with proper exercise. In Chapter 10, you will examine how your heart responds to regular physical activity.

HEART MONITORS

Bicyclist Andy Cowan puts a watch-like gadget on his wrist and straps two electrodes to his chest. With this computerized heart monitor in place, Cowan starts pedalling. By glancing at the gadget on his wrist from time to time, he knows that his heart rate is increasing. When the monitor beeps, he has reached the "target zone" — the heart rate he has pre-programmed into the monitor.

First thing in the morning before you get up, your heart rate is at its lowest, probably about 50 to 60 beats per minute. Race up the school steps and your heart will beat considerably faster.

Heart rate is a good way to measure how hard a person is working. "The heart monitor really lets you look at your physical effort," says Cowan. The monitor indicates when he is not working hard enough or when he is working his heart too much.

At a major Sports Science Training Center, Delia Roberts works with national team swimmers and speed skaters. She is an exercise physiologist, someone who studies how the body adapts to exercise. Her athletes, too, wear heart monitors during workouts. After the workout, Roberts transfers the heart rate data into her computer to analyse them and generate graphs. A rate that is too high may indicate that the athlete is ill, or is overtrained and needs some rest.

Overtrained? Is it possible to work too hard?

"Elite athletes are geared to push all the time. They tend to forget that they also have to train at the lower heart rate. That's crucial for their recovery system," says Roberts. In between heavy training sessions, the body needs recovery time to adapt, build new muscle tissue, and become stronger. If athletes always train at their upper limit, no energy is left for this recovery. A workout at

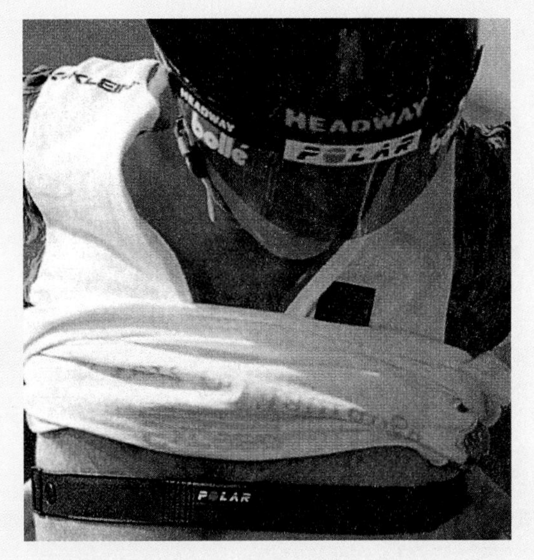

An athlete using a heart monitor during training.

about 120 beats per minute seems easy to a serious athlete, who will automatically work harder unless reminded by the heart monitor beeper to slow down.

Heart monitors are only one of the many ways computers have sharpened training methods and improved athletes' performances. Computers can record and analyse a gymnast's exact position during a somersault. They can measure how much force ski-jumpers apply during their take-off from the ramp. They can measure exactly where a swimmer is losing split seconds— on the starting dive, the recovery from the dive, or the turnaround.

For a recreational athlete like Andy Cowan, a heart monitor is not essential. However, long workouts on a rowing or bicycling machine can be very boring, and the changing numbers on his wrist give him something to focus on.

Find out how top athletes at your school train for their sport. How do they know when they are training too hard?

ACTIVITY 9C / CHANGING THE BEAT

In this activity, you will investigate how fitness affects changing heart rate.

■ **Students who are not able to take part in physical activity should not do this activity. They should work with a group to help with recording data.**

■ **Be sure to wear running shoes, and to stretch for a couple of minutes before proceeding to step 4 of this activity.**

Heart rate (beats/min)

150

100

50

0 (resting) 1 2 3 4 5

Total number of minutes running on the spot

FIGURE 9.16
Your graph will have a total of four lines. Be sure to label each one.

PROCEDURE

1. As a class, decide on three categories of fitness you will use to make up groups that represent low, medium, and high levels of fitness. For example, you may use "low physical activity" for students who rarely exercise, "medium physical activity" for students who exercise twice a week, and "high fitness activity" for students who exercise daily. Consider how long the physical activity should be for it to be counted as exercise. Work in groups of two or three students who have the same level of physical activity.

Amount of time running (min)	**My heart rate (beats/min)**	**Average heart rates (beats/min)**		
		low	medium	high
Resting				
1				
2				
3				
4				

TABLE 9.2 ***Sample Data Table for Activity 9C***

2. Copy Table 9.2 into your notebook and use it to record your heart rate measurements.

3. Sit down. Use Figure 9.8 to help you locate your wrist pulse. Use this pulse to obtain your resting heart rate as follows. Count the number of beats in 15 s, then multiply by 4 to obtain the number of beats per minute. Record this value in your chart as your resting heart rate.

4. Run on the spot for 1 min. Immediately sit down and obtain your heart rate (as in Step 3). Record this figure in your table as your heart rate after 1 min.

5. Repeat Step 4, running on the spot for 1 min. Record this figure as your heart rate after 2 min of exercise. Continue to repeat until you have values for 3, 4, and 5 min of exercise.

6. Using the data recorded on your table and the data for the other person or persons in your group, calculate the average heart rate for each exercise time for your group.

7. Share your averages with two other groups so that you have data on groups from the three fitness levels.

8. Using graph paper, plot a graph like the one shown in Figure 9.16. Show your heart rate and the average heart rates for the three levels of fitness. Use a different color for each line on your graph.

DISCUSSION

1. Why does your heart rate increase after exercising?
2. In this activity, why did you sit down and take your pulse in the same way each time?
3. What patterns are visible when you compare the lines produced by graphing the heart rates of the three groups of students?
4. Explain any differences in the changes in heart rate among the three fitness groups, based on your graph.
5. How does the line showing your heart rate compare with the three other lines on your graph? Suggest what this means.
6. Do you think the results of this activity are important for people to know about? Explain why or why not. ❖

REVIEW 9.3

1. (a) Describe the structure of your heart.
   (b) How does this structure help the heart's function of pumping blood?
2. (a) Which half of your heart pushes blood to your lungs? Explain.
   (b) Which half of your heart pushes blood to the rest of your body? Explain.
3. Why is it important to your survival that your heart rate can change so that your heart beats faster or slower as needed?

9.4 KEEPING YOUR CIRCULATORY SYSTEM HEALTHY

Many of the problems that affect a person's circulatory system can be prevented by the same steps that lead to overall good health. Not smoking is one important step to a healthy heart. A nutritious diet, regular exercise, and enough rest will help the muscles of the heart just as they help the muscles of the rest of your body. The main illnesses of the circulatory system are heart attack, atherosclerosis, and high blood pressure.

HEART ATTACK

A **heart attack** occurs when the cells of the heart do not receive enough blood. If the heart's own arteries are blocked, blood cannot reach the muscle cells of the heart, cutting off the supply of nutrients and oxygen. If this supply is cut off for too long, the cells of the heart muscle no longer work properly and some begin to die. The person experiences a heart attack. Some heart attacks are mild and others are severe, depending on how many cells die and how much of the heart is damaged.

DID YOU KNOW

■ A stroke is the result of damage to an area of the brain. It is caused when the flow of blood to that area is stopped. High blood pressure increases the risk of stroke by straining the blood vessels that carry oxygen to the brain.

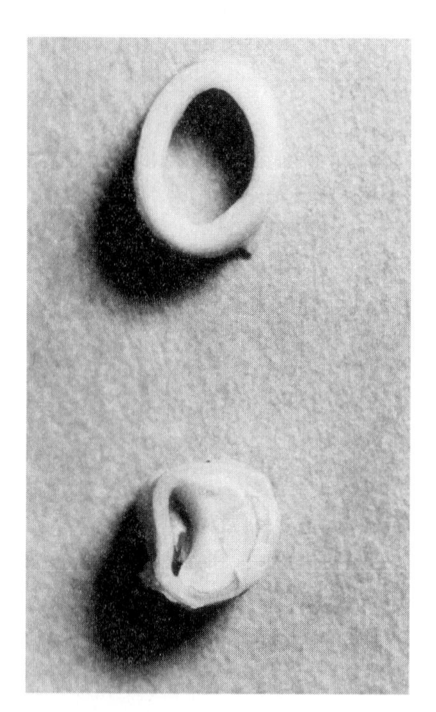

FIGURE 9.17
Atherosclerosis. The lower of the two vessels in this picture is clogged with cholesterol. What do you predict will happen to the flow of blood when blood vessels become clogged?

Most people who have had heart attacks recover. Heart muscle can repair itself, given time and rest, if it has not been damaged too badly. It is important to try to eliminate the cause of the heart attack. For example, the nicotine from cigarette smoke causes the heart to pump unusually quickly. Nicotine also closes up certain blood vessels. This strains the entire circulatory system, including the arteries that supply the heart muscle.

ATHEROSCLEROSIS

Atherosclerosis affects the blood vessels, especially the arteries. Excess cholesterol and fats stick to the inner surface of the arteries (Figure 9.17). This makes it harder for blood to move through the arteries. Eventually, an artery may be completely plugged. If this happens in an artery that carries blood to the heart muscle, a heart attack can result. You can help to prevent atherosclerosis by eating foods with less animal fat.

HIGH BLOOD PRESSURE

When a doctor takes your blood pressure, he or she is interested in how much force your heart must exert to push your blood through your arteries. Like heart rate, blood pressure normally changes from time to time. When your heart is working harder, for example, your blood pressure usually rises a bit. When you relax, your blood pressure decreases. In general, the lower a person's average blood pressure is and the more quickly it returns to a lower value after rising, the better the circulatory system is working. There are, however, medical problems with blood pressure that is *too* low.

A person with **high blood pressure** has a blood pressure that is higher than normal all or most of the time. With high blood pressure, the heart has to work much harder, and the smallest arteries and the capillaries can burst. Organs such as the heart, brain, and kidneys can be damaged. High blood pressure is one of the leading causes of heart attack. Too much salt in the diet can contribute to high blood pressure.

PERSONAL CHOICE

You might think that only older people should worry about heart attacks and other circulatory illnesses. Unfortunately, the causes of these illnesses start when people are your age or even younger. There is a definite connection between smoking and heart disease. Many people with blocked arteries are smokers or people who were frequently exposed to cigarette smoke. Fortunately, if a person stops smoking, the risk of heart disease and death will eventually decrease to the same levels as in non-smokers.

The choices you make in the next few years will make a major difference to your health in the future. You can choose a way of life that helps you keep your circulatory system in good health, reducing your chances of heart disease or heart attack. For example, regular exercise helps you develop more and larger blood vessels to supply the heart. Exercise also strengthens the heart muscle. Choosing not to smoke removes one of the major risk factors of heart disease. Reducing stress in your life may also help your circulatory system serve you better and longer.

The food you eat can make a great difference to the health of your circulatory system. The American diet often includes too much salt and fat, especially animal fat. Fast foods, such as hamburgers, french fries, and milkshakes, are usually high in fat and salt. Many packaged foods have a great deal of added salt. Look on the labels for salt (or sodium). A nutritious diet has less fat and less salt than the typical American diet.

REVIEW 9.4

1. Some people are born with a condition in which their body produces more cholesterol than it needs. This cholesterol is deposited in their blood vessels. What problem could this cause? Explain your answer.
2. What can you do to help keep your circulatory system healthy?

9.5 KEEPING THE BALANCE

There is a constant balancing act taking place inside your body. The substances carried by your blood must be present in exactly the right amounts in order for your body cells to function properly. Think of these substances as "chemical traffic" moving through your body (Figure 9.18). The liver and the excretory system act as the blood's "traffic controllers."

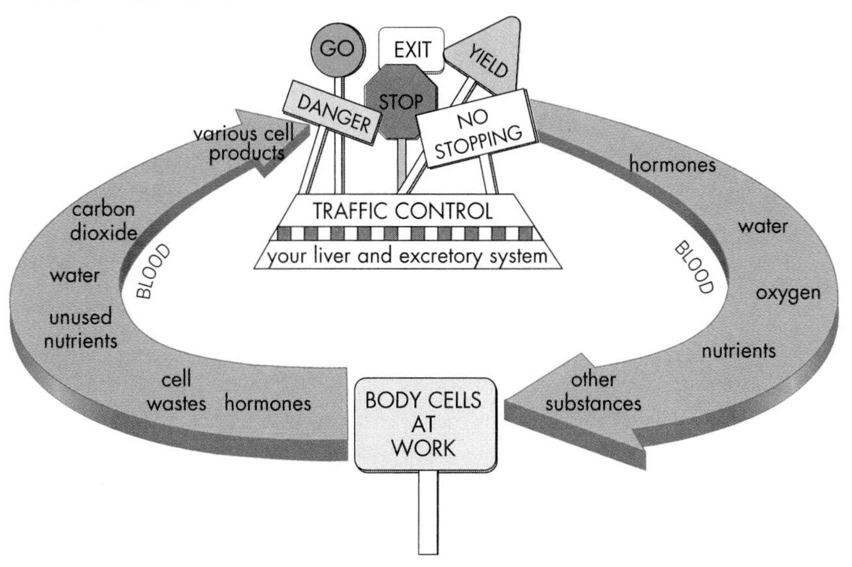

FIGURE 9.18 ◀ ***Because your blood's chemical make-up is controlled by the liver and excretory system, you can think of these body parts as a kind of traffic control for your blood.***

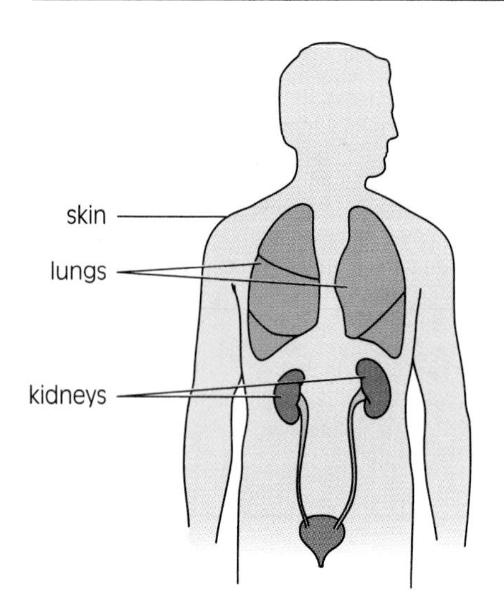

FIGURE 9.19
Your skin, lungs, and kidneys. What function do they all perform?

THE ROLE OF THE LIVER

Your **liver** is a large organ made up of cells that act on the substances found in the blood. Every substance that enters your blood passes through your liver. Your liver acts on the substances in your blood in the following ways. (These are just some of the functions of the liver.)

- Your liver can change carbohydrates into glycogen and fats if your body does not need the carbohydrates for energy immediately. These reserve energy supplies are stored until you need them.
- Dangerous substances are removed from your blood and changed by your liver to less harmful forms. For example, the ammonia produced as cells break down protein molecules is converted to a less harmful substance, urea.
- Old red blood cells are destroyed by your liver. Parts of these waste cells are used by the liver to make bile for the digestive system. Other parts are recycled into new red blood cells.

Your liver is like a processing center. It can change some of the substances that are in your blood into other forms. To remove wastes and unwanted substances from your blood, and so from your body, you rely on your excretory system.

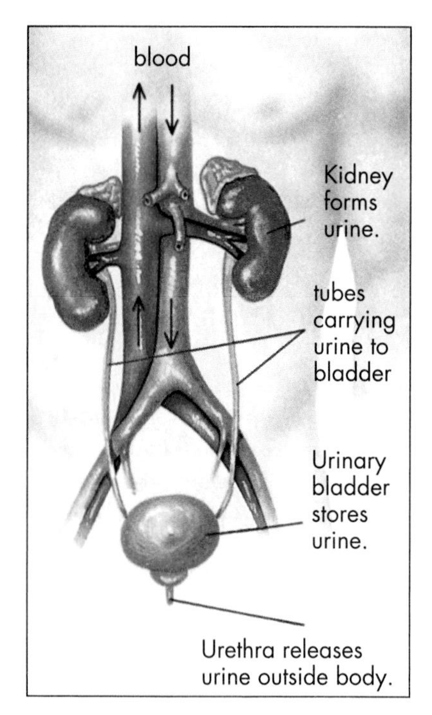

FIGURE 9.20
The kidneys. From your kidneys, urine travels to the urinary bladder, where it is stored until it can be released from your body through the urethra.

EXCRETION

Think about a balancing act again. Your body has to balance what is taken in by your cells with what is removed from your cells. When your cells perform their normal functions, they produce wastes. If these wastes were left in your blood, the balance would tip. For example, water is a waste product of cellular respiration (see Chapter 8). Your blood carries away this excess water. As it moves with your blood through your circulatory system, this water serves many functions, including dissolving salts and urea so that these waste substances can be removed from the body. But too much water in the blood is dangerous, because it affects how substances move in and out of cells. So there has to be a way to remove water from the body as well. There has to be a balance between *in* and *out*.

Excretion is the process of removing excess water, salts, and wastes from the body. Your **excretory system** includes your lungs, skin, kidneys, and associated organs (Figure 9.19). You've already learned that some water leaves your body through your lungs. This water helps moisten the air entering your body. Water and salts also leave your body when sweat evaporates on your skin. This evaporation also helps to cool your body.

Your **kidneys** are responsible for most of the excretion that takes place in your body (Figure 9.20). The kidneys are a pair of organs that filter your blood. A filter holds back some substances while letting others pass through. As blood moves through the kidneys, the kidneys keep substances needed by body cells, such as salts, carbohydrates, and most of the water. The kidneys return these to the blood (Figure 9.21). Wastes, such as excess salts and urea, pass through the kidneys' filtering system and collect together in the form of urine. **Urine** is a mixture of wastes and water.

The urine produced by your kidneys travels to a balloon-like organ, the **urinary bladder**. The bladder is able to stretch. It stores the urine until it can be released from the body through the **urethra**, shown in Figure 9.20.

FIGURE 9.21
Your excretory system balances the substances in your blood by removing some substances and recycling others.

KEEPING A HEALTHY EXCRETORY SYSTEM

Your liver and excretory system work constantly to balance the chemicals in your body. For example, after you have eaten a meal, your liver distributes the nutrients from the food to be used or stored until they are needed. When you exercise and produce more wastes than usual, your excretory system removes these wastes from your body more quickly.

This means that your level of activity affects your liver and excretory system, just as it affects the other systems in your body. It is certainly easier to feel how your heart rate changes than it is to know how well your liver or kidneys are working. Since all your body systems are interconnected with one another, you can be sure that what is proper care for one is proper care for them all.

ACTIVITY 9D / MAKING CONNECTIONS

Each of your body systems does a different job to keep you healthy. For example, your respiratory system brings in oxygen and removes carbon dioxide. Your digestive system obtains nutrients from food. Your blood links these systems together by moving substances from one part of your body to another.

PROCEDURE

1. Copy Table 9.3 in your notebook.
2. For each function in the table, write in the second column the body systems that would be involved: respiratory, digestive, excretory, and circulatory. (You can use more than one.)
3. In the third column, list all the substances that are needed or produced when each function takes place in the system. Remember, the parts of every body system are made up of living cells too.
4. Circle those substances that are carried by your blood. ❖

Body function	**Body system involved**	**Substances involved**
Removing waste	Excretory system Digestive system Respiratory system	Water, urea, salts Undigested food, fiber Water, carbon dioxide
Obtaining nutrients		
Obtaining oxygen		
Storing nutrients		

TABLE 9.3 ◀
Body Systems and Blood

REVIEW 9.5

1. The liver is one of the largest organs in your body. Why do you think this is so?
2. (a) Name three waste substances produced by your cells.
   (b) What process produces each substance?
   (c) Why must wastes be removed from your body?
3. What three parts of your body are involved in excretion?
4. How are your kidneys like filters?

CHAPTER REVIEW

KEY IDEAS

- The circulatory system consists of blood, blood vessels, and the heart.
- Blood is composed of plasma and blood cells. Plasma transports many substances, including nutrients and blood proteins, throughout the body. Oxygen is carried by red blood cells.
- Three types of blood vessels carry blood: arteries, veins, and capillaries. Arteries carry blood away from the heart and veins carry blood back to the heart. Nutrients and wastes move between the blood and body cells only through the thin-walled capillaries.
- The heart consists of two atria and two ventricles. The left side of the heart receives oxygenated blood from the lungs and sends it to the body. The right side accepts deoxygenated blood from the body and sends it to the lungs.
- The liver acts to control the substances present in blood.
- The excretory system includes the skin, lungs, and kidneys. The kidneys filter the blood, removing urea, salts, and water to form urine.
- Your circulatory and excretory systems are affected by choices, such as what you eat, how much you exercise, and whether you smoke.

VOCABULARY

blood vessel
circulatory system
plasma
blood protein
antibody
immune system
hormone
clot-forming blood protein
red blood cell
hemoglobin
white blood cell
platelet
artery
vein
capillary
valve
pulse
heart rate
oxygenated blood
deoxygenated blood
atrium (plural: atria)
ventricle
heart attack
atherosclerosis
high blood pressure
liver
excretion
excretory system
kidney
urine
urinary bladder
urethra

V1. Write the letters a to g in your notebook. Match the substance in Column A with where it is found in your body from Column B (see Table 9.4 below). Do not write in this book.

Column A	Column B
(a) Hemoglobin	Plasma
(b) Clot-forming blood proteins	Left ventricle
(c) Antibodies	Red blood cells
(d) Deoxygenated blood	Right atrium
(e) Oxygenated blood	Platelets
(f) Urea	Right ventricle
(g) Urine	Liver
	Kidney

TABLE 9.4 ***Table for Question V1***

V2. Divide a page of your notebook into four sections. In each section, write one of these terms: digestive system, respiratory system, circulatory system, and excretory system. Then, use words from Chapters 7, 8, and 9 to make a mind map that shows how these systems are connected to one another.

CONNECTIONS

C1. (a) Why do you need a transportation network inside your body?
(b) Name at least three substances brought into the body that are transported by your circulatory system.
(c) Where is each of these substances being taken by the blood?

C2. Write the letters a to f in your notebook. Match the blood component in Column A with its function in your body from Column B (see Table 9.5 below). You may need to use the same function more than once.

C3. (a) Copy Figure 9.22 into your notebook. Draw arrows at each of the points labelled with a letter to show the direction of blood flow.
(b) Give the name and function of the part of the body referred to by each letter on the diagram.

C4. (a) What are the three types of blood vessels found in your circulatory system?
(b) Explain how the structure of each part helps its function in your body.

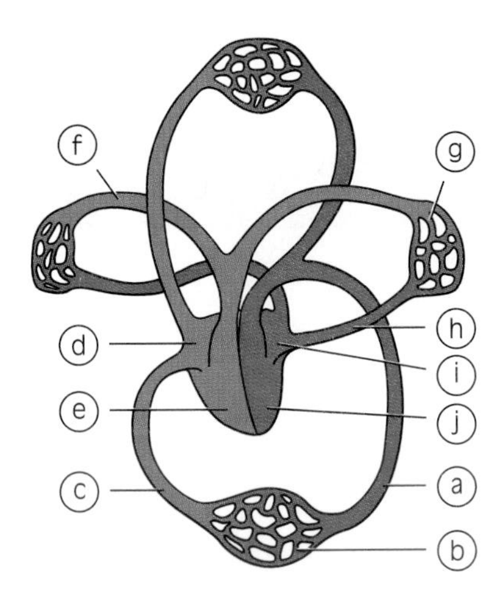

FIGURE 9.22
Use this diagram to answer Question C3.

Column A	Column B
(a) Red blood cells	Chemical signals
(b) White blood cells	Preventing blood loss
(c) Platelets	Oxygen transport
(d) Antibodies	Fighting disease
(e) Hormones	Nutrient transport
(f) Clot-forming blood proteins	Waste transport

TABLE 9.5 ***Table for Question C2***

C5. (a) What is excretion?
(b) What parts of your body are involved in excretion? Briefly describe the role of each.

EXPLORATIONS

E1. Blood tests are routinely used by physicians to detect medical problems. A sample of your blood could be used to obtain useful information about any body system. Explain why this is so.

E2. Write a 60-second commercial entitled "Choosing Now for Your Future" to dramatize the role of lifestyle choices on the future and present health of a person's circulatory system. Perform your commercial (or make a videotape) for other classes.

E3. Interview a health care professional about organ transplants or about how donated blood is used.

E4. What is cardiopulmonary resuscitation (CPR)? Find out where you can learn this life-saving technique. If possible, take a CPR or first-aid course.

REFLECTIONS

R1. Review what you wrote in your learning journal in Activity 9A. How have your ideas about how your body works and how your choices affect your body changed after reading this chapter?

R2. Describe three things you can do to have a healthy circulatory system. Be specific about how you could make healthful changes in your everyday life.

CHAPTER 10

FITNESS AND HEALTH: A WAY OF LIFE

The people you see at the starting line of the NY Marathon are all different. There are mechanics, office workers, lawyers, students, homemakers.... They range in age from about 8 to almost 80. Although their backgrounds and interests may be different, the runners all have one thing in common: a desire to be fit.

Some people keep physically fit by playing sports like soccer or volleyball. Other people prefer individual activities, such as jogging, swimming, or gymnastics. It is not necessary to be an expert athlete or to take part in competitive sports in order to be physically fit. The best way to become or stay fit is to follow a program of physical activity that suits your own interests and abilities. Following a good fitness program will help you feel and look well.

This is especially true for teenagers. Many changes occur in your body during adolescence. Some of these changes make it possible for teenagers to improve their fitness level more easily than adults can. In some ways, you have more control over your health now than you will ever have again. The habits you develop as a teenager may help you to lead a long and healthy life.

This chapter gives you a chance to examine your way of life and the role of physical fitness in it. You will then be able to make choices that may help you improve your fitness and health.

ACTIVITY 10A / WHAT IS PHYSICAL FITNESS?

What do you already know about fitness?

MATERIALS

chart paper
meter stick
drawing pens

PROCEDURE

1. Form a group with two or three other students. Use the meter stick to divide a piece of chart paper into six rectangles of equal size. Using the drawing pens, fill in these rectangles by following the instructions provided in Table 10.1
2. Post your group's chart for other members of the class to see.

A	B	C
What do you think makes a person physically fit?	List the parts of the body that you think play a role in physical fitness.	Brainstorm as many words as you can that are related to physical fitness.
D	**E**	**F**
Do you think it is important to be physically fit? Explain your answer.	List at least one link between physical fitness and the following: • the food you eat • your respiratory system • your circulatory system	List some people who you think are physically fit. Your list can include the names of individuals as well as groups of people (such as swimmers).

TABLE 10.1 ◀
Instructions for Activity 10A

3. Read the charts of other groups. In your learning journal, list 5 to 10 things you would like to learn about physical fitness.
4. How would you define physical fitness? Write your definition in your learning journal. ❖

■ 10.1 THREE FACTORS AFFECTING FITNESS

When people say that they are "in shape," they mean that they are physically fit. **Physical fitness** refers to (1) how well the parts of your body (including your heart, lungs, blood vessels, and muscles) work together, and (2) how much energy you need to make those parts work. These body parts work together efficiently in a person who is fit. Here, "efficiently" refers to how much work a person needs to do—how much energy he or she needs—to perform a certain task. If a person's body parts are working efficiently, he or she needs less energy to perform various activities than does a person who is not fit. If the runners in the photograph at the beginning of this chapter can finish the run without being very tired or feeling ill, they are probably physically fit.

In the earlier chapters of this unit, you studied several important body systems. You also learned about the nutrients your body needs in order to keep these and other systems working properly. The information in Chapters 6 to 9 will help you understand physical fitness. Physical fitness

FIGURE 10.1
Good cardiorespiratory fitness will allow you to walk up stairs without being out of breath.

includes five important factors: (1) cardiorespiratory fitness, (2) muscular strength, (3) muscular endurance, (4) body mass, and (5) flexibility. In Section 10.1 you will explore the first three of these five factors.

THE FIRST FACTOR: CARDIORESPIRATORY FITNESS

Cardiorespiratory fitness refers to how well your circulatory and respiratory systems are working. In people with good cardiorespiratory fitness, these systems, and the organs that make them up, are healthy and able to do their jobs well. For example, people with good cardiorespiratory fitness have a high vital capacity and strong heart muscle. They are able to breathe in a good supply of oxygen, and their hearts can pump oxygen-rich blood throughout their bodies with less effort.

Cardiorespiratory fitness may be the most important factor in physical fitness. Cardiorespiratory fitness affects your ability to perform any physical activity, from taking part in a strenuous sport to simply walking down the street (Figure 10.1). This fitness factor is also important because your circulatory and respiratory systems are essential to your survival. You can live without strong arm and leg muscles, but not without a heart and lungs that can do their jobs.

■ American astronaut Ed White died in a fire in his spacecraft. When doctors examined his body, they discovered that one of the major arteries leading into his heart was completely blocked as a result of atherosclerosis. Why had Ed White not had a heart attack from this blockage? It may have been because some of the other arteries near White's heart had become larger and new ones had developed as a result of the exercise he had done while training for space flight.

HOW TO MAINTAIN CARDIORESPIRATORY FITNESS

Exercise can help you improve your level of cardiorespiratory fitness. Like other muscles of your body, your heart can be strengthened by exercise. The stronger your heart, the more blood it can pump around your body each time it beats. A strong, fit heart does not need to beat as often as a weak heart to supply the same amount of blood to your body cells. A healthy heart therefore will not wear out as quickly.

Your resting heart rate is the number of times your heart beats per minute when you are sitting still. (You can measure your heart rate by taking your pulse.) A person who does not exercise regularly probably has a resting heart rate between 65 and 75 beats per minute. People who exercise regularly have lower resting heart rates. Some athletes exercise very hard and make great demands on their hearts (Figure 10.2). Long-distance runners, for example, commonly have resting heart rates of only 40 to 50 beats per minute. Their strong heart muscles can pump the same amount of blood per minute through their bodies with fewer beats than can the average person's heart.

Exercise can also help your circulatory system in another way. With exercise, you can increase the number and size of certain blood vessels in your body. More blood can then flow through these vessels, making it easier for oxygen-rich blood to reach your cells and for wastes to be carried away.

Your level of cardiorespiratory fitness also depends on how well your lungs work. Just as exercise can strengthen your heart muscle, it can also strengthen the muscles that help your lungs work. People with good cardiorespiratory fitness have lungs with a high vital capacity; they can move more air in and out of their lungs in one breath than can people with poor cardiorespiratory fitness. Their body cells are more easily supplied with the oxygen they need for cellular respiration. Also, waste products, such as carbon dioxide, are removed more easily. As a result, the body cells— and the fit person—can keep active longer without getting tired.

FIGURE 10.2
These athletes' hearts are beating fast. After the competition, when they're resting, would you expect their hearts to beat faster or slower than the heart of the average person?

RESULTS OF POOR CARDIORESPIRATORY FITNESS

In a person with poor cardiorespiratory fitness, the body is not working as well as it could be. Weak lung and heart muscles cannot supply the body with as much needed oxygen as can strong muscles. Poor cardiorespiratory fitness is a major cause of illness and death today. Over 50 percent of all deaths in the U.S. are the result of circulatory or respiratory problems that might have been reduced by good cardiorespiratory fitness.

A weak heart has to pump more often than a strong one to supply blood to body cells. In fact, a weak heart may not be able to supply enough blood to distant cells in the arms or legs if it is suddenly needed there. For example, during an emergency, a person may need to move quickly. To do this, various muscles immediately require an extra amount of oxygen and glucose. To meet this demand, the heart suddenly starts to beat very quickly. A weak heart may have difficulty with the rapid increase in beating rate and may stop beating completely. This is one cause of heart attacks. Heart attacks may also occur when a person who is not fit does a physically difficult job, such as shovelling snow. An attack can even occur when a person is resting, if the heart is too weak or has been damaged.

In Chapter 9, you read about atherosclerosis, a disease in which the arteries become blocked. This disease does not affect only adults. Studies have shown that many people younger than 20 years of age have arteries that are already partly blocked.

You may be able to lower your chances of developing atherosclerosis. For example, reducing the amount of saturated fat and cholesterol in your diet may help lower this risk. Eating less unsaturated fat may help too. Exercise is also important. A strong heart may be able to pump blood even through partly blocked arteries. In addition, because exercise can increase the size and number of some blood vessels, it may reduce the effects of a block. It may even provide new routes for blood to flow.

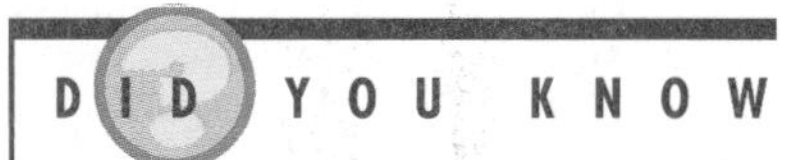

DID YOU KNOW

■ In space, where the force of gravity is greatly reduced, muscles have very little work to do. As a result, astronauts quickly begin to lose muscle strength and endurance. They must exercise with special equipment that requires them to push or pull—to use their muscles—if they are to remain physically fit.

THE SECOND FACTOR: MUSCULAR STRENGTH

A second factor related to overall fitness is **muscular strength.** Muscular strength is the amount of force (a push or a pull) that a muscle is able to exert at any one time. Strength is important for performing any activity, including walking and running, chewing and swallowing, and even breathing. Having at least some muscular strength is very important for your survival.

Muscular strength depends on the condition of the cells that make up your muscles. Although the number of muscle cells in each of your muscles does not change much over your lifetime, the size of these cells can change quite a bit. Regular exercise of a muscle can cause those cells to grow larger. The muscle itself then becomes larger and stronger; it can then perform many activities more easily. Muscles do not have to be huge to be strong. Even a very small increase in muscle size will help your body perform tasks more easily. And of course, the same exercise that increases the size of your muscles may also help decrease the amount of fat in your body (Figure 10.3).

FIGURE 10.3 ▶
Muscles do not have to be huge to be very strong. People who have strong muscles usually do not have excess fat in their bodies. Why do you think this is so?

The opposite also happens. When a muscle is not used regularly, the muscle cells tend to shrink. As a result, the muscle itself gets smaller and loses strength. If you have ever broken a bone, you may have worn a cast and been unable to move certain muscles very much. When the cast was removed, you may have noticed that these muscles were smaller than they were before you were injured. Exercise helps to build up the strength in these muscles by increasing their size.

THE THIRD FACTOR: MUSCULAR ENDURANCE

Your muscles need more than just strength. **Muscular endurance** gives you the ability to perform an activity for a long period of time. If you have good muscular endurance, your muscle cells can work longer; they get tired less easily. The hikers shown in Figure 10.4 need muscular strength to lift their packs onto their backs. They need muscular endurance if they want to carry the packs a long distance.

People who have muscles that are very large and strong do not *necessarily* have good muscle endurance. Strength depends mainly on the size of the muscle cells. Endurance depends on how well these cells work. Like all cells in your body, muscle cells need oxygen and glucose for cellular respiration. This process releases the energy the cells need to perform their jobs. Cells with a high level of endurance can carry on cellular respiration for a longer period of time and release more energy than those with a low level of endurance. High-endurance cells have two main features that make this possible.

First, such cells have many capillaries lying nearby. The blood flowing in these capillaries supplies the cells with the oxygen and glucose they need for cellular respiration. It also carries waste products quickly away. The presence of many capillaries—and thus a good supply of blood—increases the rate at which cellular respiration can occur.

Second, cells with a high level of endurance are able to make particularly good use of the oxygen and glucose that they get. In other words, muscle cells that have a high level of endurance perform cellular respiration very efficiently. Such cells can produce more of the energy the muscle needs to do its job than can muscle cells with less endurance.

You can improve your level of muscular fitness by exercising your muscles regularly. Muscular *strength* is usually increased by making muscles work very hard for short periods of time. *Endurance* increases when muscles work less hard, but for a longer period of time.

FIGURE 10.4
These hikers need to have both muscular strength and muscular endurance.

ACTIVITY 10B / TESTING MUSCULAR ENDURANCE

You can test the endurance level of muscles by seeing how long they can continue to perform a task that does not require much strength. For this activity, you should work in pairs.

MATERIALS

a 1.5 kg object or a bag filled with sand so that it has a mass of about 1.5 kg
a loop of cloth or wide ribbon that can be attached to the object or bag and hung over your wrist
clock or stopwatch

PROCEDURE

1. Copy Table 10.2 into your notebook for recording data.
2. Stand facing away from the clock and extend your right arm out in front of you, palm downward (Figure 10.5). Your partner will stand facing the clock with the object held ready.
3. Your partner will slip the loop over your wrist and release it gently so that you are bearing the full 1.5 kg mass of the object. Your partner will note the time that you started supporting the object on your own.
4. Keep your arm out straight, holding the object up until you can no longer do so. (Remember, this is a test of muscle endurance. The results do not necessarily measure the size and strength of your arm muscles.) Your partner will note and record the time when you lower your arm.

Arm	Start time	Finish time	Length of time object was held up
Right			
Left			

TABLE 10.2
Sample Data Table for Activity 10B

5. Subtract your "starting" time from your "lowering" time to obtain a measure of the endurance of your arm muscles in seconds.
6. Switch roles with your partner and repeat steps 2 to 5. Your partner should also extend his or her right arm.
7. Repeat steps 2 to 6, but testing first your left arm and then your partner's.

DISCUSSION

1. (a) Were the results the same for your right arm and left arm? If not, how different were they?
   (b) If the results for your two arms were not the same, which arm had more muscle endurance? Why do you think there might be a difference?
2. Describe some exercises that might increase the endurance of your arm muscles. ❖

■ **Release the loop gently.**

FIGURE 10.5 ▶
Do not watch the time as the object is hanging from your arm. Your partner will note the exact time you lower your arm.

▶ REVIEW 10.1

1. (a) What is cardiorespiratory fitness?
   (b) Why is it so important?
2. Suppose you decide to exercise regularly to improve your level of cardiorespiratory fitness. After you have followed an exercise plan for some time, what changes might have occurred in
   (a) your circulatory system?
   (b) your respiratory system?
3. List three effects of poor cardiorespiratory fitness.
4. (a) What is muscular strength?
   (b) What does muscular strength depend on?
5. (a) What is muscular endurance?
   (b) What does muscular endurance depend on?
6. How can you improve your muscular strength and muscular endurance?

■ 10.2 A FOURTH FITNESS FACTOR: BODY MASS

Another factor that is important for determining physical fitness is body weight, or more correctly, **body mass**. Too much body mass puts a strain on a person's body systems, particularly the circulatory and respiratory systems. Too little mass may mean that a person has small muscles without much strength or endurance. People who are physically fit have a body mass that is fairly close to the ideal, or best, mass for people of their age and height.

Body mass is an important fitness factor for teenagers to consider. Adolescence is usually a period of rapid growth, when a person becomes taller and body mass increases. Many changes in body mass during this period are quite normal. There are other factors to consider, however. When you become a teenager, your level of activity may change (increase or decrease) from what it was when you were a child. You may also gain more control over the food you eat. Understanding how these changes affect your body mass is important to maintaining good fitness.

What happens to the kilojoules of energy contained in the food you eat? Some of this energy is used to power the body processes that keep you alive, such as breathing and digestion. Another portion of the food energy is used in the "extra" activities you choose to take part in—from smiling to running. Any food energy that is not used up in these ways is added to your overall body mass.

Up to two-thirds of the food energy you use each day is for basic life processes. They include those carried out by the digestive, respiratory, circulatory, and excretory systems. The **basic metabolic rate (BMR)** is the rate at which a person's body uses energy to run such processes. The word "metabolic" is used because all of the processes going on inside a living organism are referred to as that organism's **metabolism**.

The amount of energy needed to run these basic life processes is different from person to person. It depends on a number of factors, including an individual's age, sex, and body mass. Men usually have a slightly higher BMR than women. For both sexes, the BMR is highest in young, growing individuals, decreasing when growth slows down. Another important factor controlling BMR is how much of a person's body is made up of muscle and how much is made up of fat. Muscle cells use up much more energy than fat cells do, even when a person is at rest.

You also use food energy when you take part in any activity— from reading to playing a sport (Figure 10.6). The amount of energy needed for an activity depends on many things, including the number and size of the muscles that are used, your mass, and the length of time that you take part in the activity. In general, the more active you are, the more food energy you will use up.

DID YOU KNOW

■ A simple method of determining what your mass should be is to look up the value in a table. However, the numbers in such tables are only for the "average" person of a certain age, height, and sex. These averages often do not take into account individual differences in features, such as bone density or body frame size (whether a person has a large or small skeleton relative to others of their age and sex). What problems might this cause for someone trying to determine his or her ideal body mass?

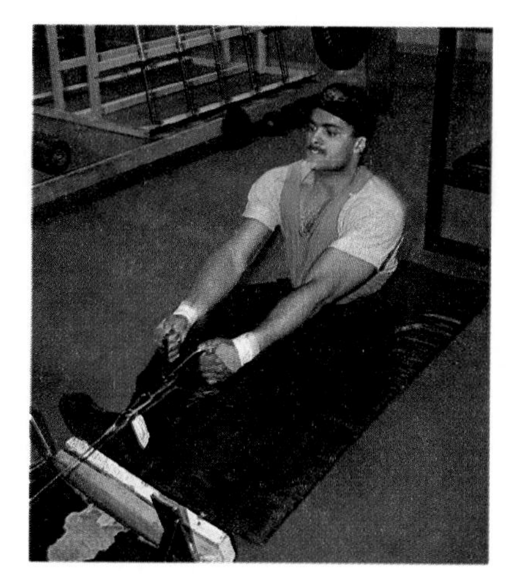

FIGURE 10.6
How is food energy used in this activity?

EXTENSION

■ The amount of energy you need for activities affects how much you should eat. What other things might affect a person's eating habits? What reasons might a person give for "overeating," or "undereating"? Do you think these are good reasons? Think about your answers to these questions and discuss them with a group of friends or classmates.

CONTROLLING YOUR BODY MASS

If you are close to your ideal body mass for your age and height, you will want to keep this mass. You do this by balancing energy input (the amount of energy taken in from food) with energy output (the amount used in basic life processes and other activities), as shown in Figure 10.7a. If you eat three meals a day and have an active lifestyle, you are probably balancing your energy input and output quite well.

People can upset their energy balance. People who are not physically active, or who often snack between meals, may have more energy input than output (Figure 10.7b). The extra energy is stored as fat in the body. In this case, a person can reach a better balance by reducing energy input or increasing output. If the person wants to lower his or her body mass, it is best to do a little bit of both. It is very important not to reduce energy input too much or to suddenly raise energy output too high. A loss of about 1 kg a week is a good average, although the amount may vary from week to week. If a person does not eat enough to supply the body's needs, he or she may feel ill or tired all the time (Figure 10.7c). If there is not enough food energy to run all the basic life processes, the person may become quite ill.

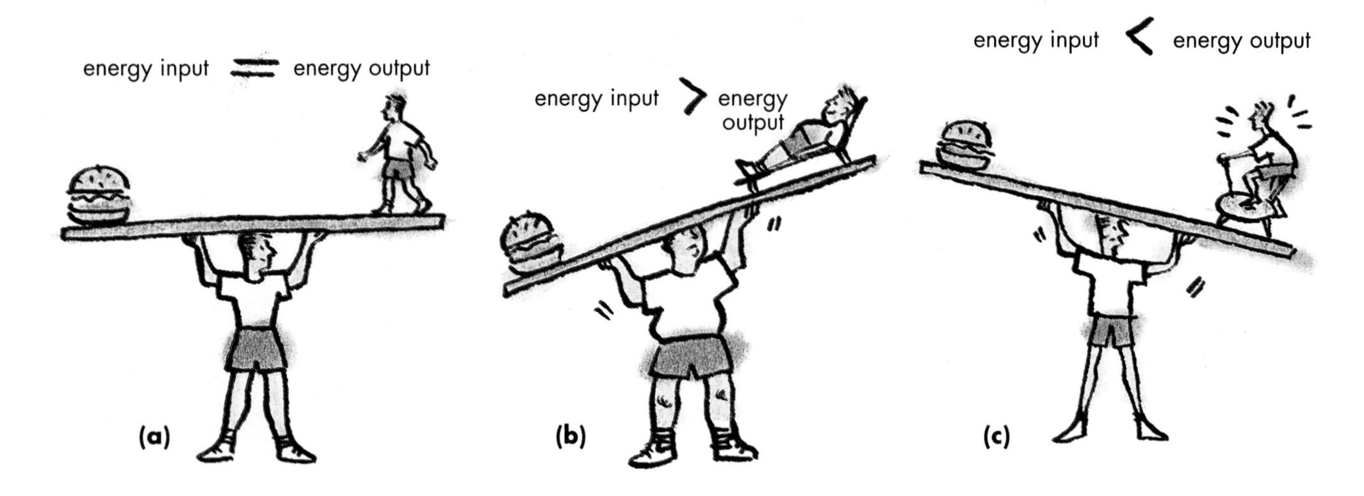

FIGURE 10.7
To maintain a suitable body mass for your height, you need to balance energy input with energy output.

ACTIVITY 10C / SNACK FOOD TRADE-OFF

Some snacks do not contain many nutrients but do contain many kilojoules of energy. Table 10.3 lists the energy in various snack foods. Examine the data given in the table, then answer the questions that follow.

DISCUSSION

1. You want a snack when you arrive home from school. You have a jelly doughnut and a cola drink.
   (a) How much energy is contained in your after-school snack?
   (b) How long would you have to jog in order to use up the energy provided by this snack?

Food	Mass (g)	Energy (kJ)	Time (in minutes) to use up the energy provided by food			
			Walk	Cycle	Swim	Jog
Cola drink (227 mL)	240	483	20	16	12	11
Milk (2%) (235 mL)	240	529	24	18	15	13
Chocolate bar	40	890	40	32	25	21
Buttered popcorn	18	344	16	12	10	8
Cream-filled chocolate cookie	12	168	8	6	5	4
Ice-cream cone	72	672	31	24	19	16
Jelly doughnut	65	949	44	34	27	23
French fries (20)	100	1150	53	41	33	27
Carrots (2 small)	100	176	8	6	5	4
Apple	150	365	17	13	10	9
Celery (two 15 cm stalks)	40	25	1	1	1	1
Banana	150	533	24	19	15	13
Orange	150	392	14	11	9	7

TABLE 10.3 ◀
Using the Energy from Your Food

(c) How long would you have to walk to use up this energy?

2. A 150 g apple provides about 365 kJ of energy. Think about what you learned about nutrition in Chapter 6. List at least three reasons why an apple might make a better snack than a chocolate bar.

3. Which of the snacks listed in Table 10.3 provides the greatest amount of energy per gram of mass?

4. Suppose you have just eaten a plate of french fries and want to "work it off." Which of the activities listed would do this most quickly? ❖

BODY MASS AND BODY FAT

If a person takes in more energy than is used up, the extra input is stored as fat. The amount of fat in your body is more of a measure of fitness level than your body mass is. For example, a physically fit athlete may have a greater body mass than someone who is the same height but who does not exercise at all. The athlete's large, well-developed muscles have a greater mass than would the same amount of fat. Being "overfat" rather than being "overweight" is the cause of many health problems.

To be healthy, everyone needs to have some fat in his or her diet and in his or her body. (Chapter 6 listed several important purposes of fat in the diet.) Although everyone needs some body fat, too much body fat contributes to a variety of health problems, including diabetes, high blood pressure, heart disease, and asthma. For people who are considerably overweight, most activities require their hearts and lungs to work harder. Such people may have trouble breathing and become tired easily if they are active. As a result, overweight people often become even less active, making their problem worse.

■ Between 5 percent and 15 percent of the total body mass of a man should be made up of fat. An amount of body fat in this range is said to be "ideal"; it is neither too little nor too much. For women, the ideal amount of body fat is 7 percent to 18 percent.

FIGURE 10.8
Why is body size not a perfect indication of body health?

BODY MASS AND GOOD HEALTH

Keeping body mass within certain limits is a great concern for many Americans. Think about the "successful" and "happy" people in magazine and television advertisements. Most of them are thin. Seeing these actors and models can make people think that everyone must be thin to be happy or successful. In fact, good health is more important than appearance; your body mass should be determined by what is healthy for you, rather than by what will make you look a certain way (Figure 10.8). Some people may need to gain mass to be physically fit. Others may need to lose mass.

If you think you have too little or too much body mass, you should talk to your doctor. She or he can tell you if it is safe to start an eating or exercise plan that is quite different from your old habits. Talk to your friends and family as well. If you are trying to lose mass, they may help you avoid eating too many high-energy foods. In addition, exercising and playing sports with family members or friends may be more fun than doing it alone. The support of others may help you stick with your plan until you have achieved your goal.

It is easy to find advertisements for "miracle diets" that promise to "melt away kilograms with ease." Sometimes these claims seem very appealing. However, the best way to reduce or control your mass is to gradually work toward a balance of energy input and energy output that will help you achieve a body mass that is suitable for your height and age.

ACTIVITY 10D / BODY MASS AND SELF-IMAGE

Body mass can affect how people think and feel about themselves as well as their physical fitness. Do you think that this could cause problems?

PROCEDURE

1. Form a group with two or three other students. Examine magazines (in the library or borrowed from home), looking for 5 to 10 advertisements related to losing weight or "getting in shape" (for example, advertisements for types of food or drink, weight loss centers, or fitness clubs). For each advertisement, describe
   (a) what is being advertised,
   (b) why, according to the advertisers, you should want to use this product,
   (c) the people shown in the advertisement (their sex and appearance).
2. If "fitness" is mentioned in the advertisement, discuss what you think the advertisers mean by this term. Do they mean all the fitness factors as described in this chapter, or something else?
3. Discuss whether you think these advertisements are realistic. In your learning journal, keep a record of your opinions and the reasons for them.
4. Imagine that your group is an advertising agency that has been hired by the government to encourage good physical fitness. In particular, your job is to show the connection between ideal body mass and fitness. Make a poster or pamphlet that you think will convince people of the importance of controlling their body mass—for the *right* reasons.

DISCUSSION

1. (a) Do you think that body mass affects the way people feel about themselves?
   (b) Do you think that body mass *should* affect a person's self-image? Why or why not?

2. (a) In general, do you think that females worry more about their body mass than males?
   (b) Do you think that females *should* worry more? Why or why not?
   (c) If your answer to part (a) was "yes," why do you think that this might be the case?
   (d) How could you and others help ensure that females do not worry more than they should about their body mass? ❖

REVIEW 10.2

1. (a) How does your body use the energy contained in the food you eat?
   (b) What happens if you eat more food energy than your body needs?
2. Would you expect Indira, who is 20, and her younger brother Sanjay, who is 14, to have the same basic metabolic rates? Why or why not?
3. Two friends are the same height, but one weighs 5 kg more than the other. Can you tell from this description which person is more fit? Why or why not?
4. Why is it a good idea to check with your doctor before making a big change in your diet or exercise habits?
5. In general, how can you control your body mass?

10.3 A FIFTH FITNESS FACTOR: FLEXIBILITY

A fifth factor that affects your physical fitness is flexibility. In order to understand what flexibility is, you must first know something about joints and muscles.

JOINTS

Any place in your body where two or more bones join, or come together, is called a **joint**. For example, the bones of your upper and lower arms meet at a joint called your elbow. Your knee is a joint at which your thigh and calf bones come together. Think of how your elbows and knees allow your arms and legs to bend. At most of your body's joints, some movement occurs between the joined bones. Your **flexibility** is the amount of movement that you have in these joins. (Figure 10.9).

How much flexibility you have at a joint depends partly on the type of joint that it is. For example, you have different kinds of joints—and different kinds of movement—at your elbow and your shoulder. The amount of flexibility also depends on the body parts, such as muscles, that control the joint's movement.

■ Do research to find out how one of the several different kinds of joints in your body works. For example, you might choose to investigate ball and socket, hinge, or sliding joints. Where in your body are joints of this type found? Why? Share your research with the class in a poster or by making a model that demonstrates how this type of joint works.

FIGURE 10.9 ▶
Many activities require a very high level of flexibility. Good flexibility can help you enjoy any kind of activity with less risk of pain or injury.

MUSCLES

The muscles that you use to move your body parts are all attached to your bones. To move a joint—and thus, a body part —one or more of your muscles must contract (get shorter). Since the muscle is attached to a bone, when the muscle contracts, it pulls the bone in the direction that the muscle has contracted. In other words, a muscle works like a spring whose ends are attached to two different objects (Figure 10.10a). When the spring contracts, or shortens, the objects move toward each other. When the spring relaxes, or gets longer again, the objects return to their original positions (Figure 10.10b). This is much the way a muscle works. Contraction of a muscle pulls the bones to which it is attached toward each other. When the muscle relaxes, the bones move away from each other.

Although a muscle can *pull* a bone, it cannot push it. For this reason, the muscles that move your body parts are often arranged in pairs, attached on opposite sides of a bone or to opposite ends of a bone. As one member of the pair contracts, and moves the bone in one direction, the other relaxes. To move the bone in the opposite direction, the second member of the muscle pair contracts, while the first relaxes (Figure 10.11). If you are to be able to move easily, the muscles must be able to relax to their full length after contracting. This allows the other muscle in the pair to do its work. The ability of your muscles to relax is part of what determines your flexibility.

Lack of activity can reduce your level of flexibility. When muscles are not used very much, the muscles get smaller and cannot stretch as far when they relax. People need special stretching exercises to overcome these problems. They must stretch their muscles and other body parts around the joints slowly, gradually, and repeatedly. This careful stretching helps to lengthen these body parts, making it possible for them to stretch farther when they relax. As a result, the joints can be moved easily.

(a)

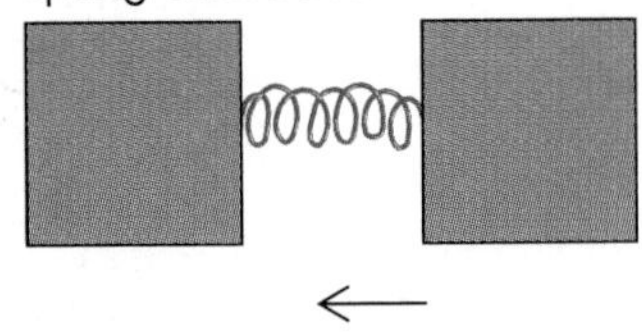

(b)

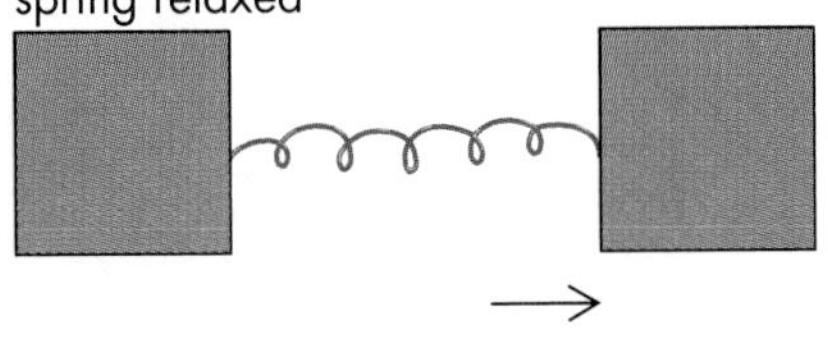

FIGURE 10.10
When the spring is contracted or squeezed (a), the objects are pulled together. What happens when the spring is relaxed?

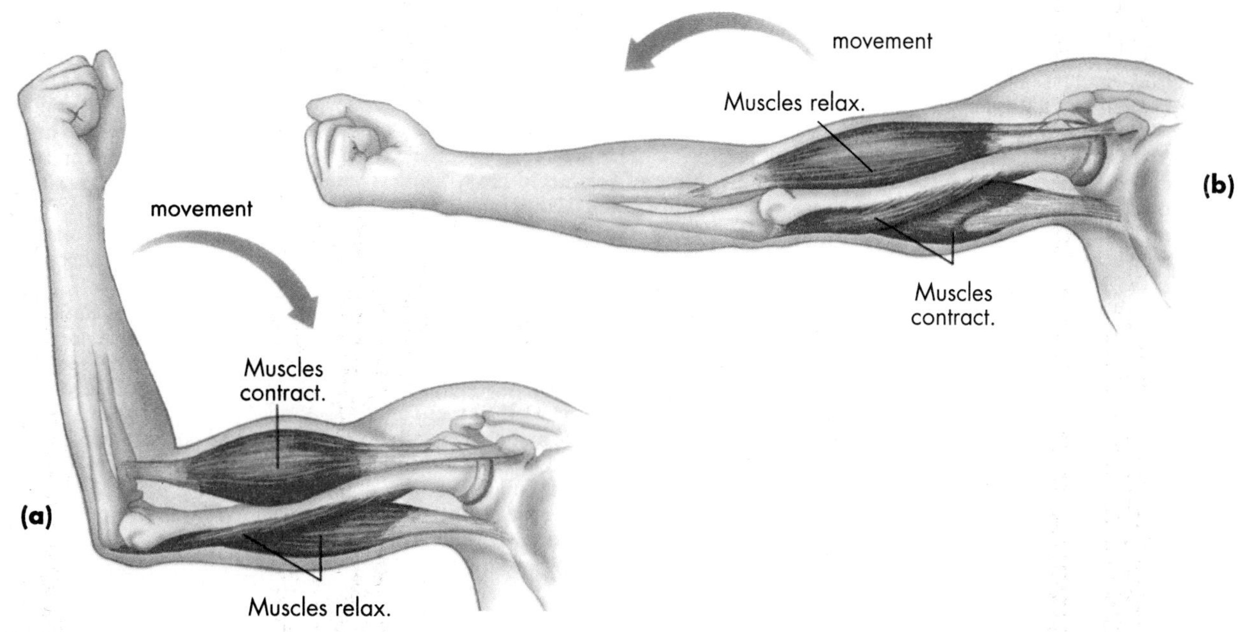

FIGURE 10.11
When you raise your forearm (a), the muscles on the top or front of your arm (between your shoulder and elbow) contract. At the same time, the muscles on the bottom or back of your arm relax. To lower your forearm (b), the muscles on the back of your arm must now contract, while the muscles on the front relax.

Being flexible helps you feel healthy—you can perform your usual daily activities without difficulty and can play sports or exercise without being stiff the next day. Flexibility also helps you avoid injuries. If you are going to do a particular activity, your body parts must be flexible enough to perform it. They must also be prepared for strenuous activity by being gradually and carefully stretched just before starting the activity (Figure 10.12). This preparation can help you avoid overstretching or tearing muscles or other body parts, which would result in stiffness or pain in the joints.

Some people are not very flexible because of bone diseases such as arthritis or serious injuries to muscles or parts of joints. Most people, however, can become or remain flexible by regularly performing proper stretching exercises.

FIGURE 10.12
This student is stretching the muscles in her legs in preparation for a run. What might happen if she didn't prepare her muscles for the stress of running?

► REVIEW 10.3

1. (a) What is flexibility?
   (b) What does flexibility depend on?
2. Why do you think it is important to be physically flexible?
3. Do you think a person could be flexible in one part of his or her body, but not in other parts? Explain your answer.
4. How could you increase the flexibility of one or more of your joints?

FITNESS INSTRUCTOR

Mary Audia runs a day center for seniors. In a recent interview she had this to say about teaching fitness to older people.

Six years ago, I was one of very few instructors offering fitness programs just for seniors. I believed that if seniors were kept more active, they could live independently longer. I wanted to design a class that would give a good workout to active seniors, but would also provide exercise for those who are frail.

I am a registered nurse, but I also have a fitness certificate with a specialty in gerontology (the study of ageing). That gives me a good background in understanding the parts of the human body.

The process of growing older affects people in different ways. I know an 80-year-old man who swims across the lake every morning. He is certainly active, but most other seniors are less mobile. If they have not been exercising, their joints are not very flexible. Their bones become brittle and can easily break. Without exercise, people lose muscle strength.

Exercise can slow down or even reverse this ageing process. Being active helps people of all ages feel better physically and mentally. One man in my class is 98.

In my classes I use gentler, slower movements than in a regular exercise program. Some seniors have problems, such as arthritis, which make sudden movements of the joints painful. I also demonstrate exercises very carefully, since seniors may have poor hearing or eyesight. We do a lot of movement of joints to increase flexibility. Many exercises emphasize relaxation and massage.

We use a chair for some of our exercises. We might walk around the chair to a beat, raising and lowering our arms to improve cardiorespiratory fitness. The chair is there so that individuals with poor balance can grab it or sit on it.

Some seniors need a lot of encouragement to keep coming to fitness classes. One thing they really like is dancing. We bring in a band and they'll dance up a storm. They enjoy the physical contact, the movement, and the sheer fun of being together!

The choice of music is really important. Music creates a mood and can be used to relax people as well as to get them moving. We have a lot of fun in our classes for seniors, and laughter is still the best medicine.

Think about some of the seniors you know. How do they keep physically fit? Could you help them become more active?

ACTIVITY 10E / ACTIVITY LEVEL AND PHYSICAL FITNESS

PART I Your Normal Routine

Think about what physical activities you did last week that you might consider exercise. If last week was unusual for you in some way (for example, if you were ill or on vacation), you might want to recall a recent week with your usual level of activity. These activities might include gym classes, team games or practices, walking or riding a bike to school, hiking on the weekend, dance classes, or jogging. Record these activities in your learning journal.

PART II Your Fitness Level

After you have read through Section 10.4, Improving Your Fitness Level: Exercise Programs, look again at your list of physical activities for the week. Do you think that your level of activity will permit you to improve your fitness, or to maintain it if you are already fit? If not, why not? If you feel you need to, how might you change your habits to accomplish a goal of good physical fitness?

Think about the kind of job you would like to have after you have finished your education. As well, consider the other responsibilities that adults have. In your learning journal, suggest how physically active you will want to be when you are an adult. You may add your observations of adults as you record your thoughts about yourself. ❖

■ 10.4 IMPROVING YOUR FITNESS LEVEL: EXERCISE PROGRAMS

You have now read about five factors related to physical fitness. All of these factors can be improved by exercise. For example, a muscle becomes stronger the more you use it. In addition, the muscle gains endurance as it is used again and again during exercise. Because your heart is a muscle, exercise helps to improve its strength and endurance and, therefore, your cardiorespiratory fitness. Exercise also allows you to use up food energy and avoid adding extra body mass. Finally, exercise helps improve your flexibility. When thinking about an activity or an exercise program to improve your fitness you need to consider the intensity, frequency, duration, and type of exercise. Before beginning any exercise program, it is wise to have a doctor check the condition of your heart.

INTENSITY

The **intensity** of the activity refers to how hard your body has to work during exercise. In order to increase your level of fitness, you make the body parts or systems you want to improve work harder than they normally do. This is called "overworking" the body parts. When performing stretching exercises to increase your flexibility, for example, you must stretch your muscles slightly more than if you were just relaxing. (You must do this carefully and slowly, however.) Exercises that improve cardiorespiratory fitness force the heart and lungs to work harder than normal. If done carefully, such overwork can help your body.

There are two ways you can overwork your body. The first is by taking part in **aerobic** activities. You may have taken part in some aerobic activities, such as running, walking, cross-country skiing, cycling, or swimming (Figure 10.13). The word "aerobic" means "with oxygen." When you perform an aerobic activity, you continue to supply your body cells with enough oxygen to allow them to carry out cellular respiration. The cells can, therefore, still obtain the energy they need to continue working. To make sure that enough oxygen reaches your cells, aerobic activities must not be too fast-paced or intense. They are carried on at a steady rate, which allows your heart enough time to pump oxygen-rich blood to your muscles.

FIGURE 10.13
What do the activities in these photographs have in common?

A second way of overworking your body is by taking part in **anaerobic** activities. "Anaerobic" means "not aerobic." Basketball, hockey, racquetball, and sprints like the 100 m run are anaerobic activities. Anaerobic activities require your body to work *very* hard (that is, with high intensity) for a short period of time. During such activities, your lungs and heart cannot constantly supply your cells with enough oxygen-rich blood: the cells use up oxygen faster than it is supplied. Therefore, while doing anaerobic activities, you have to stop to catch your breath—and give your cells time to catch up on their oxygen needs—between bursts of exercise (Figure 10.14).

When you begin an exercise program, it is best to start with aerobic activities. Performing these types of exercise for the first few weeks will get your heart ready for the more strenuous demands of anaerobic activity. Whatever type of aerobic activity you choose, it must have a high enough level of intensity: the activity must increase your heart rate quite a bit in order to help your cardiorespiratory fitness. Exercises that do this are said to have a **training effect**.

FIGURE 10.14 ◀
Many team sports involve anaerobic activity. Why do you think most teams have more players than are needed to play at any one time?

To see if an activity you are taking part in has a training effect, take your pulse right after you have stopped exercising at your highest intensity. For a training effect, the activity must cause your heart to beat much faster than its normal pace. However, you do not want to cause your heart so much stress that it is damaged. For an activity to help improve your fitness, yet still be safe, it should increase your heart rate within a certain range. You can determine this range as follows:

Lowest heart rate for training effect (in beats/min) = 170 – your age

Highest heart rate that is safe (in beats/min) = 200 – your age

For example, if you are 16 years old, to have a training effect, an activity must increase your heart rate to above 170 – 16 = 154 beats/min. The activity should not, however, increase your heart rate to more than 200 – 16 = 184 beats/min. If you find that it does, you should reduce the intensity of the exercise or take a rest.

EXTENSION

■ Different activities require the use of different muscles. Because of this, the stretching exercise that you need to perform before each of these activities may also be different. Find out what stretching exercises are usually performed before an activity of your choice, such as football, tennis, ballet, or long-distance running. Physical education teachers or coaches might be able to help you find out this information.

FREQUENCY

How often (how frequently) you exercise is called your **frequency** of exercise. Many studies have shown that three or four times a week is a good frequency of exercise. For example, to improve cardiorespiratory

■ If you ski, you should start getting in shape for the slopes a couple of months before the ski season begins. To do this you need to perform exercises that will stretch and increase the strength and endurance of the muscles you use while skiing. This will help you to ski longer without getting tired at the start of the season. It will also reduce your chances of being injured. Why do you think this is so?

fitness you should exercise to the level of the training effect at least three times a week. You need to stretch your muscles regularly to keep them flexible. You must use your muscles often in order to improve or maintain their strength.

A reasonable frequency of exercise may also help you avoid injury. If you exercise only rarely, your muscles are not kept in shape. Sudden strenuous activity may cause you to become overtired. Overtiredness can easily lead to injuries. Exercising too often may also cause tiredness, resulting in injury. Muscles and other body parts need time to recover from overwork.

DURATION

The length of an exercise period is referred to as the **duration** of the exercise. The more intense your exercise is, the shorter its duration should be. If you choose an activity that is not very intense, the duration of your exercise should be longer. For example, walking, running, and jogging are all good exercises. However, to get the same benefit from walking, you need to do it for a longer time than if you were jogging or running. Studies have shown that regularly taking part in strenuous activity (such as running or exercising in a fitness class) for 30 minutes can be very helpful. To increase cardiorespiratory fitness, you need maintain the training effect for at least 5 to 15 minutes each time you exercise.

TYPE OF EXERCISE

Although participating in any type of physical activity can be good for your health, activities differ in how much they affect your physical fitness. As Table 10.4 shows, some activities are excellent for improving one of the fitness factors but not as good for other factors. You may need to select two or three activities in order to improve all factors. The ratings for each activity also depend on the intensity with which the exercise is performed. Why do you think body mass, another fitness factor, has not been included in the table?

You should be able to enjoy any exercise program, whether it includes playing a team sport, jogging, walking, or lifting weights (Figure 10.15). If you do not enjoy the exercise, you will not continue to do it, so choose activities you like.

No matter how hard, how much, or how long you exercise, and no matter what type of activity you take part in, there are two things you should do every time you exercise. You must warm up when starting and cool down when finishing. Warming up is the best way to prepare your body gradually for strenuous activity. Always start your warm-up period with a few exercises that stretch muscles to their full length. Continue your warm-up by starting to exercise at a low intensity.

FIGURE 10.15
Why is it important to choose exercise you enjoy?

During this warm-up period, your heart should be allowed to gradually increase its rate of beating.

As your exercise period ends, you should gradually decrease the rate and intensity of your activity (Figure 10.16). This cooling down is important. If you stop exercising very suddenly, a large amount of blood may collect in the muscles since the heart was rapidly pumping the blood there just seconds before. If blood collects in your muscles, there is less oxygen-rich blood travelling to your brain. As a result, you may feel dizzy and faint.

FIGURE 10.16
After doing several laps, this runner is "cooling down" by jogging. What will he avoid by cooling down?

TABLE 10.4
Exercise Value of Certain Activities

Activity	Cardio-respiratory fitness	Muscular strength	Muscular endurance	Flexibility
Basketball	Excellent	Poor	Good	Fair
Bicycling	Excellent	Good	Good	Poor
Bowling	Poor	Poor	Poor	Poor
Calisthenics	Fair	Good	Fair	Excellent
Cross-country skiing	Excellent	Fair	Excellent	Fair
Downhill skiing	Fair	Fair	Fair	Fair
Gymnastics	Fair	Excellent	Excellent	Excellent
Ice skating	Good	Poor	Good	Fair
Jogging	Excellent	Fair	Excellent	Poor
Rope-jumping	Excellent	Good	Excellent	Poor
Softball	Poor	Poor	Poor	Poor
Swimming	Excellent	Fair	Excellent	Fair
Table tennis	Poor	Poor	Poor	Poor
Tennis	Fair	Fair	Fair	Fair
Volleyball	Fair	Fair	Poor	Poor
Walking	Good	Poor	Fair	Poor

▶ REVIEW 10.4

1. (a) What four things should you consider when thinking about starting an exercise program to improve or maintain your fitness?
   (b) Describe an exercise program that you could follow to improve your fitness level, showing how the program takes each of these four things into consideration.
2. (a) What is the difference between aerobic and anaerobic activity?
   (b) List three examples of activities that can be done aerobically.
   (c) Could the activities you listed in part (b), or activities very similar to them, be done anaerobically? Explain your answer.
3. How should the intensity of an activity be related to its duration if you are trying to improve your fitness level?
4. (a) Why should you always begin a period of exercise with a warm-up?
   (b) Why should you include a cooling-down period at the end of every exercise period?
5. Think about the activities you normally take part in. Examine Table 10.4 to see how these activities (or ones similar to them) affect your fitness level. Are there any fitness factors that your activities do not help as much as they could? What other activities might you take part in that would improve these factors?

■ Create a pamphlet, like one you might find in a doctor's office or at the local public health unit, to "sell" people on the value of being physically fit. Be creative! Make your pamphlet eye-catching and convincing. You want people to see the benefits or advantages of being fit so that they will choose a lifestyle that will help them improve their fitness.

■ 10.5 YOUR OWN HEALTH AND FITNESS

Physical fitness is very important, but it is only part of good health. What else affects how healthy you are? Here are four things to consider:

1. your past and present health
2. your family's medical history
3. the environment where you live, study, and work
4. your personal habits

Some of these are easier to control than others.

Although your past health cannot be changed, you may be able to improve your present level of health. Taking care of your body today will also help to prevent future health problems. Regular visits to a dentist are important for discovering and stopping tooth decay and gum disease. Regular check-ups can help doctors discover diseases like diabetes and cancer in their early stages. You can receive vaccinations (shots, given with a needle), to help protect you from getting certain diseases.

Your family medical history also has some effect on your own health. Some diseases are passed on to you from your parents or grandparents in the same way that curly hair or blue eyes are said to be passed on. Hemophilia and some forms of diabetes can be passed on in this way. If your family history includes certain conditions such as poor eyesight or high blood pressure, you might develop these problems too. Characteristics that help you have good health can also be passed on.

You cannot change your family's medical history. However, if you know about health problems that run in your family, you may be able to reduce your risks of developing these diseases. If there are certain types of diabetes in your family, for example, you should probably be extra cautious about gaining excess mass.

Where you live, work, or go to school can also affect your health. Although you cannot always change your environment, you can sometimes reduce some of the problems that your environment may cause. For example, living in a crowded city can be stressful for some people. Stress can lead to a variety of health problems, including high blood pressure, headaches, and more frequent colds. Such health problems are less common in people who are physically fit. Starting an exercise program may help some people reduce problems caused by stress.

Large cities may often have high levels of air and water pollution. These problems can be found even in rural areas, however. Becoming aware of such dangers in your environment is an important first step in learning how to avoid or correct them.

You have the most control over your personal habits. Decisions that you make affect your health every day. For example, smoking, drinking alcohol, and too much suntanning are behaviors that can have serious effects on your health. You must decide whether these behaviors are worth the risk. You also have some control over the food you eat each day and over how much physical activity you take part in. You can decide, for example, whether to watch television, play a computer game, or take a walk. Even very small decisions can have an important effect on your health. Using the stairs instead of an elevator whenever possible can improve your muscular and cardiorespiratory fitness. It can also add to the output side of your daily energy balance—and thus help you maintain a good body mass.

You are constantly making decisions that affect your health, whether you think carefully about them or not. Often, these decisions lead to behaviors that become habits lasting a lifetime. You can choose whether or not your habits lead to good fitness and health.

▶ REVIEW 10.5

1. How can you improve your present health and reduce the chance of health problems? Give at least one example of a way you can do this that is not mentioned in this book.

2. Teresa's parents both have had heart attacks. Does this mean that Teresa will also have a heart attack when she reaches her parents' age? Explain your answer.

3. Can being physically fit help reduce some of the health problems related to your family history or your environment? Explain your answer using examples.

CHAPTER REVIEW

KEY IDEAS

- Physical fitness refers to how well the parts of your body work together and how much energy you need to make them work. Physical fitness includes five factors: cardiorespiratory fitness, muscular strength, muscular endurance, body mass, and flexibility.
- Cardiorespiratory fitness refers to how well your circulatory and respiratory systems work. It affects all your daily activities.
- Muscular strength is the amount of force a muscle is able to exert at any one time. It depends on the size of your muscles.
- Muscular endurance refers to how long your muscles can continue working. It depends on how efficiently your muscle cells work.
- To be fit, a person's body mass should be suitable for his or her age, sex, and height. Your body mass depends on how much food energy you take in and how much you use up.
- Flexibility is the amount of movement you have at movable joints. Good flexibility helps you move easily and avoid injuries.
- Exercise can improve all five fitness factors. You should think about the intensity, frequency, duration, and type of exercise you are taking part in.
- Your health is affected by many things besides physical fitness. These include your past and present health, your family's medical history, your environment, and your personal habits. You can make choices that can help you maintain good health.

VOCABULARY

physical fitness
cardiorespiratory fitness
muscular strength
muscular endurance
body mass
basic metabolic rate (BMR)
metabolism
joint
flexibility
intensity
aerobic
anaerobic
training effect
frequency
duration

V1. Construct a mind map that shows the relationships among the words in the vocabulary list. Start with the term "physical fitness" in the center of your map.

V2. Write a paragraph describing some of the things you have done this week. Use as many terms from the vocabulary list as possible.

CONNECTIONS

C1. (a) What are five important factors of physical fitness?
(b) Which of these factors do you think is the most important to your health? Explain your answer.
(c) List one way you could improve your level of fitness for each of the five factors.

C2. Can what you eat affect your physical fitness? Explain your answer.

C3. Give five examples of personal habits (other than eating and exercising) that affect your health.

C4. Examine Figure 10.17.
(a) Why is the heart rate of the cross-country runner lower than that of the students who are not cross-country runners?
(b) What effect does eight weeks of training seem to have on the average heart rate of the students?
(c) Why might this change occur?
(d) Why is a lower heart rate a sign of good health?

C5. During the "off season," when they are not actually playing games such as football, baseball, and hockey, many athletes take part in exercise programs. What advantages would these players have over those who do not exercise regularly?

C6. (a) List five ways in which your living environment might affect your health. Remember that there are some things in your environment that may help your health as well as things that can harm it.

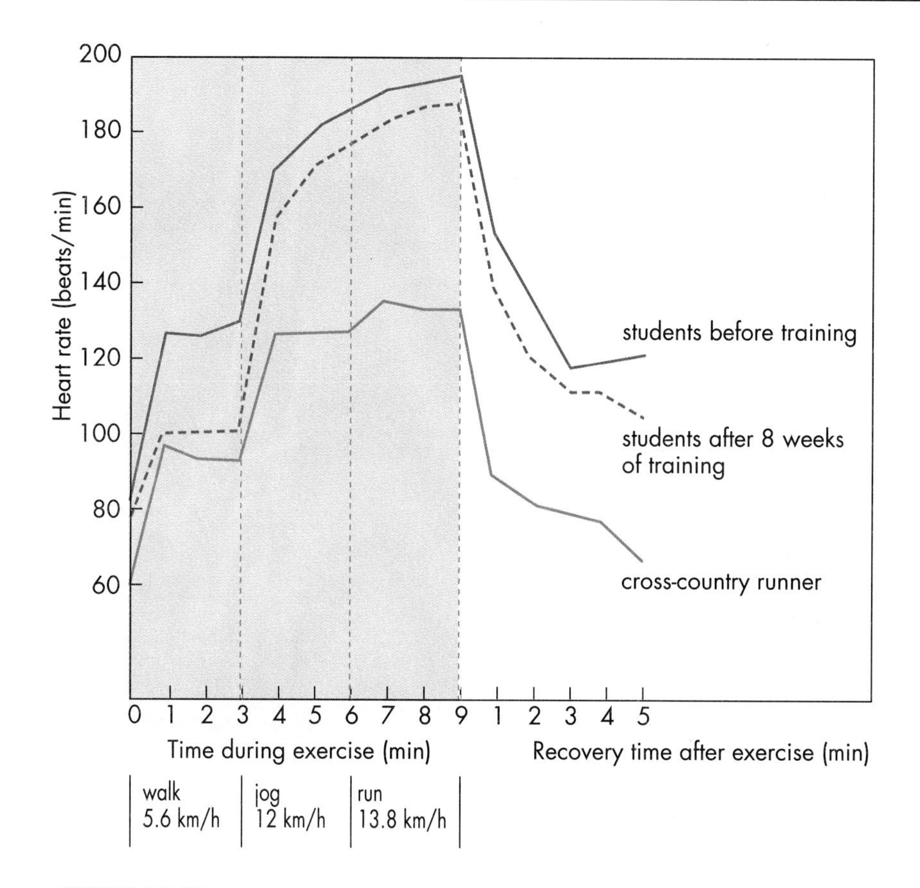

FIGURE 10.17

This graph shows the change in heart rate which occurred during and after a nine-minute exercise test. The test included three minutes of walking, followed by three minutes of jogging, then three minutes of running. The heart rate of a cross-country runner is shown, as are the average heart rates of a group of students before and after they had eight weeks of physical training.

(b) What are some changes you could make to reduce the effect of the harmful things mentioned in (a)?

C7. In what ways might a person who uses a wheelchair be more fit than one who does not?

EXPLORATIONS

E1. Some athletes take drugs called anabolic steroids, which they think will improve their performance in sports. Find out how these drugs work on the muscles, and what effects they can have on other parts of the body. Present the results of your research as an article for "World of Fitness" magazine. Let your readers know about the harmful effects of steroids as well as explaining why some athletes use them. Let your readers know whether or not it is legal to use steroids.

E2. Many people feel happy and relaxed after exercising. Some studies have suggested that this feeling is caused by chemicals called endorphins, which are released by the body into the brain during exercise. Find out more about these chemicals and how they work. You might want to write up what you find in the form of a newspaper or magazine article.

E3. Some people become too concerned about their body mass, thinking that they need to be thinner. Some of these people suffer from eating disorders, such as anorexia nervosa or bulimia. Find out more about one of these disorders. What causes the disorder, and what effects does it have? Prepare a pamphlet that would let others know what you have learned.

REFLECTIONS

R1. Look back at what you wrote in your learning journal in Activity 10A. Would you change your definition of physical fitness in any way after having read this chapter? If so, how? Did you learn all of the things you had hoped to learn from reading this chapter? Are there any new things you would now like to know about fitness or health?

R2. Think about your daily routine carefully. Where do you think you could make changes, either large or small, to improve your fitness?

EXAMINING THE EARTH

If you look at a map of the United States, you will see that the western region has many more mountains than the rest of the country. Why are there so many mountains there? Why are there so many volcanoes on the West Coast and none elsewhere? Why are there so many earthquakes in California? Is there a reason that mountains, volcanoes, and earthquakes occur together here? In this unit you will go on a "journey" to consider these questions.

Your exploration will begin in the mountains. There you will see that powerful forces have changed the shape of the Earth's surface over millions of years. Next, you will travel deep in the ocean and far underground. You will find out more about the fiery volcanic eruptions and shuddering earthquakes that can destroy whole cities. Finally, you will see how scientists have found a way to describe the structure of our planet and explain how mountains, volcanoes, and earthquakes are connected.

UNIT III

CHAPTER 11

THE RESTLESS EARTH

One day, hiking high in the Rocky Mountains, you might be lucky enough to find the fossil of an undersea animal. Fossils such as the one shown here are common on certain rocky slopes. But how did the remains of these marine organisms get there? Did the ocean level drop? Or did the mountains themselves somehow rise up out of the sea?

Amazing though it seems, direct measurements show that parts of the Cascade Mountains are, in fact, still growing higher—by as much as a meter every hundred years. These measurements and the fossils found on mountains are only two of many bits of evidence showing that mountain ranges around the world have been pushed upward. To find out more about how this happened, you can begin by looking at the types of rocks found in the mountains themselves.

ACTIVITY 11A / QUESTIONS ABOUT MOUNTAINS, VOLCANOES, AND EARTHQUAKES

Divide a page in your learning journal into three columns. Label the first column "Mountains." Under this heading, list what you already know about mountains. Draw a line under your list. Below the line, write some questions you would like answered about mountains. Label the second column "Volcanoes" and the third column "Earthquakes." In each column, list what you know and then what you would like to know about these topics. ❖

■ 11.1 THE MAKING OF MOUNTAINS

Mountains are made of one or more kinds of rocks from three groups: **igneous**, **sedimentary**, and **metamorphic** rock. You may recall that igneous rock is produced when molten rock cools and hardens, either on the Earth's surface or within the Earth's thin outer rock layer, or **crust**. Sedimentary rock is formed from small particles of rock that have been eroded by water, wind, or ice and then deposited in layers. The weight of the piled-up sediments presses the particles together and slowly forms them into solid rock. Metamorphic rock is produced when heat and pressure change the form of rock buried deep beneath the Earth's surface.

The Rocky Mountains are made mostly of sedimentary rock. Because of the way sedimentary rock is formed, a bed of sedimentary rock is usually made up of horizontal layers like a stack of sandwiches (Figure 11.1). But if you look at the tops, or peaks, of some mountains in the Rockies, such as Mount Rundle or Mount Kerkeslin, you will

FIGURE 11.1 ◀ ***What does the layered appearance of these rocks suggest about how they were formed?***

see that their layers of sedimentary rock are tilted at steep angles. In some mountain ranges, you may see layers of sedimentary rock rolling up and down like waves (Figure 11.2). Geologists call these wave-like rock features **folds**.

After looking at Figure 11.2, you might ask: Could the sedimentary layers have been *deposited* in these wavy patterns? Observations and experiments soon show that loose sediments do not stay on a steep slope for very long. Sediments are pulled downward by gravity and settle in horizontal or near-horizontal layers. The folds seen in Figure 11.2 were, therefore, most likely produced after the sediments hardened into solid sedimentary rock.

FIGURE 11.2 ▶ ***How do you think the layers of sedimentary rock became shaped in these ways?***

ACTIVITY 11B / FOLDING ROCKS

In this activity, you will consider how horizontal layers of rock may be formed into mountain peaks.

MATERIALS

two foam blocks (each about 15 cm × 30 cm × 8 cm)
felt marker

PROCEDURE

1. Using the felt marker, draw horizontal lines along the sides of each foam block.
2. Hold one end of a single block firmly against your desktop. Predict what will happen when you slowly push the opposite end of the block toward the end you are holding. Push the block, then observe and record what happens to the lines. Sketch the appearance of the pushed foam block.
3. Place two foam blocks end to end. Predict what will happen when you push on the end of one block while holding the second block still (Figure 11.3). Push the block, then observe and record what happens. Sketch and label your results.

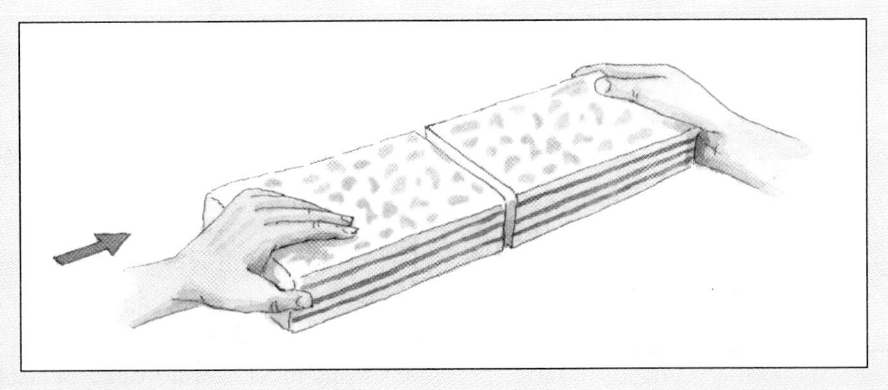

FIGURE 11.3
What happens when you hold one foam block and push the other against it?

4. Together with some classmates, arrange the foam blocks to resemble, in turn, each of the mountain peaks shown in Figure 11.4. Sketch your results. Describe what you had to do to produce each mountain.

DISCUSSION

1. Describe the movement of rock that you think produced the folding shown in Figure 11.2.
2. Suggest what might have occurred in the Earth's crust to change a horizontal section of sedimentary rock into each of the three mountain peaks shown in Figure 11.4.
3. Copy the sketch of one of the three mountains in Figure 11.14. Add arrows showing how the rock may have moved. Explain how the mountain may have been formed. ❖

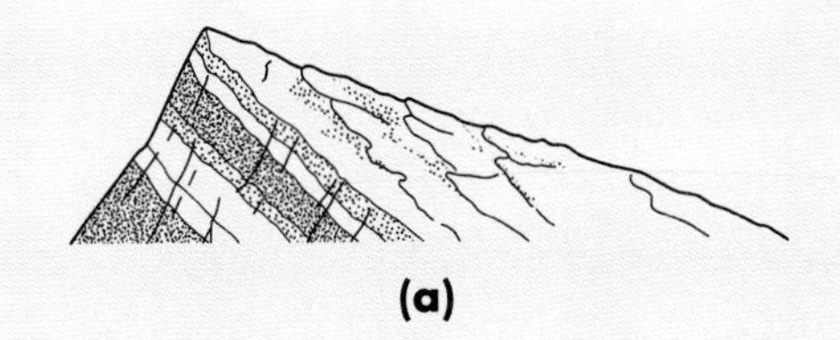

(a)

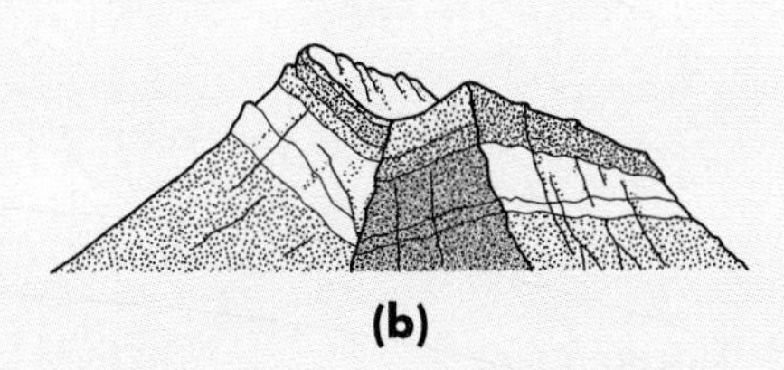

(b)

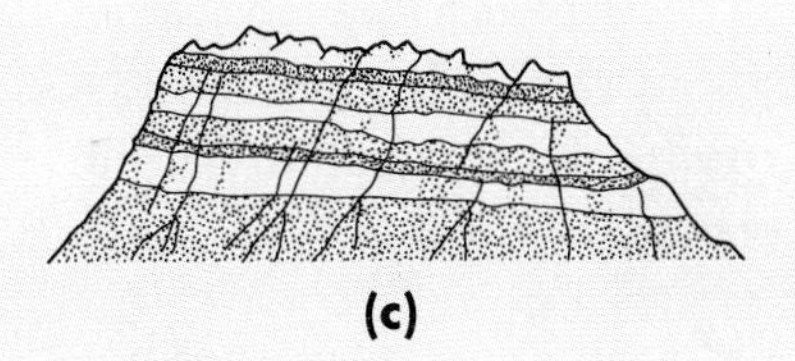

(c)

FIGURE 11.4
These three peaks are found in the Rocky Mountain Range.

FOLDS AND FAULTS

Folds occur where sections of rock have been forced to change shape. These changes take place when rock sections are squeezed from the ends like an accordion. The squeezing, or pressure, pushes parts of the rock up to form **anticlines**. Other parts are driven down to form **synclines** (Figure 11.5).

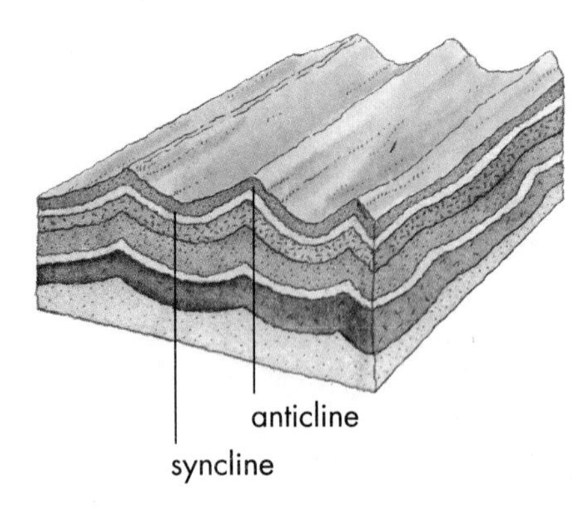

FIGURE 11.5
Synclines (downfolds) and anticlines (upfolds) are formed when a section of rock is squeezed.

In some cases, instead of folding, the rock under pressure may break along lines of weakness. Continued pressure pushes the rock sections on either side of the break past one another. A **fault** is a break in the Earth's crust along which rock moves.

Rock may move either horizontally or vertically along a fault (Figure 11.6). Vertical movement can drop one block of rock below the level of the rock on the other side of the fault. This is called uplift. If the fault runs at a sharp angle through the Earth's crust, vertical motion can also cause one block to tilt on top of another.

Uplift, along with folding, explains the appearance of the different peaks illustrated in Figure 11.4. Mount Rundle has been sharply tilted and uplifted. Castle Mountain has been uplifted while its sedimentary rock layers have remained almost horizontal. The rock of Mount Kerkeslin has been squeezed sideways to form a syncline and then uplifted.

EXTENSION

■ Whereas some processes push mountains upward, other processes wear them down. Draw a diagram showing a process that wears away rock from mountains. Some mountain ranges are high and have sharp peaks. Other mountain ranges are lower and have rounded peaks. Which type of mountains do you think are older? Write a caption for your diagram that answers this question.

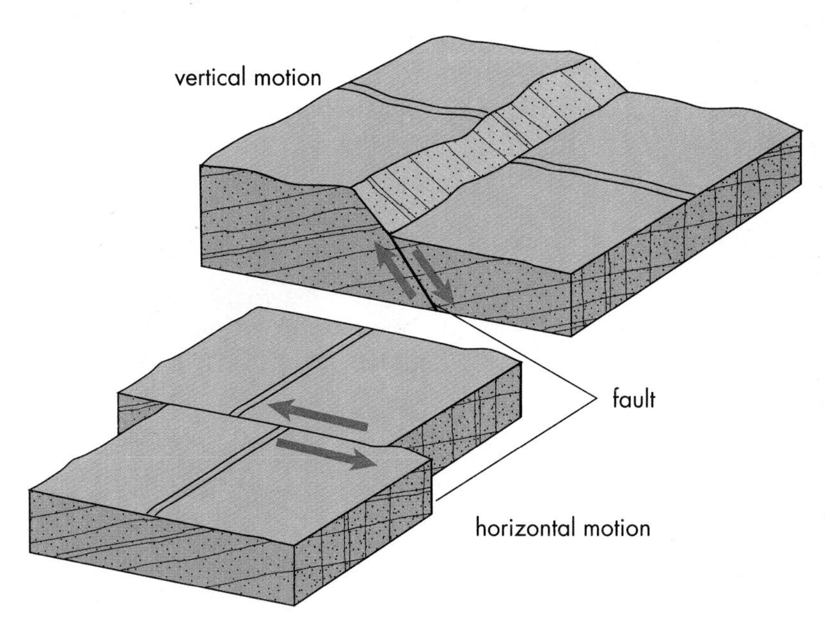

FIGURE 11.6
A section of rock may move horizontally or vertically along a fault.

► REVIEW 11.1

1. Name the three groups of rocks.

2. How is sedimentary rock formed?

3. (a) What is a fold?
(b) Sketch and label an anticline and a syncline.

4. What is a fault?

5. Name a process that may push rock upward to form a mountain. Describe a piece of evidence that suggests this process has occurred.

JACK SOUTHER

On one field trip, geologist Jack Souther had to backpack for three days with only peanut butter and Jell-O to eat. Another time, he watched an angry mother bear rip the bark off the tree he was stranded in. As one of North America's foremost experts on volcanoes, Souther has many stories to tell.

Souther's career in geology started with horses. He learned to ride as a teenager on a cattle ranch, then became a horse packer and guide in the Rocky Mountains. At that time, in the 1940s and 1950s, geologists doing field work travelled by horse. "As I took them up and down mountains," says Souther, "the geologists would point out fossils. That triggered my interest. I decided to go to university and learn more."

Now, looking back on his years as a geologist, Souther says it was "the best of both worlds. During the summer, you were out in the field doing something productive, but it was really like a holiday. Then, when you got back to your office in the fall, you were all fired up to see the results of your efforts."

A technological development that really changed geological research during Souther's time was the airplane. Souther was one of the first geologists to use light aircraft and "air drops" in his work. A small plane would drop food by parachute along a route that the geologists would later spend two weeks travelling over on foot. "This worked really well— except when the bears got there before we did," he recalls.

Another technological development that changed the way geologists work was the computer. "In the past, we took notes laboriously in longhand, came back to the office and wrote them out in longhand, and took them to a typist to type," he remembers. "Now everything is fed into a computer system that in a few seconds will do what would have taken me years."

One of the highlights of Souther's career was the 10 years he spent studying a wilderness area in the Pacific Northwest. This area contains one of the largest and youngest volcanoes in North America— Mount Edziza. It last erupted about 200 years ago.

Although he is past retirement age, Souther continues to work and to be fascinated by geology. During his career, he explored places that had never before been studied by geologists. "Now younger people are using this information and adding to it," he says. "They are looking at it with new tools and concepts. Geology today is just as exciting for the young people coming in as it was for me years ago."

Have you ever considered becoming a geologist? What qualities do you think a good geologist would have?

■ 11.2 MOUNTAINS OF FIRE

Mountains that are produced by uplifting and folding form slowly over millions of years. Other types of mountains form much faster. **Volcanoes** are fast-growing mountains made of igneous rock. They are produced when molten rock from deep in or below the Earth's crust pours out or erupts onto the surface. **Magma** is the molten rock within the crust. **Lava** is magma that has reached the surface of the Earth.

One of the youngest volcanoes on Earth began forming in 1943 in a farmer's field in Mexico. The farmer was surprised early one morning to find red-hot cinders shooting out of a hole in the ground. He tried to fill the hole, but within a day it had opened into a crater more than 2 m across. Worried villagers listened and watched each day as explosions came from the ground and the crater continued to throw out hot ashes and stones, which soon piled up to form a small dome around the hole. Within a year, the volcano had built up a cone-shaped mountain, 430 m high and nearly 1 km across at the base. The rocks and ashes from the volcano buried two small villages nearby (Figure 11.7). Nine years after they began, the eruptions suddenly ended. The volcanic mountain, named Parícutin, finally stopped growing at a height of more than 500 m.

FIGURE 11.7
Hot rock and ashes from Parícutin volcano in Mexico almost buried this church in a nearby village.

TYPES OF VOLCANOES

Because of the way they are formed, volcanic mountains are usually shaped like a cone, with a bowl-shaped hole, or crater, at the top. A few years of eruption may be followed by hundreds or thousands of years when the volcano is not active and there are no eruptions. During these quiet times, the shape of the volcano may be altered by erosion. There are some active volcanoes on the west coast of the United States. Many mountains in the area are inactive volcanoes (Figure 11.8). Some may erupt again in the future.

Volcanic eruptions are not always violent and explosive. Many volcanoes erupt gently, spilling out lava that flows slowly down the sides of the mountain. The type of lava largely determines the way a volcano erupts and the type of volcanic mountain that eventually forms (Figure 11.9).

Parícutin is an example of a **cinder cone** volcano, the smallest type. Cinder cones are produced from hot bits of material that are hurled from an opening, or vent, and pile up to produce a mound of cinder-like rock. Cinder cones are not usually found alone. More often, cinder cones grow clustered together around the sides or base of much larger volcanoes.

FIGURE 11.8 ◀
What evidence can you see that Mount Saint Helens in Washington state is, indeed, a volcano and not just a mountain?

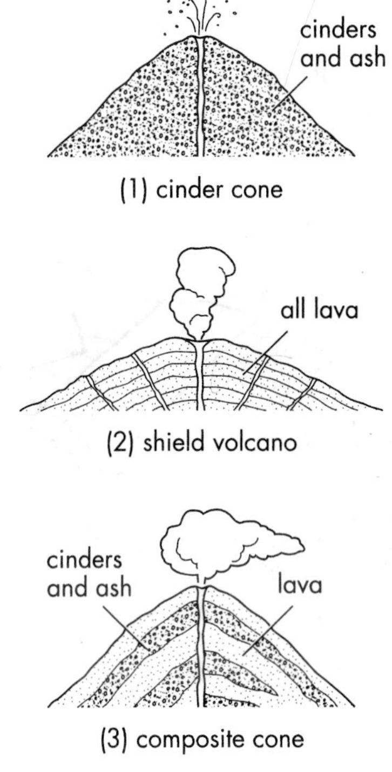

FIGURE 11.9
There are three main types of volcanoes.

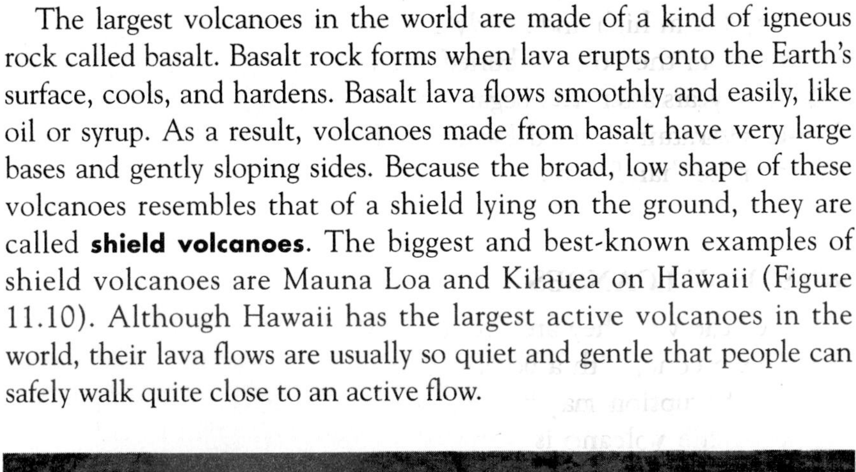

The largest volcanoes in the world are made of a kind of igneous rock called basalt. Basalt rock forms when lava erupts onto the Earth's surface, cools, and hardens. Basalt lava flows smoothly and easily, like oil or syrup. As a result, volcanoes made from basalt have very large bases and gently sloping sides. Because the broad, low shape of these volcanoes resembles that of a shield lying on the ground, they are called **shield volcanoes**. The biggest and best-known examples of shield volcanoes are Mauna Loa and Kilauea on Hawaii (Figure 11.10). Although Hawaii has the largest active volcanoes in the world, their lava flows are usually so quiet and gentle that people can safely walk quite close to an active flow.

FIGURE 11.10 ◀
Kilauea is an active shield volcano on the island of Hawaii. It rises only 1100 m above the surface of the island, but its roots go down another 5000 m to the ocean floor.

■ Composite cones are also known as stratovolcanoes ("layered" volcanoes).

FIGURE 11.11
Mount Saint Helens erupted in 1980, melting ice and snow on the peak and sending boiling mud rushing down the volcano's slopes.

Another type of volcano is called a **composite cone**. It is made from alternating layers of cinder rock and hardened lava. A composite cone has steeper slopes than a shield volcano because its lava is usually thicker and slow-flowing, like warm peanut butter. This thick, sticky lava often cools and hardens while still in the vent of the volcano. If the hardened lava blocks the vent, it traps magma below. Pressure from the build-up of volcanic gases in the magma may eventually cause the volcano to burst open, throwing gases and chunks of hot lava high into the air. Composite cones include many of the world's most famous and violent volcanoes, such as Mount Fuji in Japan, Mount Vesuvius in Italy, and Mount Saint Helens in the state of Washington (Figure 11.11).

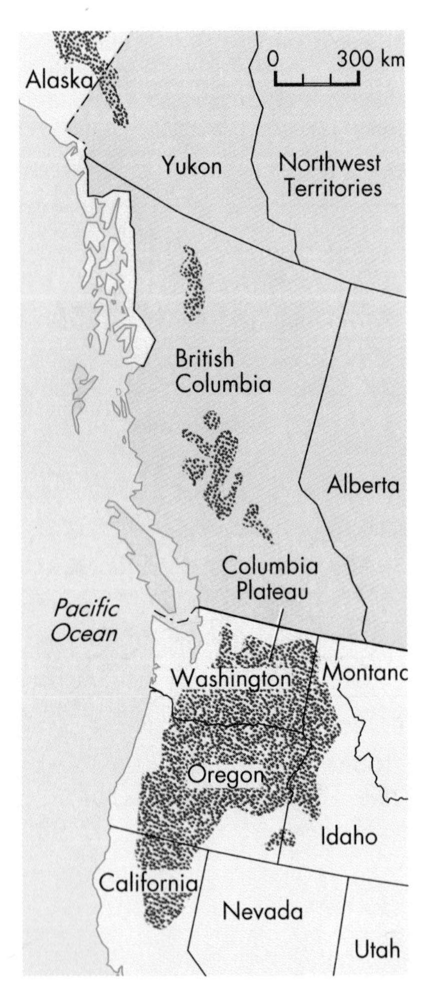

FIGURE 11.12
The shaded areas show the location of huge basalt plateaus.

TYPES OF IGNEOUS ROCK

What happens to the molten materials that pour out of an erupting volcano? They eventually cool and solidify to form igneous rock. One of the most common igneous rocks on Earth is basalt. Almost all of the sea floor is made of basalt lava that has poured out of long cracks there and cooled rapidly in the cold sea water. On land, basalt lava also pours out of long cracks. Sometimes, instead of forming a volcanic mountain, basalt lava flows out over a huge area and hardens to form a thick sheet of rock.

One of the biggest known outpourings of basalt lava on land took place in the area of the northwestern United States between 10 and 25 million years ago. This lava flood produced the Columbia Plateau, an area made up of many layers of basalt covering much of the states of Washington and Oregon (Figure 11.12). In some places, the basalt is 4 km deep. Some of the volcanoes of the Cascade Range, including Mount Saint Helens, sit on this plateau.

Whereas flowing lava produces a large, solid mass of igneous rock, exploding lava produces fragments and chunks of rock of various sizes. Some of the blocks of hardened lava shot out during volcanic explosions may be as big as an automobile. Other tiny particles of volcanic material may float up several kilometers into the atmosphere and be carried around the world by winds.

No matter what its size or final appearance, all igneous rock formed by lava cooling on the Earth's surface is called **extrusive igneous rock**. It is produced by lava that has extruded (flowed out) from a volcanic vent or crack in the Earth.

A second type of igneous rock is formed by the cooling of magma deep within the Earth's crust. This type of rock is called **intrusive igneous rock**. The magma may cool and harden to form huge underground bodies of rock. Large numbers of these together produce massive structures of igneous rock called **batholiths**, which extend over 100 km^2 or more. Batholiths may later be uplifted to form mountain ranges. Much of the Sierra Nevada Range in the United States and the Coast Range in British Columbia was produced in this way.

■ Dust and ash from large volcanic eruptions can affect the weather. A thick cloud of volcanic dust in the atmosphere blocks out some of the sun's energy and may cool normal temperatures around the world for months or even years after the explosion.

ACTIVITY 11C / CRYSTALS IN IGNEOUS ROCKS

This activity will help you learn how to recognize whether an igneous rock is intrusive or extrusive.

MATERIALS

safety goggles
apron
insulated gloves
sodium thiosulfate solution
dropper
two glass slides or two Petri dishes
container of ice
hand lens or microscope
hotplate
specimens of intrusive and extrusive igneous rocks

■ **If your teacher instructs you to use a hotplate, wear insulated gloves and handle the slides or Petri dishes carefully. High temperatures may crack glass. Be sure that the thermostat is set to low or minimum and that you wear safety goggles.**

PROCEDURE

1. Put on your safety goggles and apron.
2. Have two slides or Petri dishes ready, one cold and the other at room temperature or hot. Your teacher will help you chill and heat the slides or dishes. Your teacher will also heat the sodium thiosulfate in a water bath and have it ready for your use.
3. Remove the cold slide or dish from the container of ice and place a drop of the warm solution on it. Record the time it takes for the drop to harden. (This experiment should be done quickly, before the slide or dish warms up.)
4. Look at this slide or dish under the hand lens or microscope. Draw the pattern of crystals that you see. ➡

FIGURE 11.13
A close look at this igneous rock shows it is made up of many crystals.

5. Repeat Steps 3 and 4 with the room temperature slide or dish. You can also repeat Steps 3 and 4 with a hot slide or dish, but be sure to wear insulated gloves when heating the glass. If you are fast, you may be able to watch crystals growing on the hot glass.
6. Examine the rock specimens and look at the crystal sizes (Figure 11.13). Divide the specimens into those with mostly large grains and those with mostly small grains.

DISCUSSION

1. Why should you wear an apron, goggles, and insulated gloves when handling a hot slide or Petri dish?
2. Which slide or dish has the largest crystals? Why do you think this is so?
3. Are there any differences between the crystals other than size? If so, state what they are.
4. Assume that your samples represent crystals of rocks.
   (a) Which of these crystals do you think would form near the surface of the Earth? Which crystals do you think would form deeper down? Give reasons for your answer.
   (b) Which of the crystals would you expect to find in extrusive rocks, and which in intrusive rocks? (Hint: Think about the meaning of "intrusive" and "extrusive.")
5. Use the information you have gained from this activity to identify the rock specimens as extrusive or intrusive. ❖

DID YOU KNOW

■ Chemicals in volcanic ash are good fertilizers. Within weeks after an eruption, plants begin to grow in the empty ash-covered fields. Animals move into the area to feed on the plants. Many important farming areas are found in regions around volcanoes.

PREDICTING VOLCANIC ERUPTIONS

The clue that first told geologists to watch Mount Saint Helens more closely was a tremor (a small **earthquake**) that occurred near the mountain in 1978. Although the volcano had not erupted since 1857, the U.S. Geological Survey broadcast a warning to prepare people. The warning turned out to be two years early.

Mount Saint Helens erupted spectacularly on May 18, 1980 (Figure 11.14). Several small earthquakes shook the area, and clouds of gases and volcanic fragments shot nearly 20 km into the sky. Within minutes, hundreds of square kilometers of forest around the volcano were destroyed by the blast and the flow of volcanic fragments, hot mud, and rocks that followed. Ash settled in layers up to 7 cm thick in towns and cities far to the east of the volcano.

FIGURE 11.14
Mount Saint Helens (a) before and (b) after the 1980 eruption. What happened to the top of the mountain?

(a)

(b)

The eruption of Mount Saint Helens had serious effects on Washington state. A huge area of forest, with its wildlife, was destroyed. Rivers were filled with mud, killing fish and halting shipping. Vehicle engines and electronic equipment were damaged by the settling ash.

The danger from Mount Saint Helens did not completely end in 1980. After the big eruption, more lava began pushing up from the floor of a new crater. Throughout the summer, there were several small explosions, with steam and ash eruptions. In the past, the pattern of volcanic activity of Mount Saint Helens has been an explosive eruption followed by a period of quieter eruptions.

The eruption of Mount Saint Helens is another reminder of the tremendous forces within the Earth that are constantly changing the surface of the planet. Volcanoes that have erupted in the past are likely to erupt again. People who live near volcanoes need to know when the next eruption is likely to occur. One way to predict eruptions is by studying the history of the volcano.

EXTENSION

- Find out when and where a serious volcanic eruption occurred recently. Find out how the eruption affected the area around the volcano and whether it also affected other parts of the world. Make a display that shows and explains the results of the eruption.

STACY GERLICH: DISASTER TRAINER

Southern California experiences several hundred earthquakes every year. Fortunately, the vast majority of them are not detected by the average citizen. There have been, however, seven significant earthquakes in the area over the past ten years.

When a significant earthquake does occur, it covers a large area and causes more problems than just shaking buildings. Streets and highways are damaged or blocked by debris, making them impassable. Gas lines are ruptured, causing fires. Electrical power and telephone lines go down, making crucial communications difficult at best. This multiplicity of emergencies severely strains the conventional professional emergency agencies.

Recognizing this, the Los Angeles City Fire Department has created the Community Emergency Response Team (CERT) program. It consists of volunteer groups, comprised of ordinary citizens, assisting government agencies in the initial phase of an emergency.

Firefighter-lead paramedic Stacy Gerlich is one of eight trainers in the CERT program. Stacy had been a CERT trainer for almost four years and has worked for the L.A. Fire Department for over ten years. Gerlich and seven other instructors train approximately 1,000 people every year. To date, about 12,000 people have gone through the program.

The CERT program is a seven-week, $17\frac{1}{2}$ hour course of hands-on training. Topics covered in the course include earthquake threat, personal and family preparation, fire chemistry, hazardous materials, triage, evacuation, search techniques, and rescue training. After the initial seven-week course of instruction, team members attend quarterly $2\frac{1}{2}$ hour refresher sessions.

Gerlich says that any business group or community organization who can assemble at least twenty-five members to comprise a team may enroll in the training program. As the program continues to grow within the City of Los Angeles, it is being 'exported' to other communities and governmental agencies. Recently Gerlich and other CERT trainers were invited by FEMA (Federal Emergency Management Agency) to travel to our nation's capital to demonstrate the program.

ACTIVITY 11D / WHEN WILL VESUVIUS ERUPT AGAIN?

In this activity, you will analyze the dates when Italy's Mount Vesuvius erupted. These dates have been recorded over the past 2000 years by people living in the area. Can you use this information to predict future eruptions?

MATERIALS

2.5 m strip of paper, 6 to 10 cm wide
ruler
felt pens

A.D. **79**	**1631**	1785	1891
203	1660	1793	1900
472	1682	1804	1903
512	1689	1805	1904
685	1694	1838	1906
993	1707	1850	1910
1036	1737	1858	1913
1049	1760	1861	1926
1138	1767	**1872**	1929
1306	**1779**	1875	**1944**
1500	1784	1881	

TABLE 11.1
Dates of Eruptions of Vesuvius (major eruptions in BOLD type)

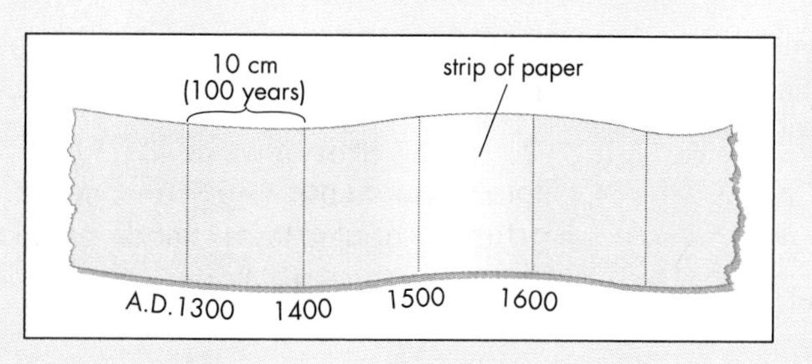

FIGURE 11.15 ◀
A portion of your time line may look like this.

PROCEDURE

1. Using a ruler, draw 21 pencil lines across your strip of paper. The lines should be 10 cm apart (Figure 11.15). Each line will represent 100 years on your time line. Label the first pencil line "0 years." Label the last line "Year 2000."
2. The dates of major eruptions of Mount Vesuvius are listed in Table 11.1 in bold type. Mark them on your paper with a thick colored line. Mark the minor eruptions with a thinner line of a different color.

DISCUSSION

1. Study your finished time line. If you see a pattern in the lines you have drawn, describe it.
2. Based on the pattern you see, predict when Mount Vesuvius will next erupt. Explain how you made your prediction.
3. When did the longest period between eruptions begin and end? How long was this period?
4. Using only the data *after* the year 1779, predict when the next eruption will occur.
5. Which prediction do you think is likely to be more accurate, the one you made in question 4 or in question 2? Explain your answer.
6. (a) Do you think a prediction would be enough of a reason to evacuate the area near a volcano? Explain why or why not.
   (b) What other information might be needed to persuade people to leave their homes? ❖

WHERE ARE VOLCANOES LOCATED?

People who live near a volcano must be prepared for an eruption. Knowing when the volcano erupted in the past can help. But volcanoes do not always erupt on schedule.

People can learn more about when and how volcanoes erupt by examining where they are located. All the volcanoes in the United States are on the west coast. There are no volcanoes in the central or eastern U.S. This distribution may reveal more about how volcanoes originate.

ACTIVITY 11E / MAPPING MOUNTAINS AND VOLCANOES

In this activity, you will examine the distribution of volcanoes and mountain ranges around the world.

MATERIALS

outline map of the world
felt pens or colored pencils
atlas or globe

PROCEDURE

1. On your map, copy the distribution of volcanoes shown in Figure 11.16.
2. Use an atlas or globe to find the location of the following mountain ranges: North American Coast Ranges, Andes, Himalayas, Alps, Urals, and Appalachians. Shade or color in the position of each of these ranges on your map.
3. Keep your map if your teacher asks you to. You may be asked to use it in another activity later in the chapter.

DISCUSSION

1. Which continents have the fewest active volcanoes?
2. In which parts of North and South America are most of the volcanoes located?
3. Does there appear to be any connection between the location of volcanoes and oceans? Give a reason for your answer.
4. List the mountain ranges where there are many volcanoes.
5. (a) In which areas are there mountain ranges without volcanoes?
   (b) What do you think is the difference between these ranges and the ones listed in your answer to question 4?
6. Do you see a pattern in the location of mountains and volcanoes around the world? If so, describe the pattern and suggest a reason for it. ❖

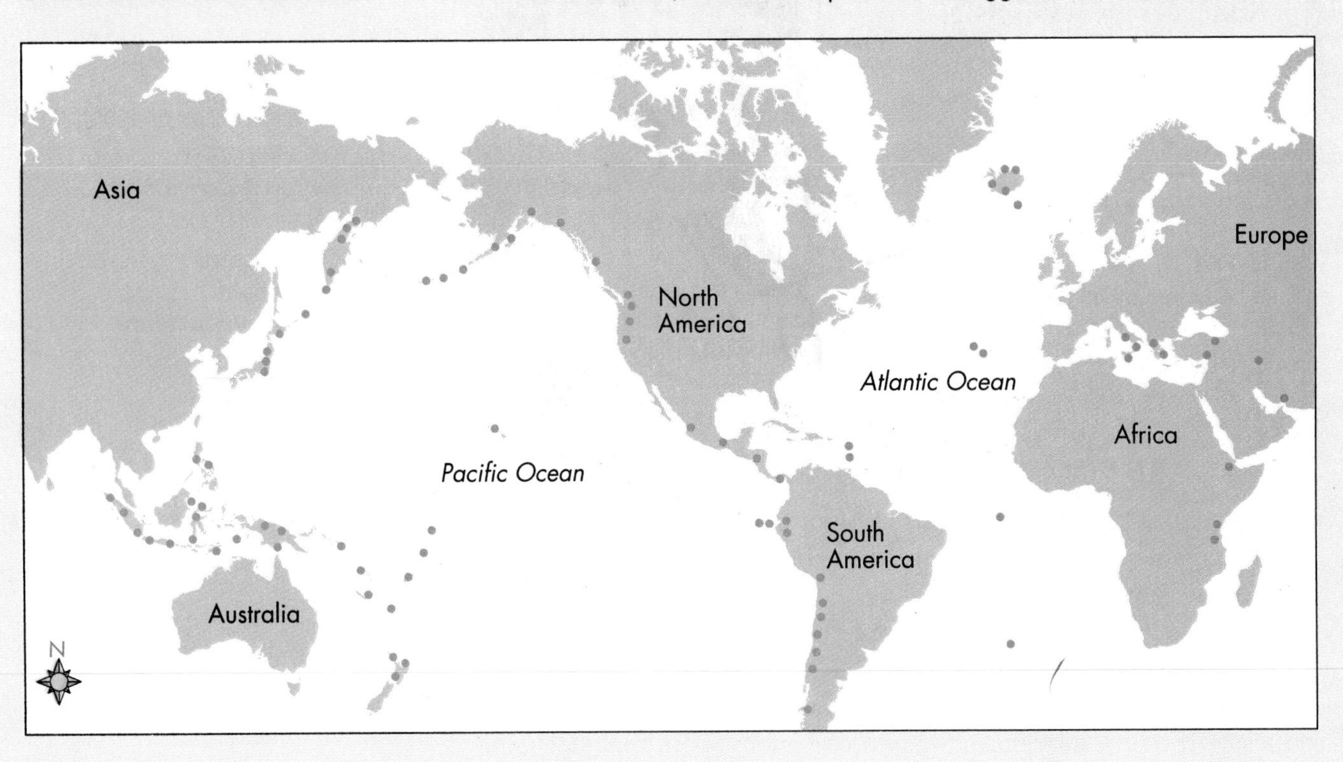

FIGURE 11.16
The dots show the location of some of the world's active volcanoes.

▶ REVIEW 11.2

1. Volcanoes with gently sloping sides, such as Mauna Loa in Hawaii, are formed from basalt. What property of basalt lava causes the shape of these volcanoes?
2. How is a cinder cone formed?
3. What is the difference between a shield volcano and a composite cone?
4. Explain why thick, sticky magma may cause a volcano to erupt explosively.
5. Aside from volcanoes, what geological features may be produced by a flow of basalt lava?
6. What is the difference between intrusive and extrusive igneous rocks?
7. Why do you think it is important to be able to predict volcanic activity?
8. How could you tell the difference between a mountain that was formed by a volcano and one that was built by folding and uplifting?

■ 11.3 WHEN THE EARTH SHAKES

The ground-shaking earthquakes that disturb parts of the world from time to time can be as awesome and powerful as an erupting volcano. The wave-like movements of the ground and the jolting shocks of an earthquake normally last for only a minute or less, but the damage they cause can be terrible.

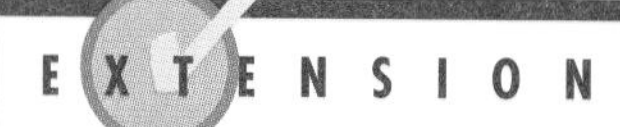

■ Plan a "volcano vacation." Describe three volcanoes you would like to visit and explain why.

FIGURE 11.17 ◀
Destruction was massive during the Alaska earthquake of 1964. After the earthquake, the residents of Anchorage rebuilt their town. How would you begin repairs on this street?

One of the most violent earthquakes in recent times occurred in Alaska in 1964 (Figure 11.17). It started without warning and lasted for seven minutes. Terrified residents of Anchorage saw the streets of their town suddenly roll and heave. Giant cracks opened in the ground, and underground gas and water lines snapped apart. Buildings twisted and collapsed onto people who were fleeing into the streets.

■ The 1964 earthquake in Prince William Sound, Alaska was one of the largest magnitudes recorded.

One end of a school building dropped 6 m, and several homes built on a cliff slipped into the churning ocean. Giant ocean waves set off by the earthquake slammed into the harbor at Valdez, 200 km from Anchorage. The waves washed away a crowd of people on the pier and lifted a 10,000 ton freighter onto the land. The effects of this massive earthquake were felt as far away as Texas, where people saw the water in their swimming pools slosh back and forth.

Major earthquakes take place every few years, but thousands of minor earth tremors are recorded by geologists every day. Although too small to be felt by people, these slight movements of the Earth are recorded on sensitive monitoring equipment. More than a million earthquakes occur each year around the world.

EXTENSION

■ Some people in California live in high-risk areas for earthquakes. Write and illustrate a report about how to prepare for an earthquake. See the front pages of your telephone book for some suggestions. Describe what to expect when an earthquake occurs and what you should do before and after. Find out if your community has an emergency plan. Describe the plan, or explain why your community does not have one.

EARTHQUAKES AND FAULTS

In the first section of this chapter, you learned how some mountains are produced by the shifting of rock along a fault or fracture in the Earth's crust. Whenever rock moves suddenly along a fault, in any direction, it releases an enormous amount of energy. The result is an earthquake.

One of the most famous faults linked with earthquakes is the San Andreas Fault, a break that runs for 700 km along the west coast of North America (Figure 11.18). The land on either side of the San Andreas Fault is slowly moving horizontally. Careful measurements have shown that the land on the west side of the fault is moving north at a rate of about 2 cm a year. During the large earthquake in 1906, which destroyed much of the city of San Francisco, parts of the land near the San Andreas Fault moved 5 m!

Many faults and earthquakes occur on the ocean floor. People on land do not usually feel undersea earthquakes, but a major one can have a devastating effect. Shock waves produced by an undersea earthquake travel outwards through the water, much like the ripples that spread from a stone dropped in a pond. The shock waves move at a speed of hundreds of kilometers an hour, and can travel through thousands of kilometers of open ocean. Far from land, these waves are no higher than other waves in the open ocean. But when a shock wave reaches the shoreline, it is suddenly forced to slow down. The wave's tremendous energy is converted into a towering wall of water, as much as 30 m high, which crashes onto the land. A huge wave caused by an earthquake on the ocean floor is called a **tsunami**. Tsunamis can cause terrible damage when they sweep ashore. A tsunami from the 1964 Alaska earthquake hit Port Alberni on Vancouver Island, wrecking cars and buildings (Figure 11.19).

FIGURE 11.18
Aerial photograph of part of the San Andreas Fault in California. Large earthquakes along this fault struck the city of San Francisco in 1906 and 1989.

FIGURE 11.19
Damage in Port Alberni, British Columbia, from a tsunami caused by the Alaska earthquake of 1964.

■ The strength of an earthquake is measured on the Richter scale. Write a brief report explaining what this scale is and how it is used.

Although scientists know that earthquakes occur along faults, they cannot accurately predict when an earthquake will occur. However, mapping the location of earthquakes around the world can help us identify regions where the risk of earthquake damage is greatest. People who live in these areas can then be prepared.

ACTIVITY 11F / MAPPING EARTHQUAKES

In this activity, you will examine the distribution of earthquakes around the world.

MATERIALS

outline map of the world
felt pens or colored pencils

PROCEDURE

1. On your map, copy the distribution of earthquakes shown in Figure 11.20. (If you completed Activity 11E, Mapping Mountains and Volcanoes, your teacher may ask you to add to the map you have already produced.)

DISCUSSION

1. Which areas of the world have the greatest numbers of major earthquakes?

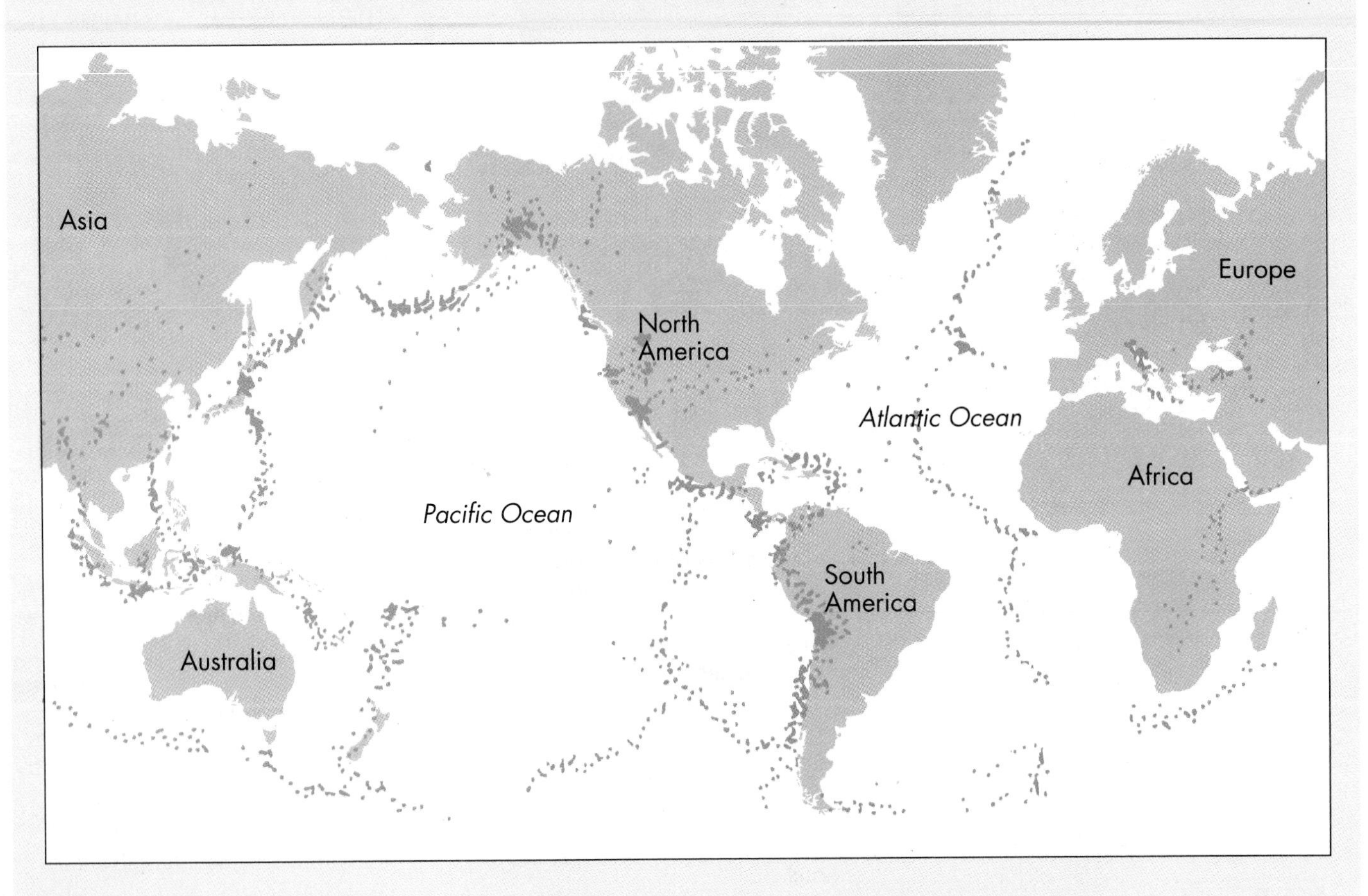

FIGURE 11.20
The dots show the locations of major earthquakes around the world. Compare this map with Figure 11.16.

2. Compare the location of earthquakes and the location of mountain ranges and volcanoes. Are they close together or far apart?

3. Do you think there is a connection between the locations of mountain ranges, volcanoes, and earthquakes? Explain your answer. ❖

MOUNTAINS, VOLCANOES, AND EARTHQUAKES

The building of mountains, the eruption of volcanoes, and the shaking of earthquakes do not happen by chance alone. You may have discovered that most mountain ranges, volcanoes, and earthquakes occur in the same parts of the world. Volcanoes and earthquakes are most active along the edges of continents where mountain ranges meet the ocean.

These and other observations about mountains, volcanoes, and earthquakes were linked together during the 1960s when geologists developed one of the most important theories in modern science. The theory describes how the Earth's entire crust is made up of several huge pieces, which move slowly over the planet. You will learn about this theory and examine some of the evidence for it in Chapter 12.

REVIEW 11.3

1. What is an earthquake?
2. "Earthquakes are rare events." Explain why you agree or disagree with this statement.
3. Explain the geological meaning of "fault."
4. What is a tsunami? Describe how a tsunami is produced.
5. Do you think that you are more likely to experience an earthquake in California or Ohio? Explain your answer.

CHAPTER REVIEW

KEY IDEAS

- Forces within the Earth cause solid rock to fold or to break along faults.
- Some mountains are produced by folding. Other mountains are produced by uplifting and movement along a fault. Folded and uplifted mountains are made mainly of sedimentary and metamorphic rock.
- Volcanoes are mountains made of igneous rock. They are produced when molten rock erupts onto the Earth's surface and solidifies.
- Small volcanoes built up from piles of rock and ash are called cinder cones. Large volcanoes made from basalt lava are called shield volcanoes. Volcanoes made from alternating layers of lava and cinder rock are called composite cones.
- Explosive volcanic eruptions occur when thick, sticky lava blocks the volcanic vent, trapping magma and gases inside.
- Rock formed by lava cooling on the Earth's surface is called extrusive igneous rock. Rock formed by magma cooling underground is called intrusive igneous rock.
- Earthquakes are caused when two large areas of rock suddenly move past each other along a fault.
- Many mountain chains, volcanoes, and earthquakes are located together in the same parts of the world.

VOCABULARY

igneous
sedimentary
metamorphic
crust
fold
anticline
syncline
fault
volcano
magma
lava
cinder cone
shield volcano
composite cone
extrusive igneous rock
intrusive igneous rock
earthquake
tsunami

V1. Construct a mind map, using as many terms as possible from the vocabulary list.

V2. Make up a crossword puzzle using at least 15 words on the list and exchange your puzzle with a classmate.

CONNECTIONS

C1. A tsunami can pass by ships on the ocean and cause no damage. The same tsunami may devastate a coastal town. Explain why this is so.

C2. Basalt is the most common rock on the Earth's crust. Why do we see so little of it?

C3. Explain how the type of lava may affect the shape of a volcano.

C4. In which type of rock(s) would you expect to find fossils: igneous, sedimentary, or metamorphic? Explain your answer.

C5. Although living near a volcano may be dangerous, farmers in many countries plant crops on the slopes of volcanoes. Why do you think they do this?

C6. List as many reasons as you can why mountains, earthquakes, and volcanoes are grouped together in certain areas of the world.

C7. Explain which you would rather live near: a shield volcano or a composite cone.

C8. Imagine that you had to live and work in either of two areas: one where earthquakes were common or one where volcanic eruptions were common. Which would you choose? Give reasons for your choice.

FIGURE 11.21
What would you like to know if you moved to the foot of this mountain?

EXPLORATIONS

E1. You have just moved to a town at the foot of a mountain (Figure 11.21). You would like to know how the mountain was formed and how old it is. List the evidence you would need to find out these things. Describe how you would find this evidence.

E2. On a large sheet of paper, make a flow chart that shows how a mountain somewhere in the world was formed. Include a photograph or drawing of the mountain.

E3. Find out where volcanoes are located in California. Contact the U.S. Geological Survey or visit your local library for information. On a map of California, mark and label areas where volcanoes are found.

E4. Mount Saint Helens erupted more than 10 years ago. Work with a partner to find out what the area around the volcano looks like now. Prepare a report with illustrations to describe the changes that have taken place since the 1980 eruption. Is the volcano still considered active?

E5. Make a model that shows how movement along fault lines can cause earthquakes.

E6. Design a brochure for people in your community to explain what an earthquake is, how they can prepare for an earthquake, and what they should do after an earthquake.

E7. Find out what makes a building able to survive an earthquake, including the construction of the building itself, the surroundings, and where the building should or should not be situated. Draw and label a building that would have a good chance of surviving an earthquake without major damage.

E8. Interview someone who studies mountains, volcanoes, or earthquakes. Prepare a list of questions before the interview.

REFLECTIONS

R1. If you did Activity 11A, reread the questions about mountains, volcanoes, and earthquakes that you listed in your learning journal. Were any of your questions not answered as you studied Chapter 11? Do you have any new questions that you would like answered? How might you find answers to these questions?

R2. How have your ideas about mountains, volcanoes, and earthquakes changed since you have read Chapter 11?

DRIFTING CONTINENTS

In 1908, a party of explorers led by Ernest Shackleton was struggling over the frozen Antarctic. These explorers wanted to be the first people to reach the South Pole, but the winter weather was harsh and their pack horses had died. The explorers were forced to turn back. Although they did not reach their goal, they did make an unexpected and important discovery.

In cliffs among the icy mountain ranges, Shackleton's team found deposits of coal. Coal is a black, rock-like material formed from plants that grew in swampy areas millions of years ago. How could coal possibly have formed in Antarctica? The coal-forming plants certainly could not have grown in Antarctica's present climate. Perhaps the climate there had once been much warmer. Or—an even more amazing idea—perhaps the Antarctic continent itself had once been located closer to the Equator!

ACTIVITY 12A / CONTINENTS CHART

Work with the members of your group to make a "continents" chart. Study a globe or map of the world, then list everything you know or think you know about the world's continents.

For example, which are the largest and smallest continents? Which continents are joined together by land? Which oceans lie between which continents?

List what you do not know and would like to find out. Add illustrations and display your chart. Refer to the class charts as you work through this chapter. ❖

■ 12.1 THE ORIGIN OF CONTINENTS

If you look at a map of the world, you may notice that some edges of the continents seem to match like the pieces of a jigsaw puzzle. For example, the coastlines on either side of the Atlantic Ocean look as though they could fit together (Figure 12.1). Some people wrote about this pattern when world maps were first drawn over 300 years ago. They wondered whether North and South America might once have been joined to Europe and Africa. Most people made fun of such an idea.

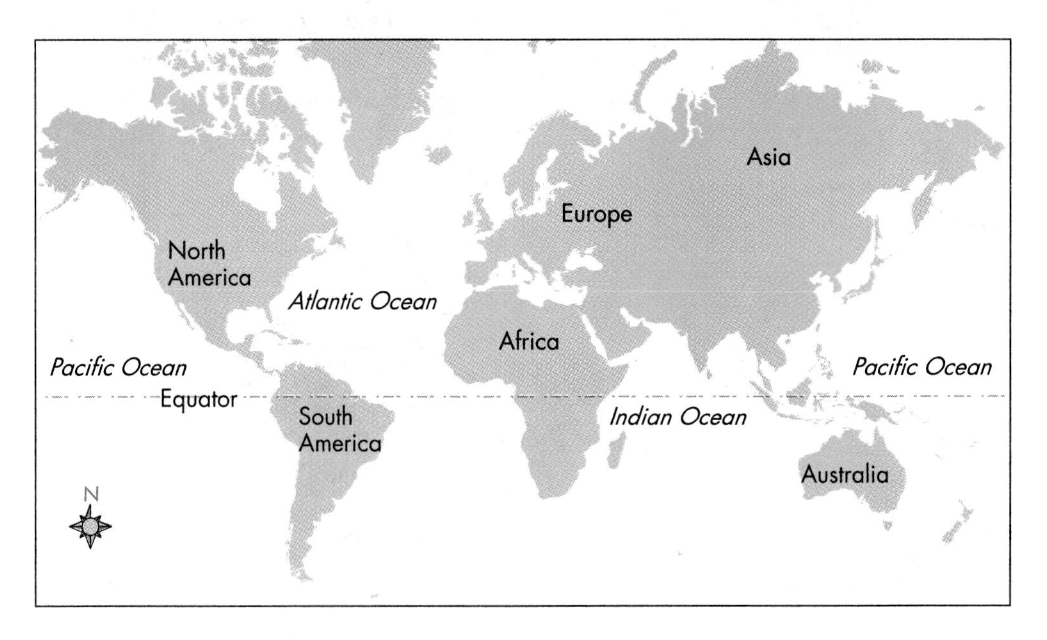

FIGURE 12.1 ▶
Examine the shape of the continents. Were the different continents of the world once joined?

Explorers and geographers later began to study the ocean bottom, or sea floor, around the continents. They discovered that the sea floor next to many coastlines is flat or gently sloping, and usually less than 200 m below the water's surface. At some distance away from the coastline, the sea floor suddenly drops away steeply to a much greater depth (Figure 12.2). The shallow sea floor near the continents is the **continental shelf.**

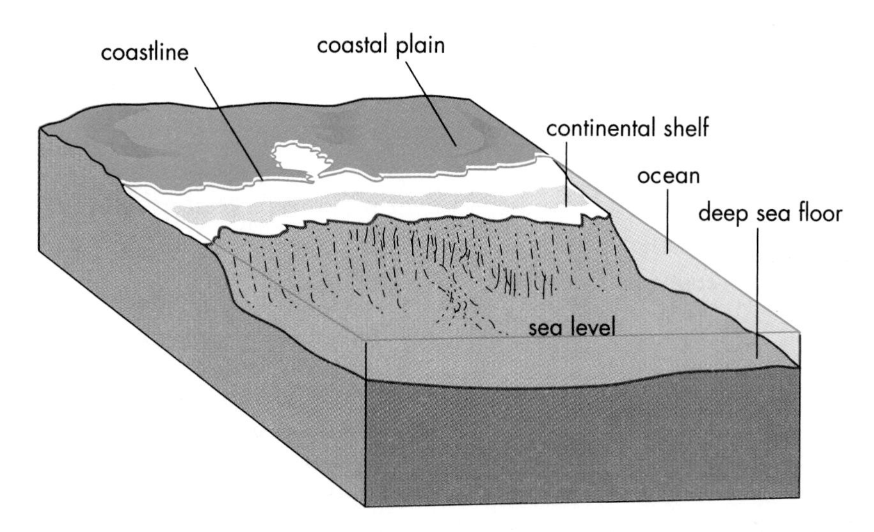

FIGURE 12.2 ◀
How could you discover where the continental shelf ends without doing deep dives?

ACTIVITY 12B / DO THE CONTINENTS FIT TOGETHER?

In this group activity, you will try joining the continents together using one of two different sets of continental shapes (Figure 12.3).

MATERIALS

a set of continental shapes with shoreline edges
a set of continental shapes with continental shelf edges

PROCEDURE

1. You will be given a set of continental shapes. Try to fit the shapes together like a jigsaw puzzle. Sketch your results, shading in areas where there are large gaps between the pieces.
2. Compare your results with your classmates' results. See how well other groups fit their shapes together.

DISCUSSION

1. Which set of shapes fits together better: the shapes with shoreline edges or those with continental shelf edges?
2. What does this suggest might have happened to the continents in the past?
3. Would you consider the continental shelf to be a part of the continent or a part of the sea floor? Explain your answer. ❖

FIGURE 12.3 ◀
How well do your continental shapes fit together?

THE THEORY OF CONTINENTAL DRIFT

In 1915, a German scientist and explorer named Alfred Wegener (1880–1930) suggested that the continents of the world had once been part of a single "supercontinent" that later broke up (Figure 12.4). The pieces then moved apart over millions of years and formed the separate continents we know today. Wegener's ideas are known as the **theory of continental drift**.

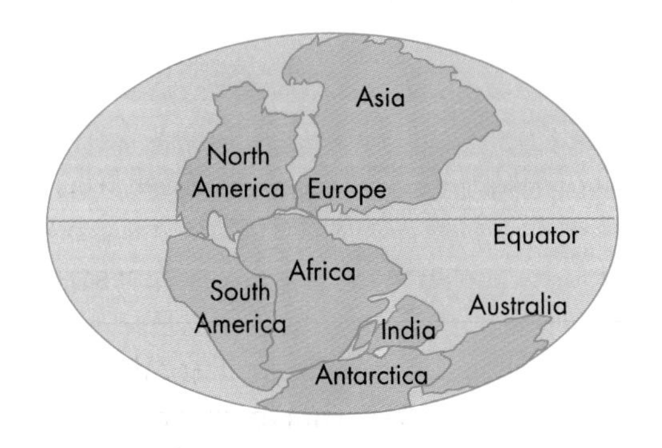

FIGURE 12.4 ◀
This map shows Wegener's "supercontinent," which he named Pangaea. In Greek, pan means "all" and gaea means "Earth."

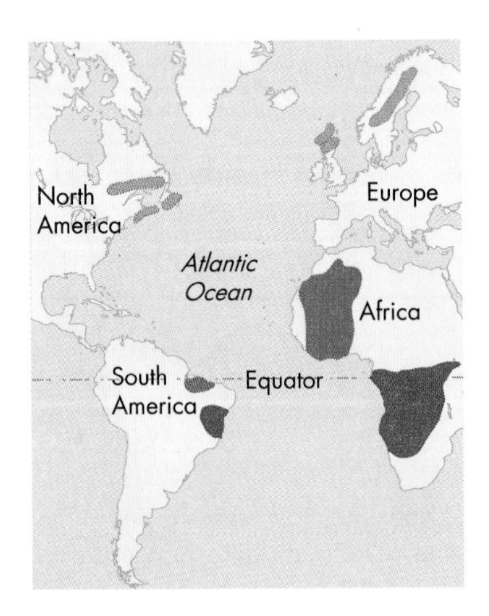

FIGURE 12.5
Similar rocks on both sides of the Atlantic are highlighted. How does the existence of these rocks support the theory of continental drift?

To find evidence for his ideas, Wegener studied the rocks on different continents. He pointed out that the rocks in the mountain ranges of western Europe are similar to the rocks in the mountain ranges of eastern North America. Similarly, ancient rocks found on the west coast of Africa match those on the east coast of South America (Figure 12.5). If the continents could be placed next to each other, the separate mountain ranges would fit perfectly together.

Wegener also looked at fossils. He discovered that fossils of a small reptile called Mesosaurus have been found in South America and South Africa, but nowhere else (Figure 12.6). These meter-long animals lived in freshwater lakes and ponds about 250 million years ago. Because it is unlikely that a Mesosaurus could have swum nearly 5000 km through the ocean from one continent to the other, South America and Africa were probably joined at the time these animals were still living on Earth.

FIGURE 12.6 ▶
Mesosaurus, an extinct reptile that lived in fresh water.

Finally, Wegener studied where certain animals and plants live today. For example, air-breathing fishes called lungfishes are found only in South America, Africa, and Australia. This evidence suggests that these continents were once linked but then moved apart.

EXTENSION

■ Living marsupials (animals with pouches) are found only in Australia, South America, and North America. Use your library to find out more about these animals. Why do scientists think they live where they do?

ARGUMENTS AGAINST CONTINENTAL DRIFT

A major problem with Wegener's theory was that it did not explain how continents might actually move. The evidence of continental shapes and of the matching rocks, fossils, and living organisms was very interesting. But it was not enough. Many geologists rejected the idea of continental drift because they did not know how areas of land the size of continents could move around on the planet. The idea of moving continents seemed so impossible, even ridiculous, that few scientists bothered to investigate it seriously.

When Alfred Wegener died in 1930, his theory of continental drift was almost forgotten. It was not until the 1960s that new evidence from the floor of the ocean made scientists think about Wegener's ideas once again.

▶ REVIEW 12.1

1. Explain why Shackleton's discovery of coal in Antarctica was unexpected.
2. Describe Wegener's theory of continental drift.
3. Summarize the evidence that Wegener obtained for his theory from
   (a) the shapes of continents,
   (b) rock formations on different continents,
   (c) fossils on different continents.
4. Explain why the distribution of lungfishes in the world is evidence for continental drift.
5. Why do you think many scientists in the first half of this century rejected Wegener's theory of continental drift?

■ 12.2 CLUES FROM THE SEA FLOOR

When Wegener was alive, people did not know much about the sea floor. They did not have the machines and equipment needed to dive deep underwater. They had few ways to map the bottom of the sea or bring up samples of rock.

As submarines and equipment to detect underwater objects were developed during and after World War II, more undersea exploration was possible. Scientists soon realized that the rocks they found on the ocean bottom were quite different from most of those on land. Bit by bit, they discovered clues in the rocks that eventually suggested how continents can move.

EXTENSION

■ Write a paragraph about sonar. Find out what the letters in the word "sonar" stand for. Describe at least three uses of sonar.

UNDERSEA MOUNTAINS AND TRENCHES

During the 1870s, scientists measured the depth of the Atlantic Ocean using ropes. They lowered a long cable with a heavy weight from the side of their research ship and recorded the length of the cable when the weight reached bottom. They took measurements at many different locations, then plotted these measurements on a graph. This research took a long time but gradually helped build up a map of the sea floor.

EXTENSION

■ The average depth of the Pacific Ocean is a little over 4 km. The deepest point on the Earth's surface is in the Marianas Trench in the Pacific Ocean near the Philippine Islands. At one place, the trench is more than 11 km deep. Compare this with the height of Mount Everest—8.8 km.

The sea floor maps clearly showed that a huge, wide mountain range, or **mid-ocean ridge**, runs underwater along the middle of the Atlantic Ocean. This craggy undersea range was named the Mid-Atlantic Ridge. It is much longer and higher than any mountain range found on land.

Mapping the sea floor was made much easier and quicker by the invention of **sonar**—a device that uses reflected sound waves to locate objects (Figure 12.7). With the help of sonar, scientists began exploring oceans all over the world. They discovered more ridges, running throughout the Earth's oceans. They found that each underwater mountain range runs roughly parallel with the coastline of the continents. They also discovered steep valleys running along the middle of each ridge.

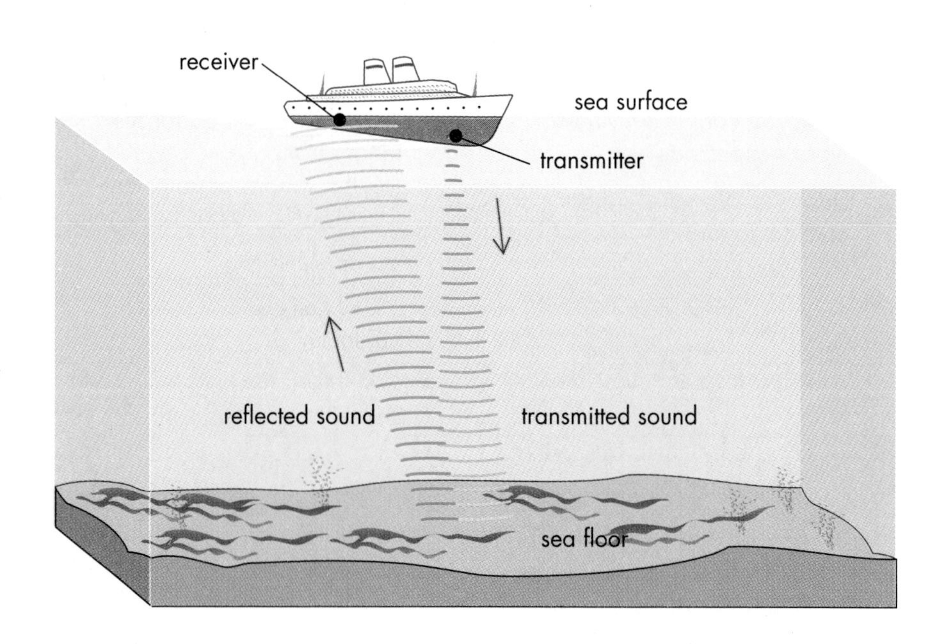

FIGURE 12.7 ▶
A sonar transmitter on the underside of a ship sends sound waves down through the water. The time it takes for the sound waves to bounce off the sea floor and travel back to a receiver on the ship is used to calculate the ocean's depth.

Sonar also revealed other unexpected features on the sea floor. Along some of the edges of the continents, just a few kilometres offshore, the ocean bed sinks into long, deep valleys called **trenches** (Figure 12.8). Some of these trenches plunge downward to several kilometers—deeper than the height of the highest mountains on Earth!

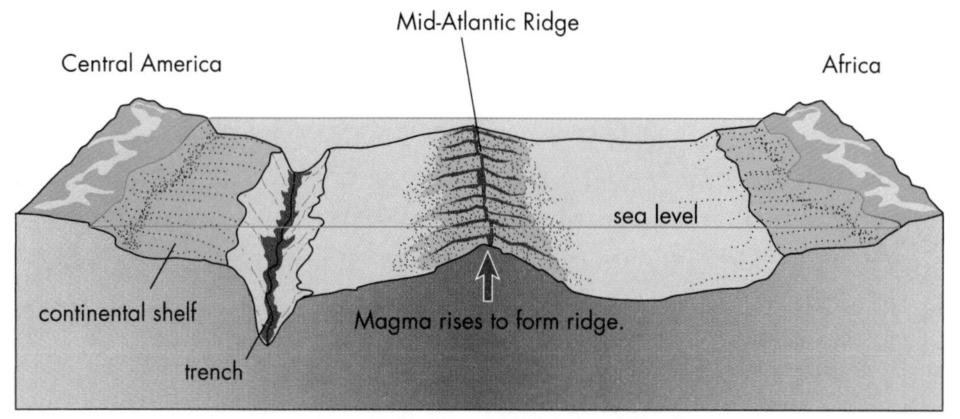

FIGURE 12.8 ◀
A cross-section of the Atlantic Ocean shows the Mid-Atlantic Ridge running down the center and a trench near the coast of Central America.

ACTIVITY 12C / INVESTIGATING TRENCHES

In this activity, you will explore the locations of undersea trenches.

MATERIALS

outline map of the world
atlas or globe
felt pens or colored pencils

PROCEDURE

1. Make a map of the Earth's ocean trenches. (If you did Activity 11E or Activity 11F in Chapter 11, your teacher may ask you to add trenches to the map you have already marked.)
2. Use an atlas or globe to find the trenches listed in Table 12.1.
3. Use a pencil to mark each trench on your map. When you are sure about all the trench locations, use felt pens or colored pencils to mark each one with a thick line.

Trench	Location
Aleutian Trench	South of the Aleutian Island Arc
Kurile and Japan Trenches	From the east side of Kamchatka Peninsula to the east side of Japan
Marianas Trench	From east of the Mariana Islands (near Japan) to east of Pelew Island (near the Philippines)
Philippine Trench	East of the Philippine Islands
Java Trench	South of the Indonesian island of Java
Tonga-Kermadec Trench	From the northeast tip of New Zealand to Samoa
Peru-Chile Trench	Along the coasts of Peru and Chile in South America
Middle America Trench	Along the lower half of the southern coast of Mexico and the coast of Guatemala in Central America

TABLE 12.1 ◀
Trench Locations Around the World

DISCUSSION

1. Describe any patterns you notice when comparing the location of trenches and the location of
   (a) mountain ranges,
   (b) earthquakes or volcanoes.
2. Suggest a hypothesis to explain any patterns you observed when studying trench locations.
3. List some questions you would ask in order to investigate your hypothesis further. ❖

MAPPING THE SEA FLOOR

The sea floor is incredibly difficult to study. In fact, we have much better maps of the moon's surface than we do of the valleys and mountains at the bottom of the ocean. Because sunlight never reaches this region of the planet, we cannot easily take photographs of it. But today scientists are slowly exposing this hidden world by using computers and sound instead of light.

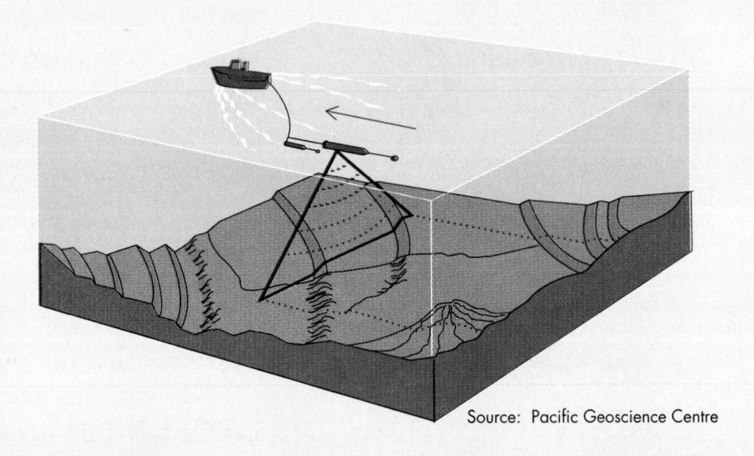
Source: Pacific Geoscience Centre

One of the best-mapped parts of the sea floor lies off the coast of British Columbia. Research ships now travel slowly back and forth in these waters, towing sonar devices below them. Each device transmits beams of sound along a broad path of sea floor up to 10 km wide. Sensors on the device then pick up the sound signals as they bounce back from the bottom of the ocean. Next, the signals are digitized—expressed as numbers—and sent to computers on board the ship. Finally, the computers sort and store the information received during a survey voyage and produce a three-dimensional contour map of the sea floor.

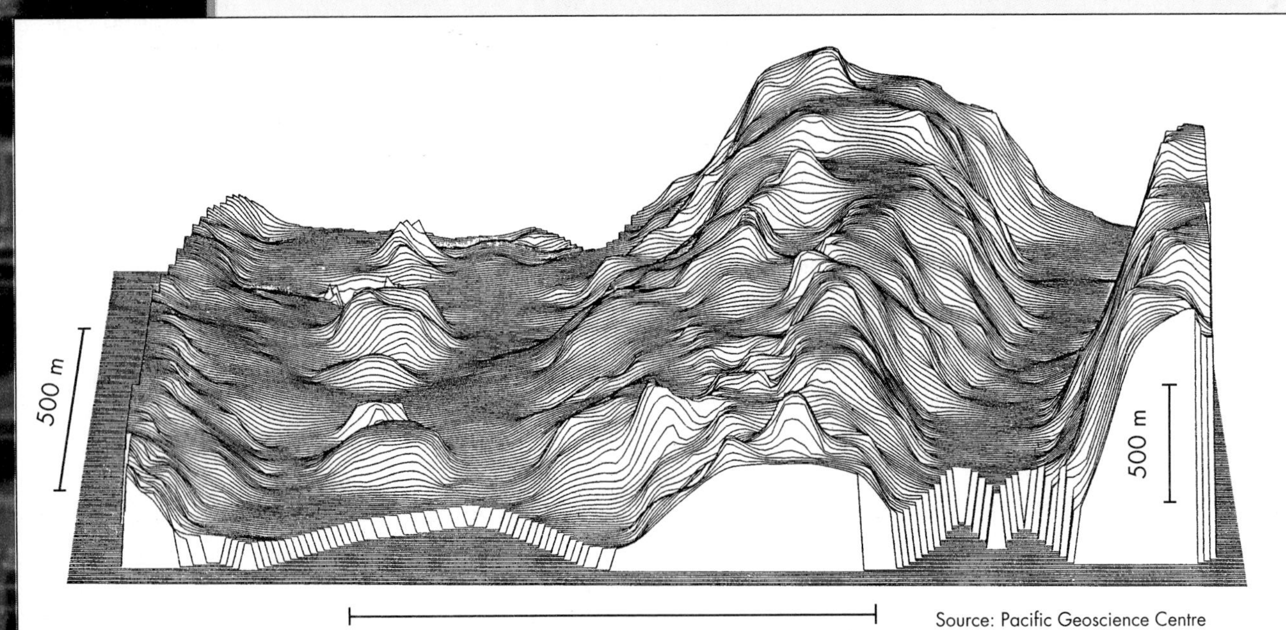

Source: Pacific Geoscience Centre

ROCK ON THE SEA FLOOR

When scientists began collecting and studying samples of rock from the sea floor, there were more surprises. They found that nearly all of the sea floor is made of basalt, a heavy extrusive igneous rock. This is quite different from most of the rock found on the continents. The continents are formed mainly of lighter sedimentary and metamorphic rock, with some igneous rock in mountainous areas.

The source of the basalt lava was not hard to find. The first clue came from undersea earthquakes. Most of these earthquakes occur along the mid-ocean ridges. The second clue came from measurements of ocean temperatures. Water around the ridges is warmer than water in other parts of the ocean. Scientists concluded that the mid-ocean ridges are volcanic. Basalt lava pours from the ridges and spreads out to form a thin layer of basalt rock over the sea floor.

Three very important differences between the rock on the continents and the rock on the sea floor had now been measured (Figure 12.9).

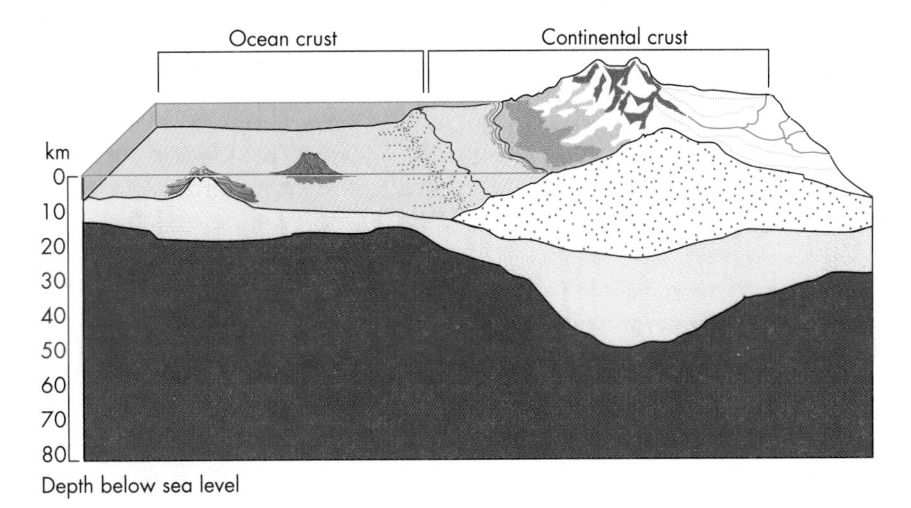

FIGURE 12.9
What differences can you see between the geology of the ocean and that of the continent?

- A volume of rock on the continents has less mass than an equal volume of rock on the sea floor.
- The rock on the continents is much older than the rock on the sea floor. (Some rocks on the continents are as old as 3900 million years, whereas no rock on the sea floor is older than about 200 million years.)
- The crust on the continents is much thicker than the crust on the sea floor. (On the continents, crust thickness averages 35–40 km, reaching 70 km under mountain ranges. Crust thickness on the sea floor averages only 6 km.)

Computer analysis of the returning sound signals reveals a lot of information. The time taken for a signal to return indicates how far it is to the sea floor. The strength of the returned signal indicates the slope and roughness of the bottom. Scientists controlling the sound image system can adjust the instruments to give different details. For example, they can alter the frequency of the transmission, or the depth at which the instruments are towed, or the speed with which they are moved through the water. Some instruments now produce many beams of sound instead of the single beam used by older equipment. This greatly increases the amount of information obtained.

The new sound image technology has revealed many smaller features on the sea floor that were missed by surveys done several years ago. Individual flows of lava have been identified and distinguished from faults. Underwater landslides have been discovered. And faults along the edges of tectonic plates have been mapped for distances up to 200 km. These and other geological features could not have been seen without the help of computers. Only a small area of sea floor has so far been studied in this detail, but the discoveries have been spectacular.

Scientists look forward to more discoveries as research continues in the oceans of the world. What exciting images from the sea floor do you think might be revealed in the future?

Scientists were astonished by the relatively young age of the sea floor. Most of the rock on the sea floor seems to have been formed by lava from the mid-ocean ridges during the past 200 million years. This means that the Atlantic, Indian, Arctic, and Antarctic oceans did not exist 200 million years ago. These ocean basins must have been formed since that time.

At last, scientists began to understand how continents could move. The continents, which are made of lighter rock, were pushed apart by new sea floor material spilling out from the mid-ocean ridges. The oceans slowly grew in size as the continents moved farther apart. More dramatic evidence to support this idea came from studies of magnetism in the rocks of the sea floor.

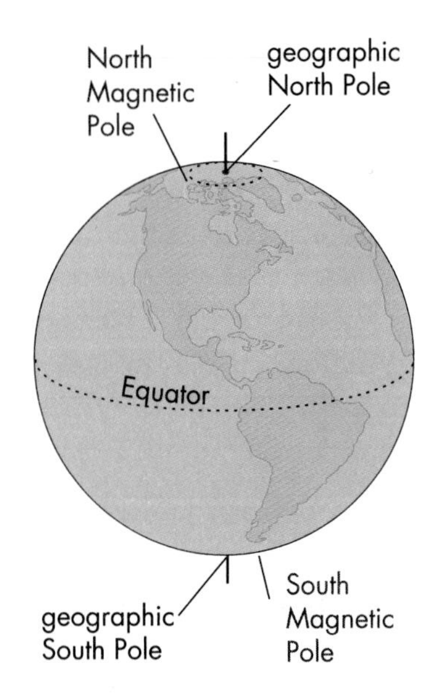

FIGURE 12.10
The Earth's magnetic poles are located a short distance from the geographic North and South Poles. If you were at the geographic North Pole, which way would your compass point?

MAGNETIC MARKERS

The Earth acts like an enormous magnet. As with a bar magnet, the areas of the Earth with the greatest magnetic force are at opposite ends, called the North Magnetic Pole and the South Magnetic Pole (Figure 12.10). You can observe the Earth's magnetism with a compass, like the type sometimes used by hikers to keep track of direction. No matter which way the body of the compass is turned, the metal needle suspended inside the compass always points north because it is attracted by the Earth's North Magnetic Pole.

Like the compass needle, many igneous rocks respond to the Earth's magnetic pull. When igneous rock cools from its molten state, particles suspended in the liquid rock material become magnetized. The suspended particles swing around like compass needles to point in the direction of the North Magnetic Pole. When the rock hardens, its magnetic particles become markers, "frozen" in solid rock. The direction in which the magnetized particles line up indicates the position of the North Magnetic Pole when the rock was formed.

ACTIVITY 12D / EXPLORING MAGNETIC FORCES

This activity will help you understand how magnetic forces affect the movement of magnetic material (Figure 12.11).

MATERIALS

safety goggles
apron
bar magnet
six small compasses
cardboard (about 20 cm × 20 cm)
iron filings
quick-drying lacquer

PROCEDURE

1. Put on your safety goggles and apron.
2. Lay the bar magnet flat on your desktop and place the six compasses in a circle around the magnet.

FIGURE 12.11
How does a bar magnet affect the compasses placed around it?

3. Draw the positions of the compass needles.
4. Turn the bar magnet 90° clockwise.
5. Draw the positions of the compass needles.
6. Place a piece of cardboard over the large bar magnet.
7. Sprinkle iron filings on the cardboard.
8. Gently tap the cardboard, then sketch the pattern formed by the iron filings.
9. Carefully spray the cardboard and filings with quick-drying lacquer to preserve the pattern. Your teacher may ask you to use a fume hood when spraying the lacquer.
10. When the lacquer is dry, lift the cardboard without turning it. Turn the magnet 90° clockwise and put the cardboard back on the magnet.
11. Predict what will happen if you sprinkle more filings onto the cardboard.
12. Test your prediction by repeating steps 7, 8, and 9.

DISCUSSION

1. Describe the positions of the compass needles when you first placed them around the bar magnet.
2. How did the compass directions change when you rotated the bar magnet?
3. Suppose the bar magnet were hidden under a sheet of paper. How could you determine the direction of the bar magnet?
4. How does the response of the iron filings to the bar magnet compare with the response of the compass needles?
5. What would have happened to the pattern of iron filings you drew in step 8 if you had turned the magnet without spraying the cardboard with lacquer? Explain. ❖

MAGNETIC REVERSALS

You might think that all igneous rocks on Earth would be magnetized in the same direction. However, this is not so. For reasons that are not well understood, the Earth's North Magnetic Pole and South Magnetic Pole have often switched positions in the past. Approximately every few hundred thousand years, these two magnetic poles suddenly change or reverse positions with one another. Scientists estimate that these sudden changes, called **magnetic reversals**, have occurred more than 170 times during the last 76 million years. As a result, rocks that formed at different times in the past have magnetized particles that are lined up in opposite directions. Some point to the north, some to the south.

To measure the magnetism of rock on the sea floor, researchers use an instrument called a magnetometer. The results of their studies show distinct bands of rock with alternating magnetic patterns (Figure 12.12).

FIGURE 12.12 ▶
Bands of rock with alternating patterns of magnetism run across the ocean floor. What does this pattern suggest?

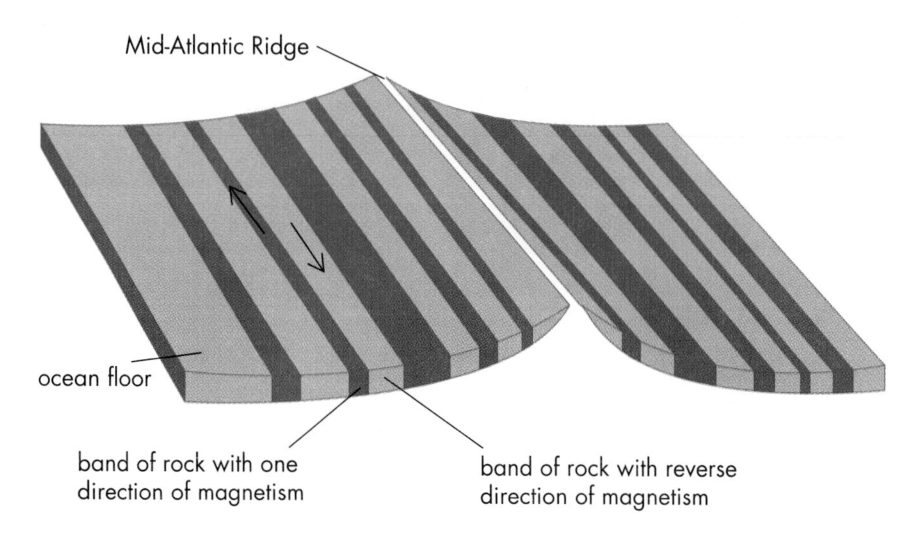

SEA FLOOR SPREADING

When scientists study the bands of rock on the sea floor, they find that the bands closest to the mid-ocean ridge were formed most recently, whereas bands of rock closer to the edges of the continents are among the oldest. In addition, the pattern of bands on one side of the ridge is a mirror image of the pattern on the other side (Figure 12.13).

FIGURE 12.13 ▶
Scientists have found that the rock of the sea floor is youngest near the mid-ocean ridge and oldest near the continents.

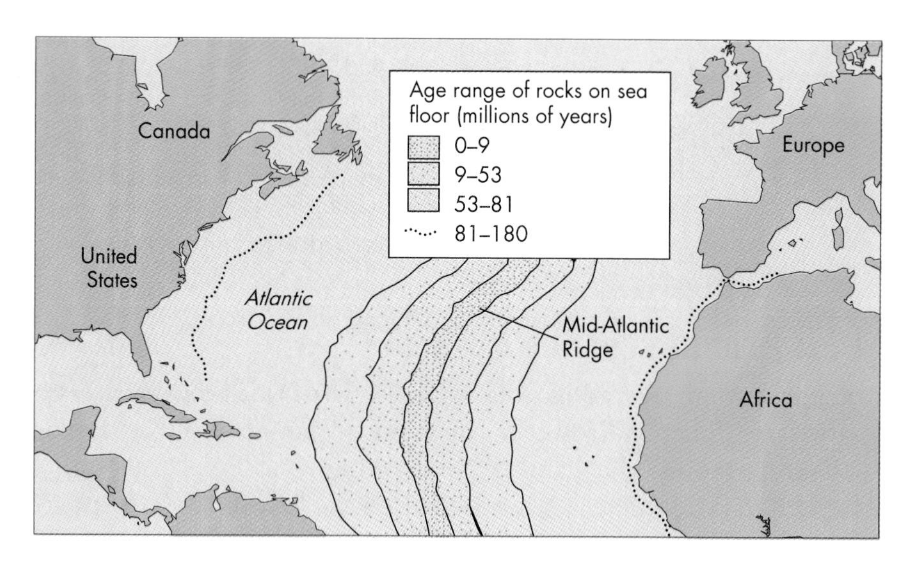

The magnetic patterns in rock on the sea floor indicate how the bottom of the ocean is formed. Basalt lava pours from the steep valley at the centre of each mid-ocean ridge and hardens into rock along both sides of the ridge. As basalt hardens, particles in the rock line up in the direction of the Earth's magnetic north. When a magnetic reversal occurs, it is recorded in the newly hardened rock. The next stream of lava pushes apart the bands of hardened rock on the sides of the ridge, carrying them outward from the center of the ridge in both directions. In this way **sea floor spreading** occurs—the rock of the sea floors is constantly moving apart at the ridges. The sea floor moves at a rate of about 1 to 10 cm every year on both sides of the ridge.

ACTIVITY 12E / THE SPREADING SEA FLOOR

In this activity, you will work with a partner to make a model that represents spreading of the sea floor.

MATERIALS

two desks or tables
two sheets of white paper (10 cm × 28 cm)
tape
colored markers
pen

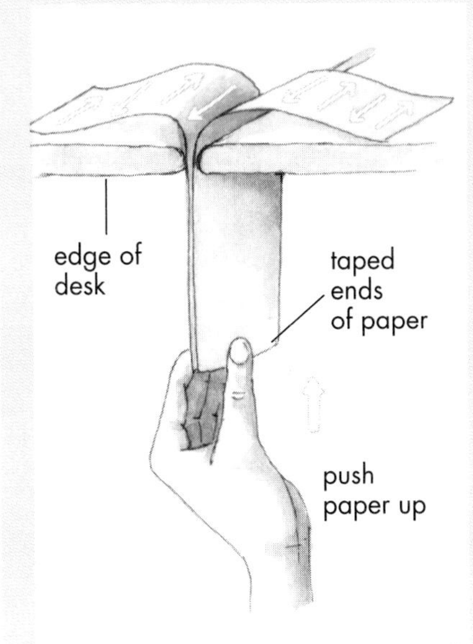

FIGURE 12.14
You can make your paper "sea floor" spread apart by pushing up on the paper as shown here.

PROCEDURE

1. With your partner's help, put two desks or tables together.
2. Tape two sheets of paper together at their shortest edges and fold along the tape.
3. Slide the folded paper into the crack between the desks, taped edge downward, until about 4 cm remain above the level of the desk.
4. Fold the ends of the paper down flat on the desktops as shown in Figure 12.14.
5. Color the ends of the paper lying on each desk a solid color or a pattern. Use the same color or pattern on each section. With a pen, draw arrows on each section of paper, parallel to the crack between the desks.
6. Push up on the paper below the desks until another 4 cm appears above the desktops. Gently pull apart the separate ends of the paper as you do this.
7. Fold the paper flat and fill in the new sections with a different color or pattern. Add arrows to each section. Draw the arrows pointing in the opposite direction to the first arrows you drew.
8. Repeat steps 6 and 7 until you have used up all the paper.

DISCUSSION

1. In your model of the sea floor, what geological features are represented by
   (a) the crack between the desks?
   (b) the paper?
   (c) the arrows?
2. Which of the colored bands on your model represents the oldest rock on your sea floor model? Explain.
3. Use your model to describe how a ridge forms on the sea floor where lava comes out.
4. How is the model you used in this activity different from sea floor spreading? ❖

▶ REVIEW 12.2

1. Name two ways that rock on the sea floor differs from rock on the continents.
2. Describe two methods used to determine the depth of oceans.
3. Magnetized particles in igneous rocks on the sea floor are not all pointing in the same direction. How does this happen?
4. Explain what the magnetic pattern in the bands of rock suggests about the formation of the sea floor.
5. On the floor of the Atlantic Ocean, where is the oldest rock located? What is the reason for this?
6. Make a cross-section (side view) drawing of the Atlantic Ocean and continents on either shore. Label the mid-ocean ridge, trench, sea floor, and continents.

■ 12.3 EXPLAINING THE MOVEMENT OF CONTINENTS

Scientists have discovered evidence that the continents are located on large, rigid (firm) pieces of the Earth's crust. These sections of the crust are called **plates** (Figure 12.15). The edges, or boundaries, of the plates are along the mid-ocean ridges. As new sea floor is formed at mid-ocean ridges throughout the world, the plates are slowly pushed apart. The moving plates carry the continents with them, like the moving conveyer belt that carries your groceries to the cashier at the checkout counter in a supermarket.

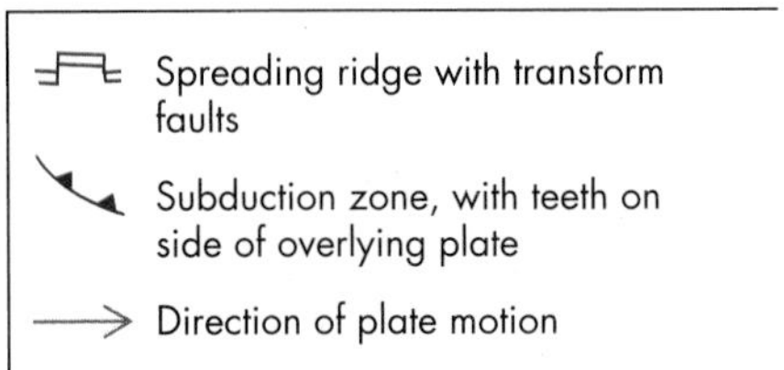

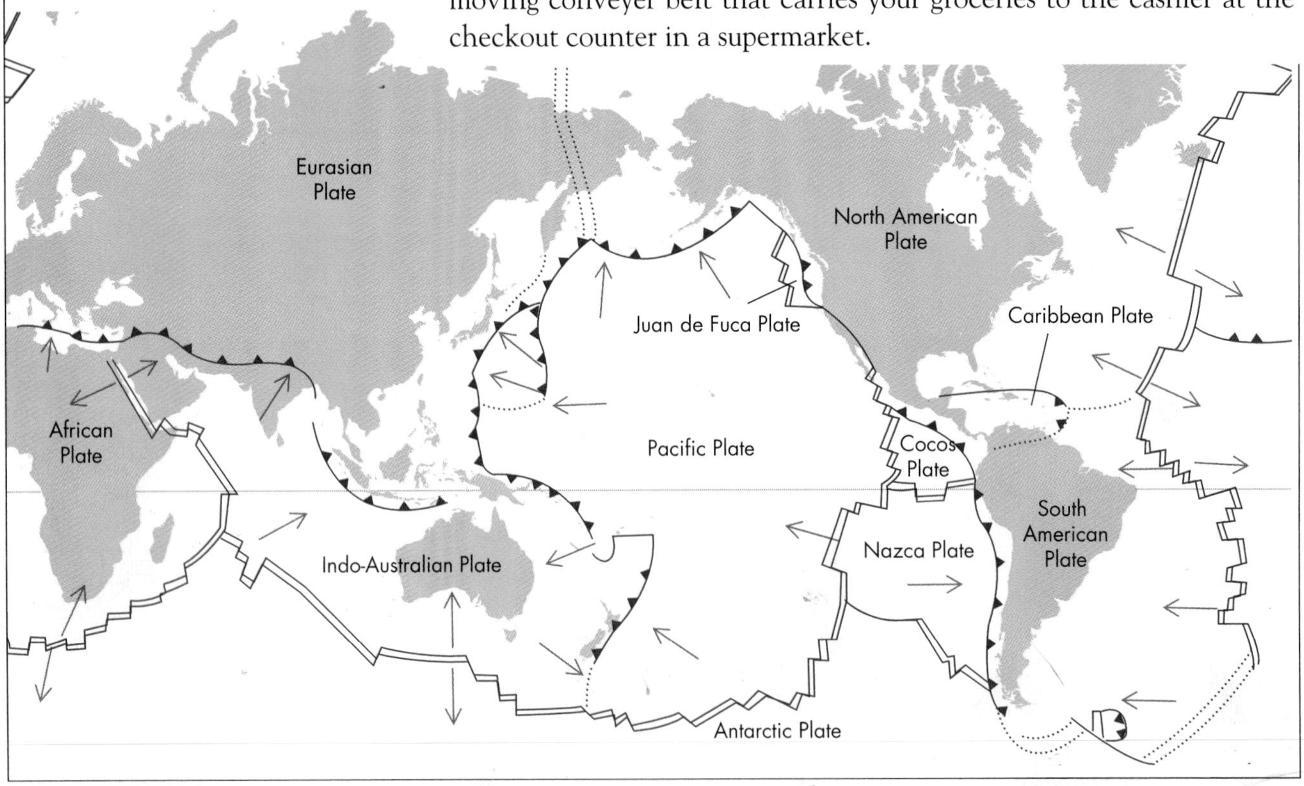

FIGURE 12.15
The major plates of the Earth's crust

If new sea floor is constantly being produced and pushed out from mid-ocean ridges, why is the Earth not growing larger? For the Earth to stay the same size, the new crust formed in one place must be balanced by crust "disappearing" somewhere else. The key to the puzzle of the disappearing crust lies in the deep sea trenches.

When trenches were first discovered, scientists noticed that the layer of sediment at the bottom of a trench was not much thicker than the layer of sediment near the top of the trench. This similarity suggested that the ocean floor at the trenches had only been formed recently.

Evidence now indicates that trenches are found at places in the Earth's crust where the edges of two plates move toward each other. As the plates are pushed together, one plate often slides beneath the other (Figure 12.16). Since the plate that carries the continents is made of lighter rock, it is always the heavier plate on the ocean side of the boundary that sinks downward. This downward motion of the plates is **subduction**.

Subduction does not occur at each plate boundary. Sometimes two plates slide past each other, with neither descending beneath the other. Breaks in the Earth's surface where this sideways movement occurs are **transform faults**.

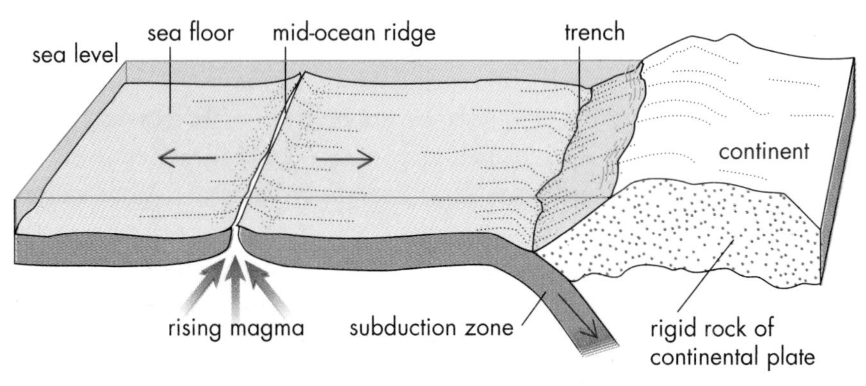

Figure 12.16
A plate descends into deeper layers of the Earth at a trench.

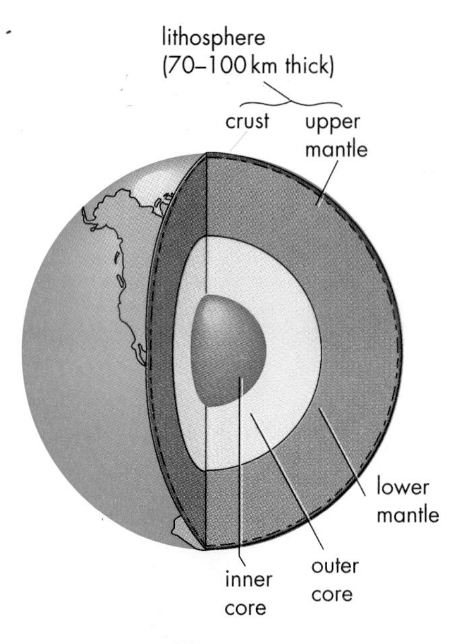

FIGURE 12.17
Heat from the Earth's molten core provides the energy that moves the plates of the lithosphere. If the Earth's core cooled, what would happen to continental drift?

WHY DO PLATES MOVE?

If you study Figure 12.17, you can see how the structure of the Earth causes plates to move. Plates are formed of the Earth's crust and the upper, rigid part of the layer below the crust—the **mantle**. Together, the crust and upper mantle form the **lithosphere**. The weight of the lithosphere presses down on the less rigid lower mantle beneath it. This enormous pressure causes the layer of rock under the plates of the lithosphere to move very slowly. This process is similar to what happens when you press your hand down on a solid-looking lump of modeling

FIGURE 12.18
Convection currents form as water is heated. How could you show that the currents exist?

clay. Under the pressure of your hand, the clay spreads sideways and your hand moves with it. In the same way, the creeping movement of rock more than 100 km below the Earth's surface carries the lithosphere—and the continents—with it.

So where does the energy that moves the rock beneath the lithosphere come from? Geologists think the energy comes from the transfer of heat from the Earth's hot interior, or **core**. This transfer of heat is carried out by convection. **Convection currents** are produced when a fluid (gas or liquid) is heated and the heated fluid rises. As hot fluid rises, cooler fluid flows in to take its place. Meanwhile, the rising fluid gradually cools and then sinks. These movements continue as long as the fluid is heated, producing a circular pattern of flow. You may observe convection currents such as this in a container of liquid being heated (Figure 12.18).

In the fluid regions deep within the Earth under the lithosphere, some areas are hotter than others. Scientists now realize that the hotter areas are where mid-ocean ridges form. At these places, hot magma moves slowly upward to form new crust. Trenches form above cooler spots, where rock descends deeper into the Earth. As the edge of a plate moves downward, the increased heat and pressure make it more fluid. The once-solid rock then becomes part of the slowly flowing rock beneath the plates.

■ Write a paragraph explaining what the word "tectonics" means when it is used by a geologist.

THE THEORY OF PLATE TECTONICS

The theory that explains how continents might move has helped scientists answer many other questions about the Earth. For example: Why are most volcanoes found along the edges of continents? Why do earthquakes often occur in the same areas as volcanoes? What causes rock to fold and produce mountains?

The theory that helps answer all these questions is called the **theory of plate tectonics**. It links together, or unifies, many different geological features that were once considered separately. The theory of plate tectonics is often called a unifying theory.

PLATE TECTONICS, VOLCANOES, MOUNTAINS, AND EARTHQUAKES

How does the theory of plate tectonics explain volcanic eruptions? According to the theory, there are three main ways volcanoes can be produced.

Some volcanoes develop above areas where subduction is taking place. As the edge of an ocean plate descends deeper below the surface, the increasing heat melts the basalt rock into magma (Figure 12.19). The magma is much lighter than the surrounding solid rock and rises up through the edge of the continental plate to build a volcano. Most of the volcanoes that form in this way are composite volcanoes.

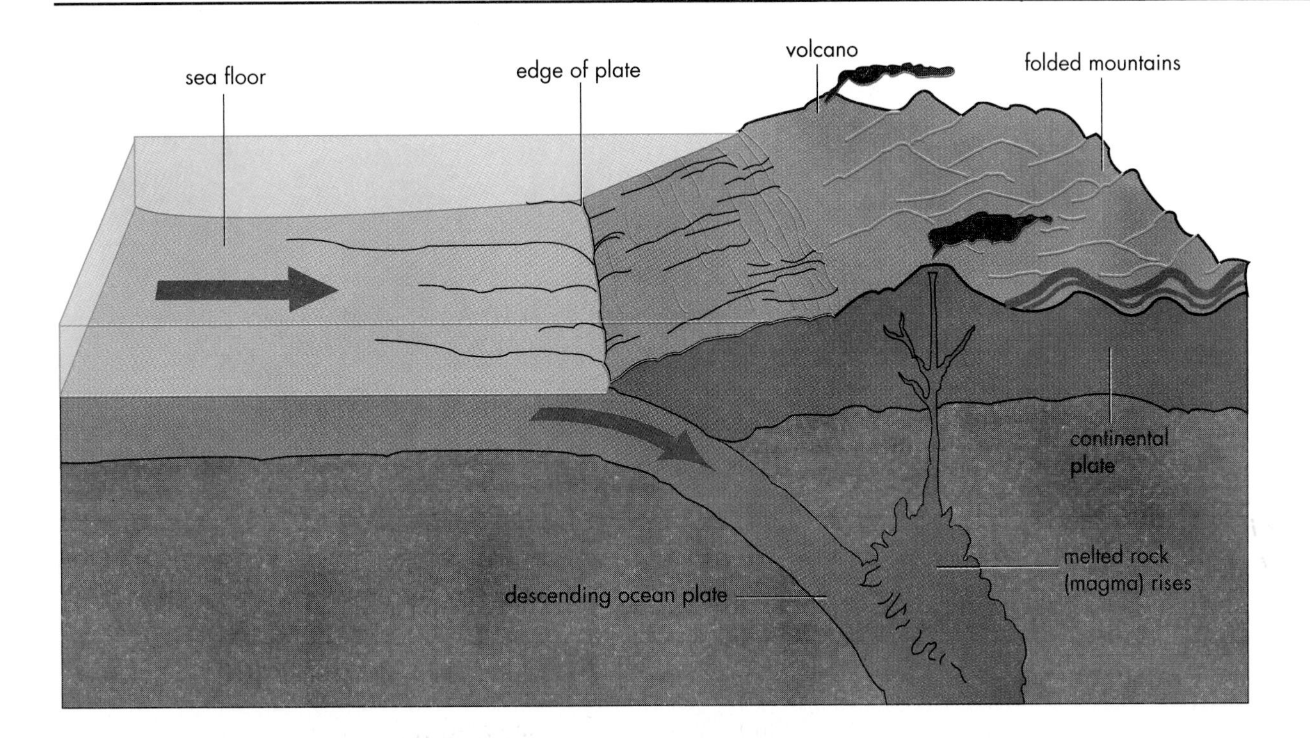

FIGURE 12.19
Volcanoes and folded mountains occur above areas of subduction. What happens to the ocean plate as it moves deeper into the Earth?

Other volcanoes are found in areas where two plates are moving apart from one another. Most of these occur on the ocean floor, along the mid-ocean ridges. Occasionally, the magma that flows from these cracks builds up volcanoes that are high enough to rise above the sea surface and produce islands. For example, Iceland and its surrounding small islands are volcanoes that have risen from the Mid-Atlantic Ridge.

The chain of Hawaiian Islands in the mid-Pacific shows a third way that volcanoes form. Research suggests that these volcanic islands were formed in the middle of a plate above an area of extremely high temperatures. This "hot spot" deep inside the Earth causes melting, producing magma that rises up to form a volcano (Figure 12.20). As the plate is pushed slowly west, it carries the volcanic island with it. A new volcano then starts to form in the part of the plate that has moved over the hot spot. This explains why the only active volcanoes are on the youngest island at one end of the Hawaiian Island chain, whereas there are no active volcanoes on the oldest island at the other end of the chain.

High mountain ranges along the coast, such as those in California, are also explained by the theory of plate tectonics. These mountain ranges are produced where two plates are moving together. The "collision" between the two plates crumples and pushes up the edge of the continental plate, much as the metal of a car is crushed and pushed into folds when the car crashes into another solid object. This force explains the folding and uplifting of rocks you studied in Chapter 11.

■ The geysers and hot springs of Yellowstone National Park in Wyoming probably represent the last stages of volcanic activity caused by a "hot spot" below that part of the North American plate.

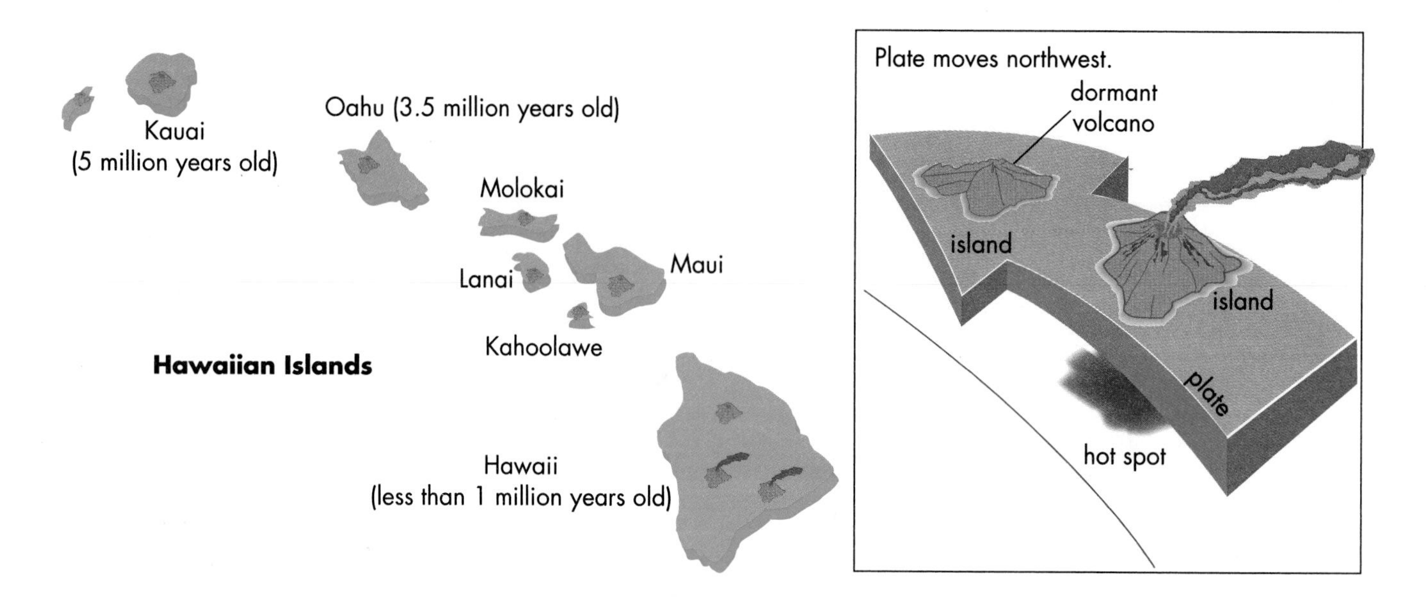

FIGURE 12.20
The theory of plate tectonics explains how the Hawaiian Islands chain was formed. Where would you expect the next volcano to appear?

Earthquakes also occur along the edges of plates. This explains why earthquakes are most common in areas with volcanoes and mountain ranges. Earthquakes are produced as the moving plates push sections of rock in the crust against one another.

ACTIVITY 12F / PLATE TECTONICS

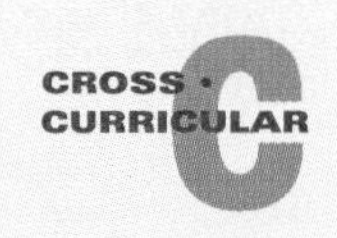

In this activity, you will explain the geology of Washington and British Columbia using the theory of plate tectonics (Figure 12.21).

PROCEDURE

1. Study the map in Figure 12.21 and review the information in this unit. Then answer the questions below.

DISCUSSION

1. Name the plates that lie next to the west coast of Washington and British Columbia.
2. Where are these plates formed?
3. In which directions are the plates moving?
4. Where is subduction taking place?
5. What is the connection between subduction and the volcanoes in the Pacific Northwest?
6. Why do earthquakes commonly occur near the Queen Charlotte Islands?
7. Explain how plate movement might be responsible for folding and uplifting in the coastal mountain ranges.

FIGURE 12.21 ▶
The location and direction of plate movement around Washington and British Columbia ❖

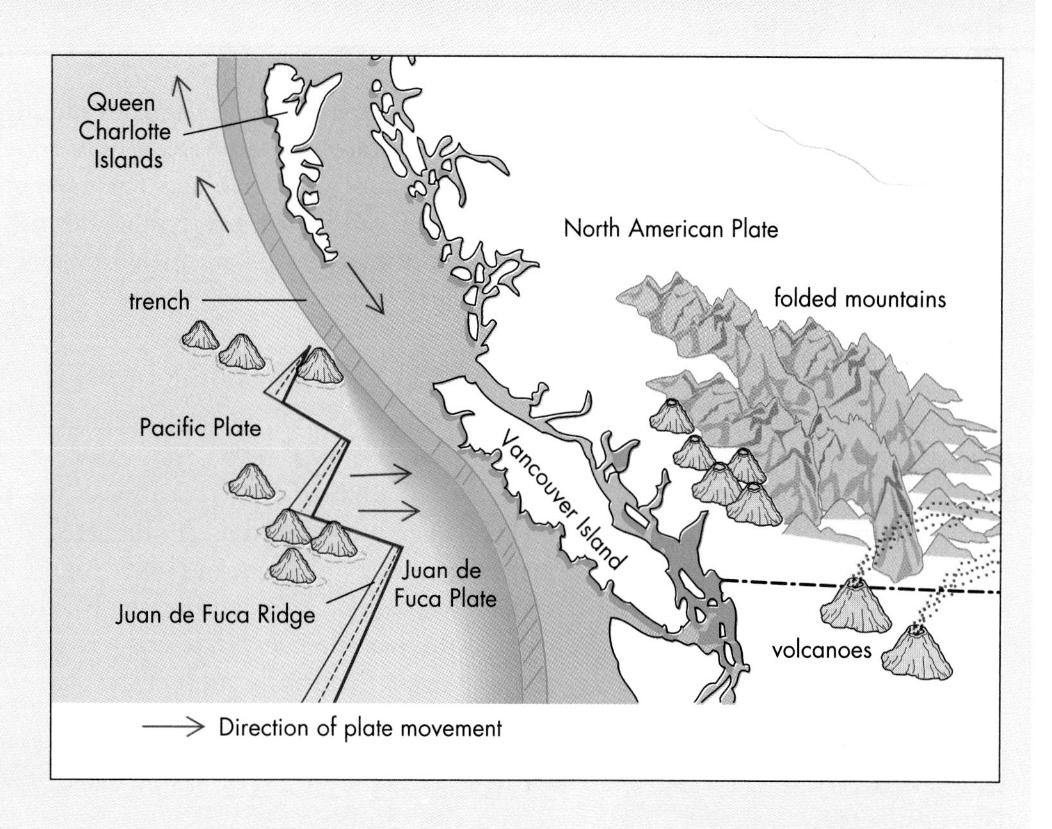

THE YOUNG MOUNTAINS OF WESTERN NORTH AMERICA

Imagine an area of ocean floor a little larger than North America. It is difficult to believe, but geologists think that this much ocean floor has descended beneath the west coast of our continent over the past several hundred million years. Part of the evidence comes from the peaks of the west coast mountain ranges, which are being pushed upward as the ocean plate is pushed beneath them.

The theory of plate tectonics links the ocean plate and the mountains. The connection between the two can be seen in the Coast Range of British Columbia. These geologically young mountains lie along the west coast of the province. They are made mostly of intrusive igneous rock. The rock was formed from magma that was produced when the descending ocean plate to the west melted deep below the crust. The magma then worked its way up through the crust, pushing up the overlying rock, but cooling before it reached the surface. The cooling rock hardened to form a batholith between 60 and 70 million years ago. (You may recall from Chapter 11 that a batholith is a large body of rock made of many individual igneous intrusions.) Erosion has since removed some of the overlying rock and made the igneous rock of the Coast Range visible at the surface (Figure 12.22).

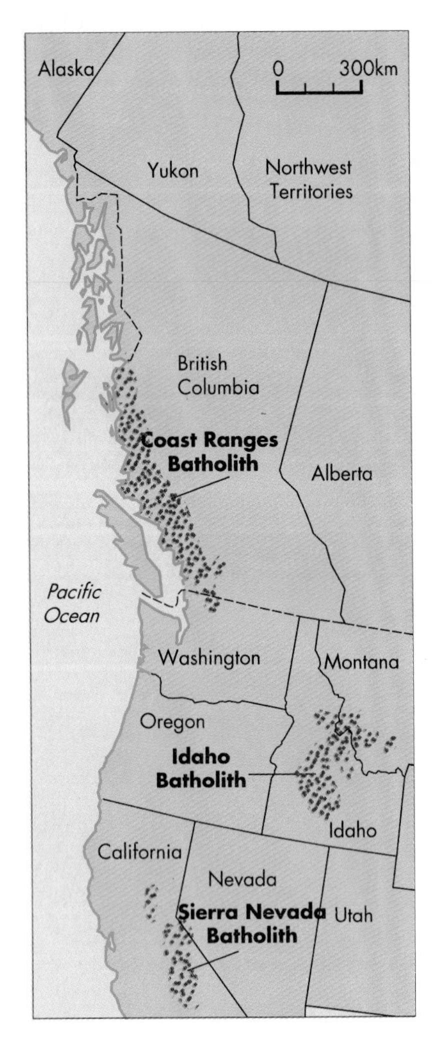

FIGURE 12.22
Location of batholiths (blocks of intrusive rock) near the west coast of North America. Which batholith do you think formed first, Coast Ranges or Idaho? How could you find out?

Unlike the Coast Range, the Cascade Mountains to the south in Washington state contain active volcanoes such as Mount Saint Helens. The Cascade Mountains are thought to be younger than the Coast Range, and have been forming during the last 25 million years. Their volcanic eruptions are fueled by magma from the descending Juan de Fuca Plate.

THE PLATES OF WESTERN NORTH AMERICA

Figure 12.23 shows how geologists think the plates of western North America have moved in the past. About 60 million years ago, when the great batholith of the Coast Ranges was moving upward, the Juan de Fuca Plate was very large. It originated at a ridge hundreds of kilometers from the west coast of North America (Figure 12.23a). As the widening Atlantic Ocean pushed the North American Plate westward, most of the Juan de Fuca Plate was forced down a long trench just off the west coast. Eventually, most of the plate descended into the trench, leaving only the small parts that remain today (Figure 12.23b).

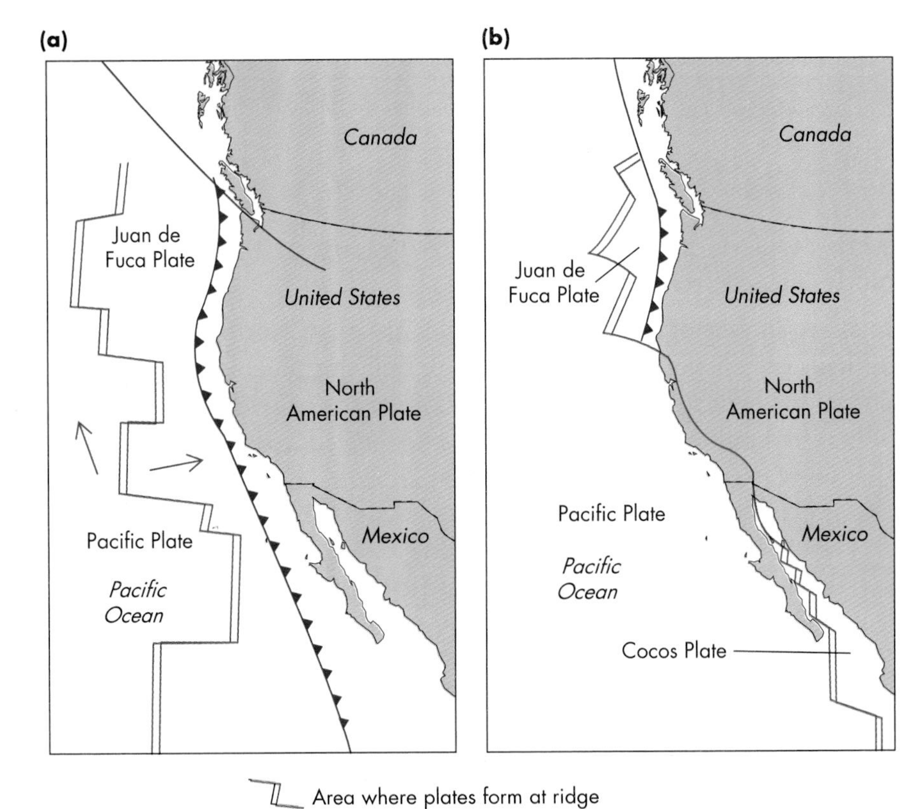

FIGURE 12.23
(a) Suggested location of plates off the west coast of North America 60 million years ago
(b) Location of the same plates today

When the Juan de Fuca Plate descended into the trench, the large Pacific Plate to its west did not follow. Instead, the Pacific Plate moved northward along the edge of the North American Plate. This movement, which continues today, produced a number of long faults, including the San Andreas Fault.

Today, the ocean ridge that produced the Juan de Fuca Plate is only a few kilometers west of Vancouver Island. The trench into which the plate descended, however, has become filled with sediment in this region. Some geologists conclude from this and other evidence that the movement of the Juan de Fuca Plate has been stalled. They predict that the volcanic activity of the Cascade Mountains will slow down and eventually end.

As scientists continue to learn how plates move, they will understand more about mountains, volcanoes, and earthquakes. They will also be able to make more accurate predictions of the volcanic eruptions and earthquakes that affect people around the world.

EXTENSION

■ The theory of plate tectonics uses evidence from geological events in the Earth's past and present. However, an important feature of a theory is that scientists can base predictions on it. Your teacher has data that you may use, together with the theory of plate tectonics, to predict the future movements of the continent of Australia.

REVIEW 12.3

1. What are "plates"?
2. What geological features would you expect to find in the following locations:
   (a) where plates meet?
   (b) where plates are formed?
   (c) where plates slide past each other?
3. What are the differences between sea floor spreading and subduction?
4. Why are continents not subducted into trenches along with the rock of the ocean floor?
5. Explain why no extremely old rocks have been found on the sea floor.
6. What conditions cause solid rock to flow like a fluid?
7. Draw a diagram showing how heat from within the Earth may cause plates to move.
8. Explain how volcanoes may be formed in the middle of a plate.
9. Explain the connection between deep ocean trenches and the mountain ranges that often occur alongside them.
10. What evidence do geologists have that the Juan de Fuca Plate is no longer being subducted?
11. Explain why the theory of plate tectonics is called a "unifying" theory.

CHAPTER REVIEW

KEY IDEAS

■ The theory of continental drift was first suggested by Alfred Wegener in 1915. Evidence for his theory included the matching shapes of continents and the distribution of rocks, fossils, and living organisms.

■ Scientists neglected Wegener's theory until the 1960s because no one could explain how continents might move.

■ Explorations of the ocean revealed mid-ocean ridges where magma erupts and forms new sea floor. This process is called sea floor spreading.

■ Rock from the sea floor is denser, younger, and thinner than rock on the continents.

■ Bands of rock with alternating magnetic patterns run across the sea floor.

■ The continents are located on rigid pieces of the Earth's crust called plates.

■ Sea floor spreading is balanced by the downward movement, or subduction, of plates.

■ The theory of plate tectonics explains the movement of plates on the Earth's surface. Plate movement produces mountain ranges, volcanoes, and earthquakes.

■ The movement of plates is caused by convection currents produced by heat deep within the Earth's core. The rigid plates move on top of a layer of fluid rock.

VOCABULARY

continental shelf
theory of continental drift
mid-ocean ridge
sonar
trench
magnetic reversal
sea floor spreading
plate
subduction
transform fault
mantle
lithosphere
core
convection current
theory of plate tectonics

V1. Construct a mind map, using as many words as possible from the vocabulary list.

V2. Write down five quiz questions, each of which can be answered by a term from the vocabulary list. Exchange your quiz with a partner and answer the questions.

CONNECTIONS

C1. Why do you think scientists consider the continental shelf to be part of the continent rather than part of the sea floor?

C2. Why has the theory of plate tectonics been more easily accepted by geologists than the earlier theory of continental drift?

C3. Rocks on continents may be much older than rocks found on the sea floor. Suggest a reason for this.

C4. Describe how bands of sea floor rock with alternating directions of magnetism might have been formed.

C5. Describe the kinds of geological activity that occur along the boundaries between plates.

C6. Look at Figure 12.15 and study the area of the Red Sea and the nearby plate boundary between the African and Eurasian plates.
(a) What is happening here?
(b) What do you think this area might look like millions of years from now? Give reasons for your prediction.

C7. Look at Figure 12.15 and find the Aleutian Islands, which form an arc running across the North Pacific from Alaska toward Siberia. Consider the plate activity in this region and suggest how this island chain may have been formed.

EXPLORATIONS

E1. From what you have learned in this chapter, explain why scientists may find it difficult to persuade other scientists to accept new ideas.

E2. Pretend you are an earth scientist. Prepare a report designed to persuade your fellow scientists (your classmates) that earthquakes are caused by the movement of plates.

E3. In Iceland, the energy of sea floor spreading has been harnessed for human use. Write a report about other parts of the world where this energy might be used. Suggest ways such energy could be harnessed.

E4. In the description of Shackleton's expedition at the beginning of this chapter, the following question was asked: "How could coal possibly have formed in Antarctica?" Imagine that you are a member of the Shackleton expedition. Write a diary entry that describes your discovery of coal on the frozen continent and uses the information in this chapter to suggest an answer to this question.

E5. Work with some classmates to make a poster that criticizes or defends the theory of plate tectonics.

FIGURE 12.24
Use these pictures as a start for question E7. How do you think each of these mountains was formed?

E6. Parts of California are located at the edge of a plate. Write a short story or a poem to illustrate why you think it is better or worse to live near the edge of a plate than in the middle of a plate.

E7. Collect some photographs of volcanoes and folded mountains in California, as shown in Figure 12.24. Label the photographs in your collection, explaining how each volcano or mountain might have been formed.

REFLECTIONS

R1. How has learning about the theory of plate tectonics changed the way you think about
(a) the area you live in?
(b) the Earth?

EXPLORING SPACE

This photograph of a small part of the sky was taken with a telescope. You can see many stars of varying sizes, with faint gases between them. In the middle of the photograph is a feature that resembles the head of a horse. This part of the sky is called the Horsehead Nebula, in the constellation Orion. What is a nebula? What is a constellation? Why is the nebula so dark compared with its surroundings? You will discover answers to these and many other questions as you study this unit.

You cannot see the stars in the photograph simply by looking up into the sky at night. Fortunately, there are many devices to help us see distant objects. Space probes are sent to various parts of the solar system to gather information about planets and other bodies. Telescopes provide magnified views of stars and other faraway objects. Cameras attached to these telescopes provide photographs that help us study what is beyond the Earth.

Although much of this unit deals with planets and stars, you will also learn about other parts of the universe. Here are more questions you will explore: How does our sun compare with other stars? What are galaxies, black holes, supernovas, and red giants? Is there life elsewhere in the universe?

Space science relates closely to other topics that you study. You will be able to use some of the ideas you may have already learned about heat, chemistry, and living things.

Why is the study of space important? Humans have always been interested in the sun, the planets, the stars, and other objects in space. Learning about objects throughout the universe helps us understand more about ourselves. Why do you think that it is important to learn about space?

UNIT IV

CHAPTER 13

THE SOLAR SYSTEM

The rings around Saturn make it one of the most beautiful planets. In the photograph of Saturn shown here, you can see shadows of the rings. People have learned much about Saturn and the other planets that travel around the sun by observing them from the Earth. But scientists have discovered much more in recent years by sending space probes that get close to the planets and their moons.

This chapter is about the solar system, which is made up of the sun and all the objects that travel around it. Our knowledge of the solar system is continually growing and changing. In this book, you will see the most recent photographs and information available, but new ideas may replace older ones even while you study this chapter. That is part of the excitement of studying space science.

ACTIVITY 13A / YOUR THOUGHTS ABOUT THE SOLAR SYSTEM

You may already know something about the sun, planets, and other objects in our solar system. Working with one or more other students, divide a large sheet of paper into four equal rectangles, as shown in Figure 13.1. Then answer the following questions, one in each rectangle. Add a title in each rectangle.

1. In the first rectangle, draw a diagram showing what you think the sun, planets, and other objects in the solar system would look like to a space traveller outside the solar system. Label your diagram, using as many words or phrases about the solar system as you can.
2. If you could travel to any part of the solar system away from the Earth, which part would you choose? In the second rectangle, describe how you would survive and enjoy life there.
3. Imagine your head as a model of the Earth. What object do you think would be a good size as a model of the moon? How far away would you place the model of the moon? In the third rectangle, draw a diagram to show your choices.
4. In the fourth rectangle, list questions you would like to have answered about the solar system after studying this chapter. ❖

FIGURE 13.1 ▶
Sharing ideas about the solar system

■ 13.1 THE SUN AND THE PLANETS

If some imaginary spacecraft were approaching the solar system from far away in the universe, what would the travellers on board see? First, they would see our sun, which is by far the biggest and brightest body in the solar system. Our sun is a star. It gives off large amounts of light and other forms of energy. This allows it to be seen from very far away.

As the spacecraft gets closer, the next objects that the travellers would be able to see are the largest planets, and then the smaller planets. A **planet** is an object that travels in a path around the sun or around any star. The Earth is one of the planets that travel around the sun.

When the imaginary travellers get close enough, they would see several kinds of objects travelling around planets. **Moons** are large objects that travel around a planet. Our Earth has one moon travelling around it.

The imaginary travellers would also see other, smaller objects. These objects are comets, asteroids, and meteors. (You will learn more about them later in this chapter.)

The **solar system** includes the sun, the planets, and all the objects that travel around the sun and planets. Figure 13.2 shows how big the sun is compared with the nine known planets in the solar system. The planets are shown in order of their average distance from the sun, starting with Mercury.

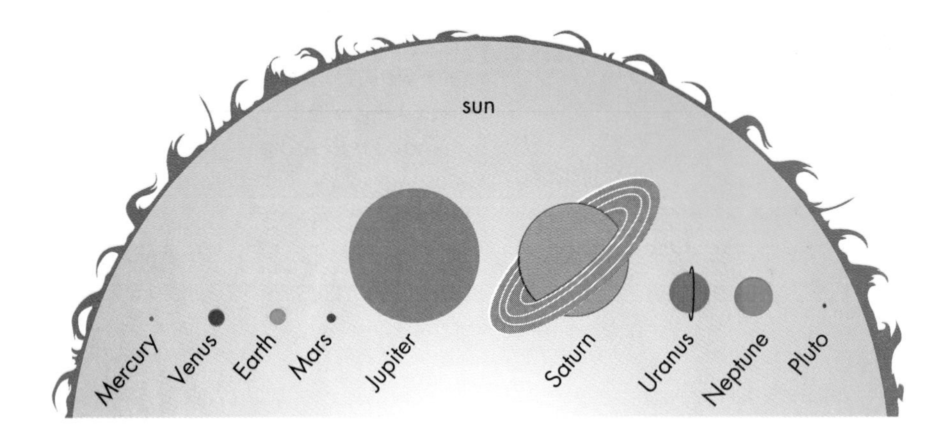

FIGURE 13.2 ▶
Comparing the sizes of the planets with the size of the sun. Can you see any trend in the sizes? If you were looking for a tenth planet, would you look for a big one or a small one?

If the travellers from outer space watched for long enough, they would also observe that everything in the solar system is in motion. One important type of motion is called **revolution**, which is the motion of one object around another. The planets revolve around the sun in paths called **orbits**. These orbits are nearly circular, with the sun at the center of each orbit. Figure 13.3 shows the approximate orbits of five of the nine known planets. How long does the Earth take to revolve once around the sun?

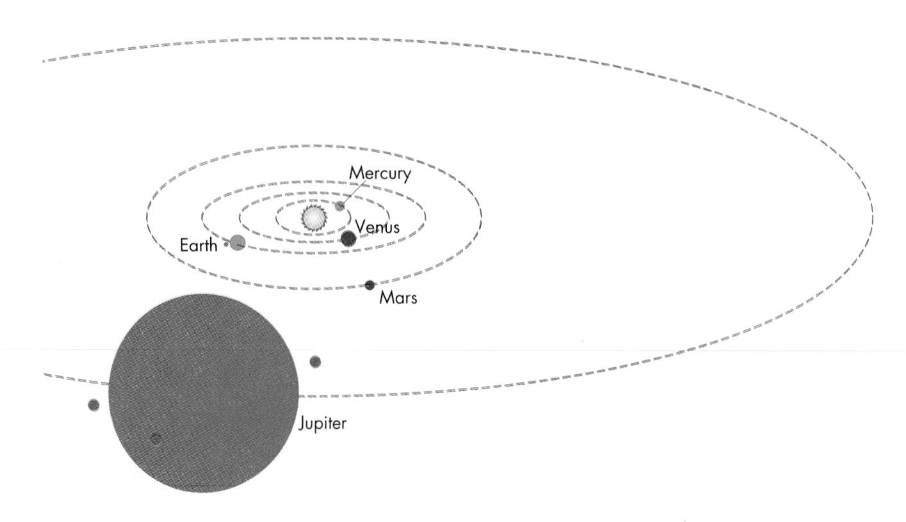

FIGURE 13.3
The approximate orbits of the five planets closest to the sun. Is there a pattern in the size of orbits as the planets get farther from the sun?

ACTIVITY 13B / A MAP OF THE SOLAR SYSTEM

A map showing the planets and their distances from the sun will help give you an idea of the tremendous sizes and distances in the solar system. Table 13.1 has the information you need to perform this activity.

MATERIALS

paper
drawing compass or other device to draw circles
meter stick
roll of paper tape
ruler
scissors
tape or glue

PROCEDURE

1. On a sheet of paper, draw circles representing the sizes of the planets. Use a scale of 1 cm = the diameter of the Earth. You will need a drawing compass to draw the larger circles neatly. Can you see from Table 13.1 that if the diameter of the Earth is 1.0 cm, the diameter of Jupiter is 11.2 cm?
2. Print the name of each planet on your diagram of that planet.
3. Using a meter stick, measure out 6 m of paper tape. Label the sun at the left-hand edge, then use the ruler to mark the positions of the nine planets using the scale 1 cm = 10,000,000 km. For example, you can see from Table 13.1 that the distance from the sun to the Earth is 150,000,000 km. To calculate the number of centimeters that represents this distance, divide 150,000,000 by 10,000,000. The answer (15 cm) can then be marked on your paper tape.
4. Use scissors to cut out the planets that you drew in Step 1, and tape or glue these planets to the paper tape at the correct locations (Figure 13.4). Tape the entire model to a wall, or hang it from the ceiling of your classroom.

FIGURE 13.4 ▶ ***Putting together a map of the solar system***

Planet	Mercury	Venus	Earth	Mars	Jupiter	Saturn	Uranus	Neptune	Pluto
Diameter of planet at its equator (Earth = 1)	0.38	0.95	1.0	0.53	11.2	9.4	4.0	3.8	0.2***
Distance of planet from sun (million km)	57.9	108	150	228	778	1427	2870	4497	5900
Time for one revolution of planet around sun	88.0 d*	224.7 d	365.26 d	687 d	11.9 yr**	29.5 yr	84.0 yr	164.8 yr	247.7 yr

TABLE 13.1 ***Planetary Data for Activity 13B***

* The symbol "d" stands for day. It means one Earth day.
** The symbol "yr" stands for year. It means one Earth year.
*** This is an estimate. Little is known about Pluto.

DISCUSSION

1. Use the information in Table 13.1 to answer the following questions:
   (a) Which planet takes the least amount of time to revolve around the sun?
   (b) Which planet takes the greatest amount of time to revolve around the sun?
   (c) How do you think the distance of a planet from the sun relates to the time it takes the planet to revolve once around the sun?
2. Which planets do you think would be called
   (a) the inner planets?
   (b) the outer planets?
   Explain your answer.
3. Create your own mnemonic sentence to help you remember the names of the planets in order of their distance from the sun. ❖

CHARACTERISTICS OF THE PLANETS

Each planet in the solar system is different from every other planet. The planets differ in their size, motion, and temperature, and in the substances they are made of.

Each planet takes a different amount of time to complete one revolution around the sun. This time depends on how far the planet is from the sun. The farther the planet is from the sun, the longer it takes to complete one revolution. The period of time for the Earth to complete one revolution around the sun is called one **year**. You can see in Table 13.1 that Pluto takes the longest time to complete one revolution, approximately 248 Earth years.

Besides revolving around the sun, all planets rotate. This **rotation** is the spinning of an object around an imaginary line called the **axis**. For the Earth, this axis joins the North Pole to the South Pole. The Earth's rotation around its axis once every 24 hours causes our day.

Surface temperatures are also different from one planet to another. The average surface temperature on a planet depends on the distance of that planet from the sun. It also depends on another characteristic of the planet: the composition of its atmosphere.

Density, too, differs among the planets. Density is a measure of how closely packed the particles of a substance are. The density of water, for example, is 1000 kg/m^3 or 1 g/cm^3. Would you expect the average density of the Earth to be less than, equal to, or greater than this value?

The planets are made up of different combinations of chemical elements, which is one reason why no two planets are the same. However, scientists have determined that four elements are more common than all the others in our solar system. These are hydrogen, helium, oxygen, and carbon.

ACTIVITY 13C / COMPARING THE PLANETS

As you look at more characteristics of the planets, think of ways that the planets can be grouped together. Especially consider which planets are most similar to the Earth.

PROCEDURE

Use Table 13.2 to help you answer the Discussion questions in this activity.

DISCUSSION

1. The length of a day from sunrise to sunrise is very different for each of the planets. Which planets have short days compared with those on the Earth?

2. (a) Name two planets that scientists think have no atmosphere.
(b) Do you think that the same reasons could be used to explain the lack of atmosphere on both planets? Explain your answer.

3. Look at the descriptions of the atmospheres on the large planets, Jupiter, Saturn, Uranus, and Neptune. Do you think that the kinds of lifeforms we have on the Earth would be able to live in these atmospheres? Explain your answer.

4. Describe what you notice about the surface temperatures on the planets.

5. (a) Which planets are most similar to the Earth in their densities?
(b) Which planets have densities much lower than the Earth's?

6. In general, which planets appear to be most similar to Earth? Explain your answer.

7. Which planets could be grouped as those that are least similar to the Earth? Why?

8. Which planet seems to fit into a category all by itself? ❖

Planet	Mercury	Venus	Earth	Mars	Jupiter	Saturn	Uranus	Neptune	Pluto
Time for one day (sunrise to sunrise)	176 d*	117 d	24 h	24 h 39 min	9 h 50 min	10 h 39 min	17 h 18 min **	15 h 40 min	153 h 18 min
Main substances in the atmosphere	None	Carbon dioxide, nitrogen	Nitrogen, oxygen	Carbon dioxide, argon, nitrogen	Hydrogen, helium, methane	Hydrogen, helium, methane	Hydrogen, helium, methane	Hydrogen, helium, methane	None detected
Surface temperature (°C)	–180 to 426	470	–85 to 65	–120 to 30	–160	–180	–210	–220	–220
Density (g/cm^3)	5.44	5.25	5.52	3.95	1.31	0.70	1.18	1.66	1.1?

TABLE 13.2 ***Characteristics of the Planets for Activity 13C***

* In this table, "d" means day, "h" means hour, and "min" means minute.
** The rotation is in the opposite direction to that of the other planets.

GROUPING THE PLANETS

The four planets closer to the sun are Mercury, Venus, the Earth, and Mars. They are all small, and their densities are all high. Because these planets resemble the Earth more closely than the other planets, they are called the **terrestrial planets**. The word "terrestrial" comes from the Latin word *terra*, which means Earth. These planets are also known as the **inner planets**.

The remaining five planets travel in the vast distances beyond the four inner planets. They are called the **outer planets**. Four of these, Jupiter, Saturn, Uranus, and Neptune, are all large, and their atmospheres consist mainly of the gases hydrogen and helium, which have low densities. For this reason, these four planets are called the **gas giants**. The last known planet in the solar system, Pluto, is neither a terrestrial planet nor a gas giant. It is so far away from us and so small that scientists know less about it than about any other planet.

THE SCIENCE OF ASTRONOMY

Throughout this chapter and the next three chapters, you will come across the words "astronomy" and "astronomer." **Astronomy** is the science that studies the composition, position, and movements of all objects in space. This includes the study of the solar system and the stars beyond the solar system. (The Greek word *astron* means star.)

An **astronomer** is a person who studies astronomy. Different astronomers study different objects in the sky. Some may study the sun, some may study the planets, while others study stars or other objects.

► REVIEW 13.1

1. (a) Describe the difference between revolution and rotation.
   (b) Which of these two motions produces a "day" on a planet?
   (c) For the Earth, what period of time is related to the other motion?
2. List at least three features of the Earth that make it different from the other planets.
3. (a) What characteristics do the terrestrial planets have in common?
   (b) What characteristics do the gas giants have in common?
   (c) How is Pluto similar to the terrestrial planets?
   (d) How is Pluto similar to the gas giants?
4. Mercury is much closer to the sun than the Earth is, yet its surface temperature can drop to −180°C at night, which is much colder than the Earth's coldest temperature. Why do you think this happens? (Hint: Table 13.2 has information that may help you answer this question.)

13.2 A CLOSER LOOK AT THE PLANETS

You learned in Section 13.1 that there are two groups of planets in our solar system. One group includes Earth and planets that are Earth-like. The other group includes planets unlike the Earth. In this section, you will explore some of the facts that astronomers have learned about these two groups of planets. Our journey will take us to all the planets in order, starting with those closest to the sun.

ACTIVITY 13D / EXPLORING THE PLANETS

You have already learned some facts about the planets in Section 13.1. In this section, you will find out more about each planet. Before you start reading, copy Table 13.3 into your notebook.

Before you read about each planet, write its name in the first column. (Do not put all the planets' names in the first column before you start, because you do not know how much space you will need in the other columns for each one.)

In the second column, write what you know about that planet. Make sure to indicate whether it is a terrestrial planet, a gas giant, or other. You may want to review Section 13.1 and Tables 13.1 and 13.2.

As you read the description of the planet in this section, fill in the last column. Then add the next planet's name to the first column.

Your table will be a useful summary of all that you learn about the planets. ❖

Planet name	What I already know	What I learned in Section 13.2

TABLE 13.3 ***Sample Data Table for Activity 13D***

TERRESTRIAL PLANETS

The terrestrial, or inner, planets are composed mainly of rocky material. They also contain metals. Besides Earth, this group consists of Mercury, Venus, and Mars, all of which could be seen by observers even thousands of years ago. Today, each one has been studied from close range by spacecraft called space probes. A **space probe** is a spacecraft with many instruments on board that is sent from the Earth to explore our solar system.

■ Find a reference book that describes the planets of the solar system in detail. Place the extra facts you discover about each planet in a fourth column in your table for Activity 13D. Title the column "What I found through research."

Mercury

Mercury is not easy to see from the Earth because it is never far from the bright light of our sun. The only times you can see it are just before sunrise and just after sunset. It is a planet of extremes. Because it is the planet closest to the sun, Mercury receives sunlight that is about 10 times stronger than what we receive on Earth. Thus, Mercury experiences scorching daytime temperatures of over 400°C. At night, because Mercury has no atmosphere to trap heat, the temperature falls to about –180°C! If you look at the data in Tables 13.1 and 13.2, you will discover that a day on Mercury (from sunrise to sunrise) is longer than its year.

Mercury was first photographed at close range when the space probe *Mariner 10* passed within 740 km of the planet in 1974. The thousands of photographs revealed a barren, rocky surface, with many holes called craters, like those on our moon (Figure 13.5). These craters were caused by chunks of rocky material colliding with the planet in the past.

FIGURE 13.5 ▶ ***This view of the planet Mercury was photographed from the* Mariner 10 *space probe. How do you think this view might change over the next several thousand years?***

DID YOU KNOW

■ The Mayas are native people who live in Central America. Over 1000 years ago they had developed ways of accurately keeping track of the movements of the planets. Many of their temples were used as observatories. One of the few Mayan books that remain from that time shows that the Mayas were able to predict the appearance of Venus so accurately that their predictions were off by only two hours in 500 years. Why do you think the Mayas were so interested in Venus that they developed such an accurate way of predicting its appearance?

Venus

Venus, our closest planetary neighbor, is one of the easiest planets to recognize in the sky. Because sunlight reflects from the thick clouds of its atmosphere, Venus is the third brightest object that we can see in the sky (Figure 13.6). (The sun and the moon are the two brightest objects that we can see.) Venus is so bright that it is sometimes called the Morning Star or Evening Star, depending on what time of the day it can be seen. Of course, Venus is not a star at all.

Venus has a thick atmosphere made up mainly of carbon dioxide. This gas acts like the glass of a greenhouse, holding in the heat. Because of this, surface temperatures on Venus are much higher than they would be without the atmosphere, and these temperatures remain the same both day and night.

Because surface temperatures are high enough to melt lead, exploring the surface of Venus is not easy. In 1975, two space probes from the former Soviet Union landed on Venus. They survived long enough to send back photographs of a rocky surface with volcanoes, canyons, ridges, and a few craters. More recently, the United States sent the space probe *Magellan* to Venus. It had special radar cameras that could "see" through the thick atmosphere. These cameras sent back detailed views of the surface of Venus (Figure 13.7).

FIGURE 13.6
Venus as seen from a distance of 720,000 km, showing cloud patterns in the upper atmosphere. Compare this photograph with Figure 13.8. In what ways are the photographs similar?

FIGURE 13.7 ◀
This view of the surface of Venus shows a volcano that is 1.7 km high. Lava from the volcano has flowed hundreds of kilometers across the surface. Before it hardens, do you think lava on Venus would flow farther, less far, or about the same distance as it would on Earth? Why?

Earth

From outer space, our Earth appears to be a bright, beautiful planet. Figure 13.8 shows a photograph of the Earth taken from a spacecraft that landed on the moon. The few astronauts who have travelled to the moon and seen this view recall long afterward how special our planet is. The astronauts who travel around the Earth in the space shuttle see a much closer view. But they, too, tell of how special the Earth looks from space. This should remind us how important it is for us to take care of our planet.

In our solar system, the only planet with an atmosphere that indicates the presence of life is the Earth. Our atmosphere consists mainly of nitrogen (78 percent) and oxygen (21 percent). The oxygen is produced by living organisms.

FIGURE 13.8 ▶
This photograph of the Earth was taken from the Apollo 17 spacecraft. What continents can you recognize? How does the Earth differ from Mercury and Venus?

FIGURE 13.9
You can see the reddish color of Mars even from a distance. The red color is due to the dusty soil. Sometimes the whole surface of the planet is hidden by huge dust storms.

More than two-thirds of the Earth's surface is covered with water or ice. Water in the atmosphere produces clouds. The water, ice, and clouds add to the beauty and color of the Earth as seen from space.

Our surface temperatures go from about –85°C to about 65°C.

We know much more about the Earth than about the other planets because we live here. But we still have much more to learn as we study features of the Earth from close up as well as from space.

Mars

Mars is called the "red planet" because of the reddish color of its soil (Figure 13.9). To observers on the Earth, Mars is one of the brighter objects in the sky, although not as bright as Venus. Mars has ice caps that change with the seasons; it has winds that reach 400 km/h; and there is evidence that it has had volcanoes, glaciers, and floods of water. It is the planet in the solar system with surface conditions most similar to those of the Earth. For this reason, scientists have been especially interested in the possibility of some form of life there, and they have studied Mars more closely than any other planet, except the Earth.

In 1976, the surface of Mars was photographed and sampled by two robot landing craft called *Viking 1* and *Viking 2* (Figure 13.10). These landing craft took many measurements and provided a huge amount of information, but they detected no life. In 1992, the space probe *Mars Observer* was launched from the Earth. It was to orbit Mars and use special cameras and other instruments to collect information about the Martian atmosphere and the planet's surface. Unfortunately, contact with *Mars Observer* was lost just as it arrived at Mars.

Astronomers think that Mars once had a denser atmosphere and liquid water on its surface, which would explain some of the surface features that we can see. Within your lifetime, it is possible that a space colony for humans will be built on Mars.

FIGURE 13.10
A robot landing craft grabs samples from the surface of Mars for analysis. Astronomers think that the frost seen on the surface is a mixture of frozen carbon dioxide (dry ice) and ice.

THE GAS GIANTS AND PLUTO

Lying in the vast outer regions of the solar system are the gas giants, Jupiter, Saturn, Uranus, and Neptune. Also far away is Pluto, a planet that we know less about than any other planet. The gas giants are composed of huge quantities of different gases and appear to lack solid surfaces. The features that we can see are clouds composed mainly of hydrogen and helium. Deep inside the atmosphere of these giant planets, the gases may become more dense, eventually becoming liquids and solids. The cores of these planets may contain metals, as those of the inner planets do.

Jupiter

Jupiter, with a diameter 11 times that of the Earth, has a greater mass than all the other planets combined. Its large size, combined with the large amount of light reflected by its surface gases, helps to make Jupiter very bright in the night sky. In spite of its large size, Jupiter's day is less than 10 hours long. This means it is rotating very quickly.

Jupiter's surface is covered with colored bands, or belts (Figure 13.11a). Perhaps its most interesting feature is a huge hurricane called the Great Red Spot (Figure 13.11b). Larger than the size of two Earths, this hurricane already existed when people first looked at Jupiter through telescopes hundreds of years ago.

Another feature of Jupiter is that it has at least 16 moons orbiting it. Using binoculars, you can see four of these moons: Io, Europa, Ganymede, and Callisto. Jupiter and its moons have been observed from close range by four space probes: *Pioneer 10* and *11*, and *Voyager 1* and *2*. One of the discoveries made by these space probes is that Jupiter has rings around it, which cannot be seen from the Earth. These rings are made up of small rocks that travel in paths around the planet.

FIGURE 13.11
(a) This photograph of Jupiter shows that its atmosphere appears to have violent storms. You can see the Great Red Spot, which is a huge hurricane.

(b) This view of Jupiter's Great Red Spot was taken from the* Voyager 1 *space probe from a distance of about 24,000 km. Compare these photos to the weather satellite photographs of Earth that you see on television.

Saturn

Saturn is the second largest planet in the solar system. It is about five-sixths the size of Jupiter, but only about one-third of Jupiter's mass. Its average density is less than that of water, so it is the least dense of all the planets. Saturn's atmosphere is cloudy, with winds that blow up to about 500 km/h. Like Jupiter, Saturn rotates quickly on its axis, so its "day" is less than 11 hours long. Saturn is farther from the sun than Jupiter, so its average temperature is lower, about –180°C.

Saturn has always been identified by its rings, which are visible from the Earth through telescopes. It is one of four planets now known to have rings (the other gas giants also have rings). Saturn's rings were seen closely by *Pioneer 11* and the two *Voyager* space probes. Photographs from these probes showed that the rings of Saturn are composed of over 1000 separate smaller rings (Figure 13.12). Astronomers are not certain whether the rings formed at the same time as the planet or are the remains of a moon of Saturn that broke up. Saturn has at least 17 moons. Several of these moons were first seen by the *Voyager 2* space probe. Saturn is featured in the photographs at the beginning of this chapter.

FIGURE 13.12
You can see the fine details of Saturn's rings in this picture taken by* Voyager 1*. The big rings appear to be made up of many tiny rings.

Uranus

Although Uranus is almost four times as big in diameter as the Earth, it is so far away that it looks like a faint star to an observer on the Earth. In fact, it was thought to be a star until its motion was discovered in 1781. In spite of the great distance of Uranus from the Earth, astronomers have gathered considerable data about this planet, mostly from *Voyager 2*, which passed near it in 1986. Uranus is unusual because its axis of rotation is in nearly the same plane as its orbit. This means that Uranus rotates on its side, as you can see in Figure 13.13. The rings of Uranus were discovered in 1977 by scientists observing light rays passing by the edge of the planet from a distant star.

The atmosphere of Uranus looks like a thick fog. Its average temperature is about –210°C. The atmosphere is made up mostly of hydrogen, with some helium and methane. It has winds that blow up to about 500 km/h.

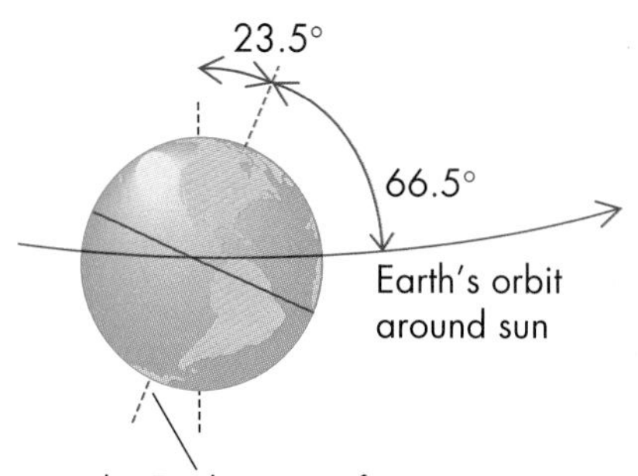

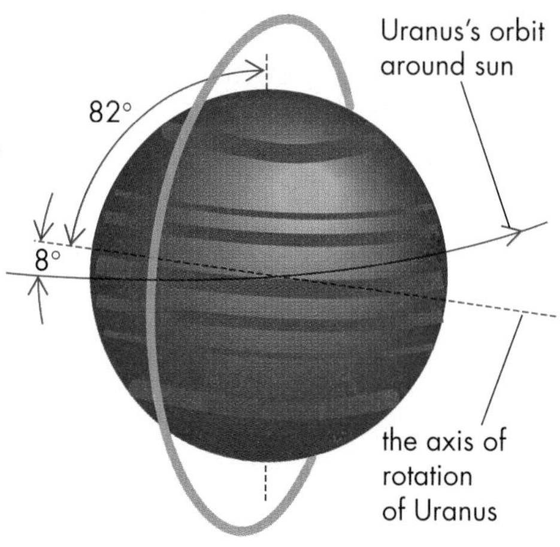

FIGURE 13.13
Comparing the rotations of the Earth and Uranus. The Earth's axis is at a large angle from its orbit. The axis of Uranus is in nearly the same plane as its orbit. How do you think the tilt of the axis of Uranus would affect its seasons?

Neptune

Today, Neptune is the farthest planet from the sun. (In 1999, Pluto's orbit will take it farther from the sun than Neptune, a situation that will last for many years before Neptune again becomes the farthest.) The story of the discovery of Neptune is one of great scientific achievement. It is so far from the Earth that it is barely visible, even through powerful telescopes. After scientists discovered that Uranus was a planet and not a star, they studied its orbit and discovered that the orbit was not a smooth path. They hypothesized that some object must be tugging on Uranus, causing its unsteady orbit. Using detailed calculations, they predicted where this hidden object must be, and a careful search began. Patient observations paid off, and in 1846 the missing planet was discovered. It was named Neptune.

Up until 1989, scientists did not know very much about this distant planet compared with the other gas giants. Then the space probe *Voyager 2*, which was launched from the Earth 12 years earlier, flew past Neptune. This space probe was expected to survive for only about five years, and had already provided exciting views of Jupiter, Saturn, and Uranus. But *Voyager 2* kept working, so scientists on the Earth reprogrammed its computers to allow it to send clearer and more detailed information about Neptune than they had thought possible. Astronomers discovered that Neptune has bright blue clouds with white sections that are moving (Figure 13.14). It has a region that was given the name Great Dark Spot, which appears to be the centre of a storm. Neptune's atmosphere is made up mostly of hydrogen, helium, and methane. Its average surface temperature is about –220°C.

DID YOU KNOW

■ If a space probe to Neptune were launched now, it would take about 30 years to get there. Why, then, did *Voyager 2* take only 12 years to travel the distance? The reason is that the gas giants were lined up in a way that allowed their force of gravity to increase the probe's speed. This alignment occurs only once every 176 years!

FIGURE 13.14 ▶
This picture of Neptune was produced from images taken from Voyager 2. You can see the Great Dark Spot clearly in this photograph.

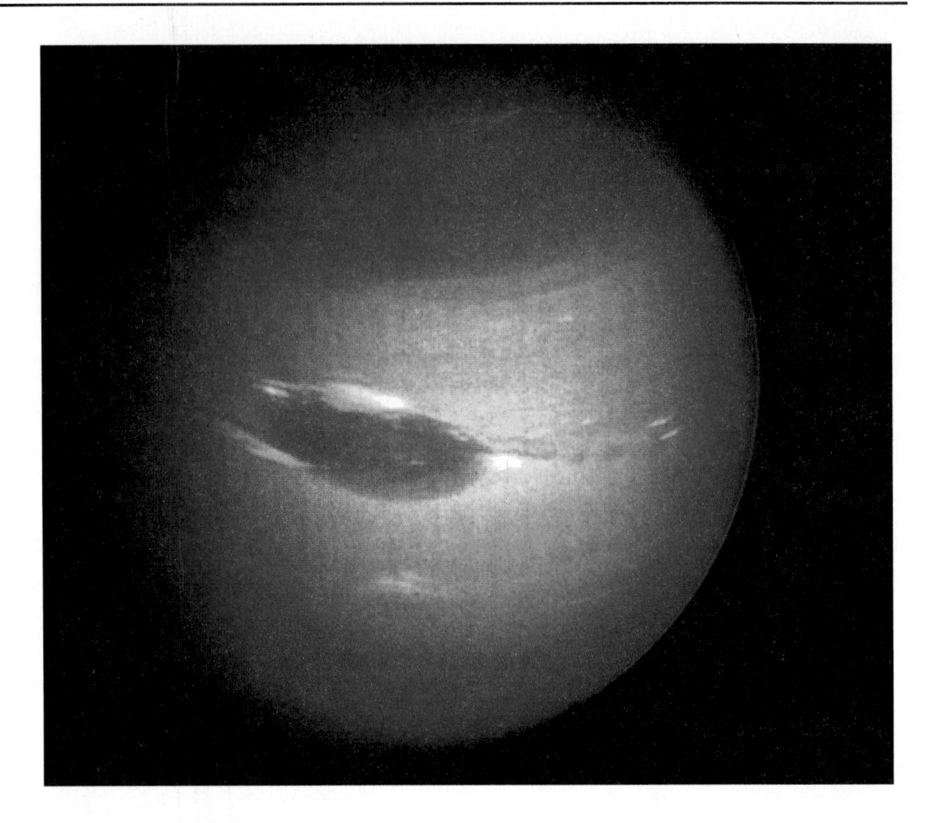

Using *Voyager 2*, astronomers also discovered moons they had not known about, bringing the total to eight known moons travelling around Neptune. They also proved that Neptune has some thin rings made up of dark, dusty material travelling around it.

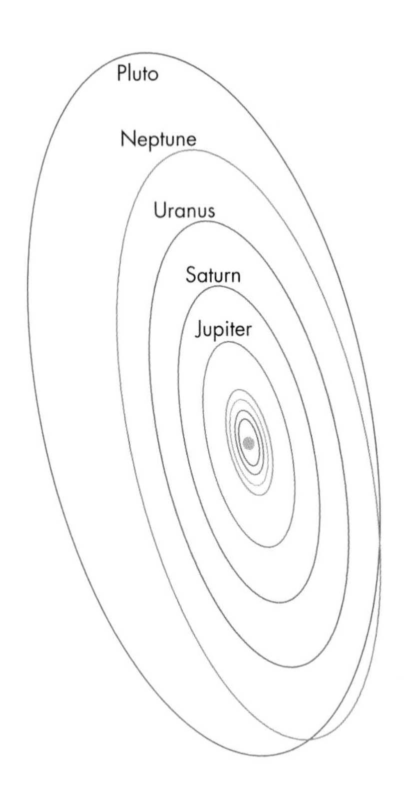

FIGURE 13.15
Pluto's orbit compared with the orbits of other planets. (The diagram is not drawn to scale.)

Pluto

Pluto is an unusual planet because it is not a gas giant and it does not seem to be terrestrial. It was discovered in 1930 after astronomers had observed that Neptune's orbit was not a smooth path. They thought that some object must be causing Neptune's unusual motion. (Recall that Neptune was discovered in this way, by observing the motion of Uranus.) But the discovery of Pluto was a lucky accident. Can you find the whole story of Pluto's discovery?

Astronomers hypothesize that frozen methane and other solids cover Pluto's surface. A moon, called Charon, was observed in 1979 by astronomers using the most powerful telescopes on the Earth.

The time it takes Pluto to orbit the sun is so great that it has completed only about one-quarter of its revolution since it was discovered. Its orbit is more oval in shape than those of the other planets (Figure 13.15). Although Pluto is normally the most distant planet from the sun, it will be closer to the sun than Neptune until 1999. Pluto's unusual orbit has led some astronomers to suggest that it may have been a moon of Neptune at one time.

TERENCE DICKINSON: SCIENCE WRITER

Terence Dickinson writes about astronomy. He has written 10 books and hundreds of magazine and newspaper articles on the subject. In an interview from his office at home, he had this to say about his work.

As far as I know, I was born interested in astronomy.

I remember at the age of five being fascinated by a bright meteor. Now, if I am away from the stars for any length of time, I get frustrated. I just have to go out and soak up some starlight. I might photograph some stars, look though my telescope, or just look at the night sky with binoculars.

That's one of the qualities I think a good science writer needs. You have to be passionate about your subject, and so curious that you would want to find out about it anyway.

At first, I was going to be a research astronomer. Then I realized that I was not the kind of person to focus on one small part of astronomy. I was fascinated by the big picture of the universe. I started working in a planetarium, writing scripts for shows about all aspects of astronomy. Later, I started writing articles for newspapers and magazines, and became the editor of *Astronomy* magazine. I finally decided to work full-time on my own as a writer.

My first really successful book was a stargazing guide called *NightWatch*. It was published in 1983. Your school library probably has one of my books.

My next book answers the 50 most-asked questions about the universe. One question is "What is a black hole?" People have heard of them, but have difficulty understanding the explanation. Another question students often ask is "What is Planet X?" They have all heard of it, this mysterious, undiscovered planet in our solar system. The fact is, there is almost no evidence for the existence of Planet X.

My work forces me to keep in touch with everything happening in astronomy: the latest spacecraft mission, next week's eclipse of the moon, the current thinking about extraterrestrial life, theories about what it was like billions of years ago when the universe was just evolving, and everything else.

I think of myself as a professional explainer of astronomy. I write as if I am having a conversation with people who want to know more. I try to introduce them to the subject easily and comfortably, and I think about the questions they might ask.

I'm very lucky I am living at a time when so much is happening in the field of astronomy. And I'm fortunate to be able to spend my days working at a job I find so fascinating.

If you were a writer, what subject in science would you like to write about? Why?

ACTIVITY 13E / PLANETARY TOURISM

■ Begin collecting information about observing planets in the night sky. You will need to know what star constellations the planets can be found in. You can see four planets without a telescope or binoculars. They are Venus, Mars, Jupiter, and Saturn. After you learn about star constellations in the next chapter, you will be able to carry out observations, and keep a record of the motion of some of the planets. Some excellent references are: the *Observer's Handbook*, published by the Royal Astronomical Society of Canada; the *Astronomical Calendar* by Guy Ottewell; and the annual January issues of *Sky and Telescope* and *Astronomy*. Ask your teacher to show you the monthly sky calendar in *Science Scope* magazine.

Imagine that you are a travel agent working for Offworld Safari Tours. You have been hired to promote tours to one of the planets other than the Earth.

In this activity, your task is to design and produce a travel brochure that will attract tourists to the planet you have chosen. Remember: you are trying to sell tours. Make your brochure exciting and interesting!

Find out all you can about the planet and the trip your customer would experience by travelling there. Your teacher will provide you with a list of books, computer software, and other reference material.

Your brochure should include the following information:

1. A colorful and creative cover page.
2. An attractive but accurate description of the planet, including the planet's size, atmosphere, temperature, weather conditions, and surface features.
3. A map showing the path to the planet. (In your trip, you may pass asteroids, comets, or other interesting objects. Read Section 13.3 to find out about these bodies.)
4. Special attractions: What are some of the features of the planet (such as craters, volcanoes, low gravity, etc.) that make the planet unique and interesting to visit? You may also offer extra trips to moons, but make sure your information is accurate.
5. Activities that could be enjoyed on the planet. The activities should be related to the special features of the planet (e.g., low-gravity sports, crater climbing, tours to special sites, etc.).
6. Things to bring and accommodations: Will the tourists need special clothing or breathing apparatus? Where will they stay?
7. Any other information that your customers might find useful or interesting. ❖

► REVIEW 13.2

1. Each of the following descriptions fits one of the planets in the solar system. Name the planet which each sentence describes.
   (a) It is, at present, the farthest planet from the sun.
   (b) It has more mass than all the other planets combined.
   (c) It has surface temperatures ranging from −180°C to 400°C.
   (d) It has an atmosphere containing oxygen.
   (e) It is neither a gas giant nor a terrestrial planet.
   (f) It has over 1000 rings around it.
   (g) It appears reddish in color.
   (h) It has a very warm surface caused by its thick atmosphere.
   (i) It rotates on its side.
2. Draw a diagram to illustrate the orbits of Mercury and the Earth around the sun. Then use your diagram to explain why Mercury is so difficult to see from the Earth.

3. Why do you think Mars is more interesting to space explorers than the other planets are? Give several reasons.
4. List the factors that affect how bright a planet in the sky would appear to you.
5. Why is Pluto listed as the last planet even though Neptune is, at present, farther away from the sun than Pluto is?
6. Astronomers predict that groups of planets similar to our solar system may be quite common in space. However, such groups of planets are very difficult to detect. Why do you think this is so?

13.3 OTHER OBJECTS IN THE SOLAR SYSTEM

Although the sun and the planets are the most important and obvious objects in the solar system, there are many other objects that are interesting to study.

ACTIVITY 13F / EXPLORING THE PLANETARY MOONS

In this section, you will read about several of the moons that travel around planets. Before you start reading, copy Table 13.4 into your notebook.

As you read about the moons of the different planets and look at their pictures, describe what you learn about each one. Write about one moon at a time so you will know how much space you will need in your table. ❖

Name of moon	Planet around which this moon revolves	What I learned about this moon in Section 13.3

TABLE 13.4 ***Sample Data Table for Activity 13F***

PLANETARY MOONS

Several planets have at least one moon. Each moon travels in an orbit around its "parent" planet. (The chunks of rock that make up the rings of the gas giants are far too small to be called moons.)

Probably the most famous moon of any planet is the Earth's moon. We know more about it than about any of our other neighbors in space. Our moon's diameter is about one-quarter that of Earth. This makes it one of the largest moons when compared with the size of the planet around which it travels. Six visits to the moon by humans from 1969 to 1972 provided much new information. The *Apollo* astronauts

■ Look up information about the *Apollo* missions to the moon. Find out what scientists discovered about the structure and history of the moon. What else did they discover? Write a newspaper article describing what you learn.

FIGURE 13.16
An Apollo 17 scientist performing tests on the surface of the moon. What is the umbrella-shaped thing beside him?

brought more than 380 kg of moon rock to Earth for detailed study. They also collected data on the composition of the moon's soil, surface conditions, and moonquakes (Figure 13.16). Our moon has no atmosphere, and its surface is filled with hills and valleys as well as craters caused by the impact of large and small objects from space.

The moons of the other planets were not discovered until after the invention of the telescope. In 1610, the Italian scientist Galileo Galilei looked at Jupiter through his telescope and observed four objects that appeared to be orbiting the giant planet. He was the first person to see some of Jupiter's moons (Figure 13.17).

Although humans have not yet been to other moons besides our own, space probes have investigated several moons at close range. For example, in 1977, *Viking* spacecraft photographed Phobos and Deimos, the two small moons of Mars (Figure 13.18). Many of the largest moons of Jupiter and Saturn were photographed for the first time by both *Voyager* space probes from 1979 to 1981. The features of some of these moons are exciting and puzzling to astronomers.

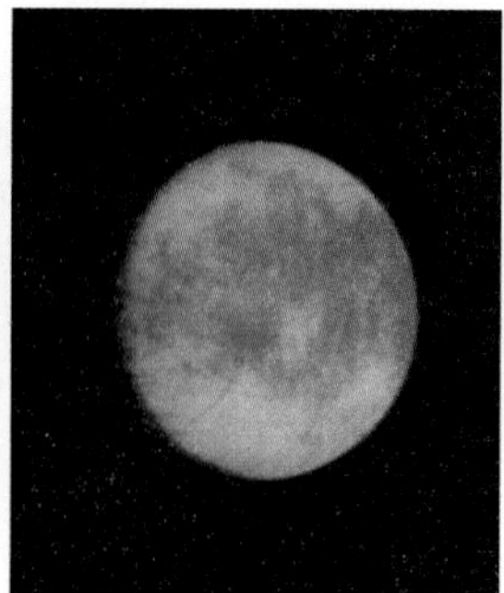

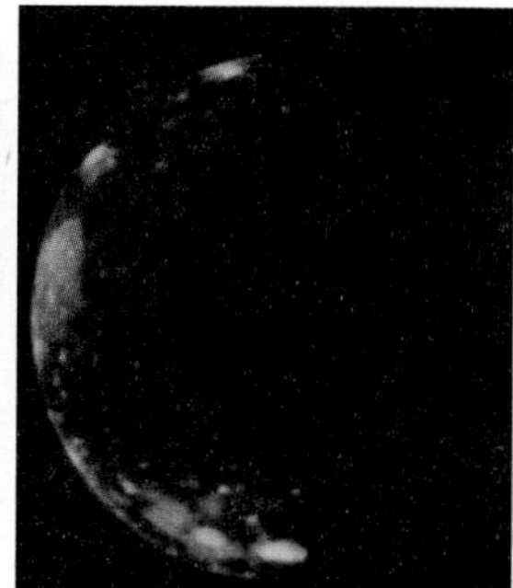

FIGURE 13.17
The four moons of Jupiter discovered by Galileo: Io, Europa, Ganymede, and Callisto. Which of these moons looks similar to our moon?

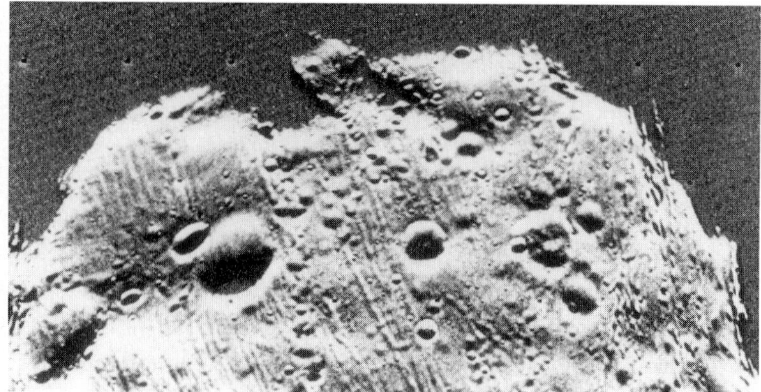

FIGURE 13.18 ***The moons of Mars***
(a) This photograph, taken from a distance of 3300 km, shows the many craters on the surface of Deimos.

(b) This is Mars's closer moon, Phobos. The photograph was taken from 880 km away. What do you think caused the craters you see on these moons and on Earth's moon?

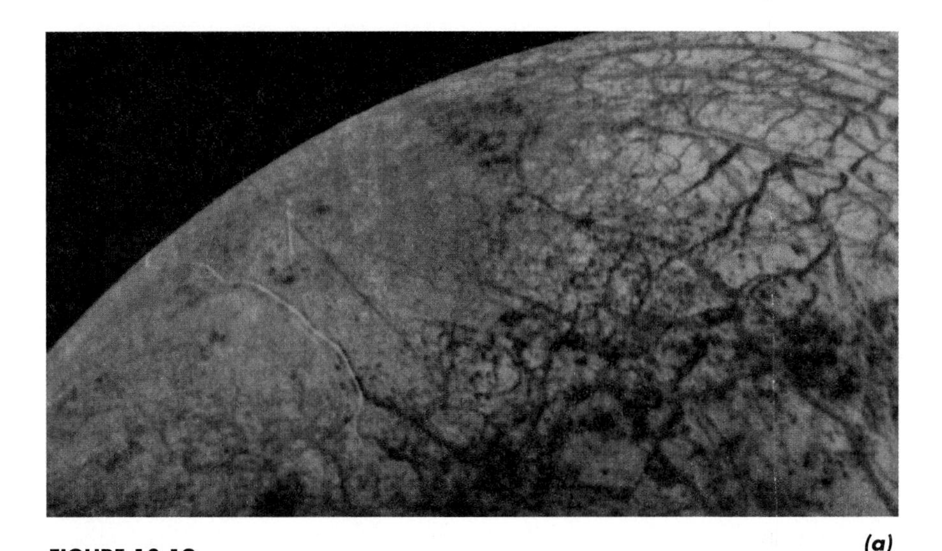

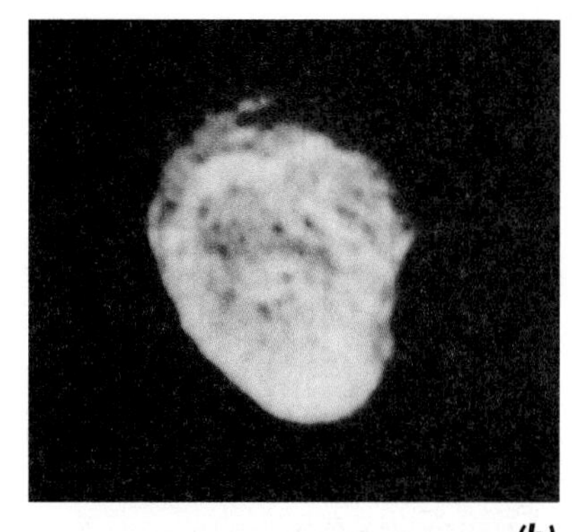

FIGURE 13.19
(a) Jupiter's moon Europa has an icy surface with very few craters.
(b) Saturn's moon Hyperion has an irregular shape. This shape could be the result of repeated collisions with large rocks from space that have broken off large portions of this moon.

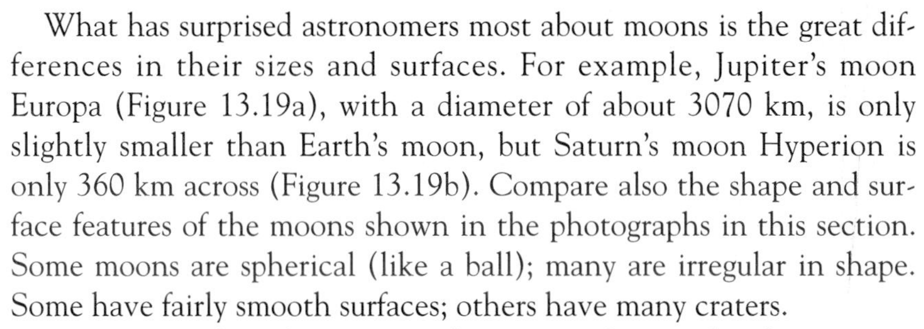

FIGURE 13.20
The volcanoes on Jupiter's moon Io are more violent than any on the Earth. The one shown here is blowing sulfur to a height of more than 100 km above the surface of Io.

What has surprised astronomers most about moons is the great differences in their sizes and surfaces. For example, Jupiter's moon Europa (Figure 13.19a), with a diameter of about 3070 km, is only slightly smaller than Earth's moon, but Saturn's moon Hyperion is only 360 km across (Figure 13.19b). Compare also the shape and surface features of the moons shown in the photographs in this section. Some moons are spherical (like a ball); many are irregular in shape. Some have fairly smooth surfaces; others have many craters.

Astronomers have been especially interested in Io, the closest moon to Jupiter. It is the only moon in our solar system known to have active volcanoes. During the seven months when the *Voyager* space probes transmitted photographs of Io, scientists discovered six active volcanoes on that moon. You can see one such volcano in Figure 13.20.

The space probe *Voyager 2* provided the most recent discoveries about moons when it passed Neptune in 1989. We now know that Neptune has at least eight moons. One of these, called Triton, is shown in Figure 13.21. This moon is about 2700 km in diameter and its surface is made up mostly of nitrogen.

Table 13.5 lists the number of known moons revolving around the planets of the solar system. Note that only the number of moons revolving around the four terrestrial planets are known for certain. Future space probes may reveal more moons.

Studying the many planetary moons in the solar system helps us understand more about the origin and evolution of the solar system. But we may find other uses for these moons in the future.

Planet	Number of known moons
Mercury	0
Venus	0
Earth	1
Mars	2
Jupiter	16
Saturn	17
Uranus	15
Neptune	8
Pluto	1

TABLE 13.5
Planetary Moon Count (1992)

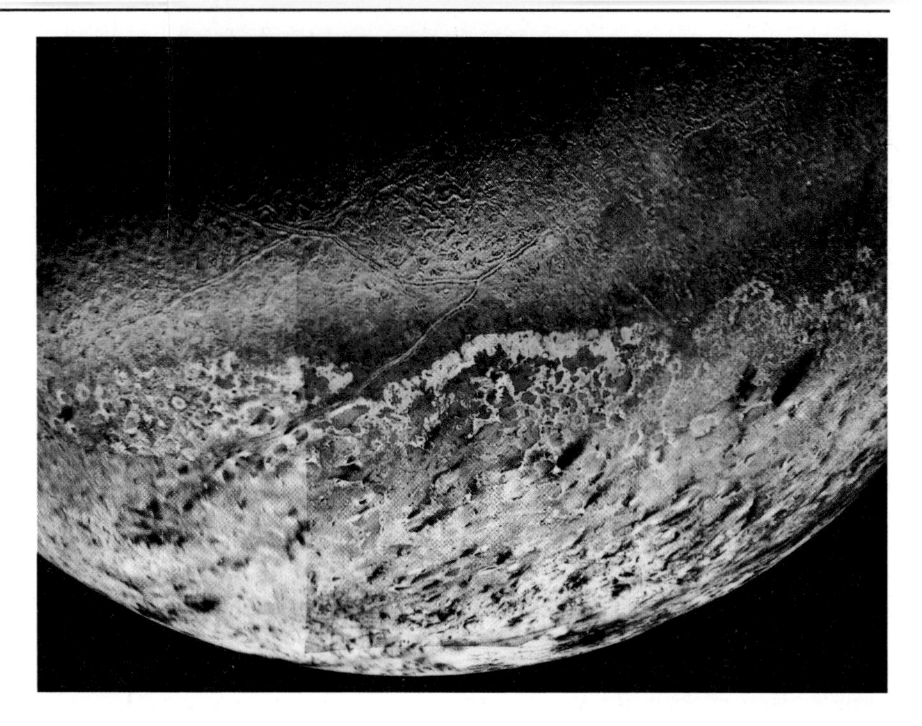

FIGURE 13.21 ▶
Neptune's largest moon, Triton. The dark spots could be liquid nitrogen shooting above the surface.

DID YOU KNOW

■ Astronomical names are taken from many sources. Ariel, Miranda, Oberon, and Titania, four moons of Uranus, are named after characters from the plays of William Shakespeare. Three craters on the Earth's moon—Plato, Aristotle, and Copernicus—are named for people who contributed to the study of astronomy. What are the planets of our solar system named after?

The moons contain huge amounts of useful minerals that humans may mine one day and use for construction on the Earth or other planets.

ACTIVITY 13G / EXPLORING SMALLER OBJECTS IN THE SOLAR STYSTEM

In the remaining parts of this section, you will read about asteroids, meteors, meteorites, and comets. To help you organize information about these objects, copy Table 13.6 into your notebook. Then complete the table as you read about the objects. ❖

Type of object	Definition	What I learned about the object

TABLE 13.6 ***Sample Data Table for Activity 13G***

ASTEROIDS

Look again at the map of the solar system you made in Activity 13B. You can see that there is a large distance between Mars and Jupiter. Would you expect a planet to be there? Scientists think that at one time a planet may have existed there, because thousands of small objects have been discovered travelling in an orbit between Mars and Jupiter. These irregular, rock objects are called **asteroids**. This word comes from a Greek word meaning "star-like," but a better name would be "minor planets." Asteroids could be "leftovers" from the time long ago when the planets formed, or they could be the result of collisions between what once was a planet and large rocks travelling through space.

Most asteroids are found in the region between Mars and Jupiter, named the **asteroid belt**. However, some asteroids follow Jupiter's orbit, whereas others travel in paths that may take them closer to the sun or the Earth. For example, an asteroid named Hermes came within 800,000 km of the Earth, only about twice the distance from the Earth to the moon. Figure 13.22 shows the orbits of some asteroids and their names.

Like the planetary moons, the asteroids are rich in minerals, which humans may someday mine. An important property of asteroids that would make mining them easy is their small size. The largest asteroid is only about 1000 km across, so it has low gravity. This means that a spacecraft would require much less energy to blast off from an asteroid than to blast off from the Earth, Mars, or even our own moon.

Jupiter
Adonis
Apollo
sun
Mercury
Venus
Earth
Mars
Eros
Icarus
Hermes

FIGURE 13.22
Orbits of several asteroids (not to scale)

METEORS AND METEORITES

A **meteor** is a lump of rock or metal that falls from space toward the Earth's surface. As it passes through the Earth's atmosphere, the meteor rubs against the molecules of the air. This rubbing, called friction, causes the meteor to become hot and burn. This produces a streak of light across the sky that you can see at night.

Most meteors burn up completely before they reach the Earth's surface. If the object is large enough to hit the Earth's surface before totally burning up, it is called a **meteorite**. One example of a meteorite is the Innisfree meteorite, named after the Alberta, Canada community where it landed in February 1977 (Figure 13.23).

FIGURE 13.23
(a) The Innisfree meteorite in flight. (b) Most of the Innisfree meteorite held just above the spot where it came to rest on the snow. Some dirt and meteorite chips are scattered around the area after having bounced off the field beneath the snow. If you lived in Innisfree, what would you suggest should be done with the meteorite?

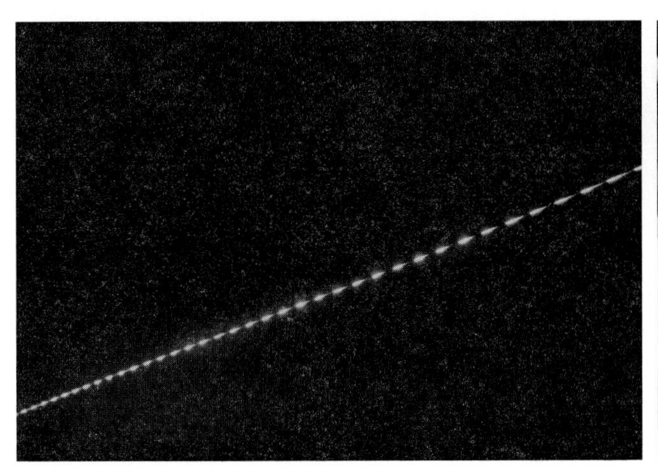

(a)

(b)

Where do meteors come from? The larger meteors probably come from asteroids that have orbits that cross the Earth's orbit. The millions of tiny meteors that produce spectacular displays called meteor showers probably come from the pieces left behind by comets.

If a large meteorite hits the Earth's surface, it can produce a crater. The best-preserved meteorite crater in the U.S. is the Barringer Crater near Winslow, Arizona (Figure 13.24).

FIGURE 13.24 ▶
Evidence gathered from local rocks suggests that the Barringer Crater was formed 20,000 to 30,000 years ago by a large meteorite. The crater is 1.2 km in diameter and 120 m deep. What is the shape of most craters? Suggest a reason for this shape.

DID YOU KNOW

■ The word "meteor" comes from a Greek word meaning "high in the air." Until about 100 years ago, the word "meteor" was also used to describe lightning, rainbows, and other things observed in the atmosphere. That explains why the science of meteorology has to do with weather, not with meteors.

■ Have you heard of "shooting stars" or "falling stars"? They are not stars at all: they are meteors. If you view the sky on a clear night, you have a good chance of seeing a meteor. The chances become very high if you view the sky during a meteor shower. The three most active meteor showers are the Perseid shower (best seen on August 12), the Geminid shower (best seen on December 14), and the Quadrantid shower (best seen on January 4). In reference books, you can find out the names and dates of other meteor showers, as well as the directions to look toward to view them.

COMETS

One of the most interesting and exciting scenes you could see in the sky is a comet with a tail that is millions of kilometers long. Some comets are so bright that you can see them even in daylight. A **comet** is a chunk of rocky or metallic material covered with ice. Comets travel in a very long oval orbit around the sun. Scientists think that the ice consists of frozen ammonia, methane, carbon dioxide, and water.

During most of its orbit, a comet is far from the sun and cannot be seen with the unaided eye. When it approaches the sun, however, the comet is warmed by solar energy and the frozen substances become gases. As the comet comes closer to the sun, these gases are pushed outward, forming a bright, glowing tail that may be over 100 million kilometers long. The glowing tail may be seen for several months as the comet travels near the sun. A comet's tail always points away from the sun because the solar energy acts like a wind sent out from the sun (Figure 13.25).

(a)

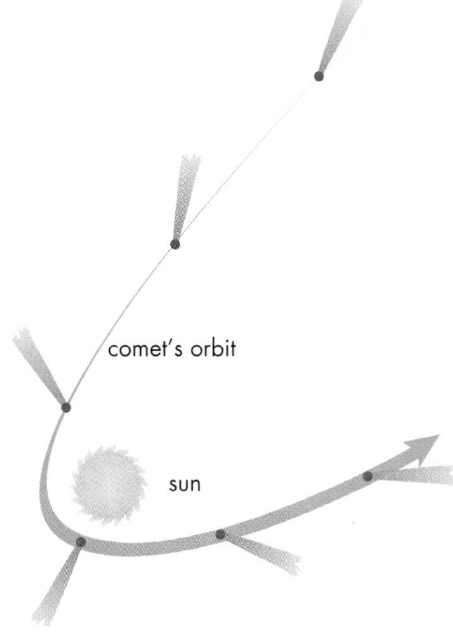

(b)

FIGURE 13.25
(a) The tail of Comet Bradfield is plainly visible in this photograph. (b) The tail of a comet always points away from the sun, even when the comet is speeding away from the sun.

One of the most famous comets is called Halley's comet, named after Edmund Halley, who lived in England from 1656 to 1743. He was the first person to predict when this comet would return close to the sun. There have been brighter comets than Halley's, but their orbits are so large that they return only once in thousands of years. Halley's comet returns every 76 years. It is also unusual because it has been visible from Earth at least 29 times (Figure 13.26). It has been seen and recorded since 239 B.C., and there is even some evidence that it was seen as long ago as 1140 B.C.

■ In 1910, people learned that the Earth was about to pass through the tail of Halley's comet and that the tail contained a deadly gas. Comet pills, gas masks, and special comet insurance were sold. But they were not necessary because the tail of a comet is not dangerous to anyone on the Earth.

FIGURE 13.26 ◀
This computer image was made from a black-and-white photograph of Halley's comet taken in 1910.

The last time Halley's comet was near the sun was in 1985 and 1986, and scientists were able to study it at close range. Professional and amateur astronomers waited eagerly, with sensitive instruments that had not existed the last time the comet had returned. The data collected from the comet's recent visit will keep researchers occupied until its next return in 2061.

DID YOU KNOW

■ In the late 1700s, after Halley discovered the regular return of the comet named after him, many people searched the skies for other comets. One famous astronomer who discovered eight comets was Caroline Herschel, who lived in England. She was the first female professional astronomer, receiving a regular salary from the king. Her brother, William, was also famous. He discovered that Uranus was a planet, in 1781.

REVIEW 13.3

1. Of all the planetary moons in our solar system, which one would you be most interested in visiting on a scientific expedition? Describe special features of this moon, and explain what you would try to discover about it while you were there.
2. (a) What is an asteroid?
   (b) Where is the asteroid belt located?
   (c) What do you think would happen to the Earth if it were struck by a large asteroid?
3. (a) Explain the difference between a meteor and a meteorite.
   (b) Meteors are fairly common. Why do you think that only a few meteorites have been found?
4. (a) What are comets made of?
   (b) What causes the glowing tails of comets?
   (c) Both comets and planets orbit the sun. How do their orbits differ?
5. Why does a comet's tail always point away from the sun?
6. Scientists think that there may be huge comets that orbit our sun, but no one has ever recorded seeing them. Why do you think they have never been recorded?

CHAPTER REVIEW

KEY IDEAS

■ The center of the solar system is the sun, which is also the largest object in the system.

■ Objects in the solar system are in motion. One type of motion is revolution around another object, and another type of motion is the rotation of an object on an axis.

■ Astronomers are scientists who study the composition, position, and movement of objects in the universe.

■ Nine known planets revolve around the sun. Four have orbits that are very near the sun, and five have orbits that are much farther away.

■ The four planets that are closer to the sun are called the terrestrial planets. They include the Earth and three planets that resemble the Earth.

■ Four of the planets that are farther from the sun are called the gas giants. They consist mainly of gases.

■ The farthest known planet, Pluto, is neither a terrestrial planet nor a gas giant.

■ Seven planets are known to have a moon or several moons travelling around them.

■ Other objects in the solar system include asteroids, meteors, and comets.

VOCABULARY

planet
moon
solar system
revolution
orbit
year
rotation
axis
terrestrial planet
inner planet
outer planet
gas giant
astronomy
astronomer
space probe
asteroid
asteroid belt
meteor
meteorite
comet

V1. Imagine that you are a space traveller from a distant group of planets, and you have just discovered our solar system. When you go back to your own world, you want to draw a mind map to show the features of the solar system. Draw this mind map in your notebook. Include as many words and ideas as you can that were used in this chapter.

V2. Make a list of all the types of objects or features on objects in the solar system that have proper names (in other words, names that are capitalized). For each type of body or feature, list as many of the proper names as you can. (Many of the names are found in this chapter, and many others can be found in reference books.)

CONNECTIONS

C1. (a) Describe the Earth's position among the nine planets of the solar system.
(b) Describe several features of the Earth that make it unique in the solar system.

C2. Choose one planet, and name three similarities and three differences between it and the Earth.

C3. (a) Name the two main groupings of planets in our solar system.
(b) For each group, describe three features that all its members have in common. (Tables 13.1 and 13.2 can help you answer this question.)

C4. Table 13.5 shows a total of 60 planetary moons in the solar system. If you checked other reference books published in earlier years, you would find the number of moons listed as follows:

Year of publication	1969	1984	1987	1991
Number of known moons	28	44	53	60

(a) Why do you think that the number of known moons has changed so greatly in such a short period of time?
(b) Do you think the number of moons will change as much in the next 25 years or so? Explain why or why not.

C5. (a) Where do you think asteroids come from?
(b) Do you think that the asteroid belt would be dangerous to travel through in a spacecraft? Explain why or why not.

C6. Why does a meteor appear as a streak of light in the sky?

C7. (a) Why do you think craters caused by meteorites are rare on the Earth?
(b) Why do you think there are many meteorite craters on a planet like Mercury?

C8. Why do you think some comets are seen much more often than others?

C9. Astronomers have observed that the brightness of a comet usually becomes fainter and fainter with each return trip around the sun. Explain this observation. (Hint: Think about the substances that comets are made of.)

C10. What are some benefits that scientists would achieve in sending robot space probes to asteroids, other planets, and moons of other planets?

EXPLORATIONS

E1. (a) Describe some of the problems that humans might face as they try to set up a settlement on Mars.
(b) Suggest one way to overcome each of the major problems you listed in (a). Explain your answer in each case.

E2. The planets are named after Roman gods and goddesses. For example, Venus (which is a beautiful planet) is named after the goddess of love and beauty, and Pluto (in the darkness and coldness of space) is named after the god of the underworld. Imagine that you have been assigned the task of renaming all the planets using modern names. Suggest a suitable name for each planet, and explain your choices.

E3. Research the story of a successful space probe, then write a report, make a display poster, or produce a video summarizing your findings. One great success story is that of the *Voyager 2* craft that passed by Neptune 12 years after it was launched. (An excellent resource for this is the August 1990 issue of the *National Geographic* magazine.)

E4. Many books are available on how to build your own telescope. Obtain one or two such references, and discover if you think you could build such an instrument. If possible, begin your project soon so you can use the telescope to view the planets (and the stars you will study in the next chapter).

E5. Make up a short story or a poem using one of these titles: "Mercuryquake," "Venusquake," "Marsquake," or "Moonquake."

REFLECTIONS

R1. Look back at the poster you worked on in Activity 13A. Think about the answers you gave to the questions in that activity. Would any of your answers be different now that you have completed this chapter? Explain.

R2. Look back at the two photographs at the beginning of this chapter. What do those photographs show you about how the study of space science and astronomy has changed since the invention of space probes?

CHAPTER 14

THE STARS

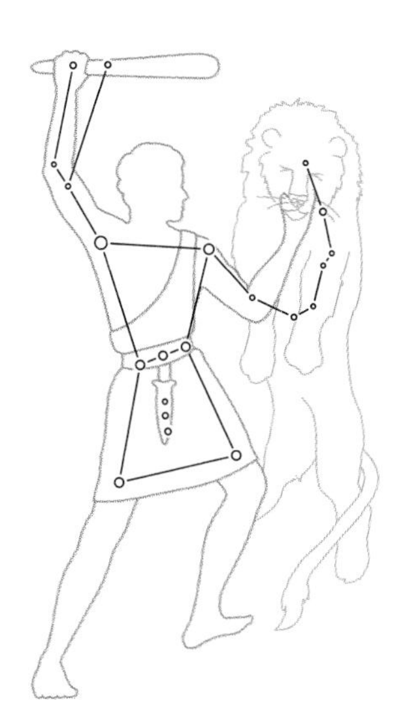

The night sky, with its stars and many other objects, has been a source of wonder for humans since the earliest times. Some stars and the patterns they seem to make have always been visible to the unaided eye. On a clear night in the winter, you can see the part of the sky shown here. Some people in ancient times imagined that they saw the shape of a hunter in this group of stars. They called the hunter Orion. They drew pictures of Orion ready to hit a lion with a club. Can you see the stars in the photograph that make up this pattern?

Stars and other objects too faint to be seen with the unaided eye suddenly become visible through a telescope. A telescope is just one of many tools used to study the sky.

This chapter is about stars. You can learn much about stars by observing the night sky, by looking at photographs taken with telescopes, and by other methods presented in this chapter. Your study begins with the star that is most important to our lives, the sun. You will also study stars in general, and discover how our sun compares with other stars.

If you are lucky to have clear weather, you can increase your enjoyment of this topic by making direct observations of the night sky.

ACTIVITY 14A / STARS AND PATTERNS OF STARS

PART I Questions About Stars

Divide a page in your learning journal into three columns with the following titles: "Question," "What I think," and "What I learned." In the first column, write the questions listed here, and add some questions of your own. When making up your questions about stars, think about what you have learned about stars in the past. Your teacher may ask you to work with one or more other students. In your group, make a list of questions about stars using everyone's ideas. Record the questions in column 1, leaving out those that are the same as your own. In the second column, write what you think are the answers to the questions. You will complete the third column after studying this chapter.

QUESTIONS

1. Why is the sun the most important star in the sky?
2. Why can you see the sun only in the day, but you can see the other stars only at night?
3. How does the temperature inside the sun compare with the temperature on the surface?
4. About how long would it take a space probe to travel to the sun?
5. About how long would it take a space probe to travel to the nearest star beyond our sun?
6. What is a constellation?
7. What is the difference between astronomy and astrology?

PART II Looking for Patterns of Stars

Your teacher will give you a copy of a map of part of the night sky. The larger dots represent the brighter stars. On your copy, outline in pencil groups of stars that seem to make patterns. The patterns can be animal shapes, geometric figures, or anything you are familiar with.

Compare your patterns of stars with those of other members of your class. Think about why it would be easier to discuss the stars in the sky if everyone saw the same patterns. ❖

■ The amount of all forms of energy that the sun produces every second is more than the amount of electrical energy that would be produced in Oregon in 1 billion years.

■ 14.1 THE SUN: AN AVERAGE STAR

By far the most important star to us is the one at the center of the solar system: our sun. The sun has been studied more than almost every other object in the sky. Learning about the sun helps us understand other stars more easily, as well as how we fit into the universe. As you read about the features of the sun, remember that we are learning more every day.

Compared with other stars, our sun is an average star of medium size. When compared with the Earth, however, the sun is huge (Figure 14.1). The sun's diameter is about 1.4 million kilometers, or almost 110 times larger than the diameter of the Earth. More than 1 million Earths could fit inside the sun. If you think of the sun as the size of a basketball, the Earth would be a tiny pearl, only about 2 mm in diameter.

The sun is the closest star to the Earth. The average distance between the Earth and the sun is about 150 million kilometers. If you travelled at a speed of 100 km/h, like a car on a highway, it would take

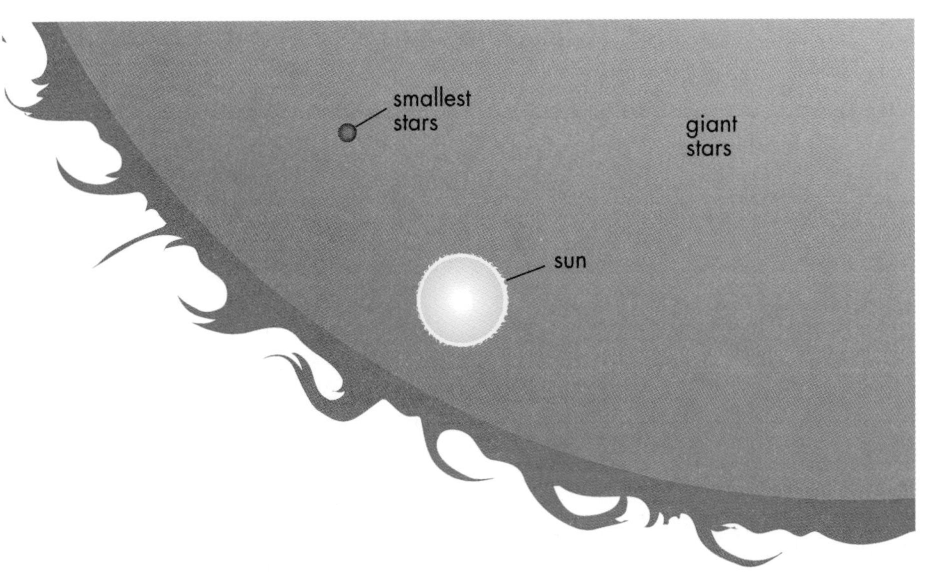

FIGURE 14.1 ◀
The sun is so large that it could hold about 1.3 million Earths. Even at this large size, the sun is only an average-sized star, and it is tiny compared to giant stars.

you about 170 years to travel from the Earth to the sun. A space shuttle, travelling at a speed of 28,000 km/h, would take about seven months to get there. This may seem like a long time, but it is short compared with the time it would take a space shuttle to travel to the next nearest star—about 180,000 years!

Because the sun is by far the closest star to the Earth, it is also the brightest object in the sky. In fact, it is so bright that you cannot see the other stars unless the sun has set.

The reason the sun is more important to us than the other stars is that it provides us with the light energy and heat needed to support life on the Earth. Where does the sun's energy come from? Like all stars, the sun produces energy through a process called nuclear fusion. Inside the sun, the temperature and pressure are so high that substances fuse (join together) to form new substances. This process produces huge amounts of heat, light, and other forms of energy (Figure 14.2). These forms of energy travel out from the sun through space. Every second, the sun makes more energy than humans have used throughout our entire history.

DID YOU KNOW

■ About one out of every five of the world's astronomers studies the sun every day. They work in observatories, government laboratories, private companies, and universities. What instruments or other devices do you think are most important to astronomers?

■ Numbers like million and billion are common when talking about stars. One billion can be written as 1000 million, or 1,000,000,000, or 1×10^9, or simply 10^9. To get an idea of how big this number is, consider the following: If you were to start counting at a rate of one number per second, it would take you about 32 years to count from one to 1 billion! How would you write out the number 5 billion?

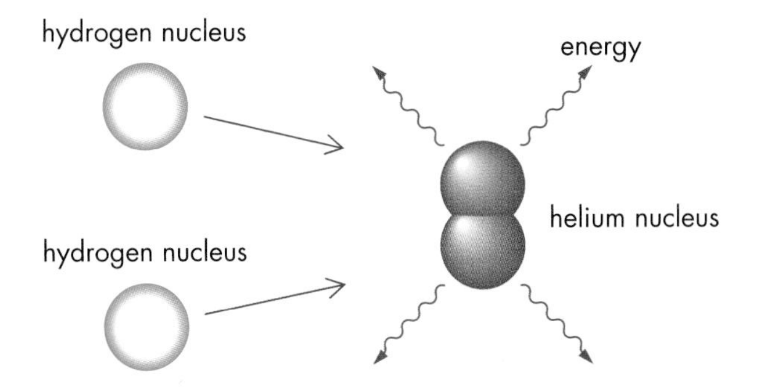

FIGURE 14.2 ◀
During nuclear fusion, substances join together to form a new substance. When this happens, large amounts of energy are released. This is the sun's source of energy.

DID YOU KNOW

■ Some particles from solar flares reach the Earth after a few days. Their energy produces the beautiful auroras seen over the Earth's North Pole (known as the Northern Lights) and the South Pole (the Southern Lights). When solar flares are strong, they can cause problems with communications systems here on the Earth. This can be a major problem, especially for people living in the far northern parts of the Earth.

Scientists have used calculations to determine that the sun has been producing energy for about 5 billion years. They estimate that it will continue producing energy for about another 5 billion years. Thus, we have no need to worry about how long the sun will live.

A CLOSE LOOK AT THE SUN

Based on many observations and calculations made by astronomers, we can draw a model of the sun. Figure 14.3 shows such a model. The sun is made up of gases: about 75 percent hydrogen and the remainder helium, with small amounts of other gases. Together, these gases form layers that are given different names. In the outer layer of the sun, the **corona**, the gases are very hot, with a temperature of about 1 million degrees Celsius. Beneath this layer is the **chromosphere**, or inner atmosphere. **Solar flares** are bursts of energy that travel outward from the chromosphere through the corona. These solar flares travel extremely quickly and last only a few minutes. **Solar prominences**, large sheets of glowing gases, burst outward from the chromosphere. They can last for days or even weeks, and they can grow as large as 400,000 km high, which is greater than the distance from the Earth to the moon. The solid black dot just outside the solar prominence in Figure 14.3 represents the size of the Earth, so you can see how large the solar prominence is.

EXTENSION

■ In 1990, the space probe *Ulysses* was launched from Earth. It was a joint European and American probe, and was the first probe intended to look closely at the poles of the sun. Use reference books or magazines to discover more about *Ulysses*. Describe what you find out about it.

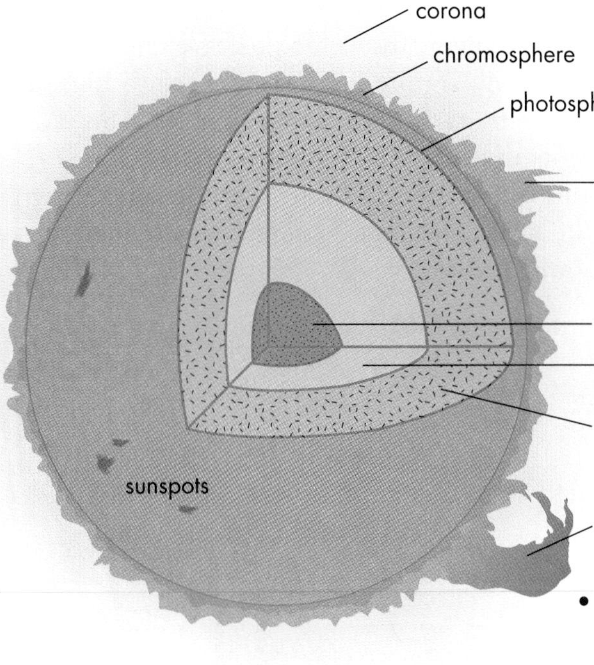

FIGURE 14.3 ▶
A model of the structure of the sun. How do you think scientists gather data to help them understand the structure of the sun?

Beneath the sun's chromosphere is the **photosphere**, made up of boiling gases. The photosphere is called the surface of the sun, although it is not a solid surface at all. Its average temperature is about 5500°C.

An interesting feature of the photosphere is that it has darker regions called **sunspots**. They look darker because they are cooler than the rest of the photosphere. Sunspots are huge. Even the smallest sunspots observed are larger than the size of the Earth. Watching them from the Earth, astronomers have discovered that the sunspots are in motion. This provides evidence that the sun rotates on its axis. The surface of the sun near the sun's equator completes a rotation every 26 days. Away from the equator, the rotation is slower.

You can observe sunspots indirectly by projecting an image of the sun through a telescope or a pair of binoculars onto a white screen. The safe way to do this is shown in Figure 14.4. The greatest numbers of sunspots occur every 11 years, with about 100 spots visible. The last time this occurred was in 1990, and the next time will be in 2001.

EXTENSION

- View the sun over a period of three or four days by projecting its image through a telescope or binoculars onto a white screen. Keep a record of your observations. Try to discover whether different regions of the sun rotate at different rates.

CAUTION!

- Never look at the sun directly with your eyes or through any instrument. You could suffer permanent eye damage in just a few seconds.

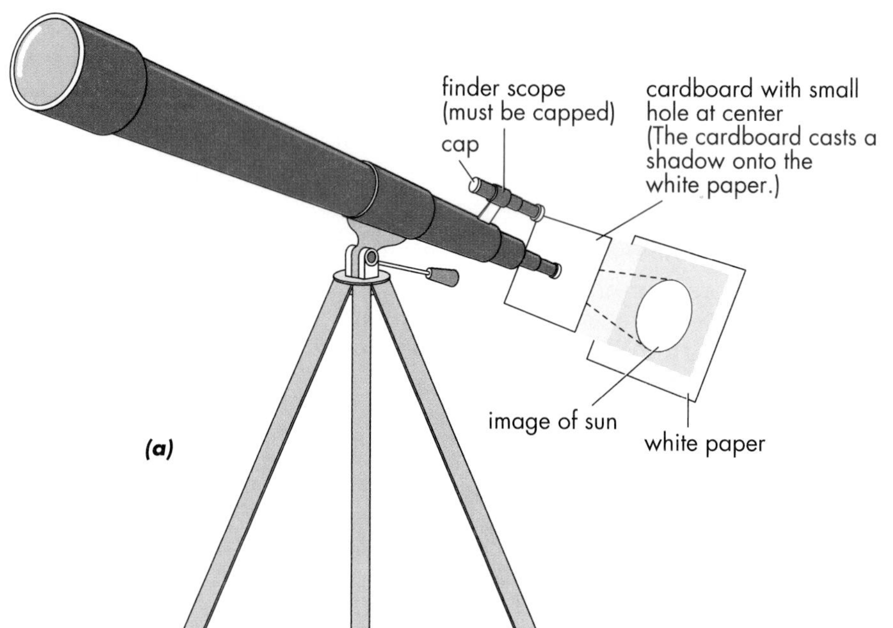

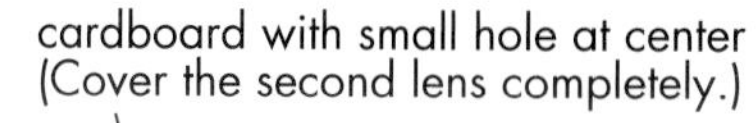

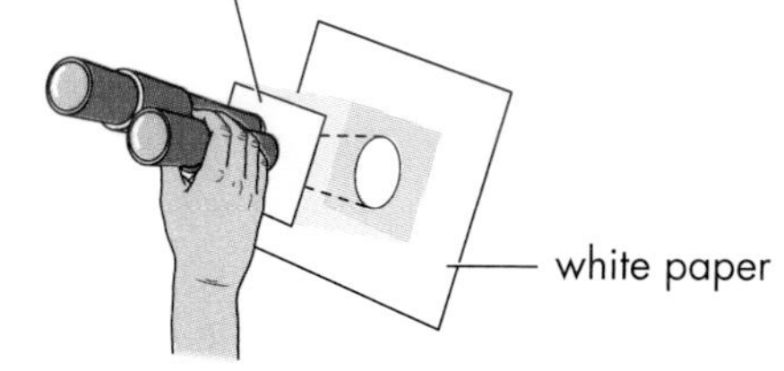

FIGURE 14.4
Two safe ways of obtaining an image of the sun. (a) Using a refracting telescope. (b) Using binoculars. In both cases, you may have to move the paper back and forth to get a clear image.

Just beneath the photosphere is a huge region of moving gases. Closer to the center of the sun, the temperature and pressure increase. At the center is the sun's core. This is the region where nuclear fusion produces the sun's energy. Temperatures here reach perhaps 15 million degrees Celsius, and the pressure is enormous.

▶ REVIEW 14.1

1. Why do we consider the sun to be the most important star?
2. Describe the differences between a solar flare and a solar prominence.
3. What feature of the sun provides evidence that the sun rotates on its axis?
4. (a) In what layer of the sun is energy produced?
   (b) What occurs there to produce so much energy?
5. You have read that the sun can continue producing energy for about 5 billion more years. Is the possible "death" of the sun at that time a problem we should worry about? Why or why not?

14.2 THE STARS YOU SEE IN THE SKY

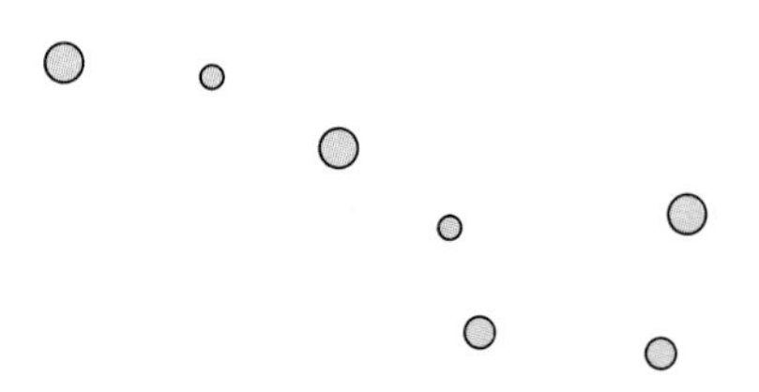

FIGURE 14.5
Of all the constellations or star patterns visible, probably the most easily recognized is Ursa Major, which contains the Big Dipper, shown here. Does the Big Dipper always appear this way up (like a saucepan)?

When you look up at the sky on a clear night, you see countless stars spread unevenly across the night sky. A long time ago, sky observers noticed that certain patterns of stars seemed to suggest the shapes of animals, mythical heroes or gods, and other objects. Groups of stars that seem to form shapes or patterns are called **constellations**. A constellation appears to move across the sky (just as the sun does) as the Earth rotates on its axis. However, the distances separating the stars of the pattern seem to remain constant for many years. Probably the most easily found constellation in the sky is Ursa Major, which contains the group of stars called the Big Dipper (Figure 14.5). Ursa Major is also called the Great Bear; *ursa* is the Latin word for "bear" and *major* is the Latin word for "great" or "large."

Even to an astronomer, many constellations do not look like the shapes that people in ancient times saw. Today, rather than showing diagrams of animals or people, many books simply show geometric shapes to represent the constellations.

ACTIVITY 14B / A BEGINNER'S STAR MAP

Books on astronomy and star maps show the patterns of stars that are recognized as constellations. In this activity, you will learn how to use star maps and identify some constellations.

MATERIALS

Star Map 1 (provided by your teacher)
ruler

PART I The Three Most Important Constellations

PROCEDURE

1. Figure 14.6 shows the shapes of three constellations that you can see in the sky all year. Near the middle of Star Map 1, locate the Big Dipper of the Ursa Major constellation. Use a ruler to draw lines joining the stars that make up the Big Dipper.
2. At the center of your star map, locate the Little Dipper. ("Ursa Minor" means "Little Bear.") Use a ruler to draw lines joining the stars that make up the Little Dipper. Label the star at the tip of the handle "Polaris." This star is

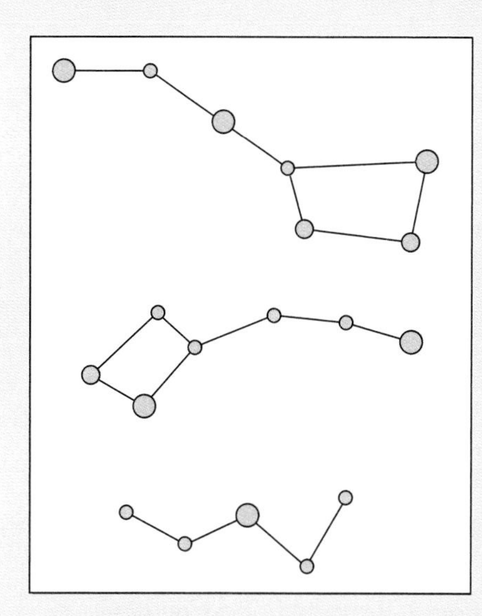

FIGURE 14.6
Three Constellations
(a) The Big Dipper
(b) The Little Dipper
(c) Cassiopeia

important. It is also called the North Star. Anybody in the Northern Hemisphere of the Earth looking toward Polaris is facing toward the North Pole.

3. Find the two stars of the Big Dipper that are farthest from the handle. From these stars, use a ruler to draw a dashed line to Polaris. This line shows how to use parts of the Big Dipper as a pointer to locate other stars or constellations. (This is important when you try to view the night sky because the Little Dipper is not nearly as bright as the Big Dipper.)
4. From the same pointer stars you used in Step 3, continue past Polaris a short distance until you reach a constellation in the shape of a spread-out *W* or *M*. This is the constellation Cassiopeia. Use a ruler to draw lines joining the stars that make up this constellation.
5. Label the three constellations you have found so far. Ask your teacher to check your star map now.

DISCUSSION

1. (a) What is another name for the star Polaris?
   (b) In what constellation is this star found?
   (c) What is the importance of Polaris to observers in the Northern Hemisphere?
2. Name the three constellations that are visible in all seasons to observers in the Northern Hemisphere.

PART II Constellations and the Seasons

PROCEDURE

1. Figure 14.7 shows constellations that you can see in the night sky during different seasons. On your star map, these constellations are found farther away from Polaris than the Big Dipper or Cassiopeia. The easiest one for you to see is the constellation Orion, with three bright stars that line up to make this imaginary man's belt.

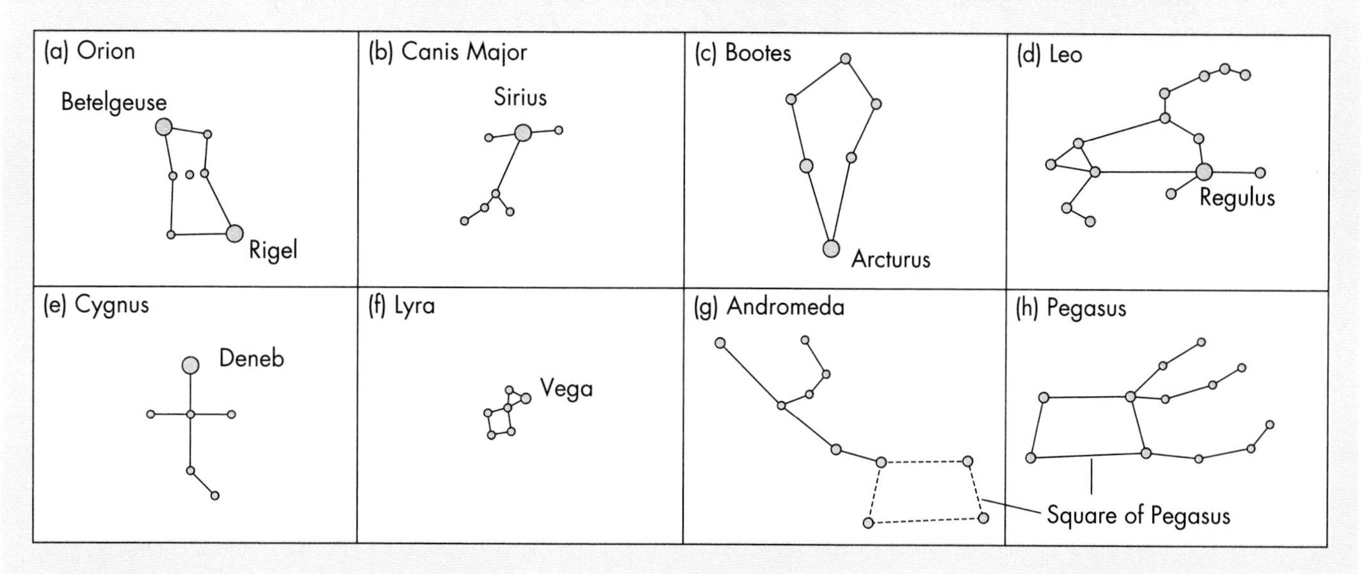

FIGURE 14.7
Eight constellations that you can see only during certain seasons. Where are they at other times of the year?

EXTENSION

■ The names of the constellations have meanings. For example, Orion is the hunter, and Cygnus is the swan. Use a reference book to look up the meanings of the names of the constellations mentioned in this activity. Choose one constellation other than Orion and draw a diagram of the shape suggested by its name. (Orion is shown in the diagram at the beginning of the chapter.)

Locate Orion on your star map. It is near the part of the map marked December. This means that Orion can be seen most easily in December, but it can also be seen in November and January. Use a ruler to draw lines joining the stars that make up Orion, and label it. Notice that one of the stars in Orion is Rigel. On your map, label Rigel as well as Betelgeuse, another star. Both are bright stars.

2. Describe how you would use the stars of Orion's "belt" as pointers to locate the brightest star in the sky, Sirius. Use a ruler to draw a dashed line on your map to show this method. Also draw in the constellation in which Sirius is found, called Canis Major (*canis* is the Latin word for "dog").
3. Locate the constellation Boötes on your map, and draw the lines joining the stars of this constellation. (Hint: The star Arcturus is part of this constellation.) Describe how you could use the stars of the handle of the Big Dipper as pointers to find Arcturus. Draw a dashed line on your map to illustrate this method.
4. Draw the constellation Leo on your map. It is located about midway between Canis Major and Boötes, and contains the star Regulus. Describe how you could use two stars of the Big Dipper as pointers to locate Regulus. Show this with a dashed line on your map.
5. Find the three bright stars Deneb, Vega, and Altair. Join them with dashed lines to show the Summer Triangle. Draw the constellations to which Deneb and Vega belong: Cygnus and Lyra.
6. Locate, draw, and label the constellations Andromeda and Pegasus. You can find these constellations by starting from Polaris and going past Cassiopeia. Also label the part of Pegasus called the Square of Pegasus.

DISCUSSION

1. Some constellations can be seen only during certain seasons. From the constellations labelled on your star map, name two constellations that you can see in:
   (a) winter
   (b) spring
   (c) summer
   (d) autumn
2. Describe the meaning and use of "pointer stars."
3. Create a data table in your notebook with two columns. In column 1, list the names of the constellations that you have labelled on your star map. In column 2, list the stars that are part of these constellations. Remember to write a title at the top of each column. ❖

EFFECTS OF THE EARTH'S MOTION

If you watch the sky at night, you will notice that the positions of the stars and planets change slowly. Like the sun, the stars appear to rise in the east, travel across the sky, and set in the west. Furthermore, you will see different stars during different seasons. You can understand these changes by thinking about the motion of the Earth.

One type of motion of the Earth is called rotation. As you learned in Chapter 13, rotation is the spinning of an object on its axis. One rotation of the Earth takes 24 hours. This motion causes the stars (as well as the sun, moon, and planets) to appear to rise in the east and set in the west. But the stars are not moving: it is the Earth that is rotating.

The Earth's axis is an imaginary straight line joining the North Pole and the South Pole. If the axis were continued northward, out into space, it would pass through Polaris, the North Star. Figure 14.8a shows why people who live in the U.S., which is in the northern part of the Earth, are able to see Polaris all year long.

As the Earth rotates, the stars near Polaris seem to travel in circles around Polaris. For observers in the U.S., the stars near Polaris include the stars in the Big Dipper, the Little Dipper, and Cassiopeia. Figure 14.8b shows a way of demonstrating that the Earth's rotation causes the stars near Polaris to appear to revolve around Polaris. Do you understand why this demonstration is not perfect?

EXTENSION

■ We are able to identify the constellations because the patterns of stars in the constellations always appear the same to us. But each of the stars within a constellation is moving on its own. If you went far back in time—for example, 100,000 years— you would see that the Big Dipper in Ursa Major would not look like a dipper at all because the stars would be in different positions. Use reference books to find out about the movement of the stars that form the Big Dipper. How fast are they moving? What will the Big Dipper look like in the future? Make a poster showing how the Big Dipper looked in the past, what it looks like today, and what it will look like in the future. On your drawing of the Big Dipper today, write below each star what its distance is from the Earth.

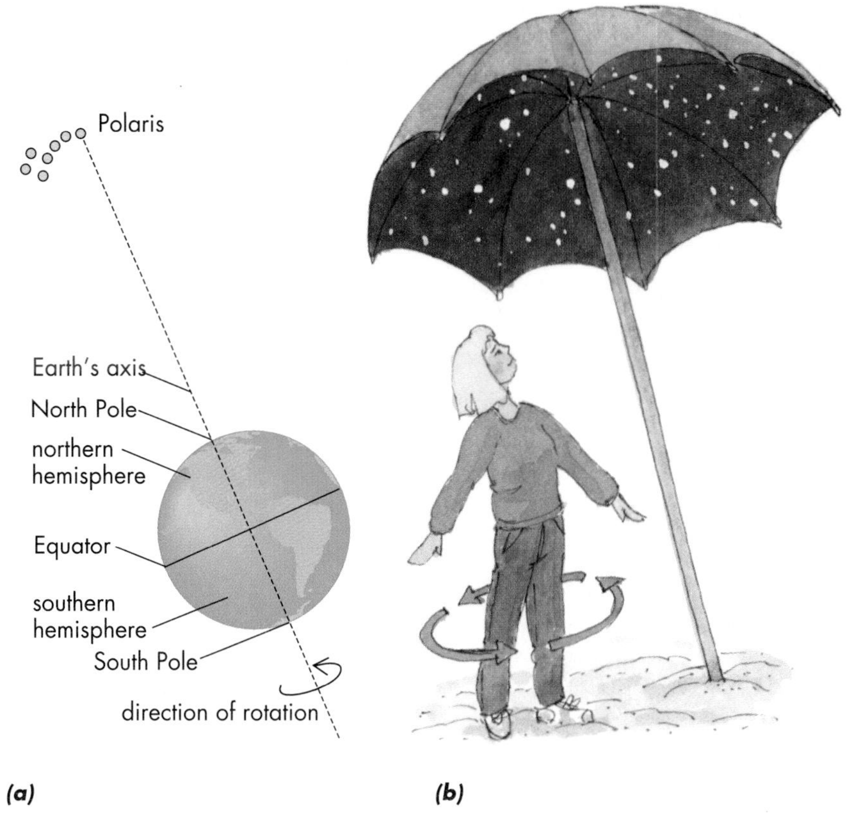

FIGURE 14.8 ◀
(a) Polaris is visible from the U.S. all year long. People south of the Equator cannot see Polaris or any of the stars near it. Are there stars that people north of the equator never see? (b) This drawing shows why the stars seem to move. The student represents the Earth. As she rotates (spins around), she sees the stars (on the inside of the umbrella) travelling in circles. Why is painting the stars on the inside of an umbrella not an accurate way to represent them?

Another type of motion of the Earth is revolution. This is the movement of one object travelling around another. The Earth travels, or revolves, around the sun once a year. This motion, combined with the angle of the Earth's axis, causes the seasons on the Earth. It also causes different stars and constellations to be visible during different seasons (Figure 14.9).

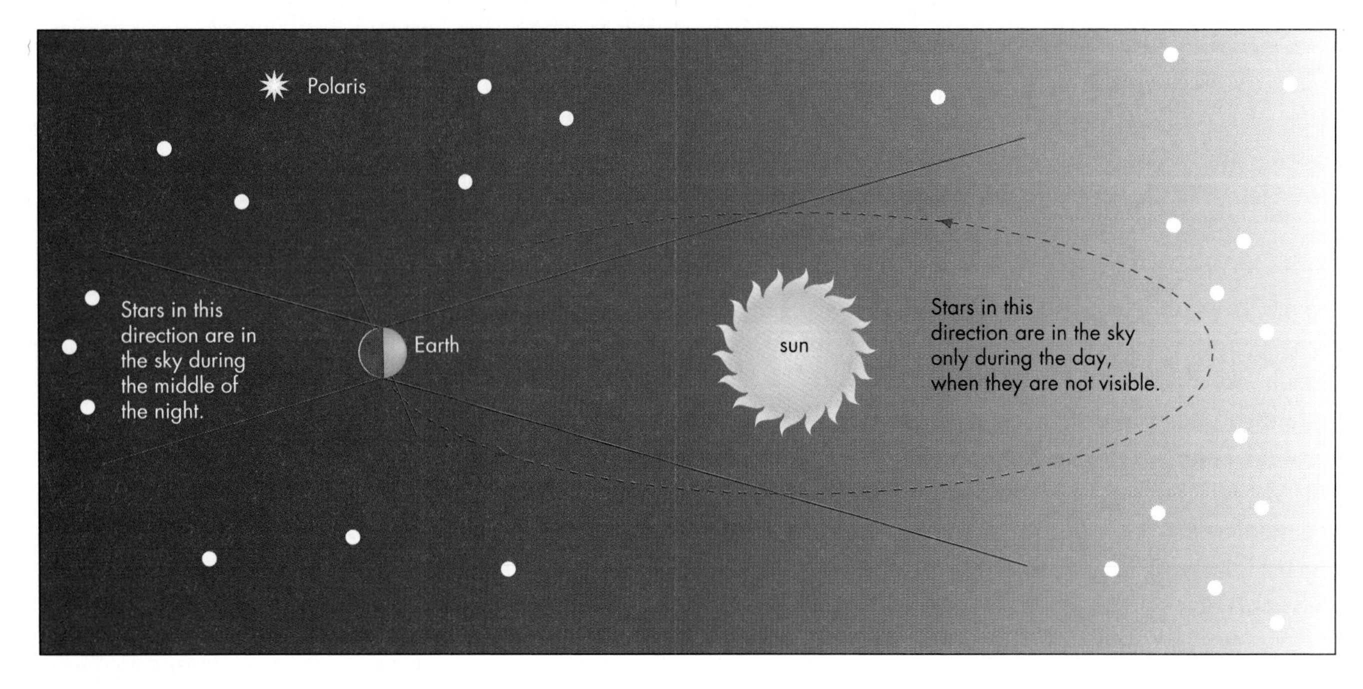

FIGURE 14.9
For observers in the Northern Hemisphere, the only constellations that we can see in all seasons are those that appear close to Polaris. We can see the other constellations only when they are not in the sky at the same time as the sun. We see stars only at night.

THE ZODIAC CONSTELLATIONS

In the past, when there were no city lights to spoil their view, people were much more aware of the night sky than most people are today. They observed that the stars and planets seemed to move in the sky. The closer, bright planets appeared to be very bright stars. Venus, Jupiter, and Mars, in particular, appear to be much brighter than the stars. As these people watched the planets, they noticed that after weeks or months the planets' positions changed when compared with the stars. They called the planets "wandering stars." (The word planet comes from the Greek word *planetes*, which means "wanderers.")

During the year, the planets appear to move through certain constellations. Many of these constellations were given names of animals. Since the Greek word for animal sign is *zodion*, the constellations were called the **zodiac constellations**.

ACTIVITY 14C / A MORE ADVANCED STAR MAP

Before beginning this activity, review Star Map 1, which you completed in the previous activity. The first step of this activity is a test of what you can remember from that map.

PROCEDURE

1. Study Star Map 2, which has all the stars found on Star Map 1 as well several more stars. Using a pencil, neatly draw in all the solid lines (constellations) and dashed lines (for example, those showing directions from pointer stars) that you can remember from Star Map 1. Use small printing to label as many of the stars and constellations as you can. Check with your copy of Star Map 1, and make any corrections necessary.

MATERIALS

Star Map 2 (provided by your teacher)
pencil

2. The only zodiac constellation found on your Star Map 1 is Leo. Find Leo on Star Map 2. There are 11 other zodiac constellations, whose shapes are drawn on Star Map 2 for you. Moving counterclockwise from Leo, label these constellations as follows: Virgo, Libra, Scorpius, Sagittarius, Capricornus, Aquarius, Pisces, Aries, Taurus, Gemini, and Cancer.
3. State what month(s) would be best for trying to see a planet that is in the constellation named. (The first answer is given.) Assume you are viewing the sky late in the evening.
   (a) Aquarius (August and September)
   (b) Aries
   (c) Taurus
   (d) Leo
   (e) Libra

DISCUSSION

1. How many zodiac constellations are there?
2. Name the zodiac constellations that have at least one very bright star in them and would be easier to identify in the sky than the other zodiac constellations. ❖

COMPARING ASTROLOGY AND ASTRONOMY

People with a knowledge of the sky can make predictions based on the regular movement of objects in the sky. In ancient Egypt, for example, when people saw the bright star Sirius appear in the dawn sky, they knew it was the time of year that the Nile River would flood. Similarly, when they saw the constellation Leo, they knew that the heat of summer was not far away. Orion is a winter constellation, and so people connected it with winter storms.

Of course, the sky changes constantly. For example, you can see the movements of the sun, moon, and the planets, the changes of the moon, eclipses of the sun and the moon, and the appearance of comets and meteors. All these changes are part of nature.

Throughout human history, some people have tried to use the motion of these objects to predict the seasons and weather conditions. Because of these attempts to link movements in the sky with events on the Earth, people came to think that events in the sky could influence events in a person's life. The belief that objects in the sky influence people's lives is called **astrology**. People who study astrology are called astrologers.

Do not confuse astrology with astronomy. As you learned in Chapter 13, astronomy is the science that studies the make-up, position, and movement of all objects in space. The findings of astronomy are tested using scientific methods such as accurate measurement and controlled experiments.

The first step in the scientific process of solving problems is observation. Early astrologers made and recorded many detailed observations of the sky. Why, then, is astrology not considered a science? It is because

■ Many native peoples used the changing patterns of the constellations to predict when the seasons were changing. They would also watch for the appearance of certain stars or planets in the early morning or evening sky. The appearance of such a star or planet would be a signal for them to begin an important ceremony. What constellations do you think native peoples might have used as guides to the seasons?

astrologers have not tested their observations through controlled experiments and other processes used by scientists. Instead, in astrology, observations are used only to support beliefs. Science provides us with a way of understanding how astronomy and astrology differ.

ACTIVITY 14D / OBSERVING THE NIGHT SKY

The information you placed on Star Maps 1 and 2 can help you observe the sky. Of course, the sky will look very different from the maps you have drawn because there are many more stars in the sky than on your maps. However, with some practice, you can become skilled in observing the stars.

A telescope or pair of binoculars will help you with your observations in this activity.

MATERIALS

flashlight
red cellophane (to cover the flashlight)
telescope or pair of binoculars (if possible)
table of observations

PROCEDURE

1. Before going outside to view the night sky, learn how to use Star Map 2. Turn the map so that the current month is at the bottom (Figure 14.10). Then the middle and lower half of the map show the stars you will see if you look south. The months on the map are labelled for viewing at midnight. If you look at the sky before midnight, the stars you expect to see will be rising in the east. If you look at the sky after midnight, the stars you expect to see will be setting in the west.

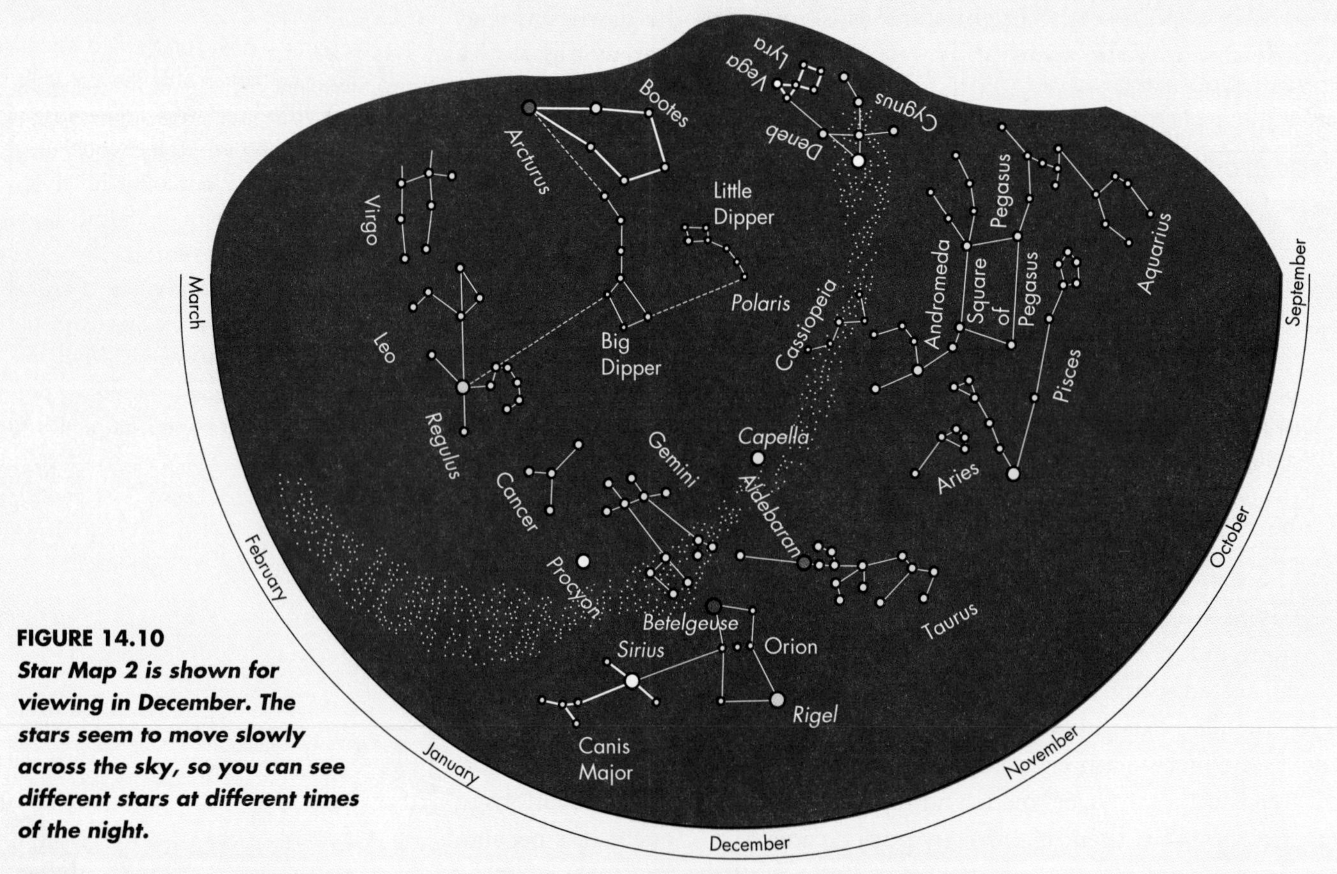

FIGURE 14.10
Star Map 2 is shown for viewing in December. The stars seem to move slowly across the sky, so you can see different stars at different times of the night.

■ **Do not go out alone to a dark area without permission from your parents. Make sure to dress warmly for night observations.**

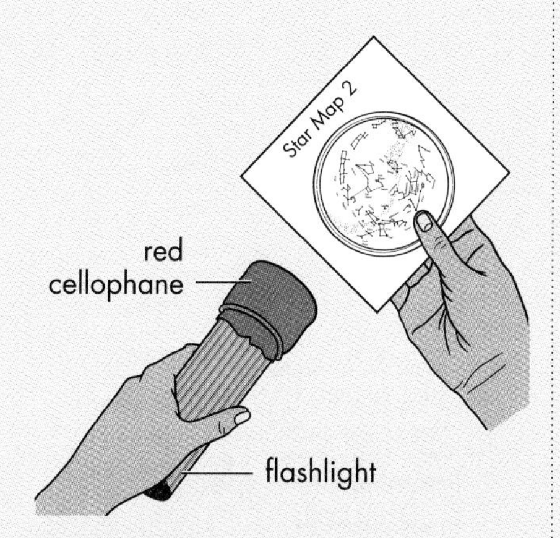

FIGURE 14.11
When using a flashlight to view a star map, it is best to cover the light with red cellophane. If you didn't cover the light, how might it affect your eyes?

2. On a clear night after the stars have appeared, take Star Map 2 to a dark location and use it to observe the night sky by following the directions below.

(a) Read your star map by using a flashlight. If the flashlight is too bright, cover the light with a thin piece of red cellophane.
(b) Allow your eyes to become adapted to the dark. It takes about 10 min for your pupils to enlarge so you can see faint stars better.
(c) Face south and hold your star map as described in step 1.
(d) Use your flashlight to locate the positions of stars and constellations on the star map (Figure 14.11). Then locate the same stars and constellations in the sky.

3. Record your observations in your table of observations. The following hints will help you make detailed observations.

(a) When observing a constellation, record any patterns of stars you see in that constellation.
(b) When you observe a planet, record its position relative to nearby constellations. (Planets do not appear to twinkle the way stars do. You can notice this even when you look through binoculars or a telescope.)
(c) Draw diagrams of what you see. They are useful in helping you recognize the patterns of stars in the constellations.
(d) Use fists and/or fingers on your outstretched arm (as shown in Figure 14.12) to determine the angle that bright stars are from the horizon or from each other. For example, you might determine that at 9:00 p.m. on a certain night, the belt of Orion was two "fists" above the horizon. You could convert this reading to degrees, then check it later with a star map in a reference book.
(e) Include any questions you have in your table of observations. For instance, you might observe a pair of equally bright stars one night, and then the next night they might not be equally bright. You could ask, "Which star has changed its brightness?" You may have other questions about the planets, the moons of planets, meteors, satellites, or other objects in the sky.

4. If possible, use a telescope or a pair of binoculars to look at a specific star, constellation, moon, or planet. Draw a diagram of what you see when you use either device. Record any other information. ➡

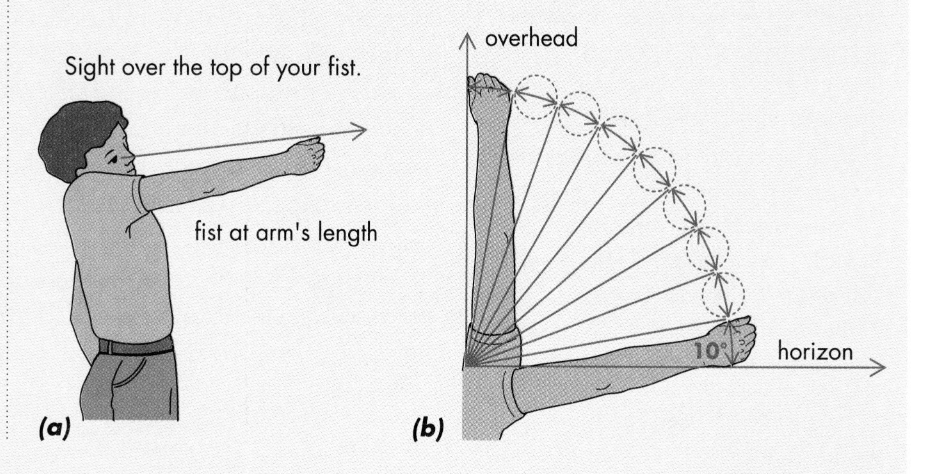

FIGURE 14.12 ▶
(a) Using your fist as a measuring tool.
(b) Each fist represents 10°. You should be able to count nine fists from the horizon to a point directly overhead.

EXTENSION

■ Use your fists to determine the angle between the horizon and Polaris. Compare the angle you get with the latitude of your location (found on an atlas or a globe). What do you conclude?

DISCUSSION

1. Why is star watching difficult in cities and towns?

2. Shawn measures the angle between Polaris and the star Sirius as seven fists. What is this measurement in degrees?

3. Describe how you would know the difference between a star and a planet when you are looking at the night sky. ❖

► REVIEW 14.2

1. (a) What is a constellation?
(b) Name three well-known constellations that are close to Polaris.
(c) Name three constellations that are located far from Polaris.

2. Explain why the constellations appear to change position from hour to hour during the night.

3. Explain why the constellations you see at a certain time one night are at different locations at the same time on another night.

4. On a clear night, what three factors affect which constellations are visible to an observer?

5. Why do you think star maps do not indicate the distances to stars?

6. Two stars appear close together in the sky. Does this mean that they are only a short distance apart in space? Explain your answer.

7. (a) What are the differences between astronomy and astrology?
(b) Why is astrology not considered to be a science?
(c) Explain why people might have developed a belief in astrology.

EXTENSION

■ Astronomers call their way of using triangulation to measure distances "parallax." Look up parallax in an astronomy reference book and see how it relates to the method of triangulation.

■ 14.3 DISTANCES TO STARS

Imagine you are standing on one side of a river and you want to estimate how wide the river is. There is no bridge across the river, so you cannot measure the distance directly using a meter stick or other measuring device. You must find a way to estimate the distance using an *indirect method.* An indirect method is one in which you use other measurements to calculate the distance, rather than measuring it directly.

You could use several methods of indirect measurement. The method discussed here is called triangulation. As you will see, this method is used to determine the distances to some stars and planets indirectly.

CURBING LIGHT POLLUTION

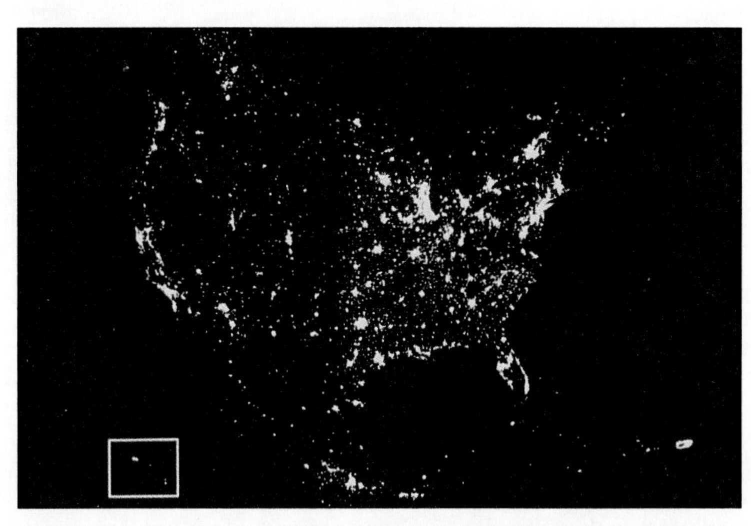

Imagine being born on a planet with an atmosphere so dense that no stars or planets are visible at night. Only the sun breaks through the haze in the daytime. You would never know that other planets exist, other star systems, or galaxies. You would have no idea how vast the universe is. How would your concept of time and your place in the universe be different if you could not walk out at night and see stars that have existed for billions of years?

If you live near a city, take a trip one night out beyond the urban glow. The darkness of the night sky will reveal a whole universe beyond, the sky the way the human race has seen it for millennia. Today, we are rapidly losing this view. Each year the city lights spread farther out into the countryside. The spreading light has even encroached on many astronomical observatories. If this continues, future generations may never be able to see the night sky in its full splendor.

This development has been given the term *light pollution.* What is the source of the problem? It can be traced to the type of lighting we use and the direction of lighting. Incandescent lights emit energy across the full range of the spectrum, and most outdoor lights are directed so that they lose much of their light up into the night sky, obscuring the stars and planets behind a hazy glow of city lights. Is there a solution?

The International Dark-Sky Association was founded in 1988 by David Crawford and Tim Hunter in Tucson, Arizona. David Crawford was a professional astronomer at Kitt Peak National Observatory and Tim Hunter is an amateur astronomer. By 1990 the organization's membership grew to 400 and by mid-1995 it grew to more than 1,700 from 49 states and 62 countries. The organization has been effective in educating the public about the adverse effects of light pollution and successful in enacting legislation to change lighting practices in many communities.

What does the International Dark-Sky Association recommend? 1) Install timers so that night lighting is used only when needed. Motion-activated lights are another alternative. 2) Make sure that lights are directed downward. It offers no greater security to shine lights up into the sky. 3) And finally, they strongly recommend the use of low-pressure sodium lights. Not only do they minimize light pollution, but they also are the most efficient lighting.

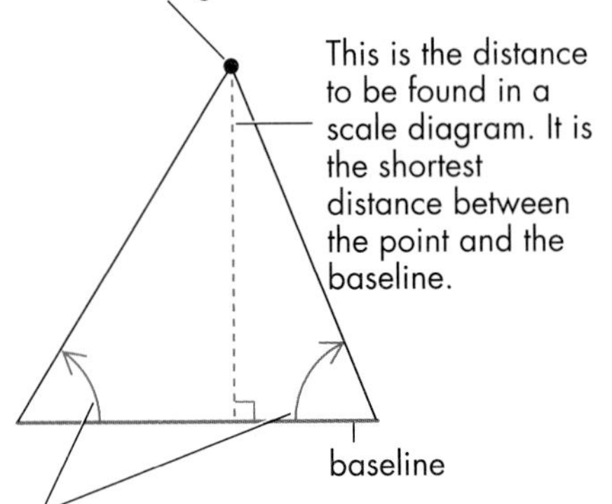
point on object to which you are measuring the distance
This is the distance to be found in a scale diagram. It is the shortest distance between the point and the baseline.
baseline
These angles are measured from the ends of the baseline to the same point.

FIGURE 14.13
The method of triangulation requires a baseline of known length. The method also requires an angle measurement from each end of the baseline to one point on the object to which you are measuring the distance.

Triangulation is the method of measuring distances indirectly by drawing a scale diagram of a triangle. The triangle has one side of known length, called the *baseline*, and two angles measured from the ends of the baseline. Figure 14.13 illustrates that the angles at the ends of the baseline must be measured to one point.

As an example, consider using triangulation to determine the distance across the river mentioned above. You mark off a 60 m baseline parallel to one side of the river. You measure the angles from the ends of the baseline to a tree on the far side of the river (Figure 14.14a). The angles are 65° and 75°.

Figure 14.14b shows the scale diagram used to solve this problem. In this case, the scale is 1 cm = 10 m, which means that 1 cm in the diagram represents 10 m in real life. (You should be familiar with number scales used to draw maps.) The width of the river is the *shortest* distance from the tree to the baseline. As you can see in the diagram, this distance is about 82 m.

FIGURE 14.14
(a) Measuring the angles from the ends of the baseline to the same tree on the opposite side of the river. (b) The length of the baseline and the two angles are drawn to scale to make a triangle. The unknown distance is measured from the point opposite the baseline (the tree, in this case) to the baseline. In this example, the scale distance is 8.2 cm, which in real life is 82 m.

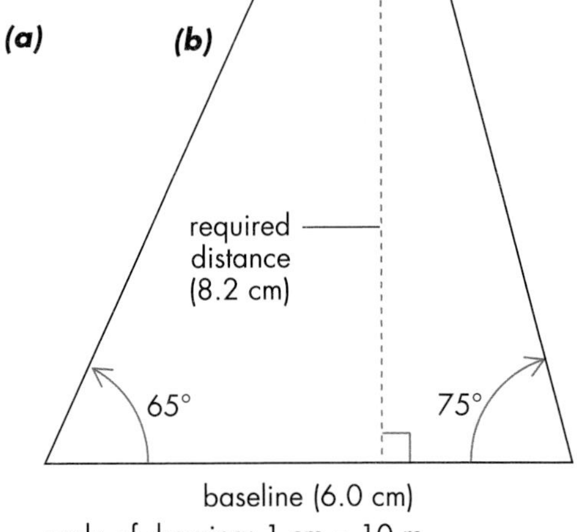
(a)
(b)
tree
required distance (8.2 cm)
65°
75°
baseline (6.0 cm)
scale of drawing: 1 cm = 10 m

ACTIVITY 14E / INDIRECT MEASUREMENT USING TRIANGULATION

GROUP ACTIVITY
G

In this activity, you will learn how to use the triangulation method by practising in or near your school. The instructions for this activity suggest that you use a meter stick and a protractor. However, you could measure the baseline distance by pacing it, and the angles required in fists, where one fist held at arm's length is about 10°.

MATERIALS

meter stick
large protractor
paper
ruler
small protractor

PROCEDURE

1. Choose an unknown distance to find indirectly. An outdoor example is the distance from the school to a distant tree. An indoor example is the distance from your desk to a light switch.
2. Mark off a baseline, then use a meter stick to measure it (Figure 14.15). For best results, use as large a baseline as possible.
3. Use the large protractor to measure the angle from each end of the baseline to one point on the object you are measuring the distance to, as shown in Figure 14.15. Be sure the protractor is placed exactly along the baseline. Record these angles.
4. Repeat the steps to find a second indirect distance.

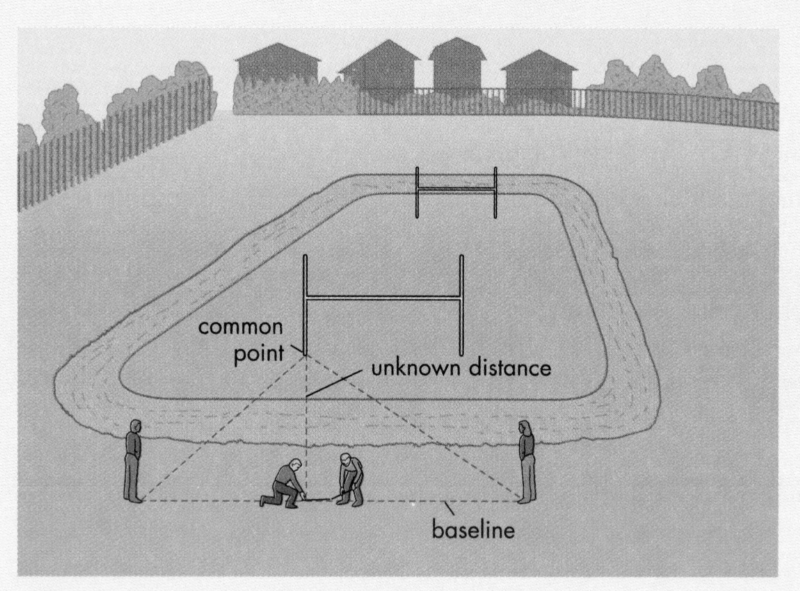

FIGURE 14.15
Choosing a baseline and measuring the angles from it

DISCUSSION

1. Decide on a suitable scale to draw a triangle on paper for the first unknown distance. Then use your ruler and small protractor to draw a scale diagram in order to determine the distance. You may want to look back at the river example and Figure 14.14.
2. Compare your results with the results of the other students who measured the same distance.
3. Describe what you could do to improve the results of your measurements.
4. What do you think is the longest baseline possible on the Earth for an astronomer using triangulation to measure distances to stars? Explain your answer. ❖

LONG BASELINE TRIANGULATION

In the method of triangulation, you can use short baselines to find short distances. But you cannot use short baselines for long distances. How can scientists obtain long baselines to measure the huge distances to stars?

One way to obtain a large baseline is to use the diameter of the Earth. Because the Earth rotates on its axis, it takes 12 hours for an observer at the Equator to move one Earth diameter from the earlier position (Figure 14.16). The baseline in this case is the diameter of the Earth, a known distance of about 12,800 km. This method could be used to determine the distance to the moon or a nearby planet.

What is the largest baseline possible to observers on the Earth? It is the diameter of the Earth's orbit (Figure 14.17). This baseline has

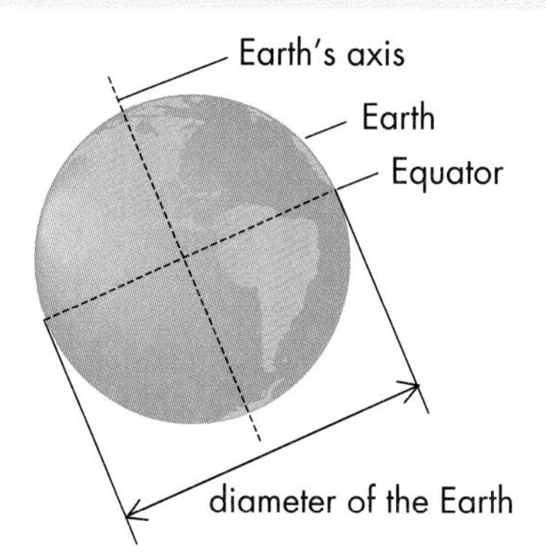

FIGURE 14.16
One example of a long baseline is the diameter of the Earth at the Equator. How much time must the observer wait between measuring angles to the object in space?

been used to calculate the distances to some of the stars nearest to our solar system. In this case, angles to the stars are taken six months apart.

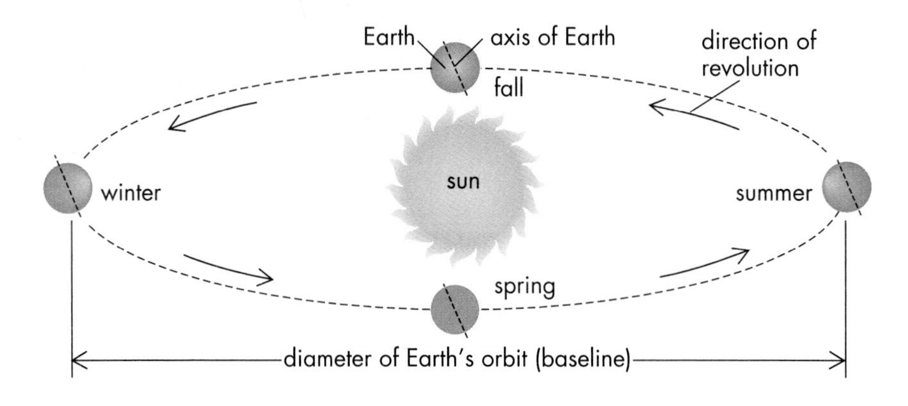

FIGURE 14.17 ▶ ***The largest baseline available to observers on the Earth is the diameter of the Earth's orbit, a distance of about 300 million kilometers. How long must an observer wait between measuring the angles from the ends of the baseline?***

■ Scientific notation is a convenient method of writing very large or small numbers. Using this notation, a number is written with a digit between 1 and 9 before the decimal, followed by a power of 10. For example, the distance of one light-year (9,460,000,000,000 km) can be written as 9.46×10^{12} km. Use this notation to write the light-year distances listed in Table 14.1.

DISTANCES TO THE STARS

The distances between stars and other objects in the universe are huge. For example, the distance from our sun to the next nearest star that you can see without using a telescope or binoculars is about 41 million million kilometers (41,000,000,000,000 km). This star is called Alpha Centauri. Most other distances in space are much greater than this. To avoid using such huge numbers, scientists have developed units of distance other than kilometers or meters.

One common unit used to measure large distances is the **light-year**. One light-year is the distance that light rays travel in one year. (Notice that a light-year is not a way to measure a time; it is a way to measure a distance.) Light travels very quickly in space, about 300,000 km/s. So in one year it can travel about 9,460,000,000,000 km. Thus, the distance to Alpha Centauri, given above, is equal to 4.3 light-years. Table 14.1 lists examples of distances measured in light-years. Notice that all the stars named are different distances from the Earth. This is true even for stars in the same constellation. For example, Betelgeuse and Rigel are both in the constellation Orion, but they are far apart from each other.

Star or object	Approximate distance (light-years)
Alpha Centauri	4.3
Sirius (brightest star in the sky)	8.8
Vega	26
Arcturus	36
Betelgeuse	700
Rigel	900
Deneb	1400
Most distant known object in the universe	16,000,000,000

TABLE 14.1 ***Some Distances of Objects from Earth***

▶ REVIEW 14.3

1. List the main steps you would follow to measure a horizontal distance indirectly using triangulation.
2. Why is the length of the baseline important when you use triangulation to measure distances?
3. Nori draws a scale diagram with the scale 1 cm = 20 m.
   (a) How long would a baseline on Nori's scale diagram be for a distance of 160 m?
   (b) An unknown distance measured in Nori's diagram is 12 cm long. What is the distance in real life?
4. A surveyor (Figure 14.18) measures off a baseline of 120 m along the shore of a river. He then measures the angle from each end of the baseline to a rock on the opposite shore. The two angles that she measures are 65° and 50°. Draw a scale diagram to determine the width of the river.
5. (a) On the Earth, what is the longest baseline possible?
   (b) How much time passes between the measurement of angles using the baseline you suggested in (a)?
6. What is the difference between a year and a light-year?
7. Explain why the distances to planets are usually given in kilometers, whereas the distances to stars are given in light-years.

FIGURE 14.18
A surveyor measures angles with an instrument called a transit. Why do surveyors need to measure exact angles and distances?

8. An astronomer uses the diameter of the Earth's orbit as a baseline to estimate the radius of Saturn's orbit. As shown in Figure 14.19, the angles to Saturn, taken six months apart, are both 84°. Use a scale diagram to find the distance from Saturn to the sun. (When astronomers use triangulation to measure such large distances, they take into consideration the movement of the distant object.)

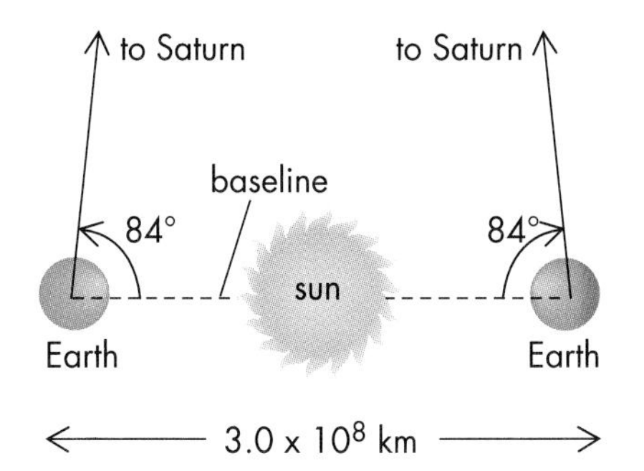

FIGURE 14.19
Diagram for Review Question 8

FIGURE 14.20
A spectroscope causes light energy to split into a spectrum of colors.

14.4 THE COLOR AND TEMPERATURE OF STARS

What color changes do you notice when you look inside a toaster as it toasts bread? When the toaster is off, the inside is usually dark in color. When the toaster is switched on, the inside becomes hotter, and its color changes to a bright red. This color change is just one example of how colors change with temperature.

Although your eye can detect a variety of colors, scientists use special devices to look closely at the light energy given off by the sun and other stars. One of the most useful instruments that astronomers use is the **spectroscope** (Figure 14.20), which is a device that splits light energy into a series of colors. The series of colors is called a **spectrum**. One example of a spectrum that you have seen many times is a rainbow. Light energy from the sun separates into the colors of the rainbow, namely red, orange, yellow, green, blue, indigo, and violet.

Scientists have found that when a chemical element is heated it gives off light energy that shows a unique spectrum when viewed through a spectroscope. Each element tested has its own spectrum, as shown in Figure 14.21.

Scientists have also used the spectroscope to look at stars. Much of what we know about stars today has resulted from using the spectroscope. The spectrum of a star can tell us some of the chemical elements that make up that star, how much of each element the star contains, and how fast the star is moving toward or away from the Earth.

FIGURE 14.21 ▶
Each element that is heated produces its own spectrum that you can see when you look through a spectroscope.
(a) Light from the sun
(b) Sodium light
(c) Hydrogen light

ACTIVITY 14F / COLORS OF LIGHT SOURCES

In this activity, you will observe the spectra of various light sources. (Spectra is the plural of spectrum.)

MATERIALS

frosted incandescent light bulb
discharge tubes containing various gases
power supply for gas tubes
spectroscope

■ **Do not touch the light bulb, discharge tubes, or electrical connections in this set-up.**

PROCEDURE

1. Your teacher will set up and identify several light sources, including an ordinary frosted light bulb and discharge tubes (connected to a power supply) containing various gases.
2. Look through the spectroscope at the first light source set up in the darkened room. In your notebook, draw and label a neat diagram of the spectrum that you observe.
3. Repeat Step 2 for all the other light sources.

DISCUSSION

1. Examine the diagrams you made after viewing the different light sources. Did the spectrum of each light source differ from all the others? Describe any differences or similarities among the different spectra.
2. How do you think astronomers might make use of the fact that each star produces its own spectrum? ❖

SPECTRAL TYPES OF STARS

The color of a hot object helps you compare its temperature with other hot objects. A dull red color means the temperature is fairly low. As the temperature increases, the color changes first to orange, then to yellow, and finally to bluish-white. Similarly, the spectrum of a hot star has more blue light than red light. The opposite is true of a cool star.

Astronomers have found that one of the best ways to classify stars is by their spectra. Using measurements from spectroscopes, scientists have grouped stars into seven color types called **spectral types**. In this classification system, each star can be ranked as an O, B, A, F, G, K, or M star (Table 14.2). Our sun is a yellow star (type G) with a surface temperature of about 5500°C.

■ Look through a spectroscope at a straight-filament light bulb that your teacher has connected to a variable power supply. Describe any changes in the light seen through the spectroscope as the voltage is increased.

■ Locate the stars listed in Table 14.2 that you can find on your Star Map 2. Label these stars and indicate their color. If possible, try to see them the next time you are sky-watching with a telescope or a pair of binoculars.

Type	Color	Temperature range (°C)	Example(s)
O	blue	25,000 – 50,000	Zeta Orionis
B	bluish-white	11,000 – 25,000	Rigel, Spica
A	white	7500 – 11,000	Vega, Sirius
F	yellowish-white	6000 – 7500	Polaris, Procyon
G	yellow	5000 – 6000	Sun, Alpha Centauri
K	orange	3500 – 5000	Arcturus, Aldebaran
M	red	2000 – 3500	Betelgeuse, Antares

TABLE 14.2 ◀
Spectral Types of Stars

REVIEW 14.4

1. Starting with red, list in order the colors of the visible spectrum that your eye is able to see.
2. How is the color of a star related to its temperature?
3. (a) Explain why a cooler star could actually appear brighter than a hotter star.
   (b) Give an example of the situation described in (a).
4. What color would you expect each of the stars described below to appear?
   (a) This star is a type A star.
   (b) This star has a surface temperature of 3,800°C.
   (c) This star has a surface temperature of 30,000°C.
5. (a) What instrument does an astronomer use to determine the spectral type of a star?
   (b) Why is the instrument you named in (a) better than a telescope to determine spectral types?
6. Make up a mnemonic sentence to help you remember the order of the spectral types.

14.5 THE BRIGHTNESS OF STARS

There are several ways of classifying stars. In Section 14.4, you learned that stars can be classified according to their color and temperature. Stars can also be classified by their size, age, distance from Earth, or brightness. In this section, you will learn about the brightness of stars. Astronomers call the brightness of a star its **magnitude**. Much of what we know about a star comes from observing its magnitude.

ACTIVITY 14G / THE MAGNITUDES OF STARS

In Part I of this activity, you will compare the brightness of stars in Orion and/or the Big Dipper. Although you can see Orion only during the winter months, you can see the Big Dipper on any clear night. In Part II, you will observe in class how the brightness of a light source depends on the distance you are from the source.

Do not go out alone to a dark area without permission from your parents. Make sure to dress warmly for night observations.

PART I Orion and the Big Dipper

PROCEDURE

1. Position yourself in an area well away from lights, with a clear view of the sky. After about 10 min, your eyes will be adapted to the dark. Make sure your flashlight, paper, and pencil are ready.
2. Locate the constellation Orion. (You may want to review Activity 14B, Part II.)
3. Sketch the stars that you can see in the constellation. Using Figure 14.22a as a reference, name as many of the stars as possible.
4. Rank the stars you see in Orion in order of their brightness to your eyes. Use 1 for the brightest, 2 for the second brightest, and so on.
5. Locate the Big Dipper and repeat Steps 3 and 4, using Figure 14.22b as a reference.

DISCUSSION

1. Compare your rankings of star brightness with those of your classmates.
2. Your teacher will provide you with a list of the rankings developed by astronomers for the stars of Orion and the Big Dipper. How do your rankings compare with these rankings on the list?

MATERIALS

Part I

flashlight covered with red cellophane
clipboard with paper, or notebook
pencil

Part II

galvanometer
solar cell
dolly or cart with wheels
ray box
graph paper

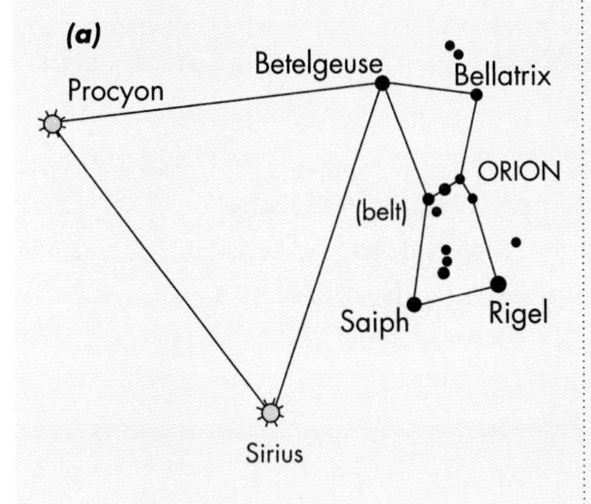

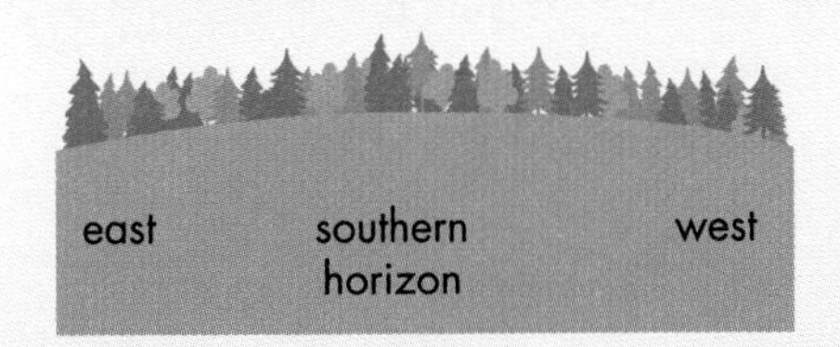

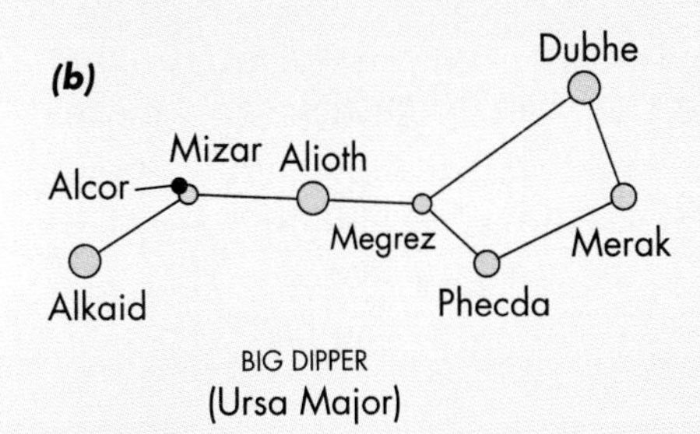

FIGURE 14.22
(a) Orion and several stars visible in the winter sky
(b) The Big Dipper

3. (a) What problems do you think there may be with your system of ranking star magnitude?
(b) Explain how you think your system could be improved.

PART II Brightness of a Light Source

PROCEDURE

1. Copy Table 14.3 into your notebook. Your teacher may suggest different distances.

2. Set up the equipment as shown in Figure 14.23. Place the ray box at one end of the room. Cover all the windows to prevent other light sources from affecting the results.

3. Place the galvanometer apparatus 1 m from the ray box bulb or at some other distance suggested by your teacher. Record the electric current, with the units indicated by the galvanometer needle.

4. Repeat Step 3 at distances of 2 m, 3 m, and 4 m from the ray box. (Your teacher may suggest different distances.)

5. On a graph that has the electric current on the vertical axis and the distance from the ray box on the horizontal axis, plot the data from your table of observations.

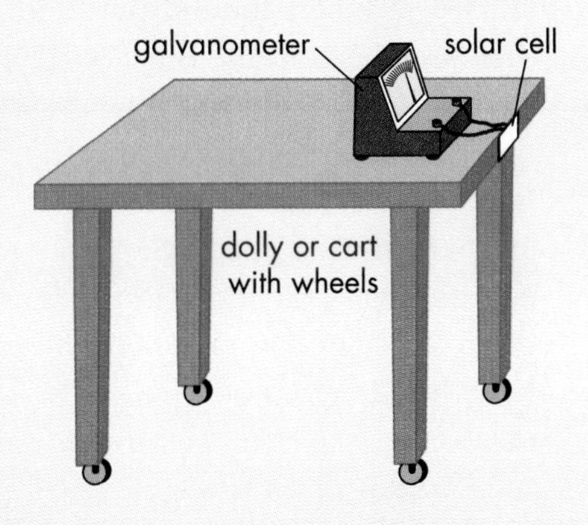

FIGURE 14.23
Apparatus for Activity 14G, Part II

Distance from bulb (m)	Reading on galvanometer
1	
2	
3	
4	

TABLE 14.3 ***Sample Data Table for Activity 14G (Part II)***

DISCUSSION

1. Examine your graph, and describe the relationship you see between the galvanometer reading and the distance from the light source.
2. Predict what the reading would be if the galvanometer were moved to 6 m from the ray box. Check your answer with your teacher.
3. Describe the relationship between the observed brightness of an object and the distance from the object.
4. Suppose that two stars, A and B, give off the same amount of light. From the Earth, star A appears to be 100 times brighter than star B. What would you conclude about the distances of these two stars from the Earth?
5. For any two stars that you observed in Part I, what possible explanations can you give for their different brightnesses as seen from the Earth? ❖

■ Astronomers are always searching for methods to help them estimate the large distances in the universe. One method for determining distances between galaxies uses the Cepheid Variables. The Cepheid Variables are stars whose brightness varies over certain periods of time. Find out where several Cepheid Variables are and how astronomers use them to calculate the distances between galaxies. Imagine that you are a scientist planning to send a space probe out to a distant galaxy. Write a short story about your plan. In your story, describe how you would use the Cepheid Variables to estimate the distance that your space probe would have to travel.

APPARENT AND ABSOLUTE MAGNITUDES

Almost 2200 years ago, the Greek astronomer Hipparchus developed the idea of classifying stars by their brightness. He decided to divide stars into six categories. The brightest stars were called first-magnitude stars, and the faintest stars were called sixth-magnitude stars.

Astronomers still use this classification system today. Since more advanced skywatching tools have improved our ability to see fainter stars, the magnitude scale set up by Hipparchus has been revised. Astronomers now use the word "magnitude" in two ways.

Apparent magnitude refers to the brightness of a star as it appears to us. This is the magnitude recorded by Hipparchus, or by you, if you are looking at the sky at night. In fact, two stars that have the same apparent magnitude can actually be giving off very different amounts of light energy. One star may simply be much closer to the Earth than the other star. The term **absolute magnitude** refers to the *actual* amount of light energy given off by a star at a standard distance. Astronomers calculate the absolute magnitude of stars by determining how bright the stars would appear if they were all the same distance from the Earth.

A simple example will show how astronomers compare magnitudes. Imagine that you are looking at two lights, one a flashlight located close to you, the other a bright floodlight that is far away (Figure 14.24a). Both lights appear to have the same brightness; this means that they have the same *apparent magnitude*. What can you do to compare their absolute magnitudes? Simply move both light sources so they are the same distance away from you (Figure 14.24b). Then you observe that the floodlight has a much brighter *absolute magnitude* than the flashlight.

Our sun has the brightest apparent magnitude of any star because it is the closest star to Earth. But our sun's absolute magnitude is only the absolute magnitude of an average star. Some stars, if they were as close to the Earth as our sun, would be nearly 1 million times brighter than the sun. Others would be only one-millionth as bright as the sun.

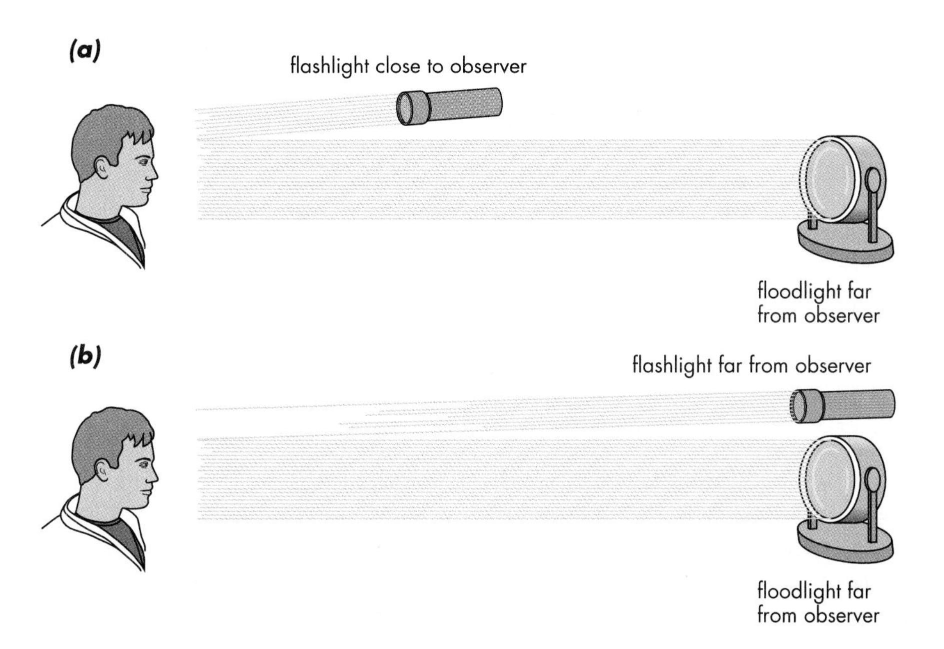

FIGURE 14.24
To the observer in (a), the flashlight and the bright floodlight have the same apparent magnitude. In (b), when the two light sources are the same distance from the observer, the floodlight is brighter. It has a much brighter absolute magnitude.

EXTENSION

■ Astronomers can determine a star's absolute magnitude from its color by using a special diagram called the Hertzsprung-Russell diagram. They can also use the diagram to estimate the distance of a star from the Earth. Find out who Hertzsprung and Russell were and how they developed this diagram. Write a script for an educational video in which Hertzsprung and Russell explain their diagram and how it is used. Include drawings of the diagram in your video. Show where our sun and at least three other stars that have been mentioned in this unit would appear on the diagram.

▶ REVIEW 14.5

1. You are in a car at night and you see a motorcycle approaching you from a distance. As the motorcycle gets closer, what do you think happens to
   (a) the absolute magnitude of the motorcycle's headlight?
   (b) the apparent magnitude of the motorcycle's headlight?
2. One night, Anna observes two stars that have the same apparent magnitude. Could these two stars be giving off different amounts of light energy? Explain your answer.
3. Describe the difference between apparent and absolute magnitudes of stars.
4. What two facts about a star must astronomers know in order to calculate its absolute magnitude?
5. What effect do you think pollution in the atmosphere would have on
   (a) a star's apparent magnitude?
   (b) a star's absolute magnitude?

CHAPTER REVIEW

KEY IDEAS

- For the Earth, the most important star is the sun. It is a star of average size and temperature.
- Like all stars, the sun gives off a huge amount of energy. If we did not receive as much of this energy as we do, life on the Earth would be impossible.
- Stars that you see in the night sky can be grouped into patterns called constellations.
- In the Northern Hemisphere, constellations appear to travel in circles around Polaris (the North Star). This occurs because of the Earth's rotation on its axis.
- Only the constellations closest to Polaris can be seen all year long in North America. The other constellations are visible only in certain seasons.
- Because we cannot travel to the distant stars in space, we must use indirect methods to measure the distances to stars. One way to measure the distance to a star is to use the indirect method called triangulation.
- The large distances to stars are often stated in light-years. One light-year is the distance light travels in one year: 9,460,000,000,000 km.
- A star can be classified by the colors of light energy it gives off as seen through a spectroscope.
- Stars can also be classified according to their brightness (both apparent magnitude and absolute magnitude), as well as other properties.

VOCABULARY

corona
chromosphere
solar flare
solar prominence
photosphere
sunspot
constellation
zodiac constellation
astrology
triangulation
light-year
spectroscope
spectrum
spectral type (of star)
magnitude
apparent magnitude
absolute magnitude

V1. Use the words from the vocabulary list and any other words you need to create a mind map about our sun and other stars.

V2. Write a clue for each word in the vocabulary list. Give your list of clues to a classmate to solve.

CONNECTIONS

C1. Why do you think the sun is considered to be an average star?

C2. (a) What is the process that produces energy inside the sun?
(b) How fast does that energy travel through space?

C3. (a) For each of the four seasons, state one constellation that you can see best during that season.
(b) Why can you see some constellations only during certain seasons?

C4. Do you think people in Australia can see Polaris? Explain your answer.

C5. If you wanted to see the planets, where would you look in the night sky? Explain your answer.

C6. Darren and Anish use an indirect method to determine the approximate distance from their school to the smokestack of a factory. The baseline they choose is one side of the football field (100 m long). Darren measures an angle of 72° from one end of the field to the smokestack, and Anish measures an angle of 82° from the other end. Use a scale diagram to find the distance from the football field to the smokestack.

C7. Using examples, explain why humans could not travel to the stars at the speeds reached by today's spacecraft.

C8. If you could travel at the speed of light, how long would it take you to travel to the following stars named in Table 14.1?
(a) Sirius
(b) Arcturus
(c) Betelgeuse
(d) Deneb

C9. (a) What information about stars can astronomers obtain using a spectroscope?
(b) Use examples to explain how the spectral type of a star gives information about a star.

C10. Explain the difference between the absolute and apparent magnitudes of stars, using our sun as an example.

EXPLORATIONS

E1. Describe how you would use "pointer stars" to locate each of the following stars or constellations in the night sky:
(a) Polaris
(b) Cassiopeia
(c) Sirius
(d) Arcturus
(e) Regulus

(Hint: Some pointer stars were described in Activity 14B.)

E2. To know their exact location at sea, sailors often used two instruments: the sextant and the chronometer. Find out when these instruments were invented, how they were used, and how effective they are in determining location.

E3. Look back at Table 14.2.
(a) What is the spectral type of the sun?
(b) Describe how you think conditions on the Earth would differ from conditions today if the sun were
(i) a type B star,
(ii) a type M star.

REFLECTIONS

R1. Look back at the table you made for Activity 14A, and read your answer to each question. If your views have changed, or if you can offer a more complete answer, write your new answers in the final column.

R2. Do you think that studying about the sun and other stars is important to a technological society like ours? Why or why not?

CHAPTER 15

EXPLORING THE UNIVERSE

When you were a young child, the only things you knew about were the things close to you. As you became older, you learned more and more about things far beyond your surroundings. In a similar way, the study of the universe has gone through stages. Early study of the universe included only those things that people could see with the unaided eye. No instruments existed to help people see distant objects better. Then, when the telescope was invented, people could see farther, and the universe suddenly appeared to be much larger. Since then, with better telescopes and many other instruments, the number and types of objects that we can detect in the universe continue to increase.

The photograph above is a view of the universe that was not possible long ago. This group of about 100 billion stars is called the Andromeda Galaxy. It is approximately the same size and shape as the Milky Way Galaxy, which is the name of the group of stars that includes our sun.

Modern telescopes and other devices have shown us that there are many galaxies of different kinds. Astronomers have also learned that there are other interesting objects in space.

This chapter is about the universe and those objects.

ACTIVITY 15A / IDEAS ABOUT THE UNIVERSE

Divide a page in your learning journal into three columns with the following titles: "Object in the Universe," "My First Ideas," and "My Final Ideas." In the first column, write the words or phrases that are listed here. In the second column, write your ideas about the object. Include drawings of these objects. These ideas can be descriptions or your own definitions. You will fill in the third column after studying this chapter.

1. galaxy
2. Milky Way Galaxy
3. open star cluster
4. globular star cluster
5. dark nebula
6. bright nebula
7. intergalactic distance
8. black hole
9. quasar
10. pulsar ❖

15.1 CHANGING IDEAS ABOUT THE UNIVERSE

Most humans are fascinated with questions about the universe. What is the universe made of? How big is the universe? What did people in ancient times think about the universe?

The **universe** consists of everything that exists, all the matter and all the energy, as well as all the space between the matter. In this section, you will learn how and why ideas about the universe have changed. And you will get an idea about how big the universe is.

ANCIENT IDEAS

Imagine that you are on a merry-go-round that is spinning clockwise. Your friends standing on the ground look as if they are moving in the opposite direction. Now imagine slowing the rate of spinning so that one rotation takes 24 hours. At that speed, it would be difficult for you to judge whether the ride were moving or the ground were moving. In ancient times, people had difficulty judging whether the Earth was moving or the objects in the sky were moving. Ancient astronomers thought that the Earth was standing still and everything else moved around it.

Everything in the sky appeared to be in motion. The sun, moon, stars, and wandering stars (planets) all seemed to rise in the east and set in the west. One early idea about the stars was that they were attached to a large glass ball that revolved around the Earth once every day. People thought that the stars were only slightly farther away than the sun, the moon, and the five planets that they could see with the human eye. In Figure 15.1, you can see one possible arrangement of the universe. This idea is called the **Earth-centered universe**.

Several ancient peoples had some excellent astronomers. For example, more than 3000 years ago the Chinese developed a calendar of

■ People all over the world have used myths (special stories) to explain events and objects in nature. Native peoples have many myths that describe the stars, the sun, the moon, and the planets. But not all native groups use the same myths. For example, the Haida call the Pleiades star cluster "a canoe bailer." Other West Coast native peoples call the Pleiades "the seven sisters." Use your library resource center to find other native myths about the universe and the objects in it. Try to find myths from different native groups about the same subject (for example, why the sun sets). How are these myths similar? How are they different? Why do you think all myths are not the same? Why are there no native myths about galaxies, quasars, or black holes? Choose one or two myths to read out loud to your class. Explain what you think the myths mean.

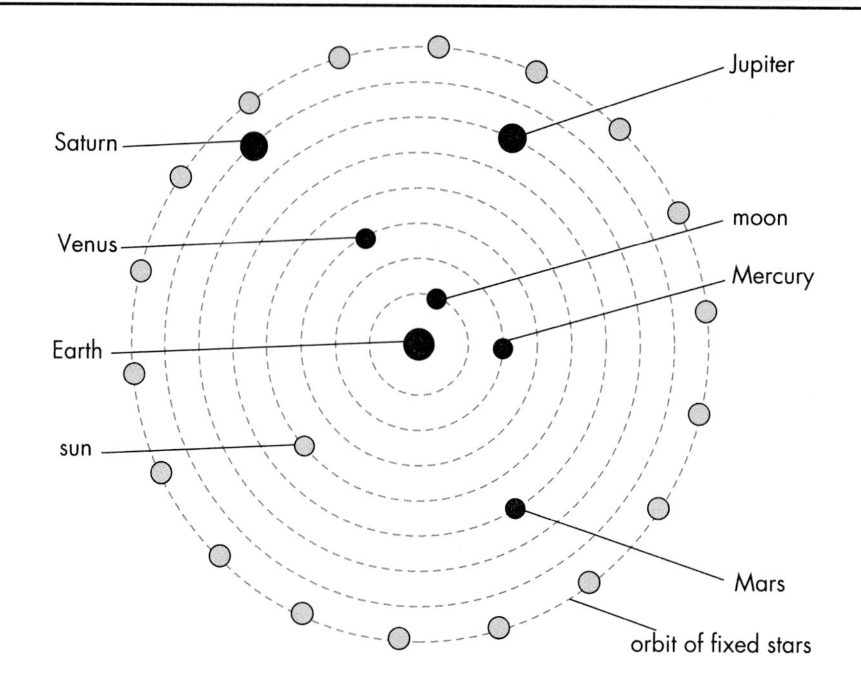

FIGURE 15.1 ▶
Many people in ancient times thought that the Earth was the center of the universe and that all the other objects revolved around it. Why did early astronomers place the sun where it is in this diagram?

FIGURE 15.2
This student is learning how to use a simple instrument to measure angles to objects in the sky. Can you figure out how to use such a device? How many stars do you think you could plot on a star map using only your eyes and this device? Ancient people plotted more than 800 stars.

365¼ days by using their observations of the sun and stars. They recorded comets, meteors, and what they called "guest stars." (We now know that these are exploding stars, which are visible for a few weeks or months.) Both the Chinese and the Greeks created star maps more than 2000 years ago showing the position and approximate brightness of more than 800 stars. They did this using only their eyes and some instruments to measure angles (Figure 15.2).

IMPROVEMENTS IN MAKING OBSERVATIONS

The Earth-centered idea of the universe was popular for thousands of years. Then around 500 years ago, some scientists began to question this idea. Scientific ideas were starting to change for two major reasons. One reason was that scientists began using experiments to learn more about nature, the Earth, and the universe. The other reason was the invention of the telescope.

The telescope was invented in Europe in the early 1600s. The great Italian scientist Galileo Galilei improved upon the invention, and he was the first person to use it to view the night sky (Figure 15.3). His first telescope made distant objects look about seven times larger than normal, which is about the same as today's inexpensive binoculars. Even with this small magnification, Galileo discovered many things never before known. For example, he saw that the planets were not merely points of light; they appeared to be circular, like the moon. He was able to see the stars, however, only as points of light. From these observations, Galileo inferred that the planets must be much closer to the Earth than the stars are.

FIGURE 15.3 ◀
In this painting, Galileo is showing his telescope to the leader of Venice.

By the time Galileo had discovered how different the sky looked through a telescope, other scientists had begun to question the Earth-centered view of the universe. Galileo's discoveries convinced him that the Earth and other planets travelled around the sun. Persuading other people that the Earth was not the center of the universe was slow and sometimes difficult. (Galileo himself was forced by the authorities to deny that the Earth travelled around the sun!) Eventually, however, the idea of the **sun-centered solar system** replaced the Earth-centered view.

TODAY'S IDEAS

Nearly 400 years have passed since the invention of the telescope. In that time, many other inventions and discoveries about the universe have been made. These have led to new and different ideas.

We know that the planets travel around the sun, and that the sun is just one of countless stars. Astronomers have observed that the sun and other stars are also moving. Stars are gathered in large groups, surrounded by gas and dust. The group of stars that our sun belongs to is called the Milky Way Galaxy. A **galaxy** is a huge collection of gas, dust, and hundreds of millions of stars. (You will study galaxies in Section 15.3.)

Beyond the Milky Way Galaxy is a vast region that appears to be almost empty. There are no stars or galaxies, except off in the distance. Far away, there are more groups of stars and other material. A simplified view of the structure of the universe is shown in Figure 15.4.

In the universe, there are countless galaxies made up of countless stars and other objects. Between the galaxies are vast spaces where few particles of matter are scattered. We do not know exactly how big the

■ With another student, research Galileo's life and his contributions to science. One student should look at Galileo's work and discoveries with telescopes. The other student should look at the difficulties Galileo had with the authorities of his time. Combine the information into a single report.

universe is or how many stars and galaxies it contains. But we do know that our knowledge of the universe is always improving. For this reason, the estimated size of the universe is getting larger.

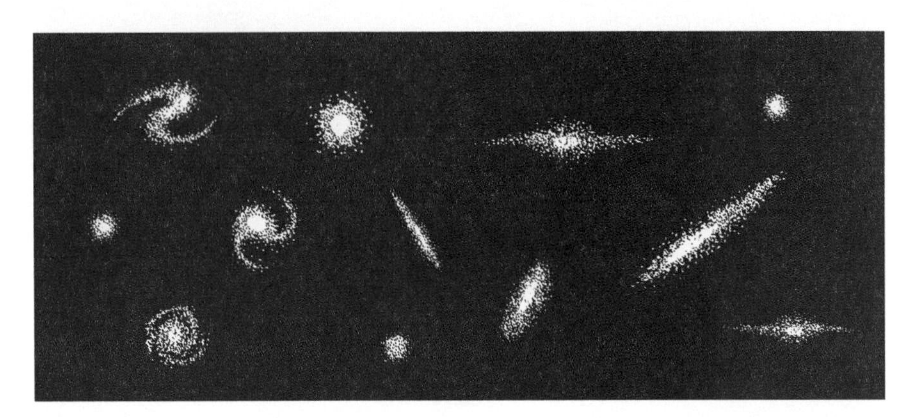

FIGURE 15.4 ▶
The universe contains huge groups of stars, called galaxies, separated by vast distances. The galaxies and everything in them are constantly in motion. Compare this concept of the universe to the concept illustrated in Figure 15.1.

ACTIVITY 15B / SCALING THE UNIVERSE

As you learned in Chapter 14, large distances in space are often measured in light-years. A light-year is the distance light travels in one year, at a speed of 300,000 km/s. Thus, a light-year is a distance of about 9,460,000,000,000 km. In this activity, the light-year is used to help you understand the size of the universe. You will use the thickness of a piece of paper to represent one light-year. To make your calculations easier, always round off your answers to easy numbers such as 10, 100, 1000, and so on.

MATERIALS

this textbook
ruler or meter stick with mm and cm divisions

PROCEDURE

1. Review the metric prefixes by copying the following questions into your notebook and replacing the question marks with your answers.

(a) 1 km = ? m
(b) 1 m = ? cm
(c) 1 cm = ? mm
(d) 1 m = ? mm
(e) 1 km = ? mm

2. Count out 50 sheets of paper in your notebook or textbook. Measure the thickness of all those 50 sheets together in millimeters (Figure 15.5). Use your measurement to calculate the approximate number of sheets of paper per millimeter (sheets/mm). Remember to round off your answer. You will need this number in the next two steps, so before going on, check with your teacher to be sure that your calculation is reasonable.

FIGURE 15.5 ***How thick are 50 sheets of paper?***

Model distance	Number of sheets thick (or number of light-years)	Example of distance in the universe
0.1 mm	1	Maximum distance of comets from the sun
0.4 mm		Approximate distance from the sun to the nearest star
Thickness of two pennies (almost 3 mm)		Approximate distance to the star Vega
Approximate length of an adult's thumb (7 cm)		Distance to the star Betelgeuse
Height of a wall in a home (2.5 m)		Distance from the Earth to the center of the Milky Way Galaxy
Length of a science classroom (10 m)		Diameter of the Milky Way Galaxy
Length of two football fields (200 m)		Distance to the Andromeda Galaxy
20 km		Distance to the Centaurus Galaxy
San Francisco to Denver (about 2000 km)		Farthest distance that people have seen in the universe
San Francisco to Chicago (about 4000 km)		Current estimated size of of the entire universe

TABLE 15.1 ***Sample Data Table for Activity 15B***

3. Assuming that the thickness of one sheet of paper represents a distance of one light-year, determine the number of light-years in:
   (a) 1 mm
   (b) 1 cm
   (c) 1 m
   (d) 1 km
   (e) 1000 km
   (f) 4000 km (the approximate distance between San Francisco and Chicago)
4. Copy the first two columns of Table 15.1 into your notebook and complete it. The first answer has been filled in using the scale from step 3 above: one sheet's thickness represents one light-year.

DISCUSSION

1. Think about the size of the Earth's orbit around the sun. Do you think it could be shown in a model of the universe using the scale in this activity? Explain your answer.
2. Imagine that you are trying to help a grade 6 student understand how big the universe is. Write a description of the explanation you would use. ❖

A MODEL OF THE UNIVERSE

If you were to draw a diagram of the universe on a piece of paper in your notebook, the Milky Way Galaxy would be so tiny that you would need a microscope to see it. You would not be able see the solar system. You can understand why you need a much larger model than a piece of paper to describe the size of the universe.

Now imagine most of the U.S. as a model of the universe (Figure 15.6). If you were to stack sheets of paper edgeways for 4000 km, from San Francisco to Chicago, you would need about 40 billion (40,000,000,000) sheets. Each thickness of paper would represent one light-year. In this model, the distance across the solar system would be less than the thickness of only one sheet. The Earth's orbit around the sun would be so tiny that it would be invisible.

The distances between different objects in the universe are given different names. Distances between planets in the solar system are called **interplanetary distances** (*inter* is the Latin word meaning "between"). Distances between the stars are called **interstellar distances** (stellar refers to stars). And distances that separate the galaxies are **intergalactic distances** (galactic refers to galaxies).

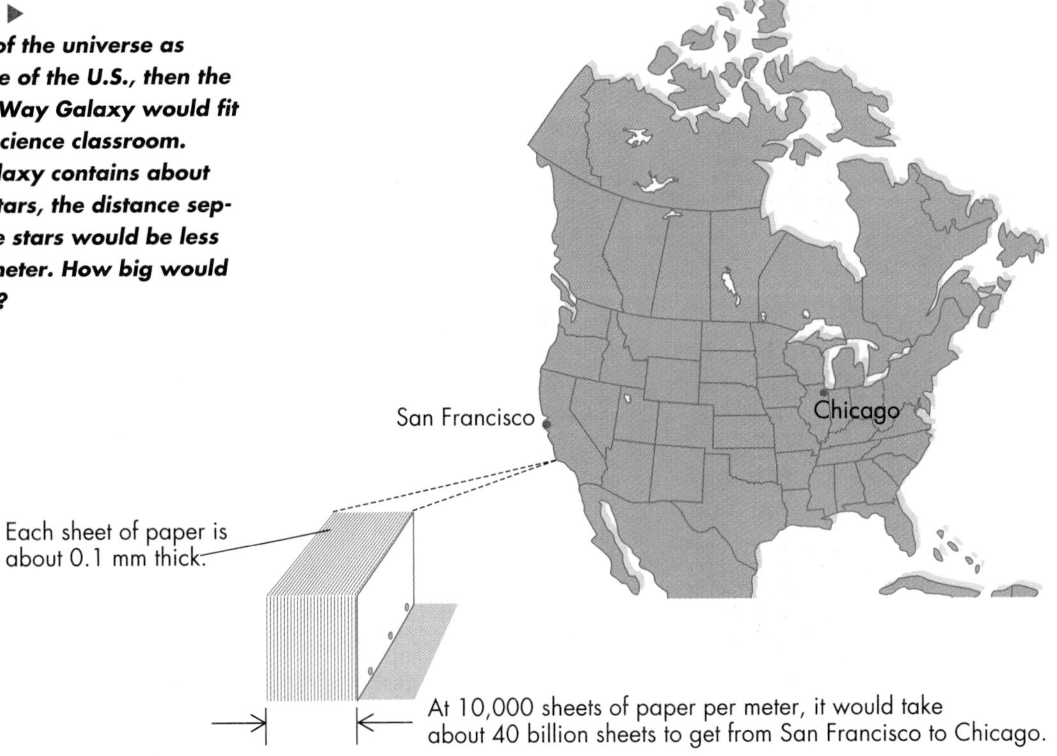
San Francisco
Chicago
Each sheet of paper is about 0.1 mm thick.
At 10,000 sheets of paper per meter, it would take about 40 billion sheets to get from San Francisco to Chicago.

FIGURE 15.6 ▶ ***If you think of the universe as being the size of the U.S., then the entire Milky Way Galaxy would fit inside your science classroom. Since our galaxy contains about 100 billion stars, the distance separating these stars would be less than a millimeter. How big would each star be?***

▶ REVIEW 15.1

1. (a) What does the expression "Earth-centered universe" mean?
   (b) What were the major reasons that caused scientists to change their ideas about an Earth-centered universe?
2. Some Chinese, Greeks, and other groups of people in ancient times were excellent astronomers, yet their star maps contained only about 800 stars. What do you think prevented these ancient astronomers from becoming more successful?
3. What evidence did Galileo use to infer that stars were farther from the Earth than planets are?
4. What is the meaning of the term "universe"?
5. (a) Has our estimate of the size of the universe increased or decreased in the past several hundred years?
   (b) Why does our estimate of the size of the universe change?
6. (a) What is a galaxy?
   (b) What is intergalactic distance?
7. Give an example of
   (a) an interplanetary distance,
   (b) an interstellar distance.

■ 15.2 TOOLS OF THE MODERN ASTRONOMER

As you have noticed in studying this unit, we can learn about space beyond the Earth by looking at photographs of planets, stars, and other objects in the universe. If you were to compare the photographs in this textbook with the photographs in astronomy books published even a few years ago, you would discover that we can now see objects that it was not possible to see then. In this section, you will learn about some of the instruments used by both amateur and professional astronomers to obtain images of objects beyond the Earth.

ACTIVITY 15C / VIEWING THE UNIVERSE

What do you already know about the instruments or methods that astronomers use to study distant objects in the universe? You will be able to answer this question by doing this activity.

MATERIALS

set of blank cards

PROCEDURE

1. By yourself or in a group, obtain a set of blank cards. On each card, name one instrument or method used to study distant objects in space, such as planets, stars, galaxies, and our sun. Try to include instruments and methods that were used long ago, several that are modern, and some that you imagine might be used in the future.
2. On each card, add some information about each instrument or method.
3. Arrange the cards in the order in which you think the instruments or methods were first used. (Start with the oldest on top.)
4. In your notebook, record the information from your cards in the order you have in Step 3 above.
5. As you read this section, change or add to the information on the cards. You may want to rearrange the order of the cards. Make changes in your notebook as well. ❖

TELESCOPES

Since the first telescope in Galileo's time, people have greatly increased the size of telescopes and improved their ability to magnify. But these telescopes still work in much the same way as Galileo's did. Galileo's telescope worked because light rays are refracted (bent) as they pass through a light-gathering glass lens, called an objective lens. This type of telescope is called a **refracting telescope** (Figure 15.7).

A larger refracting telescope provides a clearer image and allows more light to be gathered and focused. However, we cannot make telescopes with lenses larger than about 1.2 m in diameter. The glass becomes too heavy, sags under its own weight, and ruins the image. For this reason, a different telescope design is used for larger telescopes.

■ The thin atmosphere at the tops of mountains may be a great advantage for telescopes, but it is a great disadvantage for people who work there. People often become light-headed or even ill when they first arrive at a high elevation. This "altitude sickness" is caused when the brain is not receiving enough oxygen.

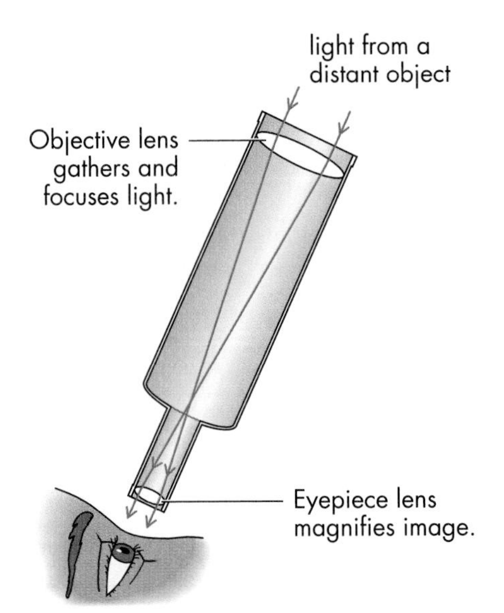
light from a
distant object
Objective lens
gathers and
focuses light.
Eyepiece lens
magnifies image.

FIGURE 15.7
How a refracting telescope produces a magnified image of a distant object

A **reflecting telescope** uses a curved mirror instead of a lens to gather light. The English scientist Sir Isaac Newton was the first person to build a working reflecting telescope (Figure 15.8). Both refracting and reflecting telescopes are called optical telescopes because they are made to help us see. The word "optical" refers to our sense of sight and is from the Greek *optos* meaning "see."

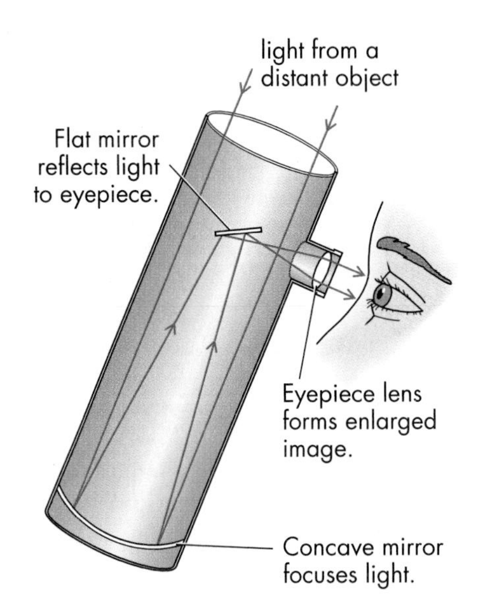
light from a
distant object
Flat mirror
reflects light
to eyepiece.
Eyepiece lens
forms enlarged
image.
Concave mirror
focuses light.

FIGURE 15.8 ◀
How a reflecting telescope works. Which kind of optical telescope do we usually find in observatories?

FIGURE 15.9
The Mauna Kea Observatory in Hawaii

Telescopes can be either portable (easily moved) or permanently located. Portable telescopes are the smaller ones. They must be placed on a special stand to help keep them steady. Large telescopes are placed permanently in buildings called **observatories**.

The Earth's atmosphere interferes with the views produced by both refracting and reflecting telescopes. To reduce this problem, most observatories today are built on top of mountains, where the air is thinner. The thin air found at the tops of very high mountains is also helpful because it absorbs and scatters less light than the denser air at lower altitudes. The Canada-France-Hawaii telescope sits at the top of Mauna Kea in Hawaii (Figure 15.9).

The most recent way that has been found to overcome the problem of the Earth's atmosphere is to place the telescope into space orbit around the Earth. An example of such a device is the Hubble Space Telescope, which was put into orbit around the Earth in 1990. Soon after its launch, scientists discovered that the telescope's main mirror was flawed. This problem has been overcome with a corrective device that was installed by space shuttle astronauts. Figure 15.10 shows that the Hubble Telescope can obtain a much more detailed view of distant objects than telescopes on Earth.

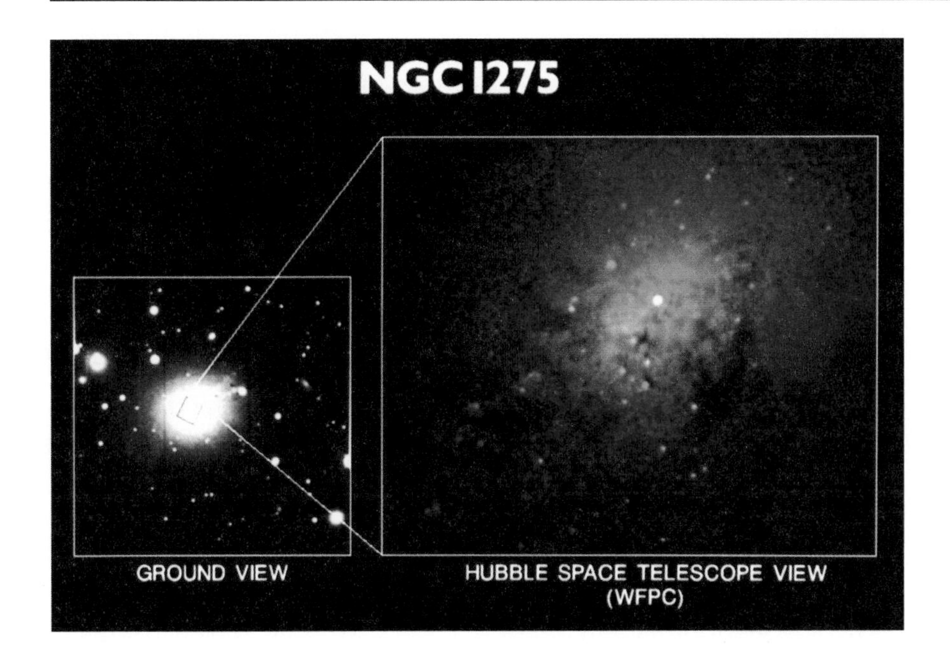

FIGURE 15.10 ◄
The image of the distant galaxy on the left was obtained by a large telescope on a mountaintop. The detailed image of the center of the same galaxy, shown on the right, was obtained by the Hubble Space Telescope, which is high above the Earth's atmosphere.

CAMERAS

About 270 years after the invention of the telescope, photography was first used for astronomy. A photograph produces a permanent image that allows astronomers to measure and analyze stars and other objects. In addition, photographic film can gather light over a period of many hours, unlike the eye, which sees an object only for a moment. As a result, we can use photographs to help us see very faint stars that our eyes cannot see (Figure 15.11).

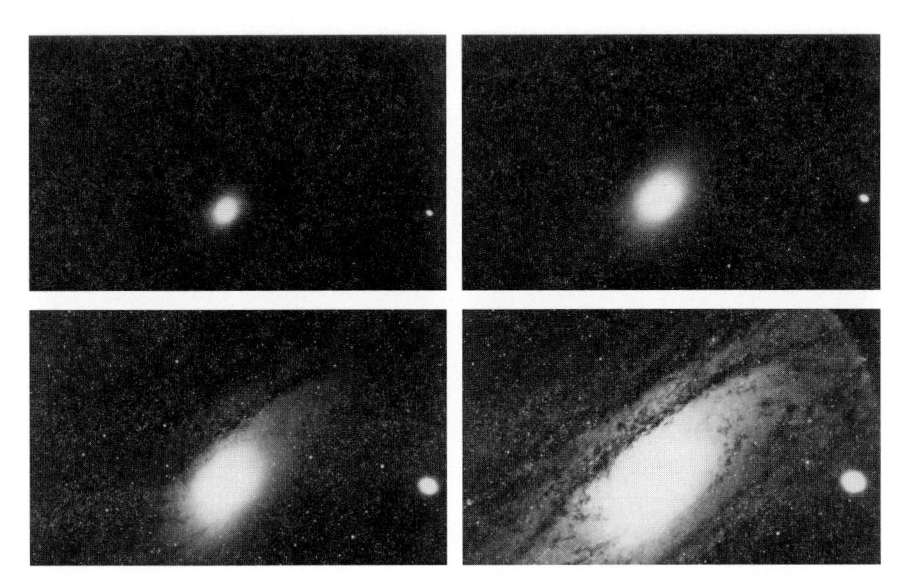

FIGURE 15.11
Four views of the same object, showing how the amount of detail increases as the time that the film is exposed to light increases. In order from top left to bottom right, these photographs show exposure times of 1 min, 5 min, 30 min, and 45 min. The film is exposed to light by leaving the camera's shutter open.

EXTENSION

■ If you have a camera with a shutter that can be held open for a long time, you can take pictures of the movement of stars. Your teacher can give you more information on how to do this.

Astronomers often combine cameras with telescopes. In fact, large astronomical telescopes are seldom used for looking directly at objects in space. Instead, they are set up to take pictures or fitted with instruments that record information about stars.

SPECTROSCOPES AND THE ELECTROMAGNETIC SPECTRUM

An important device that has greatly changed our knowledge of the universe is the spectroscope. A spectroscope is a device that separates light into a spectrum of colors. You learned in Section 14.4 that different stars have different colors. Although it is difficult for you to see slight color differences with your eyes, such differences are easy to see with a spectroscope.

The spectrum of colors that the human eye can see is called the visible spectrum. It is only a very small part of a broad band of energy called the **electromagnetic spectrum**. This spectrum consists of radio waves, microwaves, infrared rays (heat), visible light, ultraviolet rays, X rays, and gamma rays. These types of energy are emitted (given off) by stars, galaxies, and other objects in the universe. Studying these types of energy helps astronomers understand more about the universe.

RADIO TELESCOPES

FIGURE 15.12
The Arecibo radio telescope is built into the mountains of Puerto Rico. How does a radio telescope work?

What do you do when you want to receive radio signals sent out by a radio station? You simply tune your radio to receive them. What do astronomers do when they want to receive the radio waves sent out by some star or other object in the sky? They aim a radio receiver toward the object and try tuning the receiver until it receives waves from space.

A device that receives radio waves from space is called a **radio telescope**. It is able to detect radio waves that are emitted by stars and galaxies. Radio waves are easy to detect. They can pass through the Earth's atmosphere, including clouds, very easily. (You know that radio waves from radio stations can pass through the walls of a building very easily, allowing radios to work inside homes and other buildings.)

Radio telescopes often look like giant satellite dishes. They may be made of wire mesh, with large air spaces. The radio telescope shown in Figure 15.12 is the largest single radio telescope in the world, measuring more than 300 m across. It was made by placing a wire mesh in a valley in the mountains. This radio telescope receives radio signals from many different parts of the universe during both day and night, as the Earth rotates.

Radio telescopes can also be made to work together in sets called arrays. Figure 15.13 shows an array of radio telescopes in New Mexico. The radio telescope dishes work together to produce the same results as

FIGURE 15.13 ◀
An array of radio telescopes produces the same results as a much larger single radio telescope.

FIGURE 15.14
An artist's view of the IRAS in orbit around the Earth

a much larger radio telescope. Radio signals are collected from space over a period of a month and are combined, using a computer, to produce a map of objects in the sky.

SATELLITES, SPACE PROBES, AND COMPUTERS

Some parts of the electromagnetic spectrum get trapped by the Earth's atmosphere, so they cannot be detected from the surface of the Earth. To overcome this problem, scientists use satellites that are put into orbit above the atmosphere. One example of such a satellite is the Infrared Astronomical Satellite (IRAS), which can detect objects in space that emit very tiny amounts of heat (Figure 15.14). Launched in 1983, the IRAS made some of the most exciting discoveries in many years of space exploration.

Many satellite observatories and space probes have been launched. In all cases, the information sent back to the Earth is enhanced by computer. The computer is an important tool of any modern astronomer.

The remainder of this chapter looks at how all these tools of astronomy provide answers to several questions, but create many more questions to be studied.

DID YOU KNOW

■ The IRAS is sensitive enough to detect infrared (heat) radiation from a source on Pluto, over 400 million kilometers away, which gives off as little heat as a 20 W light bulb!

► REVIEW 15.2

1. State two ways in which a telescope helps astronomers to improve their visual observations of space.
2. (a) Describe the two main types of optical telescopes.
   (b) Explain how these telescopes differ from each other.
3. (a) Why is the Earth's atmosphere a problem for astronomers?
   (b) Where would you build an observatory to overcome this problem?
4. Name two ways in which photography of the night sky helps an astronomer.
5. Large astronomical telescopes are usually used together with cameras, and astronomers seldom look through such telescopes directly. Suggest reasons for this.
6. What advantages do radio telescopes have over optical telescopes?
7. Why do astronomers send telescopes and other tools of astronomy into space?
8. Look back at the four photographs in Figure 15.11. If you were to look at this part of the sky with only your eyes, what do you think you would observe? Why?

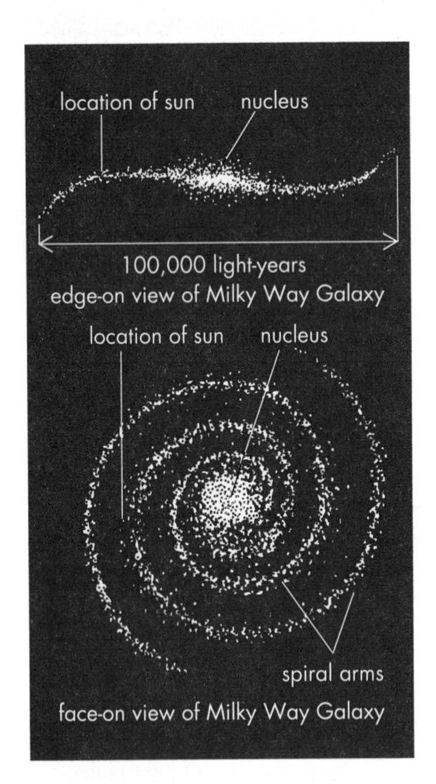

FIGURE 15.15
The spiral arms of our galaxy contain great concentrations of stars. Our sun is one of the many stars in the less concentrated regions between the spiral arms. If our sun were in the center of the galaxy, what do you think our night sky would look like?

■ 15.3 GALAXIES AND STAR CLUSTERS

If you look at a state map, you see cities, towns, and villages. These are places where people are grouped together. Between these places are large rural (country) regions. Similarly, if you look at a model of the universe, you see different-sized groups of stars. Galaxies might be compared to cities, and smaller groups of stars might be compared to villages or small towns. In this section, you will learn more about different groups of stars.

GALAXIES

In Section 15.1, you learned that a galaxy is a huge collection of gas, dust, and hundreds of millions of stars. These stars are attracted to each other by the force of gravity, and they are constantly in motion. Astronomers can see galaxies as far away as the power of their telescopes will permit.

Among the many galaxies in the universe lies the one to which our solar system belongs, the Milky Way Galaxy. The Milky Way is disk-shaped (like a compact disc), with our sun located near the outer part of the disk (Figure 15.15). The inner region of the disk is called the nucleus of the galaxy. If you look toward the nucleus of the Milky Way Galaxy, the stars are so numerous that they appear very close to one another. Yet these stars are separated by large interstellar distances. Astronomers estimate that there are about 100 billion stars in the Milky Way Galaxy.

The stars outside the nucleus of the Milky Way Galaxy travel around the nucleus. By looking at the shape of the spiral arms of the galaxy, you should be able to tell the direction of motion. Can you see in Figure

15.15 that the stars outside the nucleus are moving clockwise around the nucleus? You might observe the same type of motion if you sprinkled sawdust on the surface of water in a tub and swirled the water around in a clockwise direction.

To an observer on the Earth, the Milky Way Galaxy looks just like its name suggests: like a trail of milk spilled across the sky (Figure 15.16). The best seasons to view the Milky Way Galaxy are summer and winter, when it appears high in the sky. Through binoculars or a telescope, you can get a magnificent view, because you can see many more stars than you can with the unaided eye.

DID YOU KNOW

■ Our sun is travelling at a speed of about 70,000 km/h around the nucleus of our galaxy. Scientists estimate that to complete one trip around the nucleus takes about 225 million years!

FIGURE 15.16
This photograph, taken with a special camera, includes the whole sky. The bright band of stars across the center is the Milky Way. What part of our galaxy do we see from Earth?

The Milky Way Galaxy is called a **spiral galaxy** because of its spiral, disk shape. Figure 15.17 shows photographs of four other spiral galaxies. Some of these have shapes that resemble the Milky Way Galaxy. Notice that 15.17d shows a special type of spiral galaxy, called a **barred spiral galaxy**.

Other than our own Milky Way Galaxy, the only galaxy that you can see from the U.S. with the unaided eye is the Andromeda Galaxy, which is shown in the photograph at the beginning of this chapter.

There are other shapes of galaxies besides the spiral shape and barred spiral shape. Galaxies that are oval (egg-shaped) are called **elliptical galaxies**, and others that have no distinct shape are called **irregular galaxies** (Figure 15.18).

(a)

(b)

(c)

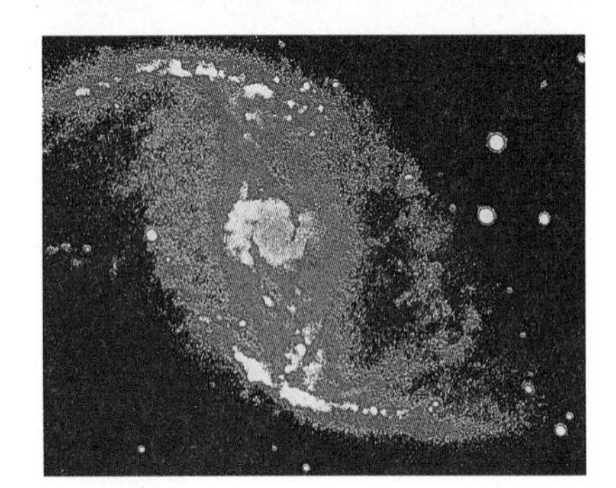

(d)

FIGURE 15.17 ◀
(a) This galaxy resembles the Milky Way Galaxy.
(b) This spiral galaxy is colored to show young giant stars (blue), which are hotter than older stars.
(c) The Whirlpool Galaxy is found in the constellation Canes Venatici. The smaller concentration of stars is another galaxy.
(d) An example of a barred spiral galaxy, colored to show the central bar. The spiral arms come out from the ends of the bar-shaped area that contains the galaxy's nucleus.

FIGURE 15.18
(a) The Large Magellanic Cloud is an irregular galaxy. It is about 160,000 light-years away.

(b) This photograph of a giant elliptical galaxy was taken by the Hubble telescope. The galaxy is emitting a jet that is 4000 light-years long.

(c) This elliptical galaxy is about 13 million light-years away. Dust between the galaxy and Earth blocks out some of the light from the galaxy.

STAR CLUSTERS

Groups of stars that are fairly close and travel together are called **star clusters**. These clusters may have as few as 10 stars or as many as a million. They have too few stars to be called a galaxy. (Recall that a galaxy has hundreds of millions of stars.) There are two types of star clusters, quite different from one another.

An open star cluster is a group of about 10 to 10,000 stars found in the main parts of the Milky Way Galaxy. These stars are bright, and are fairly close together in space.

One of the more interesting sights to observe in the sky is the open star cluster called the Pleiades, in the constellation Taurus (Figure 15.19). With the unaided eye, you can see up to six or seven stars. With binoculars, you can see many more stars, and with a telescope you could see so many stars that you would be unable to count them.

■ How good is your eyesight? An excellent test of your ability to view distant objects clearly is to see how many stars you can count in the Pleiades. Under good viewing conditions, most starwatchers can see six or seven stars with the unaided eye. You can see the Pleiades best during the winter. (Another name for the Pleiades is "The Seven Sisters.") The record number of stars seen is a total of 18!

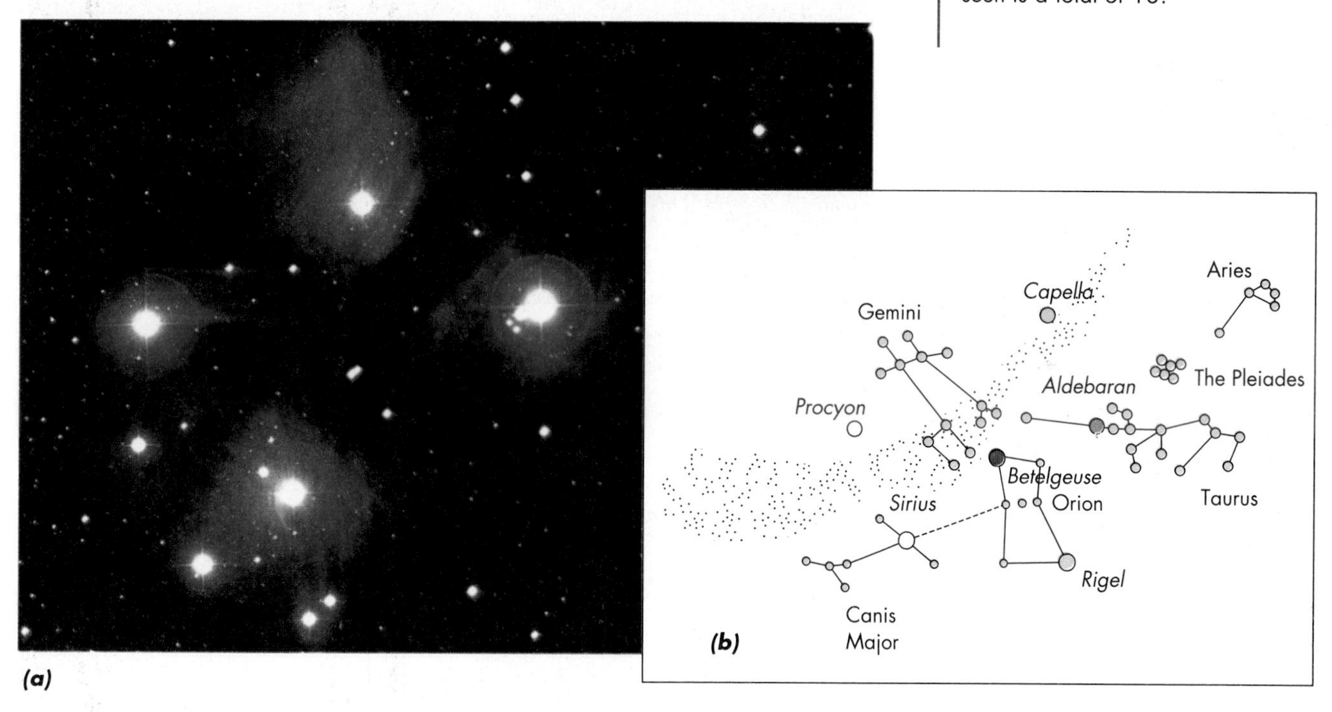

FIGURE 15.19
(a) The Pleiades is an open star cluster. (b) To locate the Pleiades, begin by finding the winter constellation Taurus. It is near the constellation Orion.

The second type of star cluster is the globular star cluster, which consists of approximately a million stars found outside the main part of a galaxy. To the unaided eye, a globular star cluster near the Milky Way Galaxy looks like a faint star, if it can be seen at all.

Through binoculars, globular star clusters are more visible, and with a telescope they appear as a fuzzy patch. Figure 15.20 shows a typical globular star cluster, and Figure 15.21 shows the location of many globular star clusters that are outside the main part of our galaxy. Some globular star clusters are also visible outside the main part of the Andromeda Galaxy, shown in the photograph at the beginning of this chapter.

Astronomers have discovered about 125 globular star clusters in or near the Milky Way Galaxy. Much of the pioneering work in investigating these clusters was done by a Canadian astronomer, Helen Sawyer Hogg. She carried out research in the Dominion Astrophysical Observatory in Victoria, as well as in other observatories. Her interesting life story is featured in the profile in this chapter.

FIGURE 15.20 ▶
This globular cluster, called Omega Centauri, is located about 17,000 light-years from the Earth.

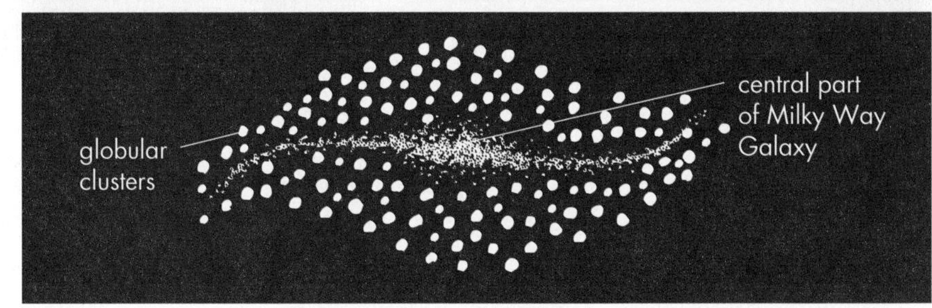

FIGURE 15.21 ▶
More than 125 globular star clusters are located outside the main part of the Milky Way Galaxy.

▶ REVIEW 15.3

1. (a) What are the names of four different shapes of galaxies?
   (b) Which one(s) do you think show the motion of stars best? Explain your answer.
2. Arrange the following in order of size, starting with the largest: globular star cluster; galaxy; universe; star; planet; open star cluster.
3. Draw a diagram showing what an observer from some distant galaxy would see when looking at our disk-shaped galaxy
   (a) from above the nucleus of our galaxy,
   (b) from the side of the disk.
4. Describe the difference between an open star cluster and a globular star cluster.
5. A certain star is located 60 light-years away from us. Which galaxy do you think this star is in?

HELEN SAWYER HOGG

Helen Sawyer Hogg remembered the day in January 1925 when she saw her first total eclipse of the sun. She was 20 years old. "We stood practically knee-deep in the snow. The temperature was well below zero. I'll never forget how cold I was, but the beauty of that eclipse was certainly one of the things that turned me on to astronomy for life."

After watching the eclipse, Hogg changed her studies from chemistry to astronomy. She then began a remarkable 60-year career. She received the Companion of the Order of Canada (Canada's top honor) for her contributions to astronomy. The University of Toronto named a telescope after her. There is a Helen Hogg Observatory in Ottawa, and an asteroid named in her honor (Asteroid 2917 Sawyer Hogg).

From an early age, Hogg was encouraged to be curious about science. Her parents allowed her to stay up and watch the stars. "I'm one of the few people who have seen Halley's comet twice in a lifetime," said Hogg. The first time was in 1910 when she was five years old. Seventy-six years later, she saw it again.

Hogg started working in the 1930s, long before the first satellite was launched or the first astronauts went to the moon. It was not an easy time for a woman to enter science and succeed at it. But astronomy became her life. Her husband was an astronomer, and together they spent long hours at the observatory. She raised three children while continuing her research (one of her sons later became an astronomer).

After her first husband died, when she was 45, Hogg took over the weekly newspaper column on astronomy that he wrote. She shared her love of astronomy though her teaching at the University of Toronto, a television series, and a book called *The Stars Belong to Everyone*. She continued working into her eighties.

Fellow astronomers remember Hogg's dedication, determination, and enthusiasm for astronomy. Her research focused on variable stars (stars that change brightness), which are found in globular star clusters. Hogg used telescopes in British Colombia, Tucson, Arizona, and Ontario.

Hogg is also remembered for her warmth and caring. One example was her habit of knitting little booties as gifts to coworkers who had babies. These booties were designed so that babies could not kick them off.

Hogg died in 1993, at age 87. Three days before her death, she was at the observatory making a video to encourage girls to study science. Even after her death, Hogg is a role model for women and for all would-be astronomers.

Can you remember an event that sparked your interest in an area of science?

15.4 NEBULAS

Some of the most beautiful sights in the sky beyond our solar system are nebulas. A **nebula** is a spread-out cloud of interstellar dust or gas. The word "nebula" comes from the Latin word for mist. (The plural of nebula is nebulas.)

Some nebulas are bright, while others are dark. In this section, you will read about both bright and dark nebulas. You can also read more about bright nebulas in Section 16.1.

Every nebula looks different in shape and color from all other nebulas. With good viewing conditions, you can see some nebulas with the unaided eye. Many are spectacular when you view them with binoculars or a telescope.

BRIGHT NEBULAS

We see an object either because it gives off its own light energy (such as a light bulb that is turned on) or because it reflects light (such as this textbook or any other object you can see). Similarly, you can see a bright nebula because it either gives off light energy or reflects it. Bright nebulas are located near stars.

An example of both types of bright nebulas is the Trifid Nebula, found in the summer constellation Sagittarius (Figure 15.22). The bright region on the right in the photograph has gases that are giving off light energy (glowing). Such nebulas often glow with a pink color

DID YOU KNOW

■ The constellation Sagittarius has so many interesting deep-sky objects that an entire book could be written about them. On a clear summer night, you can see Sagittarius near the southern horizon. It contains many nebulas, different types of stars, some star clusters, and some galaxies.

FIGURE 15.22 ▶
The Trifid Nebula has two separate regions, one with pink colors, and the other with bluish-violet colors. What do you think the dark regions in the photograph are caused by?

when they contain a concentration of hydrogen, or with a blue-green color when they contain oxygen. The region on the left of the photograph has gases that reflect the light from nearby stars. Such nebulas are often dark blue or violet in color.

Notice the dark patches blocking part of the light energy coming from the Trifid Nebula. These dark areas are caused by dust clouds located between the nebula and the Earth, somewhere in the Milky Way Galaxy.

Figure 15.23 shows examples of bright nebulas, and you can see another in the next chapter in Figure 16.3.

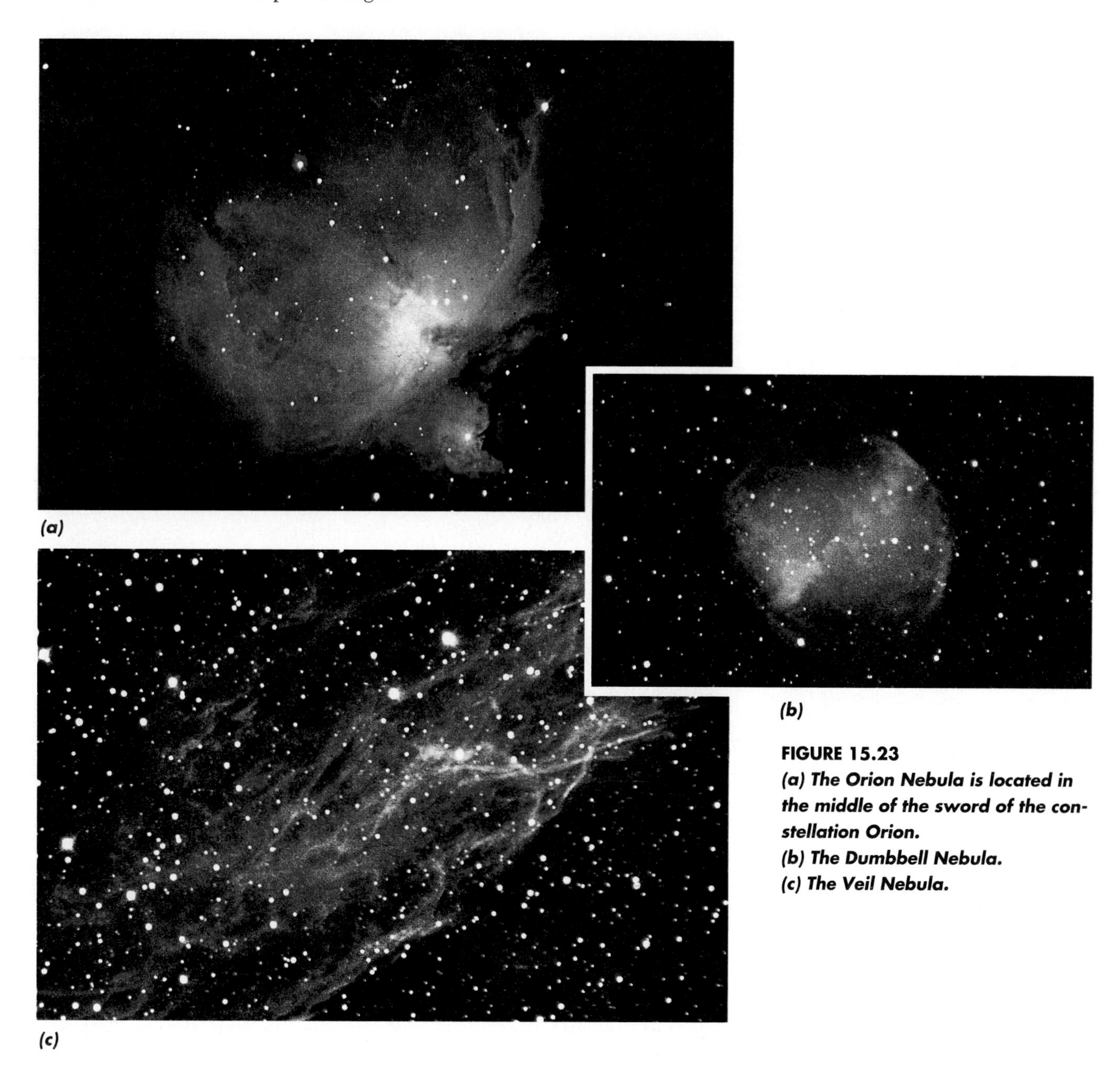

(a) **(b)** **(c)**

FIGURE 15.23
(a) The Orion Nebula is located in the middle of the sword of the constellation Orion.
(b) The Dumbbell Nebula.
(c) The Veil Nebula.

DARK NEBULAS

Dark nebulas are composed mostly of dust, so they block out the light energy from stars and bright nebulas behind them. To us on the Earth, they appear as dark patches in the sky. One dark nebula that has an interesting shape is the Horsehead Nebula, in the constellation Orion. It is shown in the photograph at the beginning of this unit.

ACTIVITY 15D / OBSERVING OBJECTS OTHER THAN STARS

You have discovered that there are many objects to view in the night sky besides stars, planets, and moons. To observe these objects firsthand, choose a clear night, preferably during the winter months, and perform this activity.

MATERIALS

Star Map 3
flashlight with red cellophane cover
telescope or good pair of binoculars
tripod
notebook and pen

■ **Do not go out alone to a dark area without permission from your parents. Make sure to dress warmly for night observations.**

PROCEDURE

1. On Star Map 3, locate and label as many constellations as possible that contain a nebula. Also label any other interesting "objects" that you have learned about but are not shown on the map.
2. On a clear night, prepare yourself for skywatching away from bright lights. Cover your flashlight with red cellophane. Allow your eyes to become accustomed to the dark. If you are going to use a telescope or a pair of binoculars, attach the device to a tripod to provide a steady support. (For specific suggestions, refer to Activity 14D and/or a reference book on observing the night sky.)
3. Locate, view, and describe as many interesting objects in the sky as you can. Draw a diagram of each object and its surroundings in your notebook. Indicate details such as date, time of your observations, size, color, location of the objects you observed, and angles.

DISCUSSION

1. Make a list of all the different types of objects in the sky that you have been able to see.
2. Of the objects named in Discussion question 1, which would you like to know more about? Why? ❖

► REVIEW 15.4

1. (a) What are nebulas made of? (b) Where are they located?
2. We can see bright nebulas for two possible reasons. What are these reasons?
3. Give one possible reason why different bright nebulas have different colours.
4. Which type of nebula, bright or dark, do you think has a higher average temperature? Why?
5. Astronomers hypothesize that interstellar space is not a vacuum. Rather, they suggest that interstellar space must contain particles of matter. How do you think nebulas provide evidence to support this hypothesis?

15.5 QUASARS, PULSARS, AND BLACK HOLES

Using modern instruments, such as radio telescopes, astronomers have discovered some unusual objects in the universe. In this section, you will learn about three of these faraway objects, called quasars, pulsars, and black holes. (You can read about how pulsars and black holes are formed in Chapter 16.)

QUASARS: NEITHER GALAXIES NOR STARS

Our galaxy has a diameter of about 100,000 light-years and contains about 100 billion stars. But there are other objects in the universe that emit up to 100 times as much energy as our galaxy does, yet they can be as small as our solar system. These massive, high-energy objects are called **quasars**. (The word "quasar" was taken from the expression "quasi-stellar radio source," which means a star-like object that emits radio waves.)

Quasars are neither galaxies nor stars, although they have characteristics of both. In a photograph or through an optical telescope, a quasar looks like an ordinary, faint star (Figure 15.24). However, radio telescopes reveal quasars to be strong emitters of radio waves. Because quasars produce such huge amounts of energy, yet appear only as faint points of light, scientists consider quasars to be the oldest, most distant, and most powerful sources of energy in the universe.

FIGURE 15.24
This image of a quasar (the red spot at the top of the picture) was created by computer using signals collected by a radio telescope. The long stream of material with the blob at the end is being fired out into space by the quasar. Quasars are all billions of light-years from the Earth.

PULSARS

In 1967, an astronomer named Jocelyn Bell was making observations of stars with a large radio telescope. During one session, she detected short, rapid pulses of radio waves. Each pulse lasted about 1/100 of a second, and the time between pulses was about one second. No star or galaxy had ever been observed that gave off signals like this. Bell and her co-workers thought they must be picking up signals from a recently launched satellite.

After two days, it was clear that the satellite explanation was wrong. The pulsing waves were coming from a source that did not move in the sky. Also, whatever the source was, it was located far from our solar system.

By the end of that year, Bell had found three more of these sources of pulsing radio waves. They were called **pulsars** because of the way they produced pulses of radio waves. Today, more than 400 pulsars have been identified. (Pulsars are also called neutron stars.)

What is a pulsar made of, and why does it send out pulses of energy? Astronomers have calculated that pulsars are very small, only about 20 km in diameter. The mass of a pulsar, however, is about the same as the mass of a star. Astronomers think that pulsars must be rotating while giving off energy. Each time a pulsar faces toward us, we detect the energy as a pulse. This is like the light from a rotating lighthouse beam. It would appear as pulses to someone far away. Thus, a pulsar is a dense, small, rotating object that emits pulsing radio waves. You would not be able to see a pulsar, even with a good telescope.

BLACK HOLES

One of the most unusual objects in the universe is called a **black hole**. A black hole is an extremely small, dense core of a star. A black hole has such a strong force of gravity that it pulls in everything near it. Its pull is so strong that even light cannot escape from it, so it cannot be seen. This explains why we call it black.

If we cannot see black holes, why do astronomers think they exist? The answer is that astronomers have discovered indirect evidence of black holes. For example, material from another star may be pulled into the black hole. Just before it falls in, this material gives off high-energy X rays that can be detected on the Earth. Astronomers have discovered some X-ray sources that they think are caused by black holes (Figure 15.25).

FIGURE 15.25 ▶
This is an artist's drawing of what astronomers think is a black hole in the constellation Cygnus. Matter drawn in from a nearby star (lower left) spirals in toward the black hole. The matter travels very quickly and gives off energy in the form of X rays. The black hole in Cygnus was discovered in 1971.

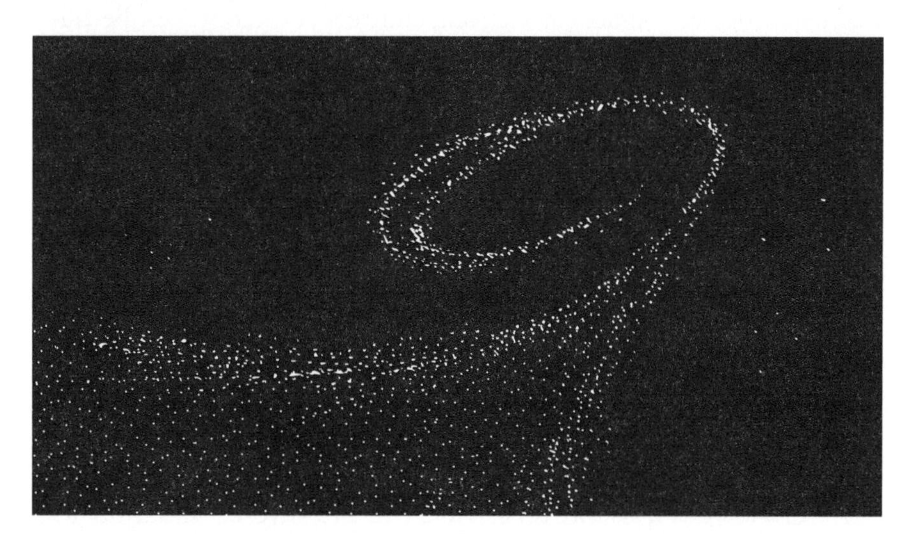

▶ REVIEW 15.5

1. What do astronomers use to detect quasars and pulsars?
2. Compare a quasar with a star.
3. In what way(s) is a quasar similar to a galaxy?
4. (a) How are quasars and pulsars similar?
   (b) How do they differ?
5. (a) Explain why astronomers cannot see a black hole.
   (b) If astronomers cannot see black holes, why do they think that black holes exist?

CHAPTER REVIEW

KEY IDEAS

- The universe consists of countless galaxies. Between the galaxies and stars are tiny particles of gas and dust.
- A galaxy is a huge collection of dust, gas, and hundreds of millions of stars.
- Our view of the universe is continually changing because the instruments that astronomers use to look into space are continually being improved.
- Galaxies can be classified according to their shape: spiral, barred spiral, elliptical, and irregular.
- Star clusters are smaller than galaxies.
- Nebulas are clouds of interstellar gases and dust. Some nebulas absorb light energy and appear dark to us. The nebulas that reflect or emit light energy are among the most beautiful objects to see in the sky.
- Quasars are massive, distant, high-energy objects that are different from both stars and galaxies.
- A pulsar is a small, dense, rotating object that emits pulsing radio waves.
- A black hole is a small, extremely dense core of a star with very high gravity that pulls in everything near it, including light energy.

VOCABULARY

universe
Earth-centered universe
sun-centered solar system
galaxy
interplanetary distance
interstellar distance
intergalactic distance
refracting telescope
reflecting telescope
observatory
electromagnetic spectrum
radio telescope
spiral galaxy
barred spiral galaxy
elliptical galaxy
irregular galaxy
star cluster
nebula
quasar
pulsar
black hole

V1. Create a mind map of the universe. Many of the words in your map should come from this chapter, but you may also use words from other chapters in this unit.

V2. Make a list of 20 words from this chapter, and check the spelling of each word. Write a description or clue for each word. Design a crossword puzzle using the words from this list. Have other students solve your crossword puzzle, then check their answers.

CONNECTIONS

C1. Write the following numbers in your notebook, then circle those that equal 1 billion. Beside your circled answers, write "= 1 billion."

(a) 1000 million
(b) 1,000,000,000
(c) 1×10^9
(d) 10^9

C2. Copy the distances listed below in your notebook. Beside each number, indicate whether the distance would be described as intergalactic, interstellar, interplanetary, or interstudent.

(a) 10 billion billion kilometers
(b) 0.001 kilometers
(c) 100 million kilometers
(d) 100 million million kilometers

C3. (a) Which theory is the one we accept today: the sun-centered theory of the solar system or the Earth-centered theory of the universe?
(b) How do these two theories differ?

C4. What are some differences between a refracting telescope and a reflecting telescope?

C5. Look at Figure 15.26, which shows one possible shape of a galaxy.

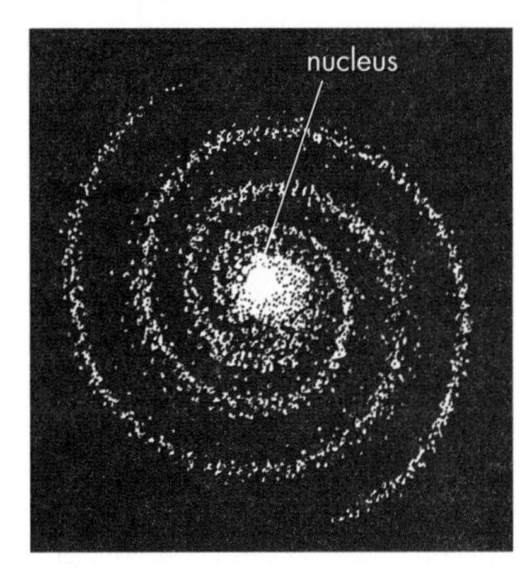

FIGURE 15.26
Is this galaxy rotating clockwise or counterclockwise around its nucleus?

(a) What type of galaxy is this?
(b) What is the galaxy composed of?
(c) What type of force keeps the parts of the galaxy together?
(d) In what direction do you think the galaxy in Figure 15.26 is rotating around its nucleus? Explain your answer.

C6. Compare a galaxy and a quasar using these titles: size; energy; distance from us; age.

C7. Explain the difference between each of the following pairs of objects:
(a) a star and a quasar
(b) an open star cluster and a globular star cluster

C8. Compare bright and dark nebulas by answering these questions.
(a) How can you identify each type of nebula through an optical telescope?
(b) Which type of nebula absorbs light energy?
(c) Suggest a reason why the shapes of bright nebulas are usually more regular than the shapes of dark nebulas.

C9. (a) How are a pulsar and a black hole similar?
(b) How do they differ?

EXPLORATIONS

E1. Obtain a copy of the supplement to the June 1983 issue of *National Geographic*, titled "Journey into the Universe through Time and Space." Write a report on the immense size of the universe, using the magazine as a guide.

E2. Interstellar and intergalactic distances may be given in light-years, but another common unit of these long distances is the parsec. Look up the meaning of parsec in a reference book or an encyclopedia. Is it longer or shorter than a light-year? Define parsec, and give some examples of distances measured in this unit.

E3. Some elliptical galaxies look like circles. Find out what an ellipse is from an encyclopedia or a mathematics reference book. Compare a circle and an ellipse. Do you think the Earth's orbit around the sun should be called a circle or an ellipse? Explain.

E4. In Star Maps 2 and 3, the Milky Way Galaxy is a strip of stars that we can see most easily during winter and summer. Describe how our view of the Milky Way Galaxy would change if our solar system were located in the nucleus of the galaxy.

E5. Suppose you owned a restaurant that had the theme of space. Write a menu for your restaurant. Will you serve "galactic burgers"? What will they look like? What other items will you serve?

E6. Write your own poem that compares the distances of objects in the universe. You may use the vocabulary list from this chapter, and you may include vocabulary from the other chapters in this unit.

REFLECTIONS

R1. Look back at Activity 15A in your learning journal. In the third column, define or describe in detail each of the 10 items. Were your first ideas about the objects close to your final ideas? Have any of your ideas changed after studying this chapter? If so, why did they change?

R2. Write at least five questions about the universe that you would like more complete answers to.

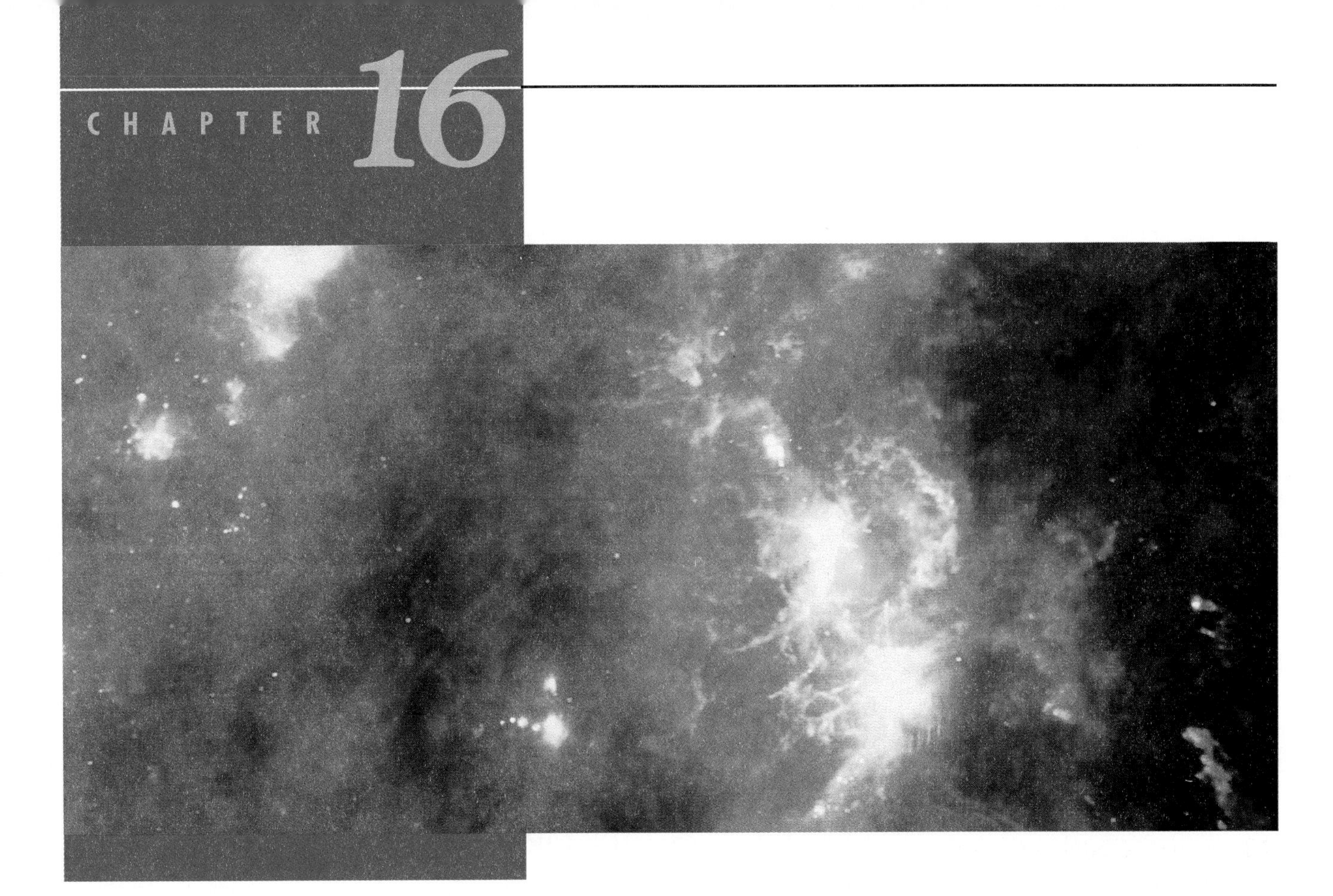

CHAPTER 16

HISTORY OF THE UNIVERSE

Are stars born? Do they grow old? During our lifetimes, most of the stars in the sky will not change. Looking at the night sky will not tell us about the lives of stars, because we do not live long enough to see them aging. But there are some clues. The photograph above shows what astronomers believe is a nursery for stars — an area where new stars are forming.

Knowing about stars and the universe will help us understand more about ourselves. Humans have always wondered who we are, where we came from, and what will eventually happen to us and our Earth. We can come closer to answering these questions by studying the many changes that are occurring in the solar system and in the universe beyond the solar system.

In the first three chapters of this unit, you learned about the planets, the sun, stars, galaxies, and other objects in the universe. In this chapter, you will look at how these objects may have been formed, and what might happen to them in the future. After completing this chapter, you should be able to answer the following questions: Do stars go through life cycles? How do scientists think the planets of the solar system formed? Did the universe have a beginning? What are the chances that life exists elsewhere in the universe?

ACTIVITY 16A / VIEWING THE UNIVERSE

Working with other students in a group, divide a poster into four equal rectangles (Figure 16.1). At the top of each rectangle, print a title to go along with the questions listed here. Then place your answers in the matching rectangle. Be creative, and include drawings that help to explain your answer.

FIGURE 16.1
Making a poster for Activity 16A

1. In rectangle A, describe at least one way that your group thinks the solar system may have formed.
2. In rectangle B, write a poem that explains the answer you wrote in rectangle A.
3. In rectangle C, suggest at least one way that humans might be able to travel to other parts of the universe.
4. Do you think there is life elsewhere in the universe? Why or why not? Write your answer in rectangle D. ❖

■ 16.1 THE LIFE OF A STAR

You can observe many cycles in nature. A **cycle** is a series of actions repeated in the same order every time. The water cycle on the Earth is one cycle you may have studied. In the water cycle, water evaporates from lakes and oceans, forming water vapor. The vapor condenses, forming clouds. Rain or snow falls from the clouds. Then the cycle begins again. Do you know of any other cycles?

Stars, too, go through cycles called life cycles. We say that stars have a "life" because they begin (are "born"), develop, and end ("die"). One major difference between the life cycle of a star and cycles we observe on the Earth is time. The cycles that we observe on the Earth may take days, months, or years. The life cycle of a star, however, takes billions of years, a length of time that is difficult for humans to imagine.

Using modern instruments, astronomers have been able to look at millions of stars at different stages in their life cycles. Scientists have put together many pieces of information, like the pieces of a puzzle, to obtain the following theory of the life cycle of a star.

All stars begin their lives in nebulas, huge clouds of dust and gases. If you have not already done so, you can learn about nebulas in Section 15.4. The gases of these nebulas are made up mainly of hydrogen and helium. The dust and gases swirl around, forming clumps. The particles of these clumps attract each other because of gravity, and the clumps become more tightly packed. Eventually, the clumps are big enough to begin producing large amounts of light and other forms of energy. They have become new stars.

New stars shine very brightly at first, and are usually blue or white in color. (You can see examples of new stars in the Pleiades star cluster, shown in Figure 15.19.) The life cycle of a star depends greatly on the star's mass. Low-mass stars, smaller than our sun, may live for 100 billion years. Medium-mass stars, like our sun, live for perhaps 10 billion years. High-mass stars have much shorter lives of only a few million years.

Eventually, a star's source of energy runs out. Low-mass and medium-mass stars then cool down and swell up into **red giants**. Their outer layers of gas drift away, and they shrink into **white dwarf stars** that are very dense. Later they cool down and fade (Figure 16.2).

Stars that begin as high-mass stars end their cycles in a very different way. After their source of energy is used up, they swell into **red supergiants**. Then they blow up with a huge explosion called a **supernova**. A supernova leaves behind a rapidly expanding nebula of dust and gases. At the center of this nebula is a small spinning star, called a pulsar, or neutron star. Pulsars are very dense, and they send out radio waves as they spin. (Pulsars were described in Section 15.5.)

Supernovas are special events. Only a small number have been recorded in history. A famous one occurred in the year 1054, and was recorded in China and India. It was so bright that people could see it with the unaided eye, even in daylight. It was visible for about 21 months. This supernova is called the Crab Nebula, and it can be seen in the constellation Taurus. The gases spreading out from the explosion form a beautiful nebula (Figure 16.3).

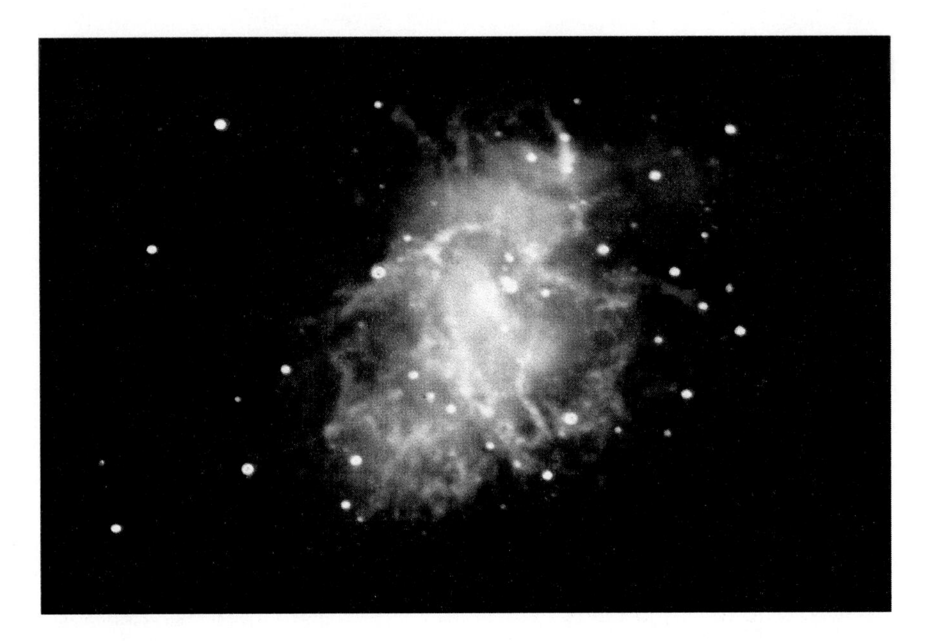

FIGURE 16.3
The supernova that created the Crab Nebula was observed in 1054. Why do you think this nebula was given this name?

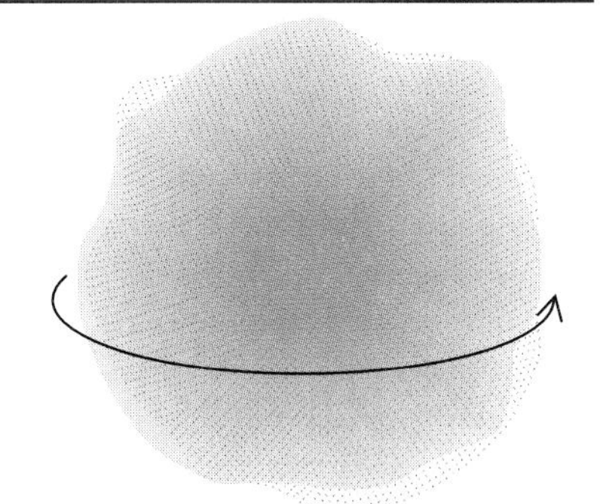

1. A rotating nebula begins contracting.

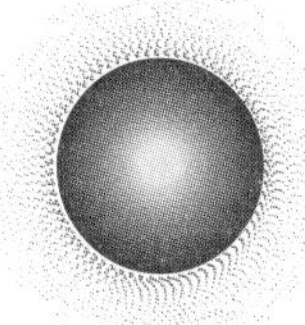

2. The nebula gradually turns into a hot, dense clump that begins producing light energy and heat.

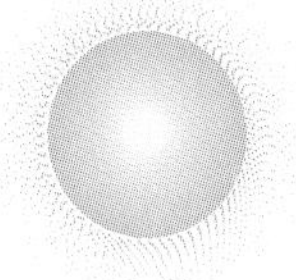

3. For a very long time, the star leads a normal life, producing energy by the process of nuclear fusion.

4. When the star can no longer produce enough energy, it swells up and its outer layers drift out into space.

5. The star ends its life cycle as a white dwarf star. Later, this white dwarf will cool down and fade away.

FIGURE 16.2
Scientists' theory of the life cycle of a low-mass or medium-mass star. At which stage is our sun?

FIGURE 16.4
The life cycles of high-mass stars: Stars with high mass form from huge nebulas. The main part of their life cycle is much shorter than that of low- and medium-mass stars. When their source of energy runs out, they explode.
(a) A high-mass star undergoes a huge explosion called a supernova, sending out a nebula and leaving behind a small, dense, spinning star called a pulsar, or neutron star.
(b) Larger high-mass stars undergo even larger explosions. They collapse past the pulsar stage, and become black holes, which are small but extremely dense.

More recently, in 1987, a supernova was discovered by a Canadian astronomer, Ian Shelton, when he was working at an observatory in Chile. His discovery is called supernova Shelton 1987A. Astronomers are observing this supernova with great care. What they learn from their observations may cause them to change their theories about the life cycle of stars.

Some high-mass stars become pulsars. At the end of their life cycles, they explode and then collapse to become pulsars. After the largest stars explode, they collapse past the pulsar stage and become black holes. (You read about black holes in Section 15.5.) The life cycles of high-mass stars are illustrated in Figure 16.4.

Just as no two clouds on the Earth are identical, no two stars are identical. However, scientists understand the life cycles of both clouds and stars fairly well. In Figure 16.4, you can see a summary of the life cycle of high-mass stars. But our knowledge of the details of how stars develop is far from complete. That is part of the excitement of studying astronomy.

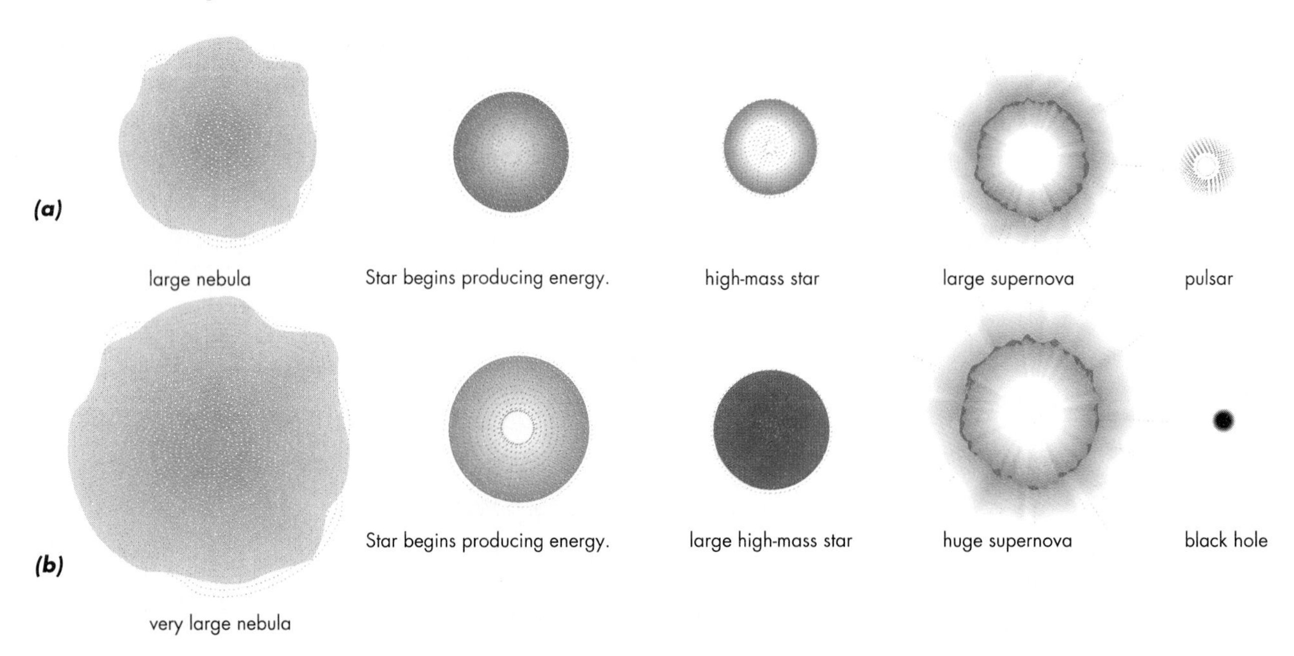

DID YOU KNOW

■ When our sun eventually swells into a red giant star, its outer layers will grow to be about 100 times as large as its size today, swallowing up Mercury, Venus, the Earth, and maybe even Mars. About how many years in the future do you think this will occur?

▶ REVIEW 16.1

1. Describe how a star forms.
2. Describe the difference between the life cycles of a low-mass star and a high-mass star.
3. How does the life cycle of our sun compare with that of other stars?
4. What is a supernova?
5. (a) Arrange the following in the order in which they occur: black hole; supernova; nebula; supergiant.
   (b) Is there more than one possible answer in (a)? Explain.

16.2 POSSIBLE ORIGIN OF THE PLANETS

Shortly after its launch into orbit, the Infrared Astronomical Satellite or IRAS (shown in Section 15.2, Figure 15.14) made some exciting discoveries. It detected a large cloud of tiny particles in orbit around the star Vega. This was the first direct evidence that solid matter exists around a star other than our own sun. Scientists thought that the clouds of particles could be planets at an early stage of development. Four months later, IRAS discovered solid material orbiting another star. These observations support scientists' theory about planets: planets probably form while a star is forming. In this section, you will explore this theory.

1. A rotating nebula starts to contract. Because of its rotation, the nebula flattens out as it contracts.

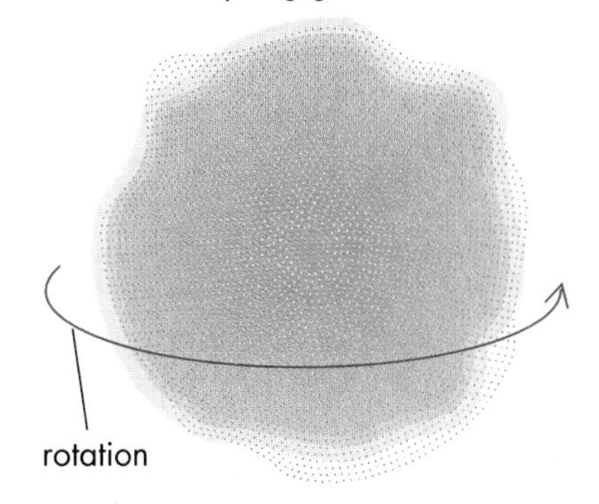

2. As the process continues, a bulge forms at the center. This bulge will later become the sun. The disk of cooler material away from the bulge will later become the planets.

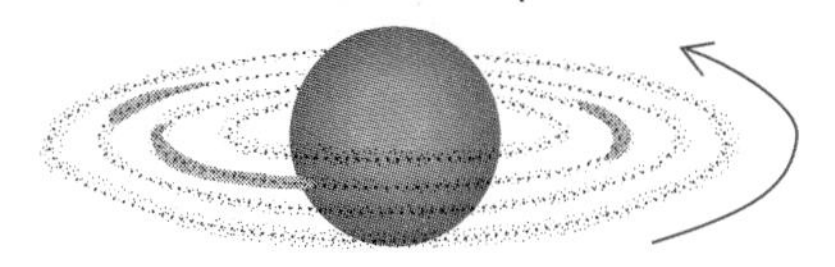

3. The disk breaks up into smaller chunks. Over a long time, these chunks form into planets. You can read the text for the entire story.

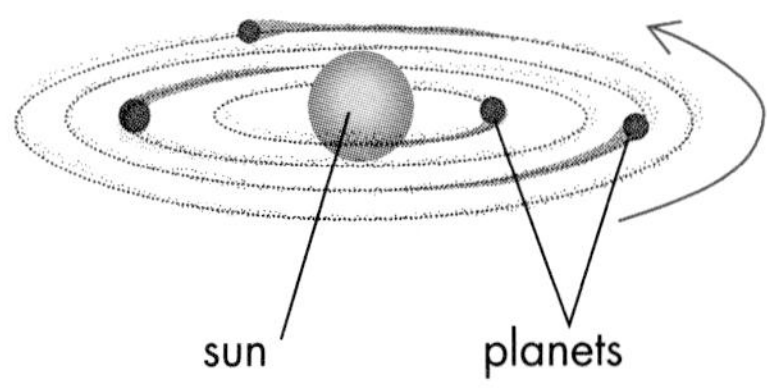

FIGURE 16.5
Possible stages in the formation of the solar system. Does this suggest to you that there might be other solar systems, just like ours, somewhere in the universe? Explain why or why not.

FORMATION OF THE SOLAR SYSTEM

Figure 16.5 shows the steps that may have occurred in the formation of the sun, planets, and other parts of the solar system. In its early stages, our solar system may have been part of a nebula consisting mainly of hydrogen gas and helium gas. Grains of solid matter such as iron, rock, and ice made up about 1 percent of the nebula. This solid matter formed from materials coming from supernova explosions, which you read about in Section 16.1.

Scientists think that our sun formed the way all stars do. Many of the particles of the nebula moved toward each other because of gravity. As the particles of the nebula came together, the pressure and temperature within the nebula increased. When temperatures were high enough, energy began to be produced. This occurred about 4.6 billion years ago, and marked the birth of our sun.

While the sun was forming, smaller clumps of matter appeared in the outer regions of the nebula. Held together by gravity, these smaller clumps of matter eventually condensed to form the planets of our solar system. This theory helps explain the formation of the planets that we call gas giants (Jupiter, Saturn, Uranus, and Neptune). The gas giants are composed mainly of hydrogen gas and helium gas, the same gases that make up much of the sun. Scientists think it is possible that these planets formed in a way similar to the sun—through the condensation of clouds of gases held together by gravity.

How do the terrestrial planets fit into this theory of planetary formation? Mercury, Venus, the Earth, and Mars are composed largely of rock and metal (iron), with little hydrogen gas and helium gas.

Scientists think that in the early stages of solar system formation, the gases in the inner regions of the solar system were too hot to condense. However, the outer regions were cool enough for the gas giants to form.

Scientists think that our young sun flared up in a sudden burst of energy. This flare-up blasted the hydrogen gas and helium gas out of

■ In 1950, the Dutch astronomer Jan Oort suggested the existence of a great collection of icy matter surrounding the solar system beyond the orbit of Pluto. Scientists think that the *Oort cloud* is the "nursery" where all comets are born. Write a report on the Oort cloud. In your report, explain why comets give scientists the opportunity to study what the solar system may have been like before the planets formed.

the inner regions of the solar system, leaving behind the heavier chunks of solid matter (mostly iron and rocky substances).

The terrestrial planets may have formed from those solid chunks. Although scientists do not know for certain how this occurred, they do have some ideas. One explanation is that as the chunks of solid matter circled the sun, they often collided with one another and stuck together. During millions of years, some of the chunks grew in size until they were large enough to have strong gravity. This gravity pulled in other bits of solid matter. The more massive chunks pulled in all the matter in the space around them. They eventually became the planets.

Over a long time (perhaps 10 million years), most of the matter in the solar system was pulled into these planets. The remaining matter makes up the asteroids, meteors, and comets, which also form part of the solar system. (You learned about them in Chapter 13.) They are called the **minor bodies** of the solar system (minor means smaller). These minor bodies are of special interest because they provide information about the early matter of our solar system, before most of the matter formed into the planets. This is just one of several possible explanations that scientists have for the formation of the planets.

THE SEARCH FOR OTHER PLANETS

As the instruments that astronomers use improve, the chances of finding planets revolving around a star somewhere in the Milky Way Galaxy increase. Learning about the formation of planets at various stages will help us better understand the development of the solar system. The goal of studying planet formation is to help us understand how life began on the Earth.

▶ REVIEW 16.2

1. What force is responsible for bringing together the particles found in space?
2. List the following stages of the formation of the solar system in order: formation of the outer planets; nebula of hydrogen gas and helium gas; formation of the sun; formation of the inner planets; flare-up of the sun, blasting hydrogen gas and helium gas away from inner regions.
3. (a) Which of the planets in our solar system are thought to have formed in a manner similar to the sun's formation?
   (b) Why could the other planets not have formed this way?
4. (a) What are the minor bodies of the solar system?
   (b) What do astronomers hope to learn by studying these minor bodies?
5. Assume that 10 years from now, astronomers have discovered planets travelling around several different stars. Do you think the planets they discover are more likely to be gas giants or terrestrial planets? Explain your answer.

16.3 POSSIBLE ORIGINS OF THE UNIVERSE

If stars go through life cycles, does the entire universe also go through some type of cycle? Scientists do not know the answer to this question, but they have gathered enough information to try to give an answer.

Through their many studies, scientists have found evidence that the size of the universe is not constant. Instead, scientists think that the universe is expanding.

The study of the origin and changes of the universe is called **cosmology**. Scientists who study cosmology are called cosmologists.

ACTIVITY 16B / A MODEL OF AN EXPANDING UNIVERSE

In this activity, you will use a simple model to illustrate the way in which most cosmologists think the universe is expanding.

MATERIALS

one balloon
black fine-point felt-tip marker
metric tape

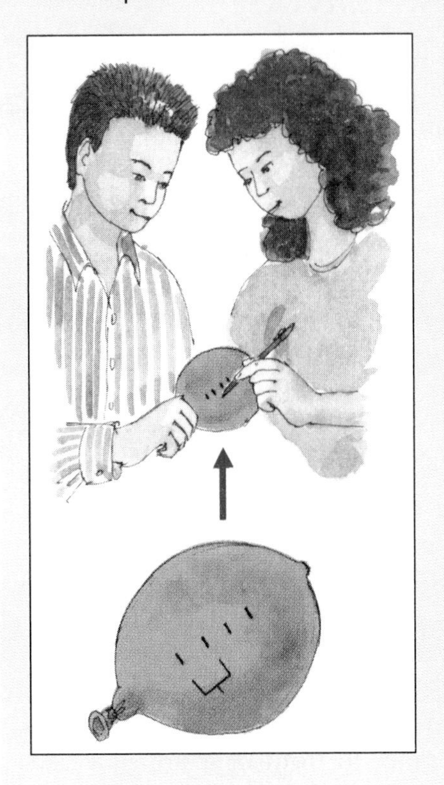

FIGURE 16.6 ◀
Labelling the balloon for Activity 16B

PROCEDURE

1. Copy Table 16.1 into your notebook. You will use it to record your results.
2. Blow up the balloon to the size of an orange. While your partner keeps the balloon at that size by pinching the opening, use the felt-tip pen to mark the balloon with four dots, each separated by 1 cm. Measure the distances between the dots with the metric tape. Label the dots in order, A, B, C, D, as shown in Figure 16.6. Record the distances between the dots in the row marked "First stage" in your data table.
3. Now inflate the balloon until it is about the size of a basketball. Measure the distances between the dots and record your results in the row marked "Second stage" in your data table.
4. Calculate the distances A to C and B to D for each stage, and add them to your table. ➡

	Measured distances (cm)			Calculated distances (cm)	
	A to B	B to C	C to D	A to C	B to D
First stage					
Second stage					

TABLE 16.1 ***Sample Data Table for Activity 16B***

DISCUSSION

1. (a) Look at how much the distances A to B, B to C, and C to D increased when you inflated the balloon to the second stage. Compare those increases.
   (b) Suppose you continued to blow up the balloon. How do you think the change in the distances A to B, B to C, and C to D would compare with the change from the first stage to the second stage? Explain what you would expect to observe.
2. (a) If you were standing on dot A while the balloon was expanding, which dot do you think would appear to be moving away from you most quickly?
   (b) Which dot do you think would appear to be moving away from you most slowly?
3. Imagine that the dots on your balloon are galaxies of stars. The separation of dots on the expanding balloon would represent the spreading of galaxies in the universe as it expands. What difference would you expect to find in how quickly close and distant galaxies are moving away from the Earth?
4. (a) Now assume that the dots still represent galaxies, and that you are on a planet in galaxy B, and a different observer is on a planet in galaxy C. Do you think the spreading of the galaxies would appear to be any different from these two locations? Explain your answer.
   (b) Assume that galaxy B is the Milky Way Galaxy. Would it be correct to say that galaxy B is the center of an expanding universe? Explain your answer. ❖

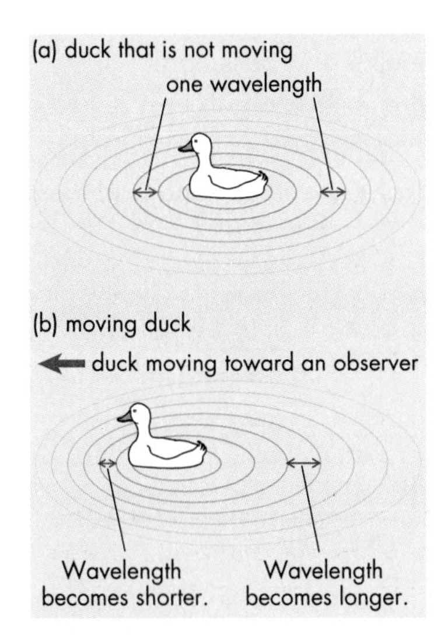

FIGURE 16.7
A model representing an energy source (the duck), energy waves emitted from the source (the water ripples), and wavelength (the distance from the top of one ripple to the top of the next).

EVIDENCE FOR AN EXPANDING UNIVERSE

A balloon is a useful model in helping us imagine how the universe expands. As a balloon expands, dots on the surface that are close together spread apart slowly. Dots that are far apart spread apart more quickly. However, as a model, an expanding balloon is inaccurate in some ways. For example, galaxies are not found on a skin or membrane; instead, they are scattered throughout the universe. How can we tell whether a galaxy is moving away from us? After all, the galaxies are very far away.

As early as 1912, astronomers observed evidence that the galaxies are moving away from the Earth and from one another. This evidence came from looking at the light spectra (colors) given off by nearby galaxies. The spectrum of a moving object in space is different from the spectrum of an object that is not moving. An example will help explain this.

Imagine a duck bobbing up and down on the surface of a pond (Figure 16.7a). Suppose the duck represents a source of energy. The ripples that spread out from the duck represent waves of energy emitted (given off) by the energy source. The distance from the top of one ripple to the top of the next ripple represents one wavelength (the length of one wave of water).

Now, imagine yourself standing on the shore, watching the duck. If the duck swims toward you, the ripples in front of it are squeezed together, and therefore have shorter wavelengths (Figure 16.7b). Meanwhile, the waves behind the duck are spread farther apart, and so have longer wavelengths. Even if you could not see the duck itself, you could observe from the changing wavelengths which way the duck was moving. Shorter wavelengths indicate that the duck is moving towards you; longer wavelengths indicate that the duck is moving away from you.

Similarly, if you observed a galaxy through a spectroscope, you would see a spectrum with lines moved toward the red region (Figure 16.8). (As you have learned, the colors in the spectrum are in the following order: red, orange, yellow, blue, green, indigo, violet.) Visible light in the red end of the spectrum has a longer wavelength than visible light in the violet end. Thus, you could infer that the light appears to have a longer wavelength than normal because the galaxy is moving *away* from you. This movement, or shift, into the red end of the spectrum is called **red shift**.

The light energy from almost all galaxies shows the red shift. From this information, we can infer that the galaxies are moving away from us, and away from each other. But what is causing this movement? Why is the universe expanding? Cosmologists have developed theories to try to answer this question.

■ The estimated density of the material that made up the universe shortly after the Big Bang is immense. One cubic centimeter of the material is thought to have had a mass of 1 billion kilograms!

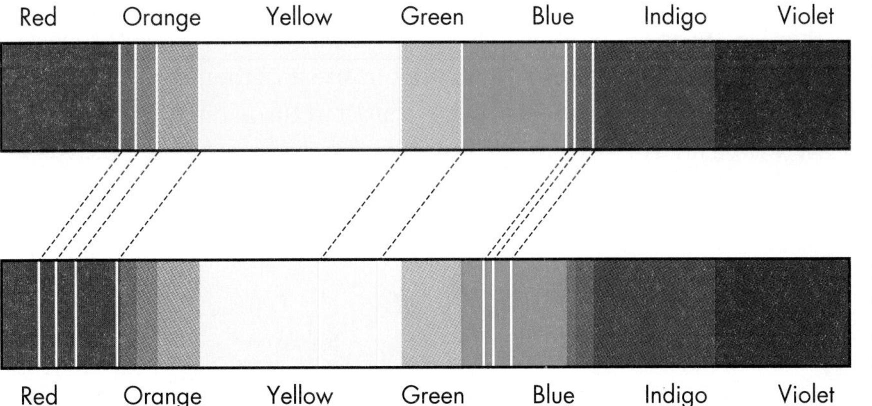

spectrum of an element on a star

Spectrum of the same element on a star that is moving away from us: All the spectral lines have shifted toward the red end of the spectrum.

FIGURE 16.8
The top diagram shows the spectrum of an element on a star. The bottom diagram represents the spectrum of the same element on a star that is moving away from us rapidly. You can see that all the lines in the element's spectrum have shifted (moved) toward the red end of the spectrum. What is this shift called?

THE BIG BANG THEORY

If the galaxies are moving apart, then at an earlier time they must have been *closer* together. Still earlier, they must have been even closer together. Continue to move backward in time in your imagination and you will reach what we call "time zero." Scientists estimate time zero as being between 13 and 20 billion years ago. At that time, the entire universe was packed together into one extremely dense, hot mass under enormous pressure. All the matter of the universe was contained within a very small volume. The explosion of this matter is known as the Big Bang (Figure 16.9). Scientists use the **Big Bang theory** to describe the beginning of the universe that we are a part of today.

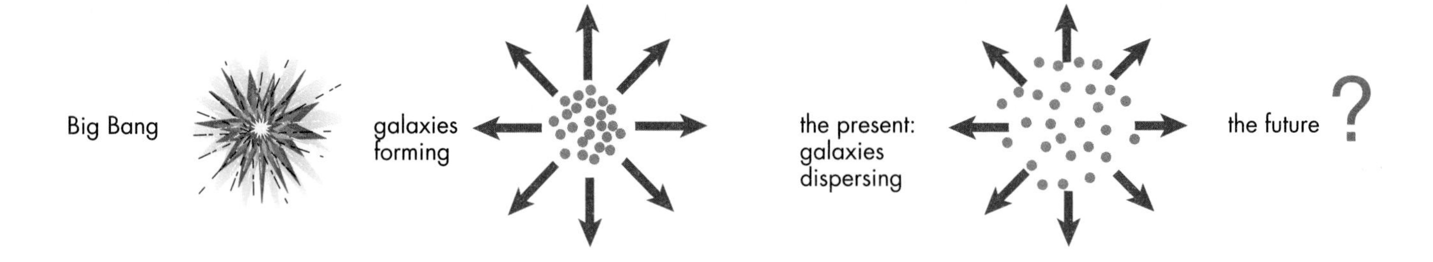

FIGURE 16.9
The Big Bang theory of the formation of the universe. How do you think this theory got its name?

Scientists have collected data that suggest that, even today, the entire universe is "glowing" from that enormous explosion. In 1965, for example, two scientists made an accidental discovery using a radio telescope. They detected unexpected radiation coming from all directions in space. After checking that there were no defects in their equipment, they concluded that this faint, unknown energy is a result of the Big Bang. The intense radiation from the tremendous explosion has faded to only a faint "whisper" after travelling for thousands of millions of years through space.

THE OSCILLATING THEORY

The Big Bang theory is not the only one that cosmologists have developed to explain the expansion of the universe. For example, the **oscillating theory** is another possible explanation. This theory suggests that the universe expands to a maximum size and then contracts (pulls back together). It contracts so far that it returns once again to the stage where another explosion and expansion occur (Figure 16.10).

FIGURE 16.10
The oscillating theory of the formation of the universe

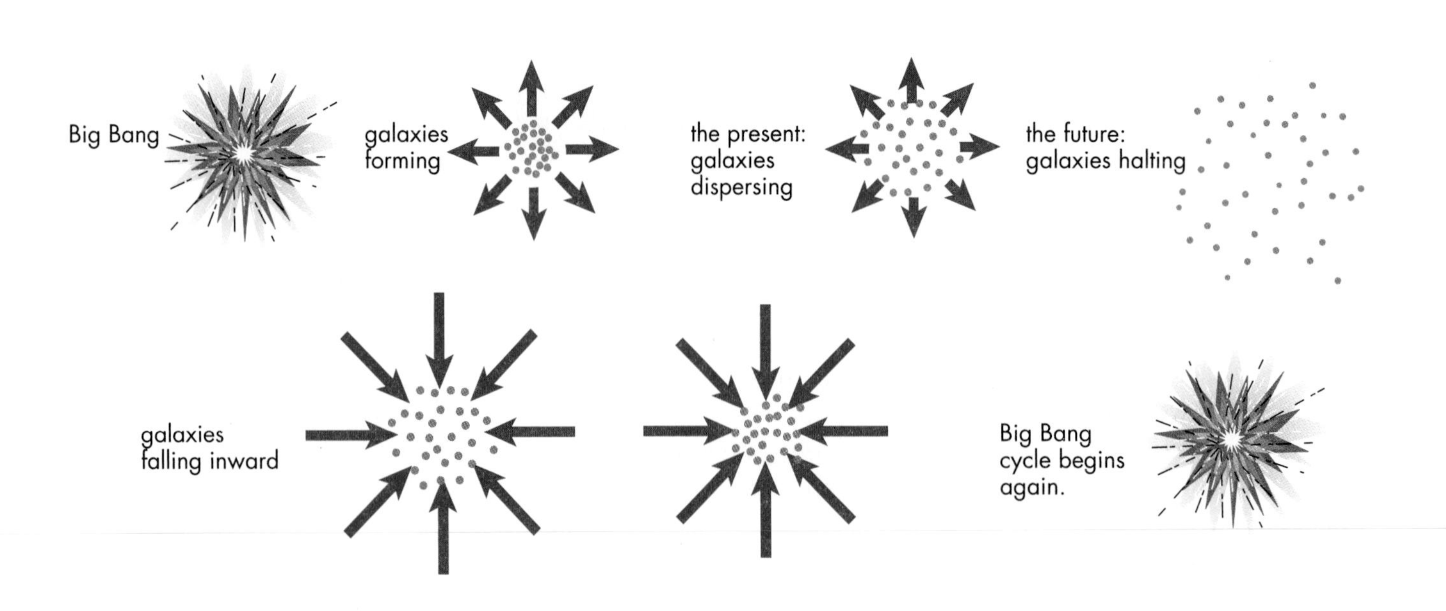

Both the Big Bang theory and the oscillating theory can be used to explain our observations about the universe. But we still know very little about how the universe began. For example, while scientists can describe what they think occurred a fraction of a second after the Big Bang, they do not know what the universe was like (or even if it existed) just before the Big Bang explosion. Changing our theories about the universe is a continuing scientific process as we try to explain the new discoveries that astronomers and cosmologists make year after year.

EXTENSION

■ Another theory about the universe is the steady state theory. According to this theory, the universe has always existed as it does today and will continue to exist. Write an article for a science magazine describing the steady state theory. How does it explain the evidence that galaxies are moving? Draw illustrations to go with your article.

REVIEW 16.3

1. (a) Copy Figure 16.11 into your notebook. Label the ripples that have shorter wavelengths, and those that have longer wavelengths. Indicate the direction of movement of the "object" causing the waves.
   (b) How do you know the direction you indicated in (a) is correct?
2. What does the term "red shift" mean?
3. Why do many astronomers think that the universe is expanding?
4. How does the Big Bang theory explain an expanding universe?
5. A round balloon has several evenly spaced dots painted on its surface.
   (a) How could the balloon be used as a model of an expanding universe?
   (b) How could it be used as a model of an oscillating universe?
   (c) Why would such a model be only somewhat useful?

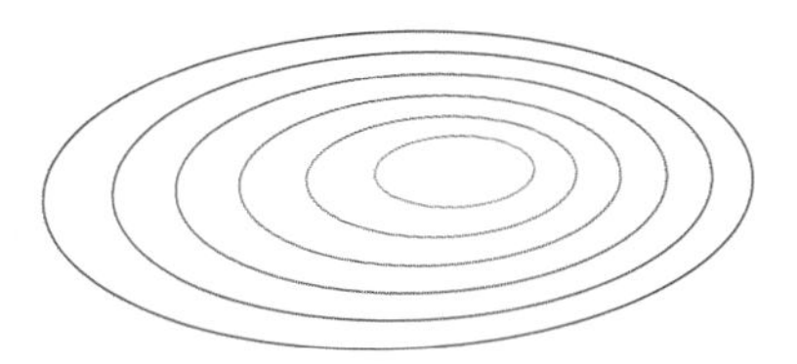

FIGURE 16.11
Illustration for Review Question 1

■ 16.4 LIFE IN THE UNIVERSE

You have studied theories about how the universe and our solar system formed. But once our planet existed, how did life begin?

THE ORIGIN OF LIFE

Many scientists think that life on the Earth developed from chemicals that filled the atmosphere and oceans of our planet in its early years of existence. Because we cannot travel back in time to see how life began, scientists must develop their ideas by observing life on the Earth today and by doing experiments.

Biologists have shown that all living organisms on the Earth are composed mainly of amino acids and other kinds of molecules. **Amino acids** combine to form the proteins that make up living things. Using this knowledge, scientists have designed laboratory experiments that have produced these molecules that are found in all living things, using only the gases that are thought to have been present in the Earth's early atmosphere.

In 1953, Harold Urey and Stanley Miller of the University of Chicago carried out the first such experiment (Figure 16.12). Urey and Miller placed hydrogen, methane, ammonia, and water in a glass flask. (It is these substances that scientists think made up much of the early atmosphere of the Earth.) Urey and Miller sent an electric spark through the mixture of gases in the flask. The spark was similar to the lightning that was probably present in the early atmosphere.

FIGURE 16.12 ▶
A simplified diagram of the apparatus used by Urey and Miller to produce amino acids. What are amino acids? What is important about the production of amino acids in this apparatus?

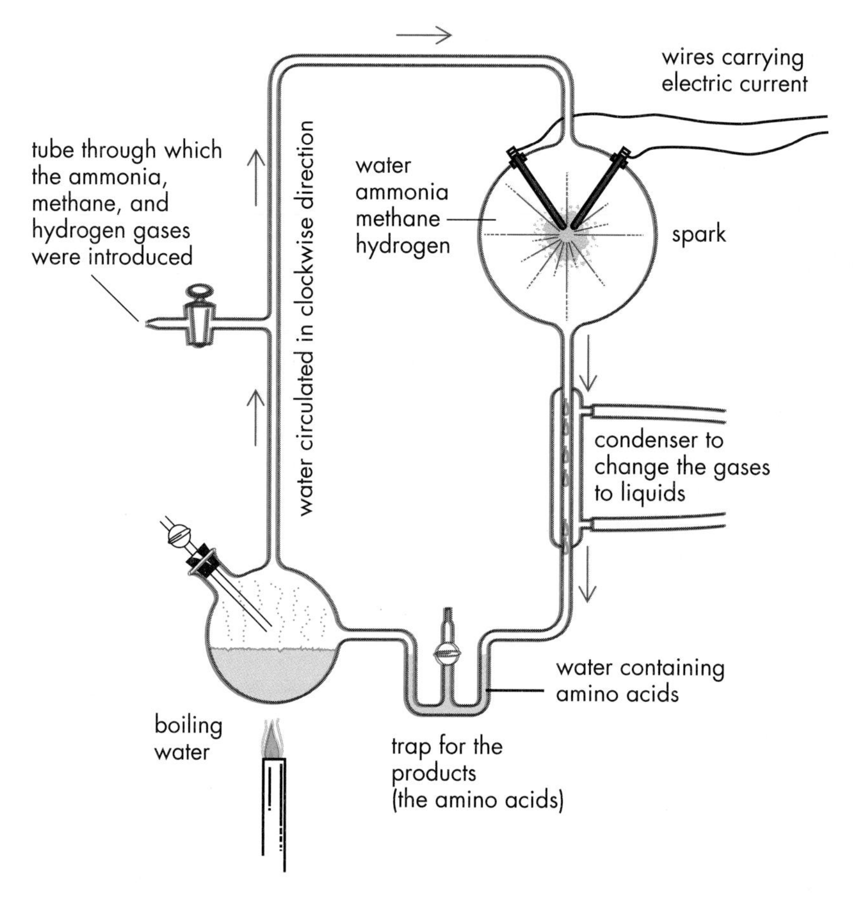

■ Scientists have repeated the Urey and Miller experiment under various conditions. For example, they have replaced the hydrogen-rich "atmosphere" of the experiment with various mixtures high in carbon dioxide and relatively low in hydrogen. These mixtures have also yielded biochemical compounds when exposed to sparks of energy.

At the end of a week, the scientists found that several different amino acids had formed in the flask. This was the first evidence that the Earth's early atmosphere could have produced the molecules for the proteins that make up living things. The results that Urey and Miller obtained excited other researchers. Similar experiments by other scientists have produced dozens of amino acids under conditions like those of the Earth's early atmosphere.

Of course, amino acids and other molecules do not by themselves make an organism. Scientists think that some of these molecules clumped together and gradually changed, forming the first cells. It probably took many millions of years before life as we know it came into being.

HOW ASTRONOMERS USE COMPUTERS

When you imagine an astronomer at work, you probably think of someone looking through a telescope. Actually, today's astronomers spend most of their time in front of computers. Without computers, most research in astronomy would be impossible. Here are some examples.

An astronomer studies a glowing image of a globular star cluster on a computer screen. The computer created the image using huge amounts of data gathered through a telescope. The astronomer can use the computer to display the image in different ways to show certain features more clearly. The computer can calculate the brightness and location of each star in the cluster.

Many objects in space emit radio waves and do not emit light energy. Astronomers use radio telescopes to detect these radio waves. For example, a radio telescope can "listen" to radio waves from an exploding star. The information is stored in a computer, which creates a radio-wave "picture" of the exploding star — what you would see if you had eyes that could see radio waves. Pictures like this give astronomers a new view of the universe.

Computers are essential in the search for extraterrestrial life. Only by using computers can scientists listen to all parts of space, at many different radio frequencies, for long periods of time. Scientists hope to detect radio signals that unknown beings might have sent from other parts of the universe.

The Hubble Space Telescope, now orbiting the Earth, sends computerized images of space to Earth. Astronomers can display these images on their computer screens, explore certain features, and make calculations. Because of an error in the construction of the space telescope, the first pictures were flawed and blurry. However, they were still better than anything we could see from Earth. A computer program was developed to correct the images. The telescope was fixed by astronauts from a space shuttle, but astronomers still use computers to improve the pictures from the telescope.

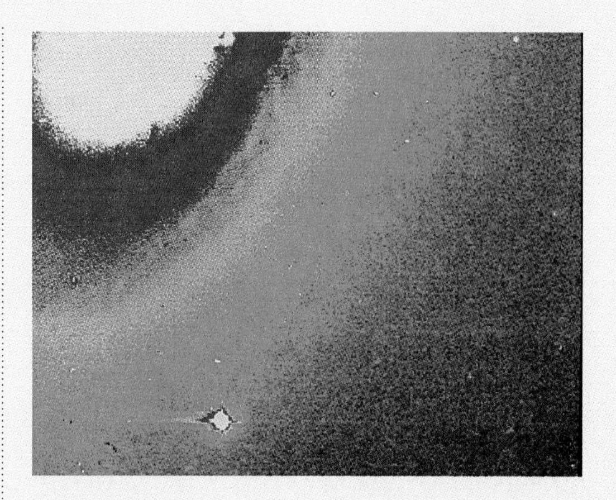

A supernova in spiral galaxy M81 is seen below and to the right of the galaxy's core. This is a false-color image, generated by computer using the image from an optical telescope.

Computers control the operation of astronomical instruments on Earth. Imagine the problem of following a galaxy when your telescope is on a planet like Earth that is rotating. In the past, astronomers had to constantly turn their telescopes in the opposite direction to the Earth's rotation. Now, computers do all the calculations and control the movement of the instrument.

Computers record large amounts of data, analyze them, create pictures, do mathematical calculations, and operate complicated astronomical instruments. But to some astronomers, the most exciting thing computers do is link scientists electronically to others all over the world. Through their computer networks, astronomers can share ideas, data, calculations, computer images, and the latest astronomy jokes.

Can you think of a disadvantage of working with computers all day?

■ Draw a time chart of life on the Earth. Include the first appearance and extinction of several species.

INVESTIGATING THE UNIVERSE

Scientists look at many factors in trying to determine if a planet can support life. They think that the Earth may be the only planet in our solar system with the right conditions to support life. But does life exist only on this one planet in the entire universe? You can find imaginary beings from other planets in everything from films to video games to cereal boxes. People are often willing, even eager, to believe that other planets in the universe are inhabited. Scientists, however, need good evidence in order to reach conclusions. What is important in science is what we observe, not what we think exists. Through investigation, scientists discover new information, which leads to new answers as well as to new questions.

ACTIVITY 16C / DESIGNING LIFE FORMS

In this activity, you will design an intelligent being that lives on a planet in some faraway part of our galaxy. Your design should show the knowledge and understanding you have gained throughout this entire unit.

PROCEDURE

1. Read about the three imaginary planets, Alpha, Beta, and Gamma, that orbit imaginary stars far away in the Milky Way Galaxy. All the planets have enough oxygen and water to support life. The stars around which they revolve are about the same mass, volume, age, and temperature as our own sun. Choose one of the planets to perform the remaining steps of this activity.

   Planet **Alpha** is about the same size as the Earth, and the force of gravity on its surface is about the same. It is closer to its sun than the Earth is to our sun, so its daytime temperatures are much higher than here. It rotates on its axis once every 100 Earth hours, and one revolution around its sun takes 200 Earth days. Its axis is tilted, just as the Earth's axis is. The atmosphere is much thicker than on the Earth, and only about 20 percent of the surface is covered with water. Conditions for life on this windy planet are harsh.

   Planet **Beta** is larger and much more massive than the Earth. Its force of gravity is much stronger, and about 80 percent of its surface is covered with water and ice. Its atmosphere is thick, and it is located somewhat farther from its sun than the Earth is from our sun. Its day is only 10 Earth hours long, but its year is about 1000 Earth days. Its axis is tilted at a smaller angle than the Earth's axis. Conditions for life on this planet are excellent in the water but harsh on the land.

   Planet **Gamma** is a small planet with a much lower force of gravity than the Earth's. Its atmosphere is about the same as that on the Earth, but water covers only about 30 percent of the planet's surface. The distance to its sun is slightly less than the Earth's is to our sun. One revolution around its sun takes 300 Earth days. Its axis is not tilted at all. One rotation on its axis takes 100 Earth hours. Conditions for life on this planet are excellent, especially for flying beings.

2. Design a creature that can survive easily and communicate on the planet of your choice. Describe the shape of the body and other characteristics of your creature. State reasons for giving your creature the characteristics it has.
3. Have someone comment on your design. If necessary, change some of the characteristics to improve your design.
4. Make a poster to show your creature and the environment in which it must live.

THE OLDEST AND NEWEST OF SCIENCES

Studying space science and astronomy helps us realize just how special our Earth is. The knowledge and understanding we have of the Earth and its place in the universe is far greater today than at any other time in history. As new discoveries about the universe are made by people such as yourself, we will learn more about ourselves and the universe. This increased knowledge should help us to learn to protect this fragile and beautiful planet as it travels in one tiny corner of a galaxy that contains 100 billion stars.

■ Design an outfit that you would need to wear when visiting the creature that you designed in Activity 16C on its home planet.

REVIEW 16.4

1. Which substances do scientists think made up the early atmosphere of the Earth?
2. (a) What characteristics of the Earth's early atmosphere did Urey and Miller try to match in their experiment?
   (b) What substances formed in the glass flask during their experiment?
   (c) Why are these substances important in the study of the origin of life on the Earth?
3. (a) In the Urey/Miller experiments, what source of energy did they assume was present in the early atmosphere?
   (b) What other sources of energy do you think might have caused amino acids to form?
4. Give reasons why life beyond our solar system would be difficult for us to observe, even if it does exist.

CHAPTER REVIEW

KEY IDEAS

■ The life cycle of a star depends on its mass. Low-mass stars have long lives that end without huge explosions. High-mass stars have shorter lives that end in explosions. They then turn into super-dense stars such as pulsars or black holes.

■ The planets of the solar system have resulted from the formation of our sun. The terrestrial planets probably formed differently than the gas giants did.

■ One important theory about the origin of the universe is the Big Bang theory. Evidence for the Big Bang comes from the red shift of the light energy that reaches us from galaxies. The red shift is the result of the expanding universe.

■ The oscillating theory is another theory about the origin of the universe. This theory describes the universe as first expanding and then contracting.

- Life on the Earth probably began when some source of energy (such as lightning) caused chemicals in the Earth's early atmosphere to join together to form amino acids and other molecules.
- Astronomy and cosmology are continually changing as scientists perform experiments, observe the universe, and make new discoveries.

VOCABULARY

cycle
red giant
white dwarf star
red supergiant
supernova
minor body
cosmology
red shift
Big Bang theory
oscillating theory
amino acid

V1. Print the phrase "Big Bang" in the middle of a blank piece of paper. Use these words as the starting point to draw a mind map illustrating the development of the universe. Include as many words from the vocabulary list as you can. Where possible, show how concepts are related by writing short descriptions on your connecting lines.

V2. As a class, make up a set of quiz questions for this chapter. To do this, each student should choose a different word or phrase from the vocabulary or another part of this chapter. The word or phrase that is chosen becomes the answer to a question made up by that student. Each question and its answer should be placed on a separate card. Organize a contest using the questions made up by the class.

CONNECTIONS

C1. (a) What is the Big Bang theory?
(b) Describe the Big Bang theory.

C2. (a) Describe what scientists think may happen to our sun over the next several million years.
(b) What effects will these changes in the sun have on the Earth?

C3. The following is a list of stages in the lives of stars: white dwarf, supernova, red giant, pulsar, black hole.
(a) Which, if any, of these stages will our sun pass through in the future?
(b) Explain why our sun will not pass through the other stages.

C4. Describe the relationship between a red giant and a white dwarf.

C5. Describe one theory of the formation of the planets of our solar system. Illustrate your answer with drawings.

C6. (a) What is "red shift"?
(b) How does the red shift provide evidence for the Big Bang?

C7. As scientists continue to observe the spectra of stars and galaxies, what do you think they would conclude if they observed the following?
(a) The red shift continued.
(b) The red shift was no longer observed for any star or galaxy.
(c) A shift was observed toward the blue end of the spectrum.

C8. (a) Name one type of molecule that helps to make up all living things on the Earth.
(b) What evidence is there that the molecule mentioned in (a) could have formed in the early atmosphere of the Earth?
(c) Describe how experiments involving this molecule have influenced scientists' ideas about the origins of life on the Earth.

C9. Suppose that a meteorite landed on the Earth and scientists discovered that it contained many amino acids. What influence might this discovery have on our theories about the origin of life on the Earth?

EXPLORATIONS

E1. Imagine you are one of five crew members on a spaceship sent to determine whether there are any planets in orbit around the star Alpha Centauri.
(a) How far is this star from our solar system
(i) in light years?
(ii) in kilometers?
(You can find both answers near the end of Section 14.3.)
(b) Estimate how long it would take your spaceship to reach Alpha Centauri. (Hint: First decide how fast your spaceship can travel in kilometers per hour.)
(c) Describe the living arrangements on the ship, and the way in

which everything you need to survive is provided for you and your children as you travel.

(d) What problems might occur during the voyage, and how might they be solved?

(e) On arriving close to Alpha Centauri, you use radio waves to send a message back to the Earth. How long will those waves take to reach the Earth?

E2. In late August 1975, amateur and professional astronomers noticed that a bright star had suddenly appeared in the constellation Cygnus. Such events, known since ancient times, are called novas (from the Latin for "new"). In fact, novas are old stars passing through an explosive stage of their lives. Using reference books, find out how novas are related to other stars, how they explode, and how they differ from supernovas.

E3. Some radio telescopes are used to try to detect intelligent life that may exist on planets revolving around other stars in the universe. What other methods could you suggest for trying to find evidence of life beyond the Earth (extraterrestrial intelligence)? Explain your answer.

E4. Under what extreme conditions can life exist on the Earth? Find out about extreme environmental conditions in which life (plants and animals) has been found. Examples of conditions to research include the coldest and hottest temperatures, the greatest ocean depths, and the highest land elevations. Design a poster to illustrate your findings.

E5. Pluto was not mentioned in the description of the possible origin of the planets in Section 16.4.

(a) Describe the orbit and basic structure of Pluto. (You may have to refer back to Section 13.2.)

(b) Suggest a theory to explain Pluto's origin.

E6. Figure 16.13 shows an incredible photograph of an event rarely observed: in this case, Supernova 1987A. A yellowish ring of gases is spreading out from what remains of the exploding star. The reddish-colored blob near the middle of the ring is the leftover material from the explosion. Astronomers who took this photograph in 1990 predicted that the leftover material, travelling faster than the gases, will overtake the gases within 1000 years. Look for a more recent photograph of Supernova 1987A (taken by the Hubble Space Telescope) and compare the newer photograph with the one shown here.

REFLECTIONS

R1. Look back at the poster your group made in Activity 16A. Now that you have finished this chapter, have your ideas about the formation of the solar system and the universe changed? If so, describe the changes.

R2. Write at least five questions about the universe that you would like more complete answers to.

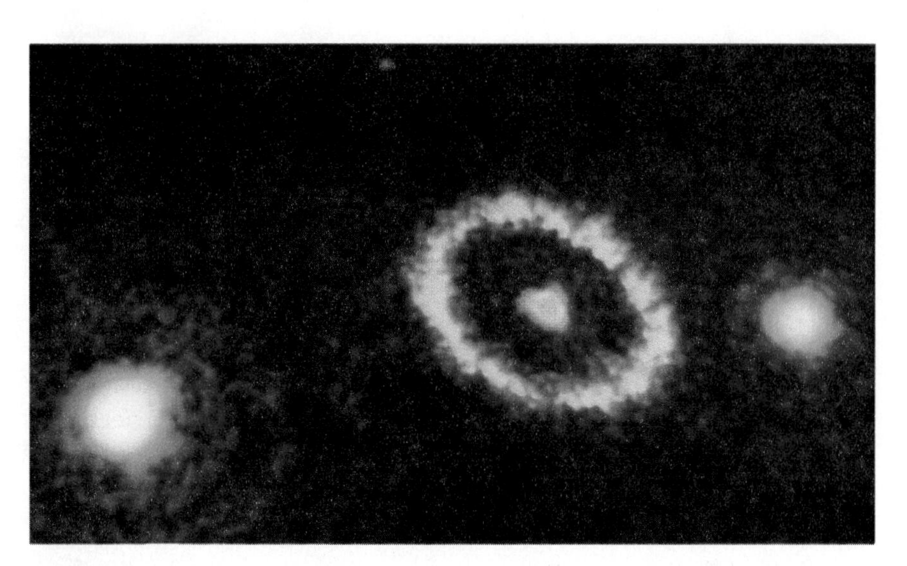

FIGURE 16.13
This image of Supernova 1987A was taken by the European Space Agency's faint-object camera about three years after the supernova was first observed.

Mobil Oil Corporation

DISCOVERING ENERGY AROUND YOU

What do you see in the photograph that has to do with your daily life? This picture of a drilling rig was taken on a sunny day. Certainly, the sun is important to you and everyone. Without the sun, life on the Earth would not exist. Oil and gas, obtained from beneath the Earth's surface, also affect your life. Many of the products that you use, including plastic objects, are made from the chemicals in oil. Oil and gas also provide energy for heating buildings. What other uses of the energy from oil and gas can you think of?

This unit is about the branch of science called physics. Physics is the study of matter and energy, and how they affect each other. In this unit, you will learn about energy. You may already know something about energy. If you have studied biology and chemistry, you may have learned that chemical energy is stored in foods, and that such energy is important for your body's needs. If you have studied space science, you may have learned that the sun and other stars give off huge amounts of energy.

In this unit, you will explore the importance and use of energy, as well as how energy changes from one form to another. You will learn how machines do work for us, and you will study in more detail the topics of heat, temperature, and thermal energy. You will consider the problem of how we can obtain the energy we use, while at the same time doing as little harm to the environment as possible. You will also find out where the energy that we use comes from. And you will learn why you should try to reduce the amount of energy that you use. Why do you think we should use less energy?

UNIT V

CHAPTER 17

ENERGY USE IN OUR SOCIETY

Think of how your life would change if you did not have some kind of energy to keep you warm in the winter. And think of how different your life would be without electrical energy to operate refrigerators, hair dryers, and other appliances.

Much of the energy we use in the U.S. comes from petroleum. The photograph above shows a refinery. In refineries, oil for heating homes and gasoline for operating cars are produced from petroleum. Where did the energy stored in petroleum come from in the first place? Should you be concerned about how much petroleum is left in the world, or by how much pollution is caused when we use it? These are some of the issues you will explore in this chapter.

As you read this chapter, keep in mind the following fact: People in the United States use more energy per person than people in any other country in the world except Canada. People in other countries, such as England, Italy, and Japan, use only about half the amount of energy per person that we use. The good news is that we Americans are becoming more aware of our energy use, and we are doing something about it. After completing this chapter, you will know more about how you can use energy more wisely.

ACTIVITY 17A / HEADLINE NEWS

In this activity, you will answer questions about titles of energy-related articles found in newspapers and magazines, such as those shown in Figure 17.1.

MATERIALS

sheet of paper with 24 headlines on it

PROCEDURE

1. Your teacher will give you a sheet of paper with 24 headlines on it. Cut up the sheet into tags so that each headline is on a separate tag.
2. Classify the tags into two different sets, one set representing problems of energy supply or use, and the other set representing solutions to problems. Arrange your tags so that the headlines you are most certain about are at the top of the list and the ones you are least certain about are at the bottom.
3. In your notebook, write the title "Headlines of Energy Problems" at the top of a page. Under the title, paste your tags in the order you arranged them in Step 2. For each headline in this list, write what you think the problem is.
4. On a new page in your learning journal, paste your second set of headlines under the title "Headlines of Solutions to Problems." For each headline in this list, write what you think the solution is.

DISCUSSION

1. Choose at least two headlines. Suggest what the articles are about.
2. Do you think newspapers and magazines give you enough information about sources and supplies of energy? Give reasons for your answer.
3. Do you think newspapers and magazines write enough about finding solutions to energy-related problems? Explain your answer.
4. Consider the following energy-related headline: "New fuel replaces gasoline — Drivers save half their fuel cost." In two or three sentences each, state what you think the article would be about if it were in
   (a) a science magazine for students,
   (b) a business or financial magazine,
   (c) a magazine about political issues,
   (d) the travel section of a newspaper. ❖

FIGURE 17.1 ▶ ***These headlines appeared recently in magazines and newspapers.***

ALTERNATIVE COULD CURB OIL DEPENDENCY

OIL RICHES ARE RUNNING DRY WORLDWIDE

WASTE AS FUEL

CONSUMERS SLOW TO ACCEPT NEW FUELS

■ Choose one of the headlines in Activity 17A, and write a magazine or newspaper article to go with it. Your article could discuss the results of an imaginary survey, be an interview with a scientist, give suggestions for a science fair project, or be your own original idea. You can be as creative as you like.

■ 17.1 WHY IS ENERGY IMPORTANT?

What does the word "energy" mean to you? A simple definition is: **energy** is the ability to make things move. In some cases, the motion is easy for you to see: chemical energy in gasoline has the ability to move a car. Electrical energy has the ability to move skiers to the top of a ski hill (Figure 17.2). However, in other cases, you cannot see the motion. For example, your heart uses energy to pump blood throughout your body, although you do not see the blood move. Cooking is another example. Energy added to food that is being cooked causes the molecules of food to move faster, although you cannot see that motion.

FIGURE 17.2 ▶ ***Energy is needed to move skiers from the bottom of a ski hill to the top. In this case, electrical energy is used. Before chair lifts and ski tows were invented, what kind of energy did the skiers use to get to the top of the hills?***

Notice that energy is the *ability* to make things move. This does not necessarily mean that something must be moving in order for energy to be there. You might be able to see the motion later when *stored energy* changes into *useful energy*. For example, the food you eat has chemical energy stored in it. That energy will become useful to your body only after you have digested the food.

Why is it important for you to study energy? One reason is that you use energy almost all the time in your daily life. Another reason is that our supplies of energy, such as petroleum, are being used up, so we must learn how to cope with this situation. A third reason is that using energy often causes pollution that harms our surroundings, and we must learn how to reduce such pollution. What other reasons can you think of for studying energy?

Energy use can be classified as either **direct energy use** or **indirect energy use**. When you ride in a car, you are using energy stored in the fuel to move the car. This is an example of direct energy use. However, you are also using large amounts of energy indirectly. For example, energy was needed to make the parts of the car. Energy was needed to put the parts together (Figure 17.3). And energy was needed to build the roads on which the car travels. These are all examples of indirect energy use when you ride in a car. So when you think of the energy you use to perform a certain activity, think of your indirect use of energy as well as your direct use.

FIGURE 17.3
How many indirect ways of using energy are involved to manufacture a car?

In your daily life, probably the most common direct use of energy involves electricity. For example, when you dry your hair with an electric hair dryer, you are using electrical energy directly. But think of all the energy that you are using indirectly. This energy was used to manufacture the hair dryer, package it, ship it, and so on.

You have seen some of the many uses of energy in the examples above. Because you use it in so many ways, energy is very important in your daily life. Energy is also important to society in general. You can see evidence of energy's importance by looking at articles in newspapers and magazines, and at books and other resources about energy.

ACTIVITY 17B / ENERGY AND FOOD PRODUCTION

CROSS-CURRICULAR

How many steps are needed to provide you with a piece of toast made from whole-wheat bread? In this activity, you will explore some of the many steps in producing food from plants. You will be asked to think about the energy used at each step of production (Figure 17.4).

PROCEDURE

1. Divide a page in your notebook into three columns. Title the columns "Steps in Food Production," "Possible Use(s) of Energy," and "Direct or Indirect Energy Use."
2. The following steps in food production are not given in order:
   - watering the plants
   - storing the dried plants
   - selling the food to customers in stores
   - preparing the soil
   - drying the plants

FIGURE 17.4
What kinds of energy are used in food production?

- processing the plants into food
- weeding the field
- transporting the dried plants to be processed into food
- cooking the food at home
- planting the seeds
- packaging the food for stores
- harvesting the plants
- distributing the food to stores

Using double spacing, in the first column of your table, list the steps in food production (given above) in the order you think they should be listed.

3. In the second column in your table, list the ways that you think energy is used at each step in food production. For each use, indicate in the third column whether the use is direct or indirect.

DISCUSSION

1. Which do you think requires more energy in the production of food from plants: direct energy use or indirect energy use? Explain your answer.
2. Is the amount of energy you need to cook rice or potatoes in your kitchen very large compared with the amount of energy used in all the other steps in producing these foods? Give reasons for your answer.
3. How do you think the number of steps in milk production compares with the number of steps listed in this activity? Explain.
4. Choose a food that is produced in the area where you live. List as many different steps in the production of this food as you can. ❖

► REVIEW 17.1

1. How do you know that there is energy in
   - (a) the food you have eaten in the past day?
   - (b) gasoline available at a filling station? (Hint: Think about the definition of energy at the beginning of this chapter.)
2. List four or five reasons why you think it is important to study energy. (Some of the reasons may be found in this chapter.)
3. (a) List at least five devices related to food in your home that use energy. For each device, state what energy source (for example, electricity or natural gas) you are using directly.
   (b) Choose one device from your list in (a) and describe several ways that energy was used indirectly to manufacture that device and bring it to your home.
4. For each example in the list below, indicate whether the source of the energy is electricity, liquid fuel (such as gasoline), solid fuel (such as wood), or the sun.
   - (a) cooling a home with an air conditioner
   - (b) operating a computer printer
   - (c) growing vegetables in a garden
   - (d) driving a car
   - (e) roasting a marshmallow at a campsite
   - (f) using a solar-powered calculator
5. Describe the energy that you are using indirectly when you use each of the products listed below.
   - (a) a gold-plated ring
   - (b) this textbook
   - (c) a bicycle
6. State why energy use is important to each of these groups of people:
   - (a) nurses
   - (b) loggers
   - (c) commercial fishers
   - (d) farmers

17.2 ENERGY RESOURCES AND TECHNOLOGY

In Section 17.1, you read that we use energy for many purposes. The energy for these purposes comes from energy resources. An **energy resource** is a material in nature that can be changed into useful energy.

Until about 500,000 years ago, food and sunlight were the only energy resources that people could use. Then they learned to make fires, so they were able to use the chemical energy stored in wood for heat and light.

Although wood is an example of an energy resource, making fires is an example of technology. Technology is the use of scientific knowledge to solve a problem. It is the tools, machines, or processes that help people use the natural world. Water wheels and windmills are examples of technology that enabled people to use running water and wind as energy resources (Figure 17.5). As people developed new technologies, they were able to use other energy resources.

FIGURE 17.5
Windmills have been used for more than 1,000 years. They have been used to pump water from the ground or grind grain into flour. This windmill is in The Netherlands.

CONTINUED GROWTH OF ENERGY USE

Since people first learned to use fire, wind, and water as energy sources, the amount of energy we use has grown greatly for two important reasons. One reason is that each person in our society is using more energy than people have ever used before. We travel more, grow and eat more varieties of food, have more comfortable buildings, do more activities that require electricity and other forms of energy, and throw away more than societies before us. Another reason is that the human population of the world is increasing, in some areas rapidly. More people need more energy. If both the energy used by each person and the world population continue to increase, the energy resources we use will be in danger of being used up.

ENERGY RESOURCES TODAY

Think of the ways we Americans heat our homes. We burn wood in fireplaces or stoves; we burn natural gas or oil in furnaces; we use electrical energy; and, less commonly, we use other resources, such as solar energy. In all cases, the energy resources can be classified as renewable or non-renewable.

A **renewable energy resource** can be renewed within an average human lifetime (about 75 years) or less. Two common types of renewable energy resources are materials from plants and animals, and water in rivers. You will read about these below. Other renewable energy resources are described in Chapter 18.

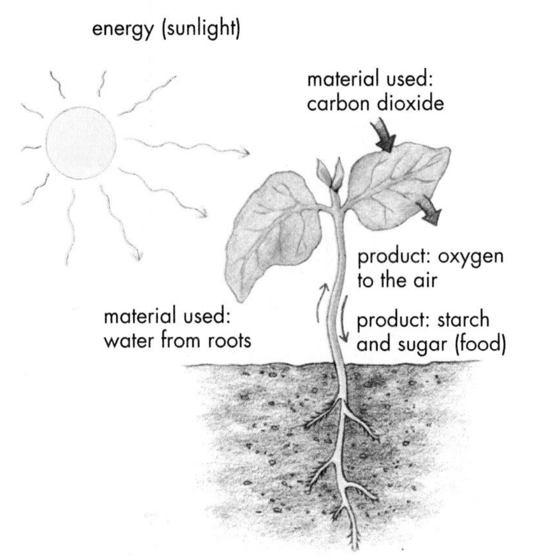

FIGURE 17.6
In the process of photosynthesis, plants use energy from the sun to make food. Where do plants store this converted energy?

A **non-renewable energy resource** takes so long to renew (millions of years) that once it is used up, we can think of it as being gone forever. The main types of non-renewable energy resources are fuels called fossil fuels, which you will read about below.

To understand renewable and non-renewable energy resources, compare them with cloth and paper napkins. After a cloth napkin has been used, it can be washed and used again. The cloth napkin is renewable. After a paper napkin has been used, it is thrown away. The paper napkin is non-renewable.

BIOMASS FUEL

Biomass fuel is made from plants and animal wastes, which contain chemical energy. The chemical energy stored in biomass comes indirectly from the sun.

You may remember that photosynthesis takes place in the green leaves of plants. Plants use solar energy to combine carbon dioxide from the air and water from the soil to form oxygen, starches, and sugars. The starches and sugars are stored in the plants and help the plants grow (Figure 17.6).

One of the advantages of biomass fuel is that it can be used in a variety of ways. To use biomass directly, we can burn wood in home fireplaces and stoves (Figure 17.7). Some industries burn biomass fuel made up of forest-product wastes, such as bark and sawdust (Figure 17.8). You can read about less common uses of biomass in Chapter 18.

FIGURE 17.7
Many people burn wood for home heating. What are the advantages and disadvantages of this?

FIGURE 17.8
This pulp mill uses wood waste as biomass fuel to produce some of the energy needed to run the mill. What are some advantages and disadvantages of this practice?

Biomass fuel has other advantages. It can be available whenever you need it. (Compare this with solar energy, which is not available at night or on cloudy days.) Also, biomass fuel is renewable. In order for wood to be renewable, however, the forests from which the wood is cut must be properly managed (Figure 17.9).

Unfortunately, using biomass fuel has some disadvantages. Burning wood and other biomass fuels produces large amounts of polluting gases and particles. As well, much of the energy is wasted when hot air goes out through the chimney. One way to reduce this waste is shown in Figure 17.10.

FIGURE 17.9
A tree planter plants young trees to replace those that have been cut.

WATER AS AN ENERGY RESOURCE

The energy of falling water can be used to produce electricity. The electricity produced by falling water is called **hydroelectricity** (*hydro* is the Greek word for water). To control the flowing water, dams are built along rivers. One example of many dams is shown in Figure 17.11. You can read about how electricity is produced from water in Chapter 18.

Using water to generate electricity has important advantages. It is the least expensive way to produce large amounts of electricity. Producing electricity this way causes little pollution, and, most important, hydroelectricity is renewable.

FIGURE 17.11
Dams built on rivers control the flow of the water. What type of energy does the generating station use to produce electrical energy?

FIGURE 17.10
Fireplaces have been designed to waste less of the heat they produce. In the heat circulator shown here, air enters through the lower vents and passes through pipes around the firebox. The heated air then rises and escapes through the upper vents into the room.

However, producing electricity from falling water also has disadvantages. Because large amounts of flowing water are needed, hydroelectricity is not available everywhere. Building a dam affects the ecology of an area and the people who live there. Often the energy must be transmitted long distances by using transmission lines (Figure 17.12).

FIGURE 17.12 ▶
Transmission lines carry electrical energy away from a hydroelectric dam. What are the disadvantages of producing energy far away from the users?

ACTIVITY 17C / THINKING ABOUT FOSSIL FUELS

You may already know something about coal, petroleum, and natural gas. In this section, you will find out more about these fuels. Before you start reading, copy Table 17.1 into your notebook.

Before you read about coal, write its name in the first column. In the second column, write what you already know about coal.

As you read about coal, fill in the last column. Leave some space before you write "petroleum" in the first column. You will want to add more information about coal as you read about where fossil fuels are found, how they are removed from the ground, and how they are transported.

Your table will be a useful summary of what you have learned about fossil fuels. ❖

Fuel	What I already know	What I learned
Coal		

TABLE 17.1 ◀
Sample Data Table for Activity 17C

FOSSIL FUELS

A **fossil fuel** is an energy resource formed from plants and animals that died millions of years ago. The fossil fuels you will learn about here are coal, petroleum, and natural gas.

Coal is a solid made up mostly of carbon, which is an element found in all plants and animals. The coal we use today as a fuel began to form in large swamps about 400 million years ago. At that time, forests of giant plants used light energy from the sun, carbon dioxide from the air, and water from the swamps to grow. When these plants died, they sank into the swamps and decayed. The result was a spongy, brownish material called **peat** (Figure 17.13). New plants grew and decayed on top of the peat, forming more layers. Over thousands of years, the layers of peat became buried far below the Earth's surface. Gradually, the peat was pressed into a more solid form that we call coal. The characteristics of the coal depend on the age of the coal and on how much material pressed down on it from above. The different types of coal (lignite, bituminous, and anthracite) are shown in Figure 17.14.

FIGURE 17.13
In some areas of the world, peat is an important energy resource. This photograph of a pile of peat was taken in Ireland. What would have eventually happened to this peat if it had not been mined?

Petroleum and **natural gas** are chemical compounds made up mostly of hydrogen and carbon. Petroleum is a liquid, and natural gas is a gas. Because they are composed of hydrogen and carbons, they are called **hydrocarbons**. Hydrocarbons began to form in oceans about 500 million years ago. Tiny ocean plants and animals died, decayed, and were buried by sediments on the ocean floor. Over millions of years, the layers of sedimentary rocks compressed and heated the decaying marine life. As this occurred, hydrocarbons were formed.

Fossil fuels took millions of years to form. This explains one big disadvantage of using them: they are non-renewable.

FIGURE 17.14 ▶
Plant and animal materials take thousands of years to turn to peat. After many millions of years, the peat has changed to coal. As it ages and more layers of sediment are added above, the coal hardens. Of the three types of coal in this diagram, which should be hardest? Which should be softest?

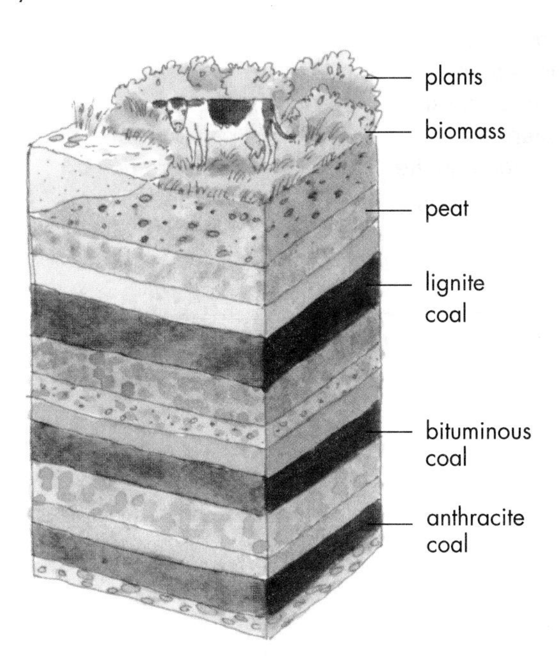

EXTENSION

■ Many states have large coal mines. Select a state and find out where these mines are, what kind of coal they contain, and what the coal is used for. To show this information, make a large map of the state on a poster. Also show how the coal is transported to the locations where it is used.

WHERE FOSSIL FUELS ARE FOUND

Coal is always found between layers of sedimentary rock. The layers of coal can be very thin (about 1 cm to 2 cm) to very thick (about 100 m to 200 m).

Petroleum and natural gas are not found in separate layers, like coal. As the petroleum or natural gas forms, it travels through sedimentary rock that contains pores (tiny spaces). The petroleum and natural gas rise until they are trapped beneath a layer of rock without pores. In Figure 17.15, you can see two examples of petroleum and natural gas trapped in sedimentary rock. In both cases, notice that water and hydrocarbons are trapped in rock with many pores (porous rock) between layers of rock without pores (non-porous rock).

Wherever sedimentary rocks exist, there is a chance of finding fossil fuels. Coal is usually found among the sedimentary layers where it first formed. Petroleum and natural gas, however, are not always found where they first formed. Over millions of years, rock layers can move, allowing the fuels to travel through pores in the rock to new layers, often a great distance away.

Figure 17.16 shows the locations of large areas of sedimentary rock in California. A very large area of the state contains sedimentary rocks that are not of marine origin.

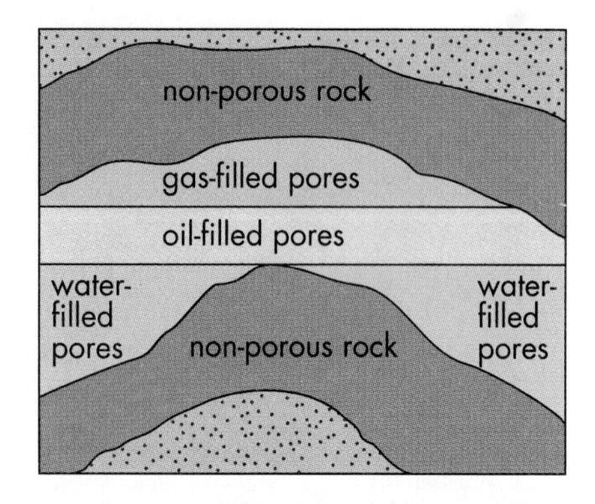

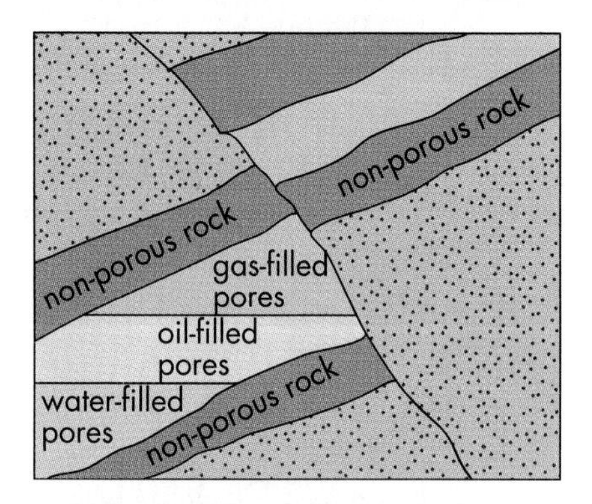

FIGURE 17.15
Examples of rock formations in which petroleum and natural gas have collected in large pockets of porous rock. The pockets are trapped between layers of sedimentary rock without pores. In the bottom diagram, the rock has shifted.

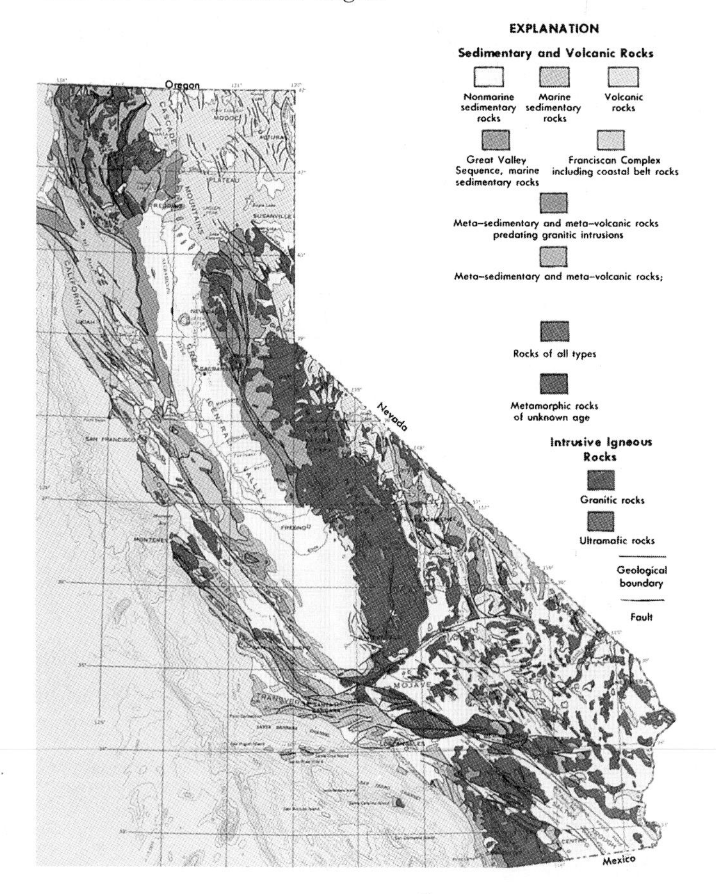

FIGURE 17.16 ▶
This map shows areas in California where sedimentary, metamorphic and igneous rocks are found.

REMOVING FOSSIL FUELS FROM THE GROUND

Two methods are used to remove coal from the ground: open-pit mining and underground mining. Open-pit mining is used when coal is near the surface (Figure 17.17). A large machine scrapes off the covering plants, soil, and rock, and digs up the coal in large chunks. Underground mining is used to remove coal that is buried deep beneath the surface.

■ In the early days of coal mining, working conditions were very poor. Men, women, and children worked in cramped tunnels for up to 18 hours a day. They had to crawl on their stomachs and backs. Cave-ins were common, the air was poor, and explosions occurred, killing people. In today's mines, safety is still a major concern.

FIGURE 17.17
Open-pit mining is used to dig up coal that is near the surface. What do you think should happen to the land after all the coal is removed?

Removing petroleum and natural gas from the ground may be either very easy or very difficult, depending on the location. If the hydrocarbon is beneath land, it may be possible simply to pump it to the surface. Petroleum and natural gas beneath the ocean are much more difficult to bring to the surface. An artificial island must be built to hold all the pumping machinery and a place for workers to live (Figure 17.18).

You will learn how energy from fossil fuels is used in Section 17.4.

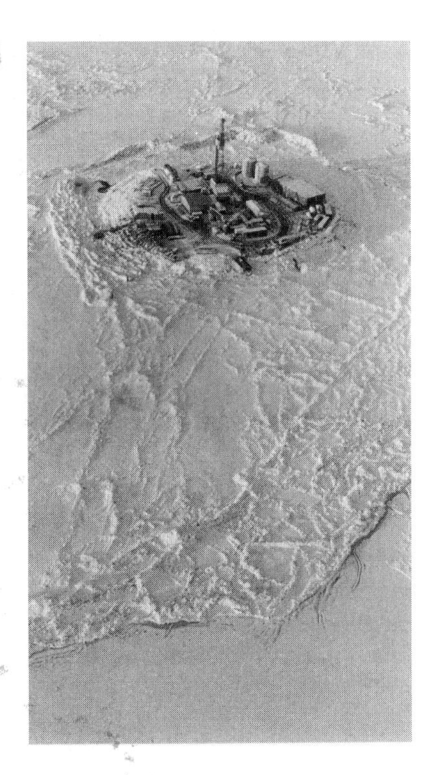

FIGURE 17.18
Drilling for petroleum in Kugmallit, an artificial island in the Arctic. What special problems face the people who drill for oil in the Arctic?

TRANSPORTING FOSSIL FUELS

Fossil fuels are hardly ever found where they are needed. Thus, they must be transported from where they are mined to where they are used. On land, coal is often moved by train, whereas on water, it is moved by barge or ship. Transporting petroleum and natural gas is not as easy as transporting coal.

EXTENSION

■ Examine the effects of oil spills on the environment. Look at what happens just after the spill takes place and at the effects years later. Use examples of recent oil spills. Discuss who should be responsible for clean-up after a major oil spill.

Petroleum from oil wells cannot be used directly. First it must be processed in an oil refinery. To get the petroleum from the oil field to the refinery, two methods are used: pipelines and oil tankers. A pipeline is easy to build where the ground is flat and firm, but it becomes costly and difficult to build in hilly or mountainous regions or across rivers. Often the ecology of a region is disturbed by a pipeline.

Using tankers to transport petroleum on water has even more problems. Since the first tankers were used, only a few decades ago, bigger and bigger ones have been built. These tankers have a bad reputation for collisions, explosions, and spills during storms (Figure 17.19). However, many oil tankers travel the oceans every day without any accidents at all.

Transporting natural gas by tankers would be very wasteful, so a different approach is used. The gas is cooled to about –160°C so that it forms a liquid (Figure 17.20).

FIGURE 17.19
An oil spill has caused oil to wash up on this beach. How will the oil affect the organisms that live near the beach?

FIGURE 17.20
The natural gas transported in tankers is in liquid form. Is natural gas transported through pipelines as a liquid or a gas? (Ask your local gas company.)

► REVIEW 17.2

1. What is the original energy source of all the energy resources discussed in this section?
2. Make a table showing the characteristics of biomass fuel using these titles:
   Definition
   Origin
   Ways of using this fuel
   Advantages
   Disadvantages
3. (a) What is hydroelectricity?
   (b) What are the advantages of hydroelectricity?
   (c) What are the disadvantages of hydroelectricity?
4. (a) How did coal form?
   (b) What is the main element in coal?
   (c) Where is coal likely to be found?

5. (a) How did petroleum and natural gas form?
(b) What are the two main elements in these resources?
(c) Where are these resources likely to be found, and in what form(s)?

6. Why are fossil fuels called non-renewable energy resources?

7. Suggest some of the environmental concerns about
(a) open-pit coal mines,
(b) drilling for petroleum in the Arctic,
(c) drilling for petroleum off the coast of the U.S.,
(d) transporting petroleum.

17.3 ENERGY FROM FOSSIL FUELS

People used coal as a fuel before they used other fossil fuels. One reason that people used coal first is that it can be lighted and burned to produce heat as soon as it is dug out of the ground. Liquid fossil fuels, however, must be refined before they can be used. **Refining** is a process to separate pure substances from a mixture such as petroleum.

As well, both liquid and gaseous fuels (such as heating oil and natural gas) need special equipment to burn safely.

The first known use of coal occurred more than 3000 years ago in China, where the coal provided heat to bake porcelain. Coal is also mentioned in ancient Greek history.

PRODUCTS AND BY-PRODUCTS OF BURNING FOSSIL FUELS

Most of the heat and light given off when fossil fuels burn results from two chemical reactions. In one reaction, oxygen from the air combines with carbon in the fuel to produce carbon dioxide, which is a colorless, odorless gas. In the other reaction, hydrogen in the fuel combines with oxygen in the air to produce water vapor. You cannot see either carbon dioxide gas or water vapor.

If there were always enough oxygen, *and* if fossil fuels were pure substances, water and carbon dioxide would be the only products. However, burning fossil fuels often releases unwanted **by-products**. These are products other than the main ones in the chemical reaction. An example is soot, which is tiny particles of carbon that escape without burning when there is not enough oxygen (Figure 17.21). Soot dirties the air and everything the air touches, including human lungs.

Other by-products result when coal, oil, and natural gas are burned because these fuels contain impurities. These by-products, such as colorless gases or thick, black smoke, can cause air pollution.

FIGURE 17.21
Soot from a burning candle. Can you think of a way to make the candle burn cleaner?

OTHER PROBLEMS FROM BURNING FOSSIL FUELS

Burning fossil fuels may cause other problems. Acid rain or snow is formed when gases such as sulfur dioxide combine with water in the air. Sulfur dioxide can be a by-product of burning fuel. Acid rain and snow harm plants, pollute lakes, and damage buildings.

Another problem is thermal pollution from the waste heat given off by vehicles and furnaces. In cities, there are so many such fuel-burning devices in a small area that the air over a city is often a few degrees warmer than the air at ground level. If a warm air layer settles over a city, it can trap cool air beneath it. This prevents polluting gases and particles from rising, and increases the air pollution in the city.

The carbon dioxide given off by burning fossil fuels may cause another kind of thermal pollution, one that affects the whole planet. When the level of carbon dioxide in the atmosphere rises above normal, the atmosphere traps heat that would usually escape from the Earth's surface, in much the same way that a greenhouse traps heat. This effect is therefore called the greenhouse effect (Figure 17.22).

Today, about 90 percent of the energy used in North America comes from burning fossil fuels.

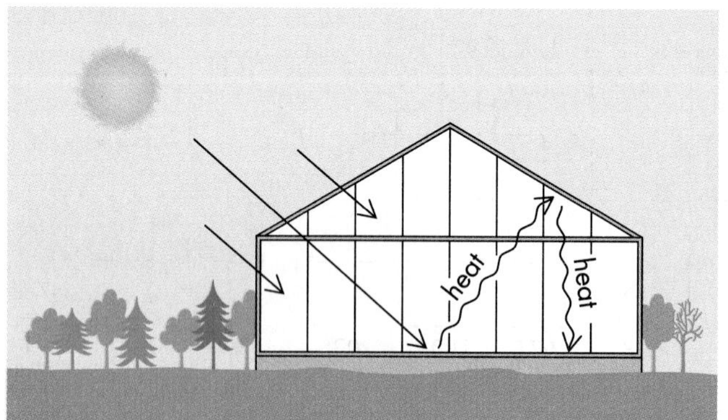

(a)

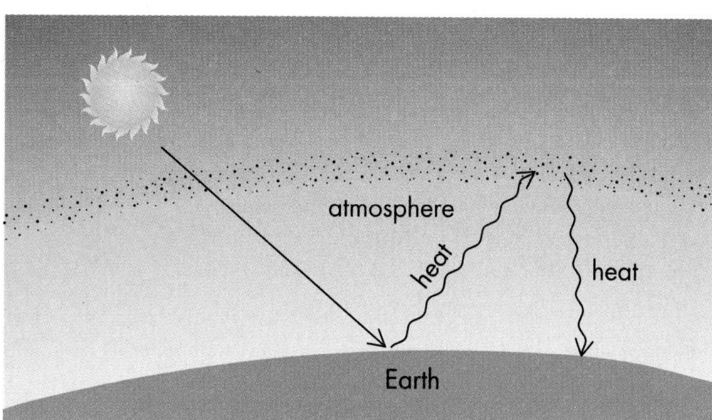

(b)

FIGURE 17.22
The greenhouse effect.
(a) The glass panes of a greenhouse allow sunlight to pass through, but prevent heat from escaping.
(b) A similar situation occurs in the atmosphere: extra carbon dioxide in the atmosphere acts like the glass in the greenhouse, trapping heat within the Earth's atmosphere.

ACTIVITY 17D / FOSSIL FUELS AND THE FUTURE

In this activity, you will investigate the production and use of fossil fuels.

MATERIALS

a map of the world or an atlas

PART I Who Has the Fuel?

PROCEDURE

Figure 17.23 shows the world distribution of known fossil fuel resources. Look carefully at the figure, then answer the Discussion questions.

DISCUSSION

1. (a) Where are most of the world's known supplies of petroleum and natural gas?
(b) Where are most of the world's known supplies of coal?

2. Which part(s) of the world do you think might have the greatest need to bring in fossil fuels from other countries? Why?

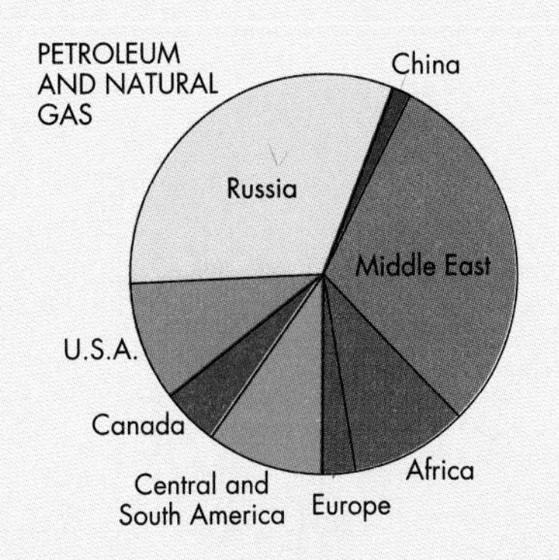

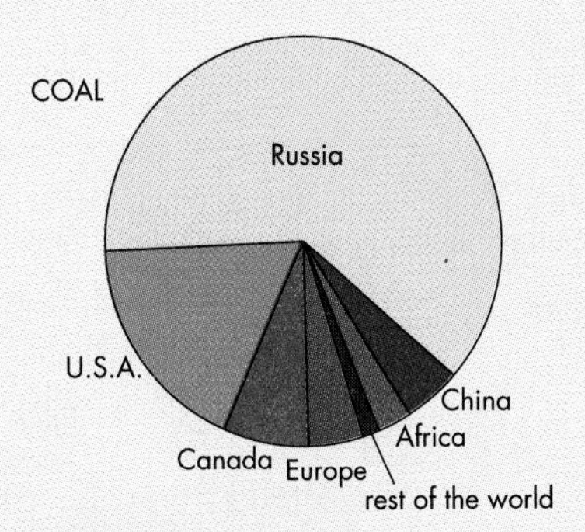

FIGURE 17.23
These pie graphs show where the known world supplies of petroleum and natural gas (top graph) and coal (bottom) are.

FIGURE 17.24 ▶
These graphs show the past and predicted future production of petroleum and natural gas (top), and coal (bottom).

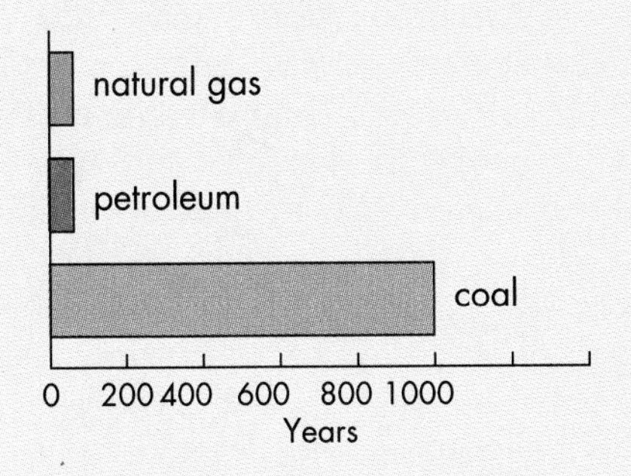

FIGURE 17.25
This bar graph shows how long the known world supplies of fossil fuels are predicted to last if we continue to use fossil fuels at the same rate as today.

PART II How Quickly Are Fossil Fuels Being Produced?

PROCEDURE

Look carefully at Figure 17.24, which shows how much petroleum, natural gas, and coal is produced each year. "Production" is the term used to describe the amount of fossil fuel removed from the ground. The solid graph lines show production up to the present; the dashed graph lines predict future production.

DISCUSSION

1. What time period (from what date to what date) is covered by
(a) the coal production graph?
(b) the petroleum and natural gas production graph?

2. Which fossil fuel do you think will be more important by the following dates:
(a) 2010 (b) 2100

3. Why do you think the production of fossil fuels is increasing?

4. Explain why scientists think that production will rise to a peak and then fall.

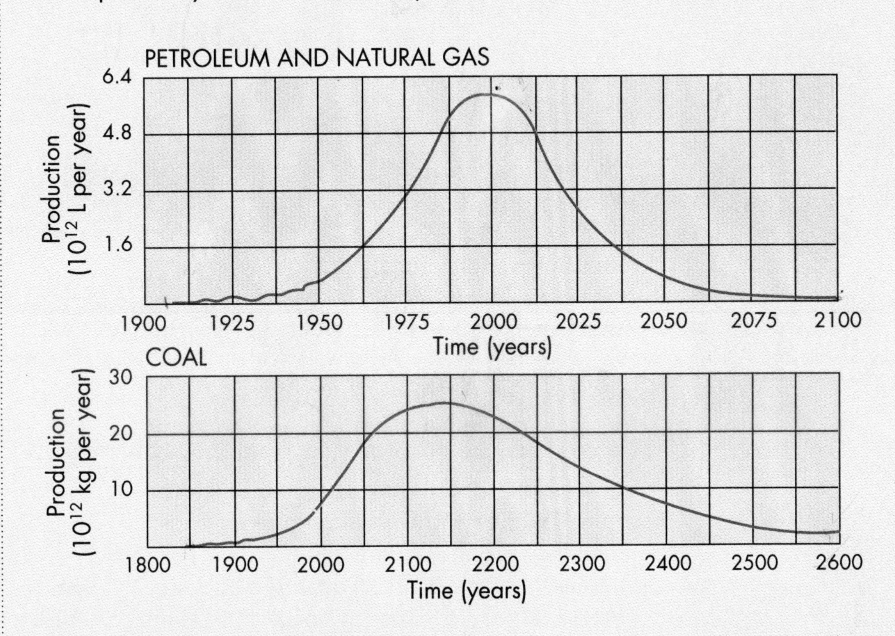

PART III How Long Will Fossil Fuels Last?

PROCEDURE

Study Figure 17.25, which predicts how long each fossil fuel will last if the rate at which the fuel is being used remains the same as at present.

DISCUSSION

1. How soon does it appear that petroleum and natural gas supplies will run out if we continue to use them at the present rate?

2. How long does it appear that coal supplies will last?

3. What do you predict will happen to the rate of using coal once all the petroleum and natural gas are used up? Explain your answer. ❖

JIM WIESE

The rollercoaster races along the track, up and down, careening around the curves. You feel your body pushed this way and that, lifted out of your seat one moment, and forced down into it the next. It's exciting and fun. Believe it or not, it's also physics.

"Amusement Park Physics" is what Jim Wiese calls it. Every year, he joins several thousand students for a day at the local amusement park. He has designed lessons to go with each of the rides. Not only do students learn about concepts such as gravity, velocity, acceleration, and centripetal force—they also experience them. "Rather than watching the experiment, they are the experiment," says Wiese.

Wiese is the kind of science teacher everyone would love to have. He spent years in the classroom doing unusual and creative things with his own students. He now works for the Surrey School Board helping other science teachers.

What's a good way to learn about vectors? Wiese suggests you climb aboard the wave swinger. It has chairs that hang by chains from a wheel that turns in a circle.

"As you start out, the only force you feel is gravity, and the chain above your chair is basically going straight down," says Wiese. "As the ride starts to turn we add another force—centripetal force." As the ride goes faster, the chairs start to swing out farther and farther, and the chain sticks out at an angle. "Now the chain that you are riding on is a vector sum."

You may be dizzy, but you are probably beginning to understand vectors. "It's a real-life experience of some very sophisticated science," says Wiese.

Wiese has designed another set of science lessons to accompany a television program called *MacGyver*. MacGyver is a character who is always getting himself out of difficult situations, often by using science. Wiese designs science experiments that students can do in class to duplicate what MacGyver does on television.

Another Wiese project is a science summer camp for students entering grades eight and nine. For a week, students work with the latest scientific and computer technology. They are encouraged to develop their own ideas.

Wiese wants students to realize that science is part of everyday life. "It's a way of viewing something you see all the time and realizing that there is science involved in it, and that it is not threatening. It's kind of fun."

Think about the fun you had the last time you went down a waterslide. What was some of the science happening during your wild descent?

▶ REVIEW 17.3

1. At one time, many homes were heated by burning coal. Why do you think coal furnaces were used long before oil or natural gas furnaces?
2. Describe how or why each of the following products, by-products, or problems are produced when fossil fuels are burned:
   (a) carbon dioxide
   (b) water vapor
   (c) soot
   (d) acid rain and snow
   (e) thermal pollution
3. How does the burning of fossil fuels add to the greenhouse effect?

■ 17.4 CONSERVING ENERGY

You have seen that using energy resources, especially non-renewable ones, can cause serious problems, such as pollution. We depend on fossil fuels, but the supply of most fossil fuels is decreasing rapidly. One way to help reduce the problem of energy use is through energy conservation. Conserving energy means not wasting it. **Conserving energy** should be an important goal for everyone in our society, including you, your family, and your friends. It is especially important in the United States and Canada, where the energy that each person uses each day is far greater than anywhere else in the world (Figure 17.26). In this section, you will explore ways in which both society and individuals can conserve energy.

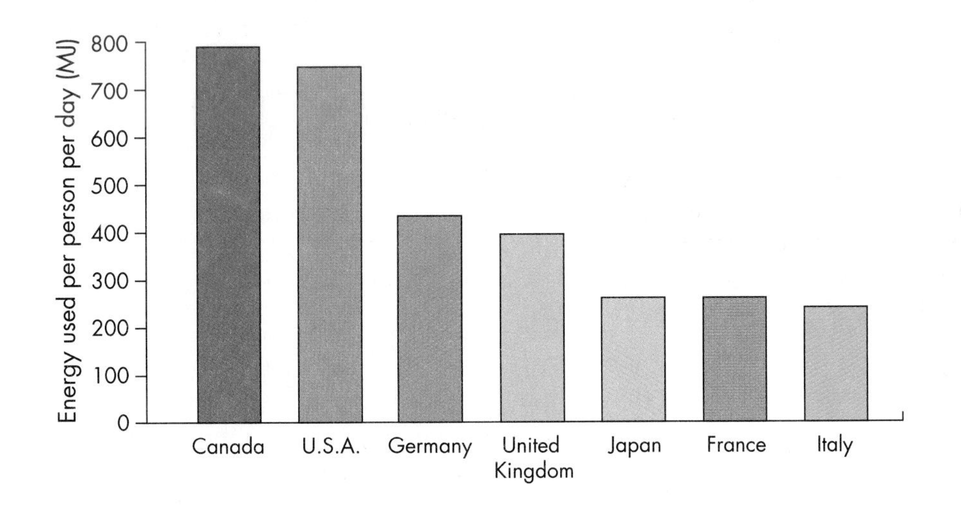

FIGURE 17.26 ◀ ***This bar graph shows the energy use (in megajoules) per person each day in seven industrialized countries. Why do you think American and Canadian energy use is so high?***

SOCIETY'S ROLE IN CONSERVING ENERGY

Figure 17.27 shows two vehicles that consume different amounts of energy. The big car shown in Figure 17.27a was built in the 1950s. At that time gasoline cost about 7¢ per liter, and people were not concerned that the car consumed about 30 L of gasoline for every 100 km. The more modern family car in Figure 17.27b is lighter and more

(a)

(b)

FIGURE 17.27
The energy consumption of two cars. (a) In the 1940s and 1950s, the average car in North America was large and heavy. Cars like this one used about 30 L of gasoline for every 100 km. (b) What features do modern cars have that allow them to use less fuel?

streamlined and uses less energy than the older car. We say it is more energy-efficient because it consumes only about 10 L of gasoline for every 100 km. The changes in automobile design are an example of how the automobile industry and the government have worked together to improve the use of energy in our society.

Citizens, government leaders, and scientists are concerned about the use of energy today and what energy will be available in the future. As a result, people in all parts of society are interested in energy conservation. Besides improving efficiency of energy use in automobiles, what else has our society done to conserve energy?

In the U.S., much energy is used to heat buildings in the winter and cool them in the summer. Since insulation helps conserve energy, government rules now require that new buildings have very good insulation. Some builders even add more insulation than the rules require so that their buildings use even less energy. Governments have provided money for people to add insulation to homes that were already built and to replace old furnaces with new ones.

FIGURE 17.28
According to the sign, only buses are allowed to use this lane during rush hour. If car drivers started using the lane, what might happen to the number of people taking the bus?

In transportation, governments have reduced speed limits and encouraged people to use car pools. Both these steps reduce energy consumption. And governments promote public transportation, especially in cities. Some lanes on city roads are for buses only so that buses can travel more quickly than cars into and out of the city. This makes public transportation more attractive to people who might otherwise take their cars (Figure 17.28). In some cities, there are also special lanes and pathways for bicycles. Outside buildings, places are set aside where bicycles can be parked safely. This is to encourage people to use bicycles for regular transportation and not only for leisure (Figure 17.29).

The government also requires new electrical appliances to be tested to find out how much energy they use. Every new appliance sold in the U.S. has an energy guide ("Energyguide") label that shows how much energy that appliance uses.

Another important way to reduce everyone's energy consumption is by public education. Radio and television advertisements remind people how to use energy wisely. You, too, can learn how to conserve energy.

FIGURE 17.29 ◀
By riding their bicycles to work, people conserve fuel, and get their exercise too!

HOW CAN YOU CONSERVE ENERGY

Every time you turn on a radio or TV, ride a bus, get a cold drink from the refrigerator, or eat a hot meal, you are using energy. Some people may not care about how much energy they use because there is lots of energy in the U.S. Perhaps they are not aware of the problems of energy use that are predicted for the future. They have not taken the first step in solving our energy problems: becoming aware.

However, having studied energy resources and use in this chapter, you are aware of possible energy problems. Now you can see how you can conserve energy.

ACTIVITY 17E / YOUR ROLE IN CONSERVING ENERGY

What are the 10 most important ways in which you can conserve energy?

PROCEDURE

1. Listed below are 13 statements about energy-related activities performed by most people. Working with a partner or a group of other students, discuss each statement.
2. For each statement, state whether you consider it to be true or false.
3. Copy each true statement into your notebook. Write an explanation of how important it is in conserving energy in our society.
4. For each false statement, explain why it is false and rewrite the statement to make it true.
   (a) In the winter, you can turn the thermostat down at home and wear a sweater to keep warm.
   (b) You can conserve energy by turning off lights and electrical appliances when you are not using them.
   (c) Incandescent lights use less electrical energy than fluorescent lights.
   (d) Toasting one piece of bread at a time in a toaster does not waste energy.
   (e) Washing dishes consumes more energy than using throwaway containers, plates, and cutlery. ➡

(f) When travelling a short distance, it is better to take a car than to walk or ride a bicycle.
(g) There are entertainment and sports activities that consume very little of our energy resources, yet they are as good for us as large energy consumers such as water skiing, motor boating, and snowmobiling.
(h) An individual can do nothing about the insulation in his or her own apartment or home.
(i) Leaving the refrigerator door open while deciding what to eat makes no difference to the amount of energy used because the refrigerator motor runs all the time anyway.
(j) The best fireplace to choose is one with a heat circulator that sends warm air to the room rather than letting all the hot air rise up the chimney.
(k) Having a car tuned up regularly makes sure that the car starts well. Tune-ups do not improve the energy efficiency of the car.
(l) A shower uses less hot water than a hot bath.
(m) Heating a full kettle of water to make one cup of a hot drink is an obvious waste of energy.

DISCUSSION

1. By using a show of hands or a chart on the chalkboard, determine which statements were considered true by all students or groups in the class.
2. Determine which statements were considered false by all students in the class.
3. Choose 8 of the 13 statements (either true or rewritten as true) that you consider to be the most important ways in which you can help conserve energy. List these 8 statements in order, from most important first to least important last.
4. List any statements about energy conservation that have not been included here but are important to you.

► REVIEW 17.4

1. Assume that a society has changed from being an energy-wasting society to an energy-conserving one. What effect do you think this will have on
   (a) the society's total energy consumption?
   (b) the need to produce energy from fossil fuels?
   (c) the predicted time before the remaining fossil fuels are used up?
   (d) the environment in the cities and the countryside?
2. For each factor listed below, decide whether it has caused an increase or a decrease in the total amount of energy consumed by automobiles in the U.S. during the past 40 years. Explain each answer.
   (a) the decrease in the average size of cars
   (b) greater streamlining
   (c) greater engine efficiency
   (d) an increase in the number of cars per family
   (e) the increase in the cost of gasoline
3. Find two examples of how industries or governments in your community are conserving energy.
4. Describe two examples that you think show that industries and governments are not doing enough to conserve energy in your area.
5. Some people think that special highway lanes in and near large cities should be reserved during rush hour for cars with two or more people.
   (a) How is this idea related to the topics you have studied in this section?
   (b) What is your opinion of these special highway lanes? Explain your answer.

CHAPTER REVIEW

KEY IDEAS

■ Energy is the ability to make things move. The things can be large and visible, such as a car, or they can be tiny and invisible, such as a molecule of oxygen.

■ Energy use can be classified as direct or indirect.

■ An energy resource is a material in nature that can be changed into useful energy. People have developed tools, machines, and processes that use energy resources.

■ A renewable energy resource, such as wood from trees, can be renewed within the average human lifetime.

■ A non-renewable energy resource, such as coal, cannot be renewed once it is used up.

■ Most of North America's energy comes from fossil fuels, mainly coal, petroleum, and natural gas. Fossil fuels were formed from plants and animals that lived millions of years ago. Our supplies of fossil fuels, especially petroleum and natural gas, are running out.

■ Using energy can create problems for our environment, such as air pollution.

■ An important goal of our society should be to reduce the amount of energy we consume.

VOCABULARY

energy
direct energy use
indirect energy use
energy resource
renewable energy resource
non-renewable energy resource
biomass fuel
hydroelectricity
fossil fuel
coal
peat
petroleum
natural gas
hydrocarbon
refining
by-product
conserving energy

V1. Create a mind map using the words in the vocabulary list and other important words, ideas, and examples from this chapter. Show how the words, ideas, and examples are related to each other.

V2. Use as many words from the vocabulary list as you can to write a paragraph about the use of energy in your own future.

CONNECTIONS

C1. (a) What is energy?
(b) How does your definition of energy apply to the gasoline in the tank of a car that is not moving?

C2. Describe an example in your school of
(a) direct energy use.
(b) indirect energy use.

C3. (a) What energy resource is used most by the people and industries where you live?
(b) Describe how your own life would be affected if supplies of this energy resource were suddenly cut off.

C4. (a) In what ways are biomass fuel and fossil fuel energy similar?
(b) In what ways do they differ?

C5. (a) In what ways are biomass fuel and energy from falling water similar?
(b) How do they differ?
(c) Could these two energy resources be used in the area where you live? Explain your answer.

C6. Classify each of the following results of burning fossil fuels as products, by-products, or problems.
(a) thermal energy and light energy
(b) soot
(c) carbon dioxide gas
(d) water vapor
(e) acid rain or snow
(f) thermal pollution
(g) the greenhouse effect

C7. How do you think each factor listed below could affect the consumption of energy in the U.S.?
(a) Our country's population is growing.
(b) Some car parts are being built with plastics to help make cars lighter.
(c) Public transportation is becoming more common.
(d) Fresh produce is trucked to supermarkets daily, even in the winter.

C8. Describe one way of conserving energy for each of the following:
(a) washing yourself
(b) using a refrigerator

(c) washing dishes
(d) washing clothes
(e) transportation
(f) home heating
(g) home air conditioning
(h) cooking food

C9. Describe ways in which your school could or should conserve more energy.

EXPLORATIONS

E1. Investigate the difference in energy use between the past and present. Interview a parent, aunt, uncle, grandparent, or any older person and find out what appliances, tools, and machines did not exist during his or her childhood. Which inventions since that person's childhood do you think have changed people's lives most dramatically? Explain your answer.

E2. Write a story or newspaper article titled "What Happened in My Area When the Power Went Off."

E3. Set up a class discussion or a debate to consider one or more of the statements given below. You may need to do some research to find more background information than you have in this chapter.

(a) The U.S. should sell its natural resources to any country that is willing to pay for them.
(b) It is wise to increase the rate at which we are removing fossil fuels from the ground in order to provide cheaper energy for the world's population.

E4. Design and make a poster advertising at least one important way that you think people can reduce their energy consumption. You can be as creative and original as you like.

REFLECTIONS

R1. Look back at the work you did for Activity 17A. How have your ideas about the problems of energy use or the solutions to the problems changed? Explain your answer.

R2. Use the knowledge you have gained in this chapter to describe your opinion of the following statement: "The United States does not need to worry about wasting energy; Canada consumes almost 10 times as much energy per day as the United States."

CHAPTER 18

ENERGY TRANSFORMATIONS AND ALTERNATIVE ENERGY RESOURCES

You learned in Chapter 17 that our main energy resources, fossil fuels, are being used up rapidly. How will we meet our energy needs in the future? One way is by finding new ways of producing energy. Wind turbines that provide electrical energy are just one example of these new ways. You will explore several other possible ways in this chapter.

Who will need new sources of useful energy? Who will be responsible for developing them? The answers to both questions are the same: you, the students of today, will be the energy consumers of tomorrow, and you will become the scientists and technologists who develop new methods of producing energy.

As you study this chapter, remember that everybody's ideas about how to solve energy problems are important. Any ideas you have can help. And ideas that may not seem possible today may help in the future.

ACTIVITY 18A / NEW SOURCES OF ENERGY

Fossil fuels are used more than any other energy resources. In this activity, you will think about other energy resources that could be used.

In your notebook, list several energy resources that you think could be used to provide energy for your area. Beside each energy resource in your list, state what the energy from that resource could be used for. (For example, energy from the wind could be used to pump water for growing plants.)

Get together with another student and compare your lists. From your lists, choose an energy resource that you would both like to see developed in your region.

Working with your partner, make a poster showing the energy resource you selected and how it could be used to do some interesting work for you. Label your drawings as much as possible. ❖

■ 18.1 FORMS OF ENERGY AND ENERGY MEASUREMENT

In Chapter 17, you learned that energy is the ability to make things move. For example, you need energy to make your muscles move to help you walk or perform other activities. Where does that energy come from? It is stored in the food you eat, and it changes into energy of motion by processes that occur in the cells in your body.

Do you recognize the forms of energy involved in the paragraph above? The energy you get from food is an example of chemical energy. The energy that your bicycle has when you are pedalling it is an example of mechanical energy. Besides these forms of energy, there are many other forms that you may remember from earlier studies. These forms of energy are shown in Figure 18.1.

FIGURE 18.1
Can you find all the examples of the ten different forms of energy in this scene?

Electrical energy is the energy of moving electrical particles. Electrical energy runs your stereo, your toaster, and the light bulbs in your lamps.

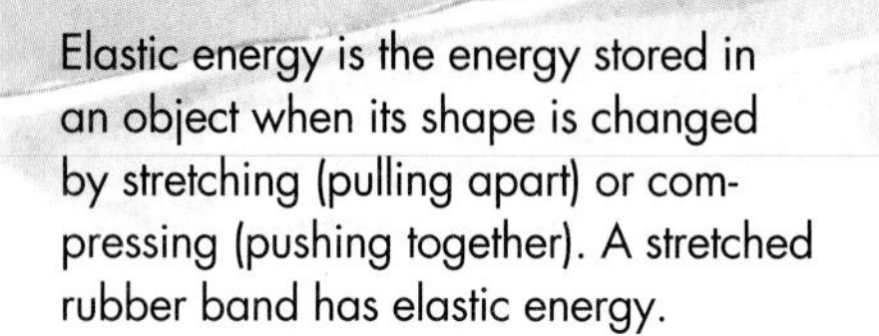

Elastic energy is the energy stored in an object when its shape is changed by stretching (pulling apart) or compressing (pushing together). A stretched rubber band has elastic energy.

Thermal energy is the energy of the particles in an object. The greater the movement of particles, the more thermal energy the object has.

Gravitational energy is the energy that an object has when it is above the surface of the Earth. Any object that can drop, roll, or slide down has gravitational energy.
Magnetic energy (magnetism) is the form of energy that causes some kinds of metal to attract (pull together) or repel (push apart) some metal objects.
Sound energy is the form of energy that you can hear. It travels as vibrations through materials and through air.
Chemical energy is the form of energy stored in matter, such as the fuel for your family's car and the food that you eat.
Mechanical energy is the energy of a moving object, such as a ball or a person in motion.
Nuclear energy is the energy stored in the central part of an atom, called the nucleus. This energy is used for producing electricity.
Light energy is the form of energy that you can see.

ACTIVITY 18B / COMMON FORMS OF ENERGY

You may remember the different forms of energy from your studies in an earlier grade. In this activity, you will match the forms of energy with their definitions, and give examples of them from Figure 18.1 and from your own experience.

Copy Table 18.1 into your notebook.

In the first column, list the following forms of energy, leaving three or four lines between each one: sound energy, chemical energy, electrical energy, mechanical energy, gravitational energy, magnetic energy, thermal energy, elastic energy, nuclear energy, light energy.

The definitions for all these forms of energy are given in Figure 18.1. Use these definitions to fill in the second column of your table.

Now look carefully at Figure 18.1 to discover examples of each form of energy. List two or three examples of each form in the third column in your table. Add one or two examples of each form of energy from your everyday life. ❖

Form of energy	Definition of this form	Examples of this form
Sound energy		

TABLE 18.1 ◀
Sample Data Table for Activity 18B

CLASSIFYING ENERGY

You can think of stored energy—like the energy in your food— and the energy of motion as classes of energy. There are two classes of energy: potential energy and kinetic energy. **Potential energy** is stored energy. **Kinetic energy** is the energy of motion. The operation of a roller coaster is an example of both classes of energy (Figure 18.2). In a roller coaster ride, the first hill is the highest. At the top of this hill, the coaster has

FIGURE 18.2 ▶
A roller coaster is an example of both potential energy and kinetic energy. At the top of a hill, the coaster has the greatest potential energy. As the coaster moves down, it gains kinetic energy. The next time you ride a roller coaster, check for signs of an engine being used. Is the engine working throughout the ride, or only when the roller coaster is going up the first hill?

its greatest amount of potential energy (caused by gravity). As the coaster rolls down the slope, it gains kinetic energy. Think about the different forms of energy and how they can be classified as potential or kinetic energy.

EXAMPLES OF POTENTIAL ENERGY AND KINETIC ENERGY

Several different forms of energy can be classified as potential energy. For example, the chemical energy in food and fossil fuels is potential energy. The elastic energy stored in a stretched guitar string is potential energy. The gravitational energy in raised objects such as the water at the top of a waterfall is potential energy.

There are also some forms of energy that can be classified as kinetic energy. All forms of energy that involve motion can be classified as having kinetic energy. For example, sound energy causes the back-and-forth motion of the molecules of the medium (gas, liquid, or solid). These molecules cause your eardrum to vibrate back and forth, allowing you to hear. Mechanical energy is the energy of a moving object, such as a ball or a person in motion.

FIGURE 18.3
It would take a little more than one hour of cycling to use up the energy from one slice of pizza. Can you find this out from the data in Tables 18.2 and 18.3?

MEASURING ENERGY

You have learned that energy is measured in units called joules, represented by the symbol J. One joule is a very small amount of energy. The average person needs about 15,000 J of energy to walk quickly for one minute.

If the amount of energy were given only in joules, very large numbers would often be needed. Therefore, we sometimes give amounts of energy in larger units, such as kilojoules (kJ) and megajoules (MJ). For these units, 1 kJ = 1,000 J and 1 MJ = 1,000,000 J. Using these units, you can write the amount of energy you would need to walk quickly for one minute as 15 kJ rather than 15,000 J. Table 18.2 shows the amounts of energy a 50 kg student would use to do some familiar activities (Figure 18.3). Table 18.3 shows the amount of energy stored in some familiar foods.

Energy in kilojoules (kJ) needed for 50 kg person to perform activity for 1 min (values are approximate)	
Rowing in race	50
Swimming (fast)	30
Walking (fast)	15
Cycling (average speed)	15
Washing dishes	13
Writing	5

TABLE 18.2 ***Typical Energy Amounts for Activities***

Nutrition Facts
Serving Size 1 Box (21g)

Amount/serving	
Calories	80
Fat Calories	0
	% DV*
Total Fat 0g	**0%**
Saturated Fat 0g	**0%**
Cholesterol 0mg	**0%**
Sodium 170mg	**7%**
Total Carb. 15g	**5%**
Fiber 1g	**4%**
Sugars 2g	
Protein 4g	
Vitamin A 10% • Vitamin C	15%
Calcium 0% • Iron	30%
Vitamin D 8% • Thiamin	20%
Riboflavin 20% • Niacin	20%
Vitamin B_6 20% • Folate	15%
Phosphorus 4% • Magnesium	4%
Zinc 15% • Copper	4%

* Percent Daily Values (DV) are based on a 2,000 calorie diet.

FIGURE 18.4
Where would you find information about the amount of energy that a serving of breakfast cereal contains?

Energy (kJ) provided by average serving of food (values are approximate)	
Pizza (slice)	1000
2% milk (glass)	540
Chocolate fudge	470
Chicken drumstick	370
One apple	290
Butter (5 mL)	150

TABLE 18.3 ***Typical Energy Amounts in Several Foods***

You can find out about the amount of energy in some of the foods you eat by reading the labels on containers (Figure 18.4). You might discover, for example, the difference between the amount of energy in a breakfast cereal that contains sugar and one that does not.

ACTIVITY 18C / FOOD ENERGY AWARENESS

In this activity, you will learn the energy content of some of the foods you eat.

MATERIALS

food products in unopened containers
poster paper

PROCEDURE

In a group, design and create a colorful, attractive poster that illustrates the amount of energy in various foods that you eat. In some cases, such as breakfast cereals, you may be able to cut out the information and paste it onto your poster. In other cases, such as drinks in metal cans, you may have to copy the information. On your poster, also indicate what else you discover about the amount of energy in food products.

DISCUSSION

1. What is the meaning of "a single serving"?
2. Which types of products do not indicate the amount of energy they contain?
3. Why do you think some products do not list the amount of energy they contain?
4. Do you think the way the information on energy is given is useful to consumers? Explain your answer. ❖

▶ REVIEW 18.1

1. What form (or forms) of energy can you find in each of the following examples?
   (a) After walking across a rug in the winter, Hiroshi touches a metal doorknob and receives a shock.
   (b) Propane gas is often used in taxi cabs.
   (c) You use a magnet to help find a sewing needle that has fallen onto a rug.
   (d) Janine hears a bird chirping.
   (e) Narmutha's metal soup spoon becomes hot soon after being placed in hot soup.
   (f) Maria hits a baseball with a bat.
   (g) Anders fully winds the spring of a wind-up toy.
   (h) Rosa raises an axe above a block of wood.
2. Change each of the energy quantities in the following descriptions to an easier unit to use, such as kilojoule or megajoule.
   (a) One gram of coal contains about 28,000 J of chemical potential energy.
   (b) One kilogram of propane gas contains about 43,000,000 J of chemical energy.
   (c) Each Stone Age human used about 22,000,000 J of energy per day obtained from food and wood fires. Today, each American uses about 1,000,000,000 J of energy per day, mostly from fossil fuels.
3. Based on the information in Table 18.3, how much energy is in three average slices of pizza? Based on the information in Table 18.2, how many minutes would this energy allow a 50 kg person to
   (a) walk quickly?
   (b) write?
4. Choose three activities that you like to do. Then use Table 18.2 to estimate how many kilojoules of energy you would use in performing these activities. (Be sure to think about how many minutes you would perform each activity as well as your mass in kilograms.)

■ 18.2 ENERGY TRANSFORMATIONS

Think about all the different forms of energy that you use every day. You can use some forms of energy directly. Visible light, for instance, is a form of energy that you can see. But you cannot use other forms of energy directly for many of your needs. For example, it would be difficult to cook a meal directly using gravitational energy. If we want to have energy in a useful form, we often must change, or transform, the energy into another form. An **energy transformation** is a change in energy from one form to another form.

There are many examples of energy transformations. As you read about energy transformations, keep in mind that everything you do and see in the world around you is possible because of energy transformations constantly taking place.

BOUNCING A BASKETBALL

What energy transformations do you think occur when you bounce a basketball? As the ball is moving downward toward the floor, it has mechanical energy. As soon as the ball hits the floor, the part of the

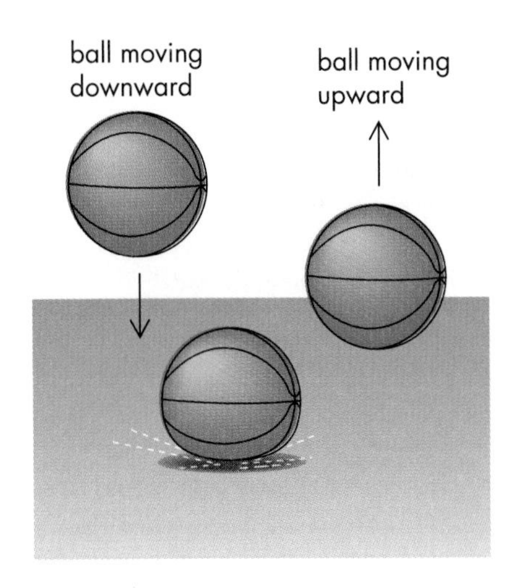

FIGURE 18.5
When a basketball is moving toward the floor, it has mechanical energy. When it hits the floor, the ball gains elastic energy. After bouncing from the floor, the ball has mechanical energy. Can you think of other transformations that might be occurring?

ball that touches the floor gets pushed inward a small amount. This gives the ball elastic energy. Very soon, this elastic energy changes back to mechanical energy as the ball moves upward (Figure 18.5). This set of energy transformations can be written using the following word equation:

mechanical energy → elastic energy → mechanical energy

PORTABLE SOUND SYSTEMS

What are some of the energy transformations that occur when you use a portable cassette or compact disc player (Figure 18.6)? The player has a battery that contains chemical energy. When you turn on the player, this chemical energy changes to electrical energy, which operates the cassette or compact disc. The operation of the cassette or compact disc causes the speakers to vibrate; vibrations are a form of mechanical energy. The vibrations cause sound energy that you can hear. There are three energy transformations in this example, as shown in the following word equation:

chemical energy → electrical energy → mechanical energy → sound energy

FIGURE 18.6 ▶
In changing from chemical energy to the sound energy that comes from a portable player, three main energy transformations occur. What other transformations occur when the sound reaches your ear?

APPLIANCES IN THE HOME

How many electric appliances do you have in your home? How many non-electric appliances do you have? As you read about the energy transformations in electric appliances, think of how you would write word equations to summarize them.

Many appliances transform electrical energy into thermal energy. Examples are a clothes dryer, a toaster, and a curling iron. What other examples can you think of?

Several appliances transform electrical energy into mechanical energy. An electric can opener, a blender, and a fan are examples of such appliances. What are others?

Some non-electric devices are shown in Figure 18.7. In all these devices, the resulting form of energy is mechanical energy. What is the source of energy in each case?

FIGURE 18.7
Not all appliances used in the home are electric. What energy transformations occur when you use each of these devices?

ACTIVITY 18D / DEMONSTRATING ENERGY TRANSFORMATIONS

In this activity, you will set up a demonstration of an energy transformation and observe others set up by members of your class.

MATERIALS

safety goggles
apron
rubber gloves (if using chemical batteries)

(see note on materials on page 394)

■ **Use only materials and tools that are safe and approved by your teacher.**

■ **Handle chemical batteries very carefully. The acid used in them is very corrosive.**

PROCEDURE

1. Put on your safety goggles and apron. Also put on your rubber gloves if you will be handling chemical batteries.

2. Think of at least one way to demonstrate in class how each of the sources of energy listed below can make something move. The motion can be bouncing up and down, spinning around, travelling in a straight line, or some other motion. Try to think of interesting demonstrations. One example of using flowing water to raise an object is shown in Figure 18.8.

Sources of energy:

(a) Water in a raised container
(b) Water (from a tap) flowing through a hose
(c) Moving air (wind) from an electric fan
(d) Heated air moving upward through cooler air
(e) Electrical energy from a rechargeable battery
(f) Solar cells producing electricity
(g) Hot water moving upward through cooler water
(h) Steam from boiling water
(i) Waves moving up and down on the surface of water 

MATERIALS

The materials you use will depend on which demonstration(s) you choose. Some possible apparatus and materials are:

chemical batteries
toys
solar cells
light bulbs or floodlights
sources of heat, such as a hot plate or a heat lamp
water in a container and running water
containers to hold water
hoses or tubes
fans (electric and non-electric)
rubber bands
springs
metal masses
mirrors
wood, cardboard, paper, and other materials
tools
electrical outlets

FIGURE 18.8 ▶
Flowing water causes a fan and an attached shaft to spin. As the shaft spins, the string wraps around it and pulls up the attached toy. How many energy transformations do you observe in this situation?

3. Choose your most interesting and suitable demonstration idea. Discuss with your teacher whether it is safe, original, and easy to set up. If it is approved, write a report about your demonstration using the following titles:
   Purpose
   Materials
   Safety Precautions
   Procedure (Include diagrams if they will help.)
4. In your notebook, set up a table to summarize your observations of the energy transformations demonstrated by you, your classmates, and your teacher. The table should have three columns with these titles:
   Title of Demonstration
   Word Equation
   Description of Energy Transformation
   Describe the transformations in detail in the final column.

DISCUSSION

1. From the demonstrations you observed, describe two examples of energy transformations that involve only one energy change.
2. Describe two examples of energy transformations that involve two or more changes in energy.
3. Which of the energy transformation demonstrations that you observed could be most useful in the future to help reduce our use of fossil fuels? Explain your answer. ❖

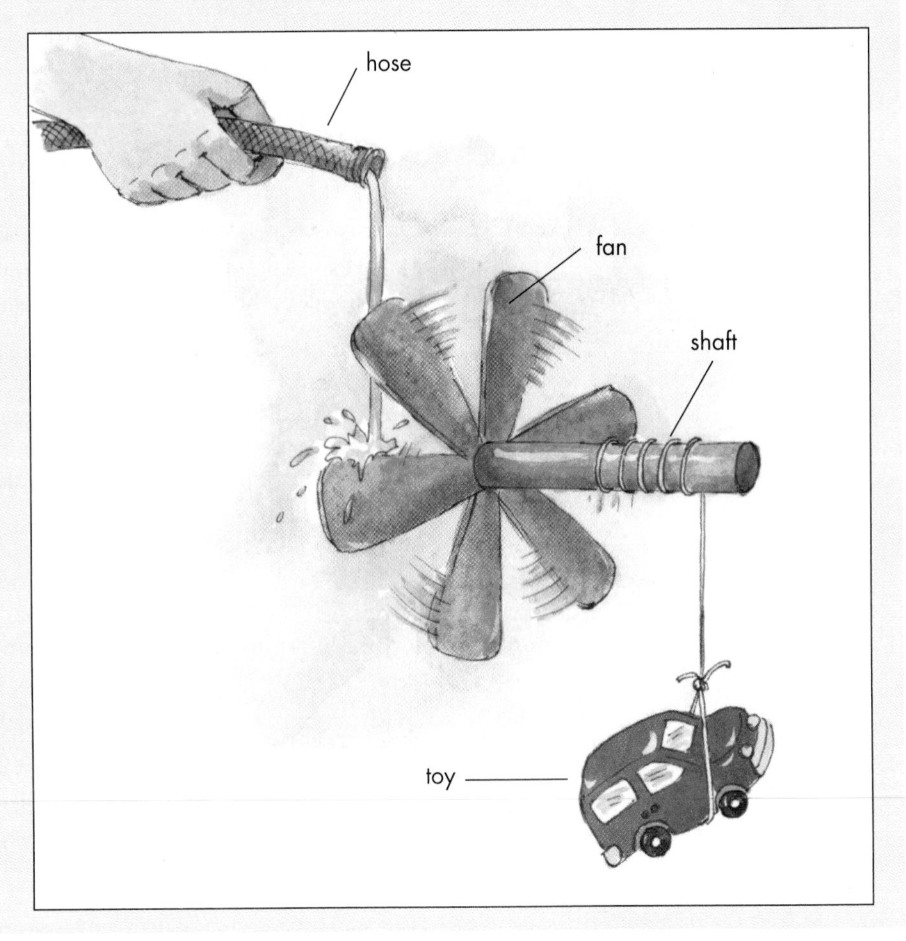

PHYSICAL THERAPY TECHNOLOGIES

It was a good game of beach volleyball, until you lunged for the ball and landed on your foot the wrong way. Now your ankle is swollen and painful. The doctor says you have sprained it, and he has sent you to a physical therapist. You wonder what comes next.

Physical therapy is a way of treating physical injuries without using drugs. A physical therapist uses physical methods of treatment. These include exercising the injured area, soaking it in a whirlpool bath, and applying heat or cold packs.

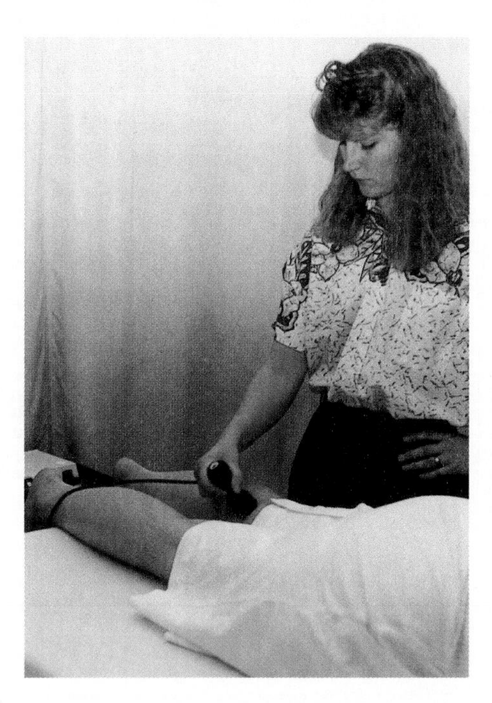

The physical therapist uses a laser to help heal the soft tissues of an injured leg.

Physical therapists may also use some complex equipment. For your ankle, the physical therapist might use an interferential machine. It sends an electrical current through your ankle, which helps reduce the swelling. You feel only a vibration.

Or, your physical therapist might use a laser machine. A laser is a beam of light that is all one color. Laser light can be focused to one small point. The physical therapist directs the laser to the appropriate spot on your ankle for two or three seconds to treat the swelling.

In a few days, as the swelling goes down, the physical therapist may use yet another piece of equipment. The ultrasound machine sends out high-frequency sound waves. You can't hear them, and you can't see anything, but these sound waves help the soft tissues in your ankle to heal.

The way these different forms of energy (electrical, light, and sound) can help heal injured areas seems amazing. Scientists believe that these forms of energy work on the walls of the cells in your body, allowing nutrients from the blood to move in and out of cells in the injured area. The new blood brings both nutrients and more oxygen to help the healing process and reduce swelling.

These new treatments do not replace more traditional methods of physical therapy. For instance, soon you'll be starting on some exercises to get your ankle working properly again.

Not all physical therapists use these new technologies. More research needs to be done to determine exactly how the technologies work and when they should be used. However, the new methods can speed up treatment time, making physical therapy much more efficient.

Heat is a traditional physical therapy tool. Think about how your body reacts to heat. What happens to your body when you sit in a hot bath? Now imagine getting into a very cold bath. How does your body feel?

▶ REVIEW 18.2

1. (a) Define the term "energy transformation."
   (b) Describe an example of an energy transformation from the room where you are now.
2. "Energy transformations are basic to life as we know it." Do you think this statement is true or false? Why?
3. Write a word equation for each energy transformation described below. In some cases, more than two forms of energy are involved.
   (a) Niko uses a rechargeable battery to set off a camera's flashbulb.
   (b) Kay uses gasoline to operate a motorcycle.
   (c) Light strikes your solar-powered watch.
   (d) Peter uses an electric frying pan to cook some scrambled eggs.
   (e) An alarm bell rings on your wind-up alarm clock.
   (f) Your cousin cooks a roast in an oven.
   (g) Erica focuses sunlight through a magnifying glass to start a campfire.
   (h) An electric motor operates a sewing machine.
   (i) Annika reaches a high speed before jumping over a pole-vault bar.
   (j) Your toy steam engine moves along a track.
4. At a restaurant, you order a hamburger and salad. Starting with energy from the sun, which of these two foods do you think requires more energy transformations in order to be served to you? Explain your answer.
5. Exercise bikes are installed in a school's exercise room. The room has no fans or air conditioning, and the students find that the room is too hot to exercise for long. The students want to find a way of transforming their "muscle power" into a method to help keep them cool. Suggest how they could do this.
6. (a) Name three electric appliances in your home other than the ones mentioned in this section.
   (b) For each appliance that you named in (a), write a word equation for the energy transformation.

18.3 THE LAW OF CONSERVATION OF ENERGY

Figure 18.9 shows the energy transformations that occur when a pitcher throws a ball. In Figure 18.9a, the pitcher's muscles have transferred 200 J of mechanical kinetic energy to the ball. What happens to this mechanical energy when the catcher catches the ball? Most of this energy is transferred to the catcher's glove and pushes the glove backwards. Some of the energy, however, is transformed into sound and thermal energy. Figure 18.9b shows that the total energy before and after the ball is caught is the same.

Think about the energy transformations that you have studied earlier in this chapter. Do you think the amounts of energy present before and after these transformations would match exactly the way they do in Figure 18.9b?

Scientists have investigated energy transformations in thousands of experiments. They made careful measurements to account for all the forms of energy involved. They performed both simple experiments

mechanical kinetic energy = 200 J
pitcher
catcher
(a)
160 J of mechanical kinetic energy are transferred to the catcher's glove and hand.
10 J of mechanical kinetic energy are transformed into sound energy.
30 J of mechanical kinetic energy are transformed into thermal energy that warms up the glove.
(b)

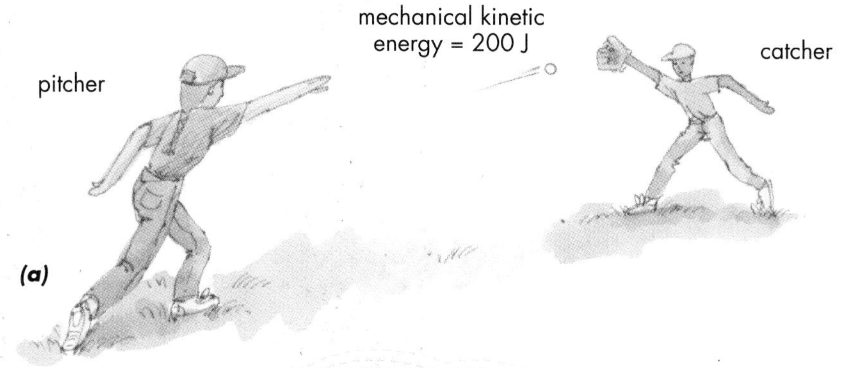

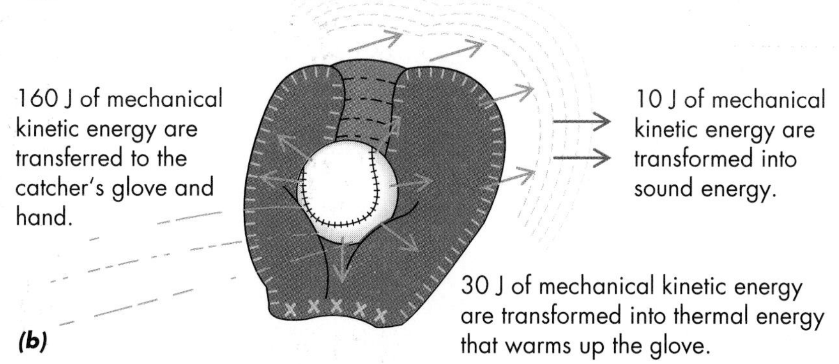

FIGURE 18.9 ◀ ***(a) When the catcher catches the ball, it stops moving. What happens to the ball's mechanical energy? (b) The total of 200 J of energy can be accounted for when all the forms of energy involved are considered.***

and complicated ones. In all cases, the experimenters found that the total amount of energy before a transformation is always exactly equal to the amount of energy after the transformation.

When many experiments lead to the same conclusion with no opposing evidence, scientists summarize their findings in a statement known as a law. The statement that sums up these findings about energy is called the **law of conservation of energy**:

Energy can neither be created nor destroyed; it can only change form.

The law of conservation of energy is one of the most important ideas in science. In this case, "conservation of energy" means that the total amount of energy remains the same. (Sometimes, the expression "energy conservation" may be used to mean not wasting energy.)

ACTIVITY 18E / ENERGY TRANSFORMATIONS OF A BALL

When you drop a ball on the floor, several energy transformations take place. In this activity, you will try to identify all of them.

PROCEDURE

1. Put on your safety goggles.
2. Allow the ball to fall from a height of 100 cm several times, and observe its behavior.
3. Use your observations in Step 2 to design a method to measure the highest point that the ball reaches on the first bounce, the second bounce, ➡

MATERIALS

small "super ball"
meter stick
graph paper

■ Determine the percentage of its gravitational potential energy that the ball regains after each bounce by using the following calculation:

$$\frac{\text{height after bounce}}{\text{height before bounce}} \times 100\%$$

What percentage of its potential energy does the ball seem to "lose" each time it bounces?

and the third bounce. Repeat the measurements at least three times to obtain the best possible results. Record these height measurements.

4. Prepare a sheet of graph paper as shown in Figure 18.10. Make a bar graph using the data collected in Step 3.

DISCUSSION

1. (a) Name the class of energy and the form of energy that the ball has just before it is dropped.
   (b) What happens to this energy as the ball falls?
   (c) What happens to this energy when the ball hits the floor?
   (d) What happens to the remaining energy after the ball bounces up from the floor?
2. Why might the ball's behavior make some people think that the law of conservation of energy is not always true?
3. How would you explain that energy really is conserved when a ball bounces? ❖

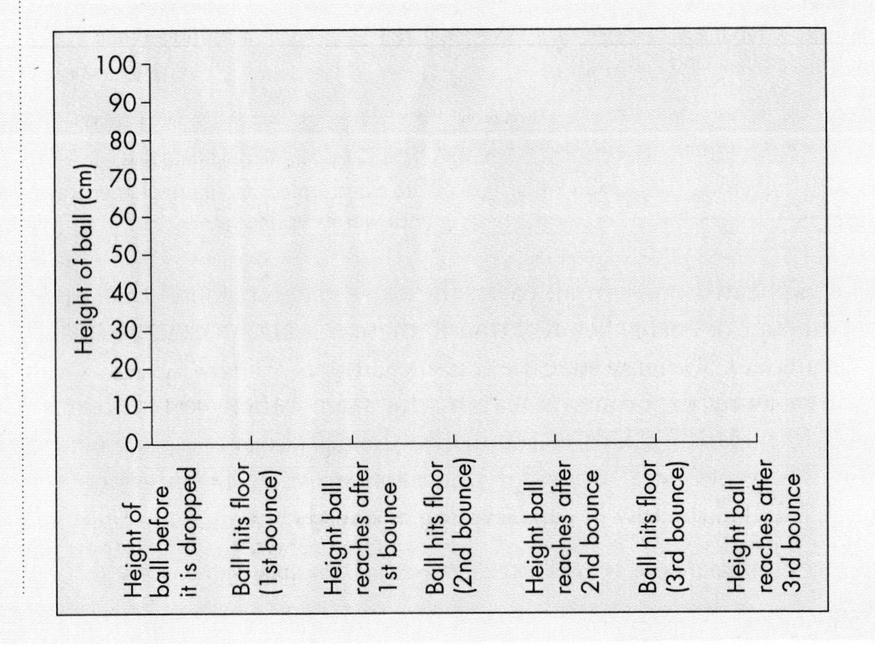

FIGURE 18.10 ▶ ***Copy this outline on graph paper. Plot the height of the ball as marked on the vertical axis at each stage shown on the horizontal axis.***

■ Cooking an egg in its shell in a microwave oven can be dangerous. As the egg becomes hotter, the moisture in the egg turns to steam and the pressure inside the shell builds up. The egg can explode even *after* it is removed from the oven. People have suffered serious eye damage caused by such explosions. Anyone trying to cook an egg this way should remember to first poke a hole in the shell.

A CLOSER LOOK AT THE CONSERVATION OF ENERGY

Two important parts of the law of conservation of energy affect all energy transformations. First, energy cannot be created. This means that you cannot get energy from a device unless you put energy into it. This is true for a ball that falls toward the floor. Before the ball can gain mechanical kinetic energy, you first must raise it above the floor to give it gravitational potential energy. The situation is similar for an electric blender. The blades of the blender rotate only as long as the blender has a steady supply of electrical energy (Figure 18.11).

Sometimes, energy can be stored for later use. For example, energy is stored in the foods we eat and in the fossil fuels we burn. In both cases, the energy stored is chemical energy.

The second important part of the law of conservation of energy is that energy cannot be destroyed. In other words, energy does not "disappear"; it simply changes form. An interesting example of this is found in microwave cooking. Experienced cooks know that the food will continue to cook even after the microwave oven is no longer operating. How does this occur?

The microwaves produced in the oven cause the water molecules in the food to vibrate easily. The vibrating water molecules cook the surrounding food molecules. Even after the microwaves have stopped striking the water molecules, those molecules continue to vibrate for a while. Thus, they continue to help cook the food.

In many energy transformations, some of the energy does not end up being useful. Such energy is called "wasted energy." This wasted energy does not disappear; it simply changes to some other form of energy, such as sound or thermal energy. The wasted energy is still present, but it is not in a form that is useful. This is a big problem with using fossil fuels as sources of energy. Much of the chemical energy in these fuels is wasted as thermal energy when the fuels are burned.

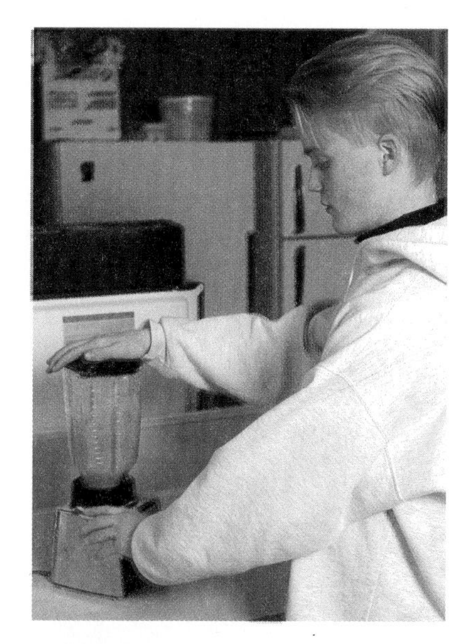

FIGURE 18.11
The mechanical energy of an electric blender doesn't exist all the time. What must you do to put that mechanical energy into the blender?

▶ REVIEW 18.3

1. (a) State the law of conservation of energy.
   (b) Explain how this law applies to the energy transformations that occur in an incandescent light bulb.
2. Describe a situation in your daily life when energy seems to be "lost" or to have "disappeared." Explain what has really happened to the "lost" energy.
3. An inventor claims to have invented a device that transforms all the energy supplied into useful work, without any energy being lost. Is this possible? Explain how you know.
4. Describe two examples of energy transformations in which some of the "wasted" energy becomes sound energy.
5. Describe two examples of energy transformations in which much of the "wasted " energy becomes thermal energy.

■ 18.4 ALTERNATIVE ENERGY RESOURCES

Much of the energy we use comes from fossil fuels. These fuels help generate electricity, operate vehicles, and heat buildings. However, using fossil fuels has important disadvantages. They are non-renewable, they harm the environment, and they will not last forever.

To reduce these problems, our society is learning how to use energy more efficiently. For example, our homes are better insulated and our cars use less gasoline than ever before. Even so, we will always need to use energy, so new energy resources must be found. Since these new energy resources give us an alternative to fossil fuels, they are called **alternative energy resources**.

What features would the best alternative energy resource have? It would be renewable, plentiful, easy to use, clean, harmless to the environment, and available all the time. It could be used for producing electrical energy, and for operating cars and other vehicles. And it would be inexpensive.

Can any one resource have all these features? Perhaps not. Every type of energy resource has disadvantages. To choose alternatives wisely, we must consider which ones have the most benefits and the fewest problems.

Some alternative resources, such as running water and biomass fuel, were described earlier in this unit. After looking at these resources, and as you look at other alternative resources, think about their advantages and disadvantages for the entire world and for the region where you live.

ACTIVITY 18F / COMPARING ALTERNATIVE ENERGY RESOURCES

In this activity, you will use information from Chapter 17 and from the following subsections to compare alternative energy resources. Try to determine which ones are likely to be best for the future of both the world and your state.

PROCEDURE

1. Before you begin reading, copy Table 18.4 into your notebook.
2. Based on the information in Section 17.2, and on your knowledge of the area where you live, give each energy resource a plus or a minus sign by considering each of the questions from (a) to (i). Add one or two of your own questions as (j) and (k).
   (a) Is it renewable?
   (b) Is it plentiful?
   (c) Is it available all the time?
   (d) Is it non-polluting?
   (e) Is it easy to use?
   (f) Is the technology that is needed to use this resource available?
   (g) Is it safe for the environment?
   (h) Can it be used to operate automobiles?
   (i) Is it inexpensive to use?
3. As you read this section, fill in the plus or minus signs for questions you were not able to answer the first time. You may want to change some of your answers as a result of your reading.

Non-fossil fuel energy resource	Questions (+)(−)											Total
	a	b	c	d	e	f	g	h	i	j	k	
Hydroelectricity												
Biomass fuels												
Wind												
Solar energy												
Tides												
Geothermal energy												

TABLE 18.4 ***Summary Table for Activity 18F***

4. After you have finished reading and answering the questions in Step 2, add up the total (+) and (–) scores for each resource, and place the totals in the last column of the table. Then answer the Discussion questions.

DISCUSSION

1. Which of the questions asked in Step 2 do you think is (or are) most important? Why?
2. Which question do you think is least important? Why?
3. Do you think this rating system is detailed enough? Explain your answer.
4. List the resources in order of most desirable to least desirable. Is one resource obviously better than the others? Explain.
5. Do you think that only the resources with the most plus signs in this activity should be researched? Explain.
6. Which alternative resources do you think will be most important in the future of
   (a) the world? (b) your state?
   Explain your answers.
7. Which resource might provide much of the energy needed for transportation in the future? Explain. ❖

ENERGY FROM WATER

In Chapter 17, you read about hydroelectric generating stations where huge dams along rivers hold back water. The water behind the dam has gravitational potential energy. As the water falls, it gains mechanical kinetic energy and causes turbines to spin. A **turbine** is a wheel-shaped device with blades that help it rotate. The turbines are connected to generators that produce electrical energy (Figure 18.12).

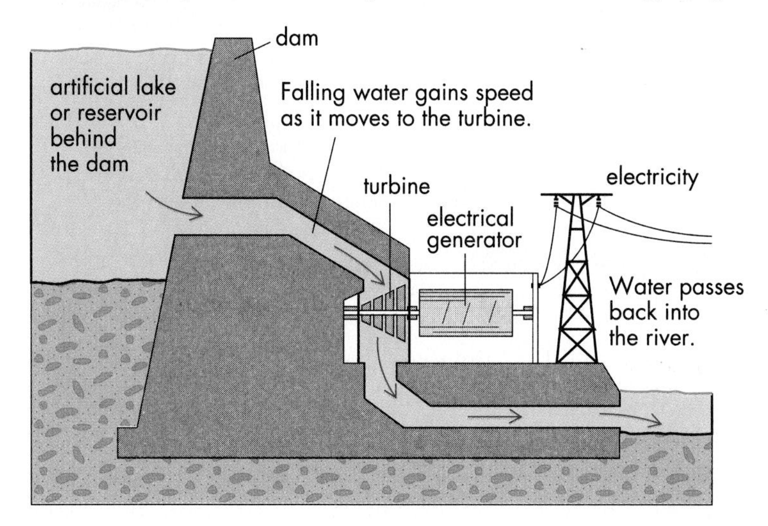

FIGURE 18.12 ◀
At a hydroelectric generating station, the gravitational potential energy of the water is transformed into mechanical kinetic energy as the water falls. The moving water forces the blades of the turbine to spin, which also results in mechanical kinetic energy. The turbine is connected to a generator that produces electrical energy. Write the word equation that describes these energy transformations.

Using falling water to produce electricity is the least expensive way of generating large amounts of electrical energy. It causes little or no pollution. It is renewable (or it should be, unless the river dries up or changes its course). Using dams to store water makes this resource available all the time, even during the non-rainy season.

EXTENSION

■ Some energy experts predict that by the year 2010 much of the gasoline used in automobiles will not come from petroleum because that resource is being used up so fast. Rather, they say, it will come from "synthetic oil" produced by making coal into a liquid. This process produces liquid fuels from the liquefied coal. Find out more about synthetic oil, and describe its advantages and disadvantages.

However, hydroelectricity is not ideal. Often the generating stations are built far from where most consumers live. Dams cause large areas of land to be flooded. This flooding harms plant and animal life, and it forces people to leave behind their homes and property. Also, hydroelectricity is not available everywhere because fast-flowing rivers are needed. In fact, most possible sites for hydroelectric dams near populated areas in the U.S. have already been developed.

ENERGY FROM BIOMASS

In Section 17.3, you learned about biomass fuels that come from plants and animal wastes. But these fuels are not always easy to use because they must be chopped up for use, and they require a lot of room for storage. However, energy can also be produced from biomass fuels that are easier to use.

One method of producing these biomass fuels involves strongly heating biomass, such as garbage, without oxygen. This method forms methane gas, which is a fuel that can be burned (Figure 18.13).

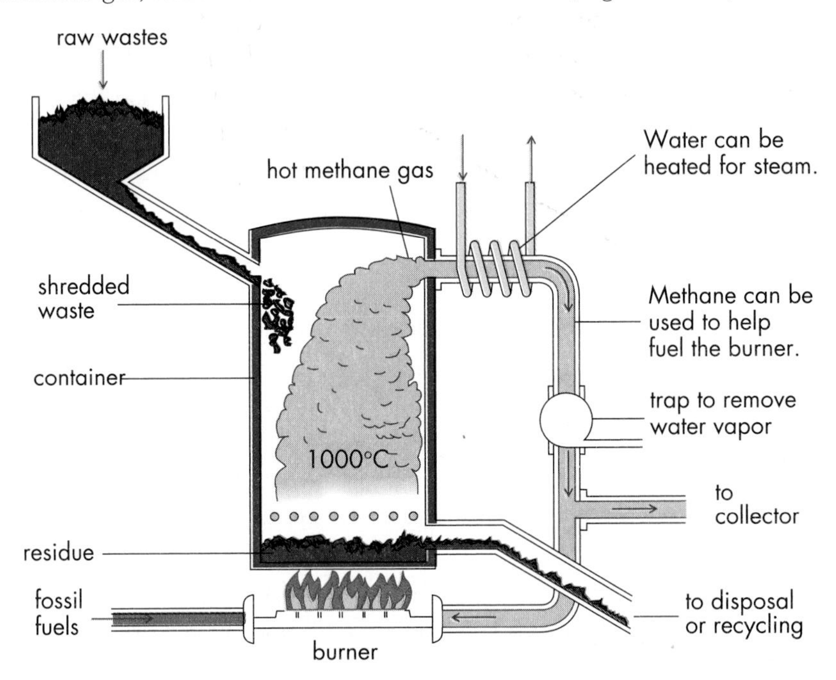

FIGURE 18.13 ▶ ***Producing biomass fuels from garbage. Some of the methane gas produced can be used as fuel for the burner, and the rest is collected for heating, cooking, and producing electricity. What could the steam be used for? What could be recycled from the residue (the solids that remain after burning)?***

Another chemical method involves the use of microscopic organisms, such as yeast or bacteria, to act on biomass in a place where there is little or no oxygen. This method produces methane gas from garbage and animal dung (Figure 18.14). The method can also be used to produce ethanol from corn or sugar, and methanol from wood. Ethanol and methanol are two types of alcohol that can be used as fuel. Ethanol or methanol can be mixed with gasoline to produce gasohol. This mixture can be used in car engines. Leftovers from this method can be used as crop fertilizer.

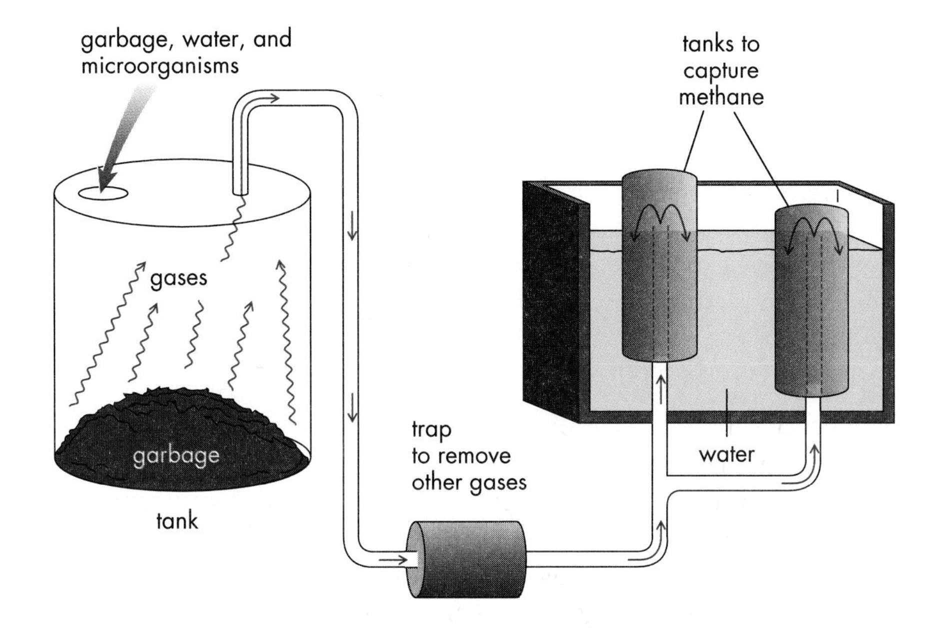

FIGURE 18.14
Using microscopic organisms to produce methane gas from garbage

Biomass fuels such as methanol are convenient forms of biomass. They can be stored and transported more easily than wood, garbage, or other forms of biomass. However, all biomass fuels have disadvantages. Like fossil fuels, biomass fuels release large amounts of polluting gases, carbon dioxide, and waste thermal energy into the air when they are burned. In addition, many biomass fuels are made from plants grown on agricultural land that could be used to produce food.

ENERGY FROM THE WIND

Wind energy is energy from moving air. Without the sun, this energy resource would not exist. Wind occurs when energy from the sun heats the air on Earth. The ocean and land surfaces heat up at different rates. These differences cause the air to move, creating wind.

For thousands of years, people have used energy from the wind to move their sailboats. On land, they used the wind to help them pump water and grind grain into flour. But wind energy can also be used to spin turbines that are connected to electric generators. Wind energy is renewable, and it does not pollute the environment. In windy areas, it is plentiful.

But there are some problems with using wind energy. These can be overcome with modern technology. For example, to obtain the most energy from the wind, the turbine must face toward the wind. But the wind does not always come from the same direction. In the modern wind farm shown on the first page of this chapter, computers help control the direction of the turbine blades. Researchers have designed a different type of turbine, called a vertical-axis turbine, that operates whatever the wind's direction (Figure 18.15).

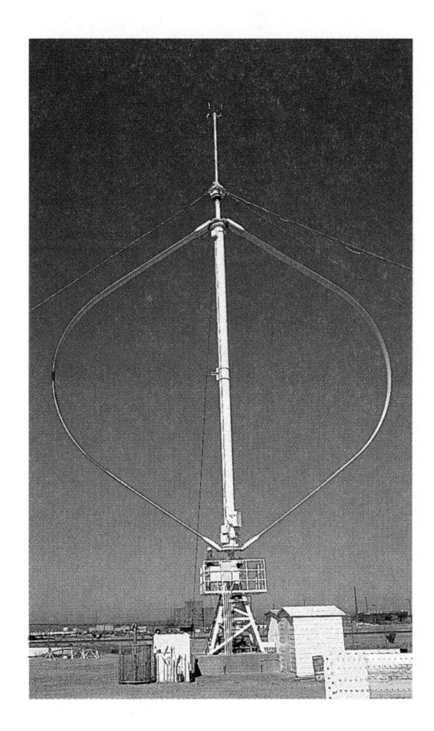

FIGURE 18.15
This vertical-axis wind turbine spins, no matter what direction the wind is coming from. Into what form of energy is the wind power transformed?

DID YOU KNOW

■ Some fishing boats are saving fuel by using sails that assist their engines whenever the wind is suitable.

Another problem with wind energy is that the strength of the wind continually changes. Energy from the wind would be easier to use if the energy produced could be stored for later use. One method being investigated involves storing electrical energy in special batteries.

Using wind energy is not cheap. Researching and developing methods of transforming this energy into electrical energy makes this resource very expensive. As modern wind-driven generators become more common, these costs will probably decrease.

ENERGY FROM THE SUN

Energy from the sun is called **solar energy**. This resource is both renewable and plentiful. In fact, in only about 12 minutes, enough solar energy reaches the Earth's surface to supply the entire world's energy needs for a year! However, only a tiny fraction of this energy is being used directly by humans.

Solar energy is not available all the time. Clouds can block out some of it. Shorter days in winter reduce the amount we receive. It is not available at night. Even so, solar energy can be used directly for heating homes. Many homes and other buildings have been designed with large south-facing windows, and thick walls and floors made of concrete (Figure 18.16). As sunlight enters the windows, it warms the inside air to a comfortable temperature. At the same time, the concrete absorbs heat, storing it as thermal energy. During the night, as outdoor temperatures fall, thermal energy in the walls and floors is slowly released and keeps the indoor air warm.

Solar energy can be used to heat water for washing clothes, washing dishes, and bathing, by using a device called a solar collector (Figure 18.17). Water heated this way can also be stored in a tank to warm indoor air by releasing heat.

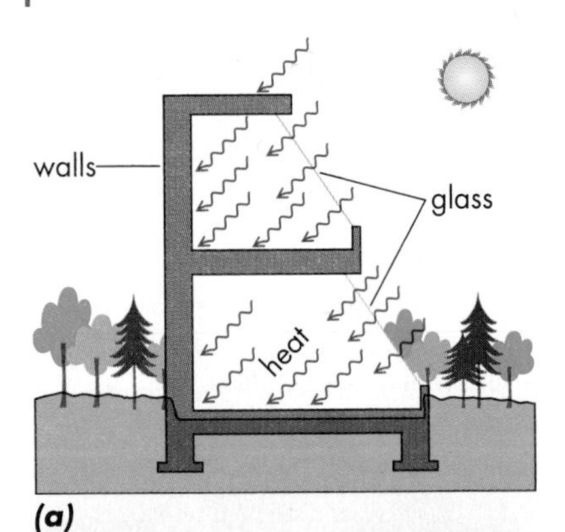

(a)

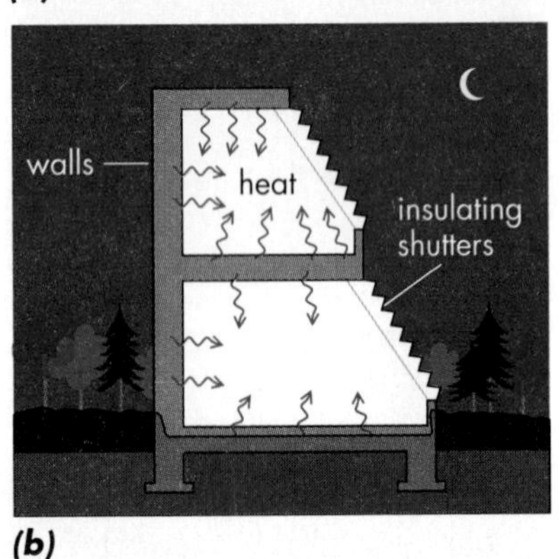

(b)

FIGURE 18.16
Cross-section through a passive solar-heated house, showing the walls and the large, south-facing windows, (a) during the day, (b) at night. What could the owners do during a hot night? During a dull, cold day?

FIGURE 18.17 ▶
This solar collection system provides heat and hot water. Hot water is stored in insulated containers for the winter and is occasionally reheated by oil-fired burners.

A problem is that solar energy cannot bring water to a high enough temperature to produce steam to operate an electric generator. This problem can be overcome by technology. For example, mirrors can be used to focus solar energy. The solar furnace shown in Figure 18.18 creates temperatures high enough to produce steam, which can be used to turn a turbine connected to an electric generator. So far, very few solar furnaces are being used. This is partly because they require a large area in a sunny location, but it is mainly because they are very expensive to build.

Solar energy can also be transformed to electrical energy directly through the use of a solar cell (Figure 18.19). Wires carry the electricity from the cell to another device to do work.

Since solar energy is so plentiful, energy experts hope to make better use of solar cells in the future.

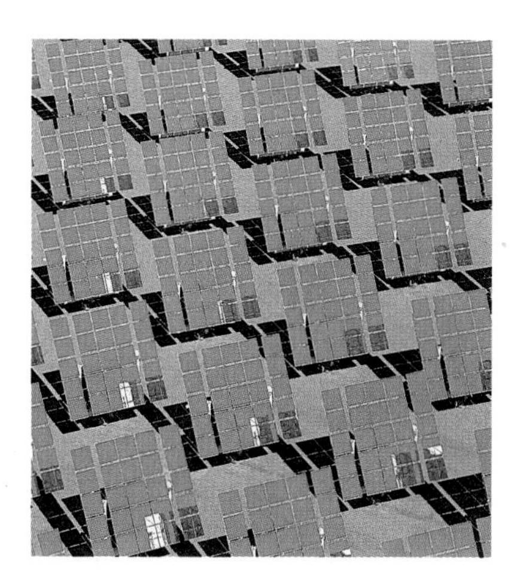

FIGURE 18.18
At this solar-powered generating station in California an array of flat mirrors tracks the sun and reflects its light on to a central tower. In the tower, the gathered heat powers a steam turbine. Where might this type of generating plant be possible in your area?

(a)

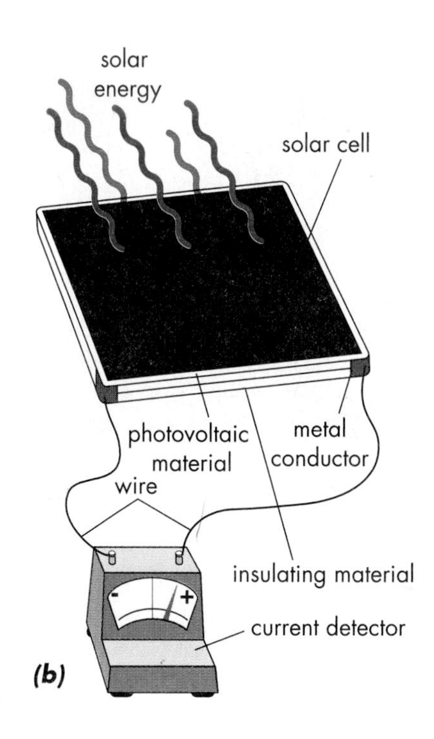

(b)

FIGURE 18.19
A solar cell depends on the properties of special materials such as selenium. When solar energy strikes a solar cell, electricity is produced. (a) Space satellites use solar cells to operate electrical devices. (b) A solar cell. Why are solar cells more effective in space?

ENERGY FROM THE OCEAN

Approximately twice a day, huge amounts of ocean water surge up over the rocks of Hornby Island, then rush out to sea again (Figure 18.20). Tides like these occur along shorelines all around the world. **Tidal energy** results from the gravitational pull of the moon and the sun on the oceans. It is an example of an energy resource that does not rely on solar energy.

FIGURE 18.20 ▶
The ocean's tides have enough energy to erode and smooth this sandstone on Hornby Island. Could people who live near this beach harness the tidal power for their own use?

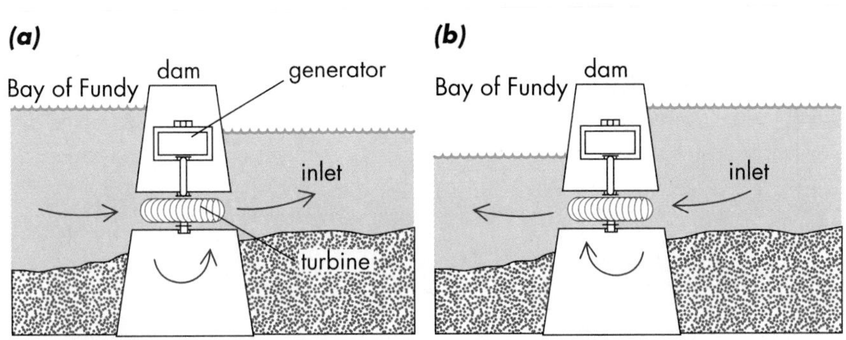

FIGURE 18.21 ▶
As the sea level rises, water passes through the dam, turning the turbines connected to the electric generators. When the water level outside the dam becomes lower than in the inlet, the direction of flow is reversed, and electricity is again produced. (a) High tide. (b) Low tide.

To use tidal energy and transform it into electricity, a dam must be built across a bay or inlet with a strong tidal flow. Two-way turbines are used, so that they can operate the generator whether the tide is coming in or going out (Figure 18.21).

For a tidal energy project to be practical, there must be a large difference in height between low and high tides. The Bay of Fundy on Canada's Atlantic coast has the largest tidal difference in the world, as much as 15 m. North America's first tidal generator began operation on a small inlet of this bay in 1985.

Tidal energy is clean and renewable. Its major disadvantages are the costs of construction and that there are few locations where it can be used.

FIGURE 18.22
What powers this geyser?

ENERGY FROM THE HEAT OF THE EARTH

Geysers and hot springs are evidence of the thermal energy beneath the Earth's surface (Figure 18.22). Heat that comes from the Earth's interior is called **geothermal energy**. For many centuries, people have used water heated by geothermal energy for washing clothes or for bathing. More recently, this energy resource has been used for heating many homes and other buildings in Iceland. There are also more than 300 geothermal electrical generating stations in the world.

Geothermal energy's advantages are that it is always available and easy to use. Like all energy resources, it has some disadvantages. Hot underground water often has an unpleasant odor caused by dissolved gases or minerals. The dissolved minerals also make the hot water undesirable for washing dishes and clothes, and they can corrode pipes.

The use of geothermal energy is limited to the few places where it is readily available. Geothermal generating facilities have been used in these regions for many years. In Reykjavik, Iceland, most people heat their homes with water piped from volcanic hot springs. Geothermal energy also is used to produce electricity in California and in such countries as Italy, Mexico, and New Zealand. In Hawaii government officials hope that by the year 2007 geothermal energy will totally replace fossil fuels for generating electricity.

EXTENSION

■ Several methods of using alternative energy resources now available or being researched have not been described in this section. These include nuclear fission, nuclear fusion, chemical cells or fuel cells, heat pumps, piezoelectricity and electrical generators that use wave energy. Choose one of these alternative resources, find information about it, and prepare a poster or written report about your findings.

REVIEW 18.4

1. (a) What do you think the features of the perfect energy resource should be? List them in order of importance.
(b) What energy resource used today do you think comes closest to matching your perfect energy resource?

2. How do you know that wind and running water contain energy? (Give evidence from your own experience as well as from this unit.)

3. (a) Explain why wind and running water are renewable energy resources.
(b) Energy experts often refer to wind and running water as "indirect solar energy." Why do you think this term is used?

4. Can energy from wind and falling water be used as energy resources for transportation? Explain.

5. (a) Describe one method used today for transforming solar energy to a more useful form.
(b) What kind of new technology is needed to help people use more of the available solar energy?

6. (a) What causes the tides?
(b) What technology is needed to make use of tidal energy?

7. Define geothermal energy, and give an example.

CHAPTER REVIEW

KEY IDEAS

■ There are many forms of energy: sound energy, light energy, mechanical energy, gravitational energy, magnetic energy, thermal energy, chemical energy, nuclear energy, electrical energy, and elastic energy.

■ There are two classes of energy: potential energy and kinetic energy.

■ Energy is measured in joules, but it can also be stated in kilojoules and megajoules.

■ An energy transformation is the change of energy from one form to another.

■ The law of conservation of energy states that energy can neither be created nor destroyed; it can only change form.

■ In most energy transformations, some energy is wasted. The most common form of waste energy is thermal energy.

■ Energy resources that can be used in place of fossil fuels are called alternative energy resources. Many of them are important for our future energy needs.

■ Electric generators, which operate by spinning turbines, transform various sources of energy into the very useful form of electrical energy.

VOCABULARY

potential energy
kinetic energy
energy transformation
law of conservation of energy
alternative energy resource
turbine
wind energy
solar energy
tidal energy
geothermal energy

V1. Draw a mind map by listing as many forms of energy as you can. Where possible, connect the various forms with examples of transformations. For instance, you could connect chemical energy to light energy with the transformation burning wood.

V2. List as many examples as you can of how to produce electrical energy. Beside each example, write a word equation summarizing the energy transformation(s).

CONNECTIONS

C1. (a) List all the forms of energy that you have used or observed in the past 24 hours.
(b) How did you use each form of energy that you listed in (a)?

C2. The energy content of a single serving of a certain breakfast cereal, served with milk, is 750 kJ.
(a) What form of energy does this cereal contain?
(b) According to Table 18.2, about how many minutes of writing by a 50 kg student would consume this amount of energy?

C3. Give an example of each of the following energy transformations. Try to use examples other than those described in this chapter.
(a) chemical energy to mechanical energy
(b) mechanical energy to thermal energy
(c) electrical energy to light energy
(d) mechanical energy to elastic energy
(e) mechanical energy to electrical energy

C4. Write the word equation for each energy transformation listed below. If more than one change is involved, include all the steps.
(a) Lari cooks a hamburger on a propane barbecue.
(b) You stir-fry vegetables in a metal wok (a curved frying pan) on the burner of an electric stove.
(c) Alia sounds the horn on her car.
(d) Energy from the sun is stored in plants.

C5. A chef heats some water in a metal pot on a gas stove. The amount of energy supplied by the fuel is 5 MJ, yet the amount of energy gained by the water is only 3 MJ.
(a) What might have happened to the 2 MJ of "missing energy"?
(b) Does this example mean that the law of conservation of energy is incorrect? Explain.

C6. Name 10 energy resources described in this chapter and the previous one. Classify these resources as those that depend on the sun and those that do not depend on the sun. Divide the resources that depend on the sun into those that depend directly on the sun and those that do not. (One example of a resource that does not depend directly on the sun is the wind. Solar energy heats up air at different rates, causing winds, but the sun does not need to shine for the wind to blow.) You may want to use a table to organize your information.

C7. The United States has many, many miles of coastline. The windy conditions along many coastlines provide good conditions for using wind to generate electricity.
(a) Look at the photograph at the beginning of this chapter. What are some advantages and disadvantages of building such a "wind farm" on the coast?
(b) In the country of Denmark, many farms in isolated locations have their own wind turbines to generate electricity. Do you think people in isolated regions of Alaska would be interested in using this type of wind turbine? Is cost important?

C8. (a) Which of the energy resources described in this chapter form products other than energy when they are used?
(b) Are these products useful, harmful, or neither? Explain.

C9. For each purpose listed below, decide which of the following alternative energy resources would be most practical: wind energy, geothermal energy, tidal energy. In each case, give reasons for your answer.
(a) heating a home in winter
(b) heating a swimming pool
(c) cooking food
(d) pumping water
(e) providing electricity for home lighting

C10. Which of the alternative energy resources described in this chapter are plentiful in the U.S.?

C11. (a) What resource is used to supply energy for most of America's electricity?
(b) Suppose the population of the U.S. grew so large that the resource named in (a) could not supply the demand for electricity. What other energy resource do you think would most likely be used? Explain.

EXPLORATIONS

E1. Suppose you are in charge of designing wood-burning stoves for use in small homes and cottages. Your responsibility is to create a design that provides the greatest amount of energy for the quantity of wood burned. Describe the features of such a design, and draw a diagram to illustrate your design.

E2. Investigate local weather patterns to decide if your community could use energy from wind to produce electricity. If it could use wind energy, determine where the best location for the wind turbine would be.

E3. Suppose that you live in a small town located near hot springs. Make a labeled sketch showing how your town could use geothermal energy to
(a) heat all the houses,
(b) heat water for bathing, and for washing clothing and dishes (assume that the underground water contains too many dissolved minerals to be used directly), and
(c) produce electricity.

REFLECTIONS

R1. Refer to the list of energy resources you made for Activity 18A.
(a) How have your ideas about energy resources changed? Add energy resources to your list that you think are appropriate.
(b) Look again at the poster you made in this activity. Describe the changes you would now make to your poster.

R2. What energy resource do you think will play the greatest part in meeting the future overall energy needs of each of the following? (You may choose from resources described either in this chapter or the previous one.) Explain your answer. (Be sure to consider all the purposes for which energy is used.)
(a) your state
(b) the U.S.
(c) the world

CHAPTER 19

SIMPLE MACHINES

Think of the many ways in which bicycles are used: for transportation, physical fitness, and racing. You may not have thought that bicycles can also be used to study machines. A bicycle may seem to be a simple device. However, it consists of many parts, each with a special function. You may have found that a bicycle allows you to do things that you would otherwise find very difficult. This is a property of machines: they allow us to do things more easily.

How do you make a bicycle go forward? You use force on the pedals by pushing downward on them. Forces are involved in the operation of all machines. Before you study machines in this chapter, you need to know about forces: What can they do? How are they measured? How are they important in your life?

ACTIVITY 19A / THINKING ABOUT MACHINES

In this activity, your group will create a poster to show what you know about machines. Working with three other students, divide a large piece of poster paper into six boxes. Fill in each box as shown here.

With your group, present your poster to the rest of the class. Later in this chapter, you can compare your ideas about machines with scientific descriptions of machines. ❖

List as many different kinds of motion as you can think of.	Draw a mind map of things that move that affect your life every day.	Print the names of at least 10 machines.
Draw large pictures of at least three of the machines that you listed in the upper right box.	Write a short story about how life on Earth would be different without machines.	Print at least five questions that you would like to know the answers to about machines and what makes them move.

19.1 FORCES

A **force** is a push or a pull. Each time you pull on your boots, lift your feet, pick up a book, and push open a door, you are using your muscles to exert a force. When we use a force we say that we are "exerting" a force. All forces may cause change when they act on an object or material. Figure 19.1 illustrates four ways that a force may cause change. These are called force effects.

(a)

(b)

(c)

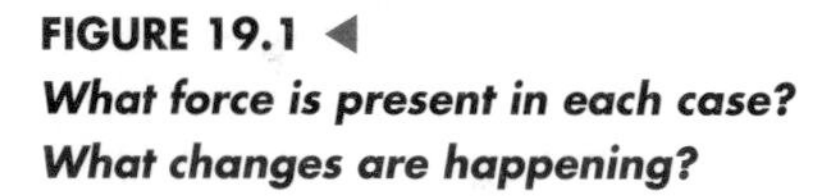

FIGURE 19.1 ◀
What force is present in each case? What changes are happening?

(d)

FIGURE 19.2
The force of friction causes this bicycle to slow down. How can the rider use friction to stop the bicycle quickly?

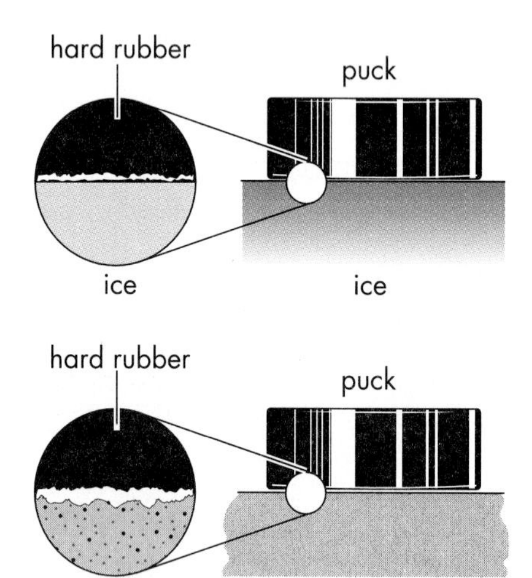

FIGURE 19.3
The puck slides more easily on ice because the ice surface has fewer tiny bumps than the concrete surface.

FORCE, MOTION, AND FRICTION

Forces are needed to make objects move. This motion depends on the size of the force and the mass of the object being moved.

Once an object starts moving, it will continue to move until a force acts on it in the opposite direction. On Earth, all moving objects are affected by forces that oppose their motion. For example, if you stop pedalling a bicycle on a level path, it will continue to roll for a while, but eventually it will stop (Figure 19.2). The force slowing down the bicycle is called friction.

Friction is a force that resists motion whenever one material rubs against the surface of another. The rougher the surface, the greater is the friction. For example, a hockey puck can slide much farther along smooth ice than across a concrete parking lot. You can see the rough, uneven surface of the concrete, but *all* surfaces are uneven, no matter how smooth they may seem. For example, both the hockey puck and the ice have tiny bumps that you can see when they are magnified (Figure 19.3). When the puck slides on the ice, the bumps on the puck and the ice catch on each other. The result is a force of friction that slows down the moving puck and eventually stops it.

THE FORCE OF GRAVITY

If you stop pedalling a bicycle uphill, the bike will soon stop moving forward. In this case, two forces are acting against the bike's motion: friction and gravity. **Gravity** is a force of attraction between all objects. Its effects on motion are easy to see and feel. Climbing a hill is tiring because you must exert force against the downward pull of gravity.

Gravity can also speed up objects. For example, you can coast down a steep hill on your bicycle without exerting any force on the pedals of the bicycle. Can you see how this is similar to what you learned about gravitational potential energy? Moving up a hill increases your gravitational potential energy. Then this energy changes to mechanical kinetic energy when you move down the hill.

For the force of gravity between two objects to be noticeable, at least one of the objects must have a very large mass. The mass of the Earth is so large that it attracts your body strongly enough to keep you on the ground. Gravity is also affected by distance. The closer the objects are to each other, the greater is the gravitational attraction between them.

MEASURING FORCE

The unit for measuring force is the **newton** (N). It is named after Sir Isaac Newton, a great scientist who did many early studies on force and motion. He lived in England from 1642 to 1727.

To get an idea of the size of a one-newton force, imagine you are holding an orange, as shown in Figure 19.4. The pull of gravity on an average orange is about 1 N. This means that your hand must exert an upward force of 1 N on the orange in order to hold it.

You can measure force using a spring scale, which is marked in newtons to indicate the amount of force pulling on the spring.

FIGURE 19.4 ◀
If gravity pulls downward with a 1 N force on an orange, you must exert an upward force of 1 N to hold up the orange. If you hold a small mass in one hand, and a larger mass in the other, which arm gets tired first? Why?

ACTIVITY 19B / THE FORCE OF GRAVITY

In this activity, you will learn how to measure forces in newtons.

MATERIALS

spring scale (0 N to 10 N or 20 N)
set of masses with hooks attached (100 g, 200 g, 500 g, and 1,000 g)
graph paper
bathroom scale (marked in kilograms)

PROCEDURE

1. Copy Table 19.1 into your notebook. Convert the mass amounts from grams to kilograms (1 g = 0.001 kg), and enter the kilogram values in the middle column of your table.
2. Hold the spring scale straight up, and check to see if it reads zero newtons. If it does not, ask your teacher to help you adjust the scale.
3. Use the spring scale to measure the force of gravity on each of the masses. Record your readings in the final column of your data table.
4. Graph the data from your table. Plot the force of gravity (in newtons) along the vertical axis (y-axis). Plot the mass (in kilograms) along the horizontal axis (x-axis). Starting from the point (0,0), draw a straight line that best fits the points on your graph. ➡

TABLE 19.1 ▶
Sample Data Table for Activity 19B

Mass (g)	Mass (kg)	Force of gravity (N)
100		
200		
500		
1,000		

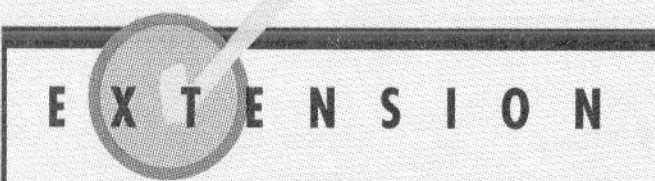

EXTENSION

■ Describe how you would make a spring scale to measure forces in newtons. If the materials are available, build the scale and try it out.

5. Use the bathroom scale to measure your own body mass in kilograms.

DISCUSSION

1. From the graph you made in Step 4, determine how many newtons of gravitational force pull down on each of the following:
   (a) 0.30 kg of mass
   (b) 0.70 kg of mass
   (c) 750 g of mass
2. How much mass would be needed to exert a downward force of 1 N on your hand?
3. (a) How much gravitational force pulls down on 1 kg of mass?
   (b) Use your answer in (a) to calculate how much gravitational force pulls down on:
      (i) 10 kg of mass
      (ii) 20 kg of mass
      (iii) your body ❖

▶ REVIEW 19.1

1. (a) Define the term "force."
   (b) Describe three examples of forces that you have experienced today.
2. Examine Figure 19.1 closely, and answer these questions for each of the four photographs.
   (a) What do you think is exerting the force?
   (b) On what object or material is the force acting?
3. In a grocery store, Jamie and Jitendra are pushing grocery carts with the same force. Jamie's cart is empty, and Jitendra's cart is filled with groceries. Which cart will reach a safe top speed first? Explain why.
4. (a) Explain what causes friction.
   (b) Describe three examples in which friction slows down a moving object.
5. From your own experience, give one example in which gravity speeds up a moving object, and one example in which gravity slows down a moving object.
6. (a) What type of instrument is used to measure force in the laboratory?
   (b) What is the unit of force that you use this instrument to measure?
7. What is the force of gravity acting on each of the following items?
   (a) a school bus of mass 5,000 kg
   (b) a piece of cake of mass 150 g

■ 19.2 FORCE AND WORK

Who is doing more work, physical education students having a game of tennis or mathematics students learning a new equation? The answer depends on what you mean by "work." Many people would say that tennis is play, whereas studying is work. The scientific definition of work is more precise. In its scientific sense, **work** measures the transfer of energy to an object. Work is done on an object only when a force moves that object through a distance. This means you are doing work on a tennis ball when you hit it, and you do work on a book when you turn its pages. When a force moves an object, the amount of work done on the object depends on two factors:

(a) the amount of force exerted,
(b) the distance that the object moves.

These two factors are used to calculate the amount of work done on an object, as shown in the following equation:

$$\text{work} = \text{force} \times \text{distance}$$

This same equation can be written using symbols:

$$W = F \times d$$

Is it possible to exert a large force without doing any work? Imagine pushing as hard as you can with your hands against the wall of the school. No matter how long you push, the wall does not move. Without motion, no distance is covered, so no work is done on the wall (Figure 19.5).

When you use the equation for calculating work, it is important that you use the correct units of measurement. Distance is measured in meters (m), and force is measured in newtons (N), which you learned about in Section 19.1. The measuring unit of work is the newton meter (N•m). The newton meter has been given the special name **joule** (J). It is named after James Joule (1818–1889), a physicist who did many experiments on work and energy. One joule is the amount of work done when a force of one newton pushes or pulls an object a distance of one meter (1 J = 1 N•m). Notice that the unit of work is the same as the unit of energy. This is because work is energy that has been transferred from one object to another.

The example below will help you learn how to use the equation to calculate work.

FIGURE 19.5
In science, if an object does not move when a force is applied to it, no work is done on it. Use the equation W = F × d, to show that the above statement is true.

Example

To raise a box of books from the floor to a shelf, Cheryl exerts a force of 200 N. The box moves a distance of 1.5 m. How much work did Cheryl do on the box?

Solution

(a) Write down the work equation.

$W = F \times d$

(b) Substitute the values given above for the force and the distance. Be sure to include the units.

$W = 200\text{ N} \times 1.5\text{ m}$

(c) Multiply the numbers.

$W = 300\text{ N}\bullet\text{m}$

DID YOU KNOW

■ Sir Isaac Newton (1642–1727) stated three simple laws to explain his observations of moving objects. Newton's second law of motion describes the mathematical relationship between force, mass, and acceleration. What do Newton's other two laws describe?

(d) State your answer in joules.

$W = 300\ \text{J}$

(e) Write your answer as a sentence.

Cheryl did 300 J of work on the box in lifting it 1.5 m.

In the example above, the work Cheryl does is used to push upward against the force of gravity on the box of books. If Cheryl used the force to push the box along the floor for some distance, she would still be doing work. However, that work would not be done to overcome gravity; it would be done to overcome friction.

Sometimes when calculating the work done to overcome gravity, you may know the mass of the object being raised, but you do not know the force needed. When this happens, use the observation in Activity 19B that the Earth pulls down on each kilogram of mass with a force of about 10 N. Thus, if a babysitter needs to lift a 4 kg baby, he or she must use an upward force of about 40 N.

ACTIVITY 19C / HOW MUCH WORK IS DONE?

In this activity, you will do work against gravity and against friction.

MATERIALS

spring scale (in newtons)
piece of string (if needed)
brick with holes
meter stick

■ **Bricks are heavy. When lifting or moving a brick, make sure you have a strong grip on it.**

PART I Doing Work Against Gravity

PROCEDURE

1. Using the spring scale, measure the force that you must use to raise the brick slowly against gravity. If necessary, tie a piece of string onto the brick so the spring scale can hold it (Figure 19.6). Calculate the amount of work done (in joules) raising the brick from the floor to a height of 1.0 m.
2. Measure the distance from the floor to the top of the lab bench. State this distance in meters, using a decimal. Then calculate the work done raising the brick from the floor to the top of the lab bench.
3. With the help of your teacher, measure the distance (in meters) from the top of the lab bench to the ceiling. Then calculate the amount of work you would have to do on the brick to raise it from the top of the lab bench to the ceiling.
4. Hold the brick steady for 5 s in your hand without raising or lowering it. Determine the amount of work you have done on the brick during the 5 s.
5. One student in the class will measure the height of the stairs nearest your classroom. Calculate the amount of work you do each time you climb those stairs. (Hint: You calculated the force of gravity on your body in Activity 19B.)

DISCUSSION

1. In which step (2, 3, or 4) was the amount of work done on the brick the greatest? Explain why.
2. In which step was the amount of work done the least? Explain why.

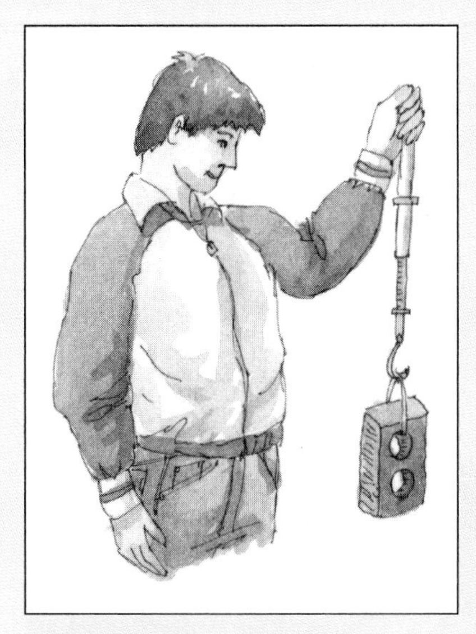

FIGURE 19.6
Measuring the force needed to oppose the downward pull of gravity

3. Use your calculations in Steps 2 and 3 to determine the amount of work needed to raise the brick from the floor to the ceiling.
4. Assume the mass of an orange is about 100 g. In which of the following cases do you think the amount of work being done on the orange is closest to 1 J? Explain your choice.
   (a) You roll the orange across a lunch table to a friend.
   (b) You pick up the orange from the floor and put it on the table.
   (c) You raise the orange from the floor of your classroom to the ceiling.
5. (a) On what object is work being done when you climb stairs?
   (b) If you climbed one set of stairs twice every school day, how much work would you do in an average school year of 190 days?

PART II Doing Work to Overcome Friction

PROCEDURE

1. Measure the force needed to pull the brick at a slow, constant speed across the floor. Try to obtain a reading that is as accurate as possible.
2. Calculate the amount of work needed to pull the brick a distance of
   (a) 1 m (b) 2 m (c) 3 m.
   Show all your calculations.

DISCUSSION

Note: In questions where you need to calculate work, be sure to show the steps that were given in the example earlier in this section.

1. (a) Suppose you need a horizontal force of 80 N to push a desk across a floor at a slow, constant speed. If you move the desk 2 m, how much work do you do?
   (b) What does this work do?
2. In Question 1, if the friction between the floor and the desk becomes less, what do you think would happen to the work needed to move the desk the same distance? Explain why. ❖

► REVIEW 19.2

1. (a) Use an example to explain the difference between force and work.
   (b) How can you tell if work has been done on an object?
2. State the work equation, and give the meaning of each symbol, as well as the units used for each part of the equation.
3. Laurie needs a force of 40 N to pull her younger brother on a toboggan. If she pulls him a distance of 150 m, how much work has she done?
4. Philip picks up an apple having a mass of 100 g.
   (a) How much force does Philip need to pick up the apple?
   (b) How much work does he do lifting the apple a distance of 50 cm?
5. Returning from a hike, Josephine places a 12 kg backpack on the ground.
   (a) How much force does Josephine need to pick up the backpack again?
   (b) How much work must she do on the backpack to raise it a distance of 0.3 m?

19.3 DOING WORK WITH MACHINES

Look at Figure 19.7. Did you recognize immediately that the truck shown is a machine? Did you know that the ramp is also a machine? How can something so simple be a machine?

A **machine** is a device that helps people do work more easily. For example, to lift heavy furniture straight up onto the back of a truck, you would probably need several friends to help you. With a ramp, however, you can do this work more easily.

FIGURE 19.7 ▶
A large force would be needed to lift this couch straight up from the street into the moving van. With a ramp, the movers use a much smaller force.

ACTIVITY 19D / DOING WORK WITH A RAMP

In this activity, you will compare three different methods of raising an object to the top of a platform.

MATERIALS

cart
spring scale (0 N to 20 N)
platform without ramp
(0.25 m high)
short ramp (0.25 m high; sloping surface 0.5 m long)
long ramp (0.25 m high; sloping surface 1.0 m long)

PROCEDURE

1. Copy Table 19.2 into your notebook.
2. Method A: Hang the cart from the spring scale, as shown in Figure 19.8a. Measure the force needed to raise the cart slowly and steadily until it can rest on top of the platform. Record this number in the column titled "Force" in your table.
3. Set up the platform and short ramp as shown in Figure 19.8b. Measure the vertical distance from the bottom of the ramp to the top of the platform. Record this number in meters in the "Distance" column for Method A.

TABLE 19.2 ▶
Sample Data Table for Activity 19D

Method used to do work	Force (N)	Distance (m)	Work done on load (J)
Method A no ramp			
Method B short ramp			
Method C long ramp			

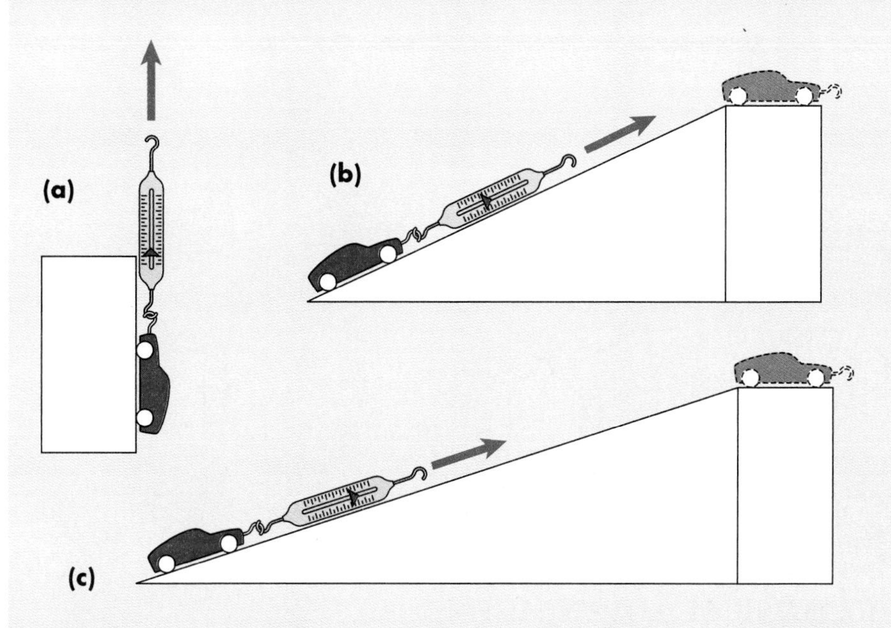

4. Method B: Place the cart on the short ramp, as shown in Figure 19.8b. Measure the force needed to pull the cart at a slow, steady speed up the sloping surface. Record this force in your data table.
5. Measure the length of the ramp. Record this number in meters in the "Distance" column for Method B.
6. Method C: Repeat Steps 4 and 5, using the long ramp as shown in Figure 19.8c. Record the force required and the distance travelled.
7. Using the work equation ($W = F \times d$), calculate the work done to move the cart to the top of the platform for each method. Enter your results in the final column in your table.

FIGURE 19.8
Apparatus for Activity 19D. (a) Measuring the force needed to lift the cart straight up against the force of gravity. (b) Measuring the force needed to pull the cart up the short ramp. (c) Measuring the force needed to pull the cart up the long ramp.

DISCUSSION

1. Consider all the data gathered in this activity. Compare methods A, B, and C in each of the following:
   (a) the size of the object that was raised
   (b) the height reached by the raised object
   (c) the force needed to raise the object
   (d) the distance required
   (e) the amount of work done
2. (a) Does using a ramp reduce the amount of work needed to raise an object? Explain your answer.
   (b) Why do you think people use ramps to do work?
3. Describe some examples in everyday life when people use ramps to help them do work. ❖

FUNCTIONS OF MACHINES

Every machine performs at least one of the following machine functions:

(a) A machine may transfer forces from one place to another (e.g., the chain of a bicycle transfers the force from the pedals to the rear wheel).
(b) A machine may transform energy from one form to another (e.g., an electric generator transforms mechanical energy into electrical energy).
(c) A machine may change the direction of a force (e.g., a pulley on a flagpole changes the direction of a force, as shown in Figure 19.9a).
(d) A machine may multiply speed or distance (Figure 19.9b).
(e) A machine may multiply force (Figure 19.9c).

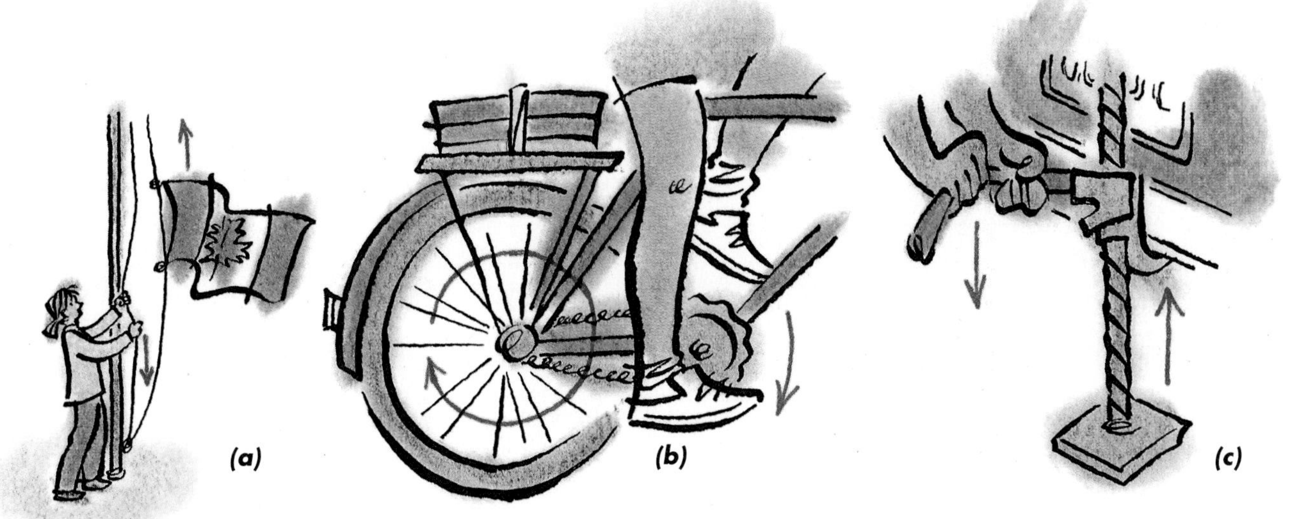

FIGURE 19.9
Examples of machine functions.
(a) Changing the direction of a force: you can lift a flag up by pulling down on the rope. What is at the top of the flagpole to make this action possible?
(b) Multiplying speed: during one revolution of the pedals, the rim of the bicycle's wheel travels much faster than the pedals. Why?
(c) Multiplying force: a small force applied to the handle causes the jack to exert a force large enough to lift a car.

MECHANICAL ADVANTAGE

The amount that a machine can multiply a force is called its **mechanical advantage**. You can calculate the mechanical advantage of any machine by using the following equation:

$$\text{mechanical advantage} = \frac{\text{load force}}{\text{effort force}}$$

Load force is the force needed to move the object without the machine, and **effort force** is the force needed to move the object with the machine.

For example, the effort force required to push a person in a wheelchair up the ramp shown in Figure 19.10 is 100 N. If the person and wheelchair had to be lifted straight up against gravity, the load force would be 900 N. You can calculate the mechanical advantage by using the equation:

$$\begin{aligned}\text{mechanical advantage} &= \frac{\text{load force}}{\text{effort force}} \\ &= \frac{900\ \text{N}}{100\ \text{N}} \\ &= 9\end{aligned}$$

This means the ramp multiplies the effort force by 9 times.

Using a ramp may seem as if we are "getting something for nothing." But that is not what actually happens. Although the load in the wheelchair example can be raised with only 1/9 as much effort force, the effort distance becomes 9 times as long. You do not save work by using a ramp instead of lifting the load directly. A ramp simply makes the work easier for you to do by multiplying the force. It also makes the work more convenient, because pushing the wheelchair by its handlebars is less awkward than trying to lift it straight up. This is true of all simple machines. They do not change the amount of work that you do; they only change the force that you need to use.

FIGURE 19.10 ◀
The ramp is a machine because it can multiply force.

▶ REVIEW 19.3

1. Indira pushes Justin in a wheelchair up a ramp. The length of the ramp is 14 m and the top of the ramp is 2 m above the bottom. The combined mass of Justin and the wheelchair is 70 kg.
   (a) Find the force of gravity in newtons acting on Justin and the wheelchair.
   (b) Calculate the work required to raise Justin and the wheelchair vertically to the height of the top of the ramp.
   (c) How much work does Indira do in using a 100 N force to push the wheelchair all the way up the ramp?
   (d) What is the advantage of using this ramp?
   (e) What is the disadvantage of using this ramp?
2. Why is the ramp in Question 1 called a machine?
3. Can a stairway be called a machine? Explain.
4. What is the main function of a ramp?
5. Lifting a friend in a wheelchair onto a porch requires a force of 900 N. With a ramp, the effort force is only 150 N. Calculate the mechanical advantage of the ramp.
6. To have repairs made on its underside, a 12,000 kg railway car is pulled up a ramp and onto a raised platform. The effort force needed is 1,500 N.
   (a) Calculate the load force, which is the force of gravity on the railway car.
   (b) Determine the mechanical advantage of this ramp.

■ 19.4 THE SIX SIMPLE MACHINES

All machines, no matter how complex, are made up of one or more of the six **simple machines** shown in Figure 19.11. Whether you use them alone or together, the simple machines make your work easier and more convenient. In this section, you will study the features of these six simple machines.

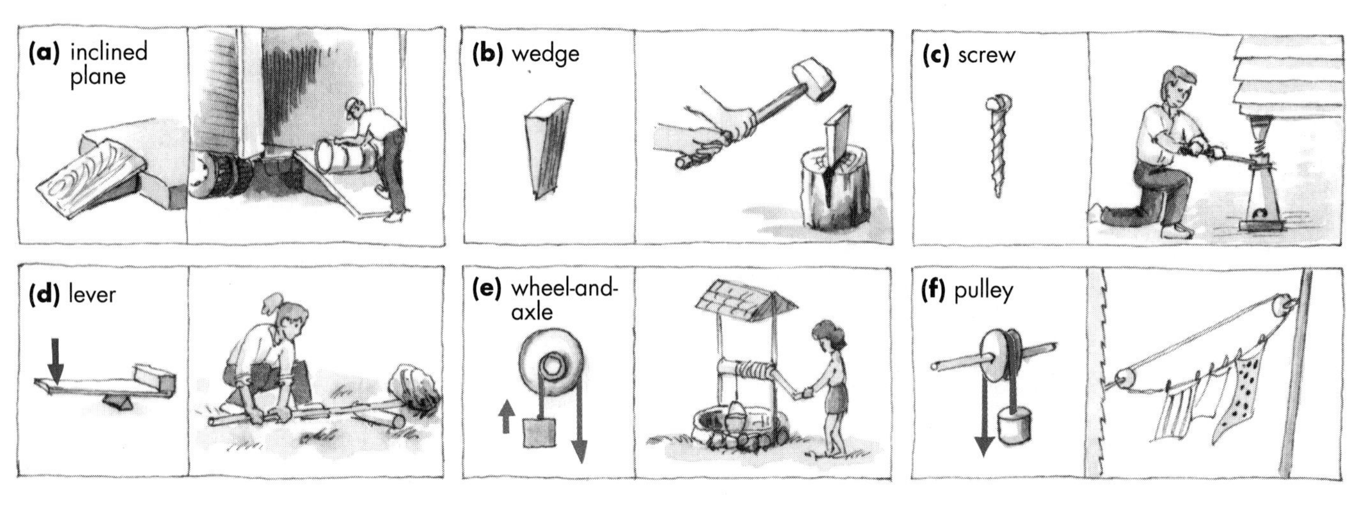

FIGURE 19.11
The six simple machines, with an example of each machine. What is being done by each machine shown here?

ACTIVITY 19E / LEARNING ABOUT SIMPLE MACHINES

Copy Table 19.3 in your notebook. Make enough space for six simple machines.

As you read the following section, write the names of the machines in the first column. In the second column, write the main function of the machine. In the last two columns, give examples of each machine mentioned in this section and others not mentioned. ❖

Simple machine	Function	Examples from text	Other examples
Inclined plane			

TABLE 19.3 ***Sample Data Table for Activity 19E***

THE INCLINED PLANE

An **inclined plane** is a sloping surface along which a load is moved. A ramp is an example of an inclined plane. The gentler the slope of the inclined plane, the greater is its mechanical advantage. The function of the inclined plane is to multiply force. For example, mountain paths sometimes zigzag several times to provide a series of inclined planes (Figure 19.12). These gentler slopes allow a hiker to climb a mountain with a smaller effort force than if the path were shorter but steeper. Although a gentler slope makes the hike easier, the hiker must walk a longer distance, so the hiker's body does the same amount of work in moving up the mountain. Another example of the inclined plane is a wheelchair ramp.

THE WEDGE

A **wedge** is a simple machine shaped like an inclined plane, but it does work on an object by being pushed into the object. A wedge may be used to split objects or prevent an object from moving (Figure 19.13). The function of the wedge is to multiply force. The longer and narrower the wedge, the greater is its mechanical advantage. This means that you can use less force when using a longer wedge. For example, you need much less effort to sew with a sharp, thin needle than with a blunt, thick one. Other examples of wedges include cutting tools such as knives, axes, razor blades, and your front teeth.

FIGURE 19.12
Two mountain tracks: On which track will the hiker need to use a smaller force? On which track will the hiker travel the greater distance?

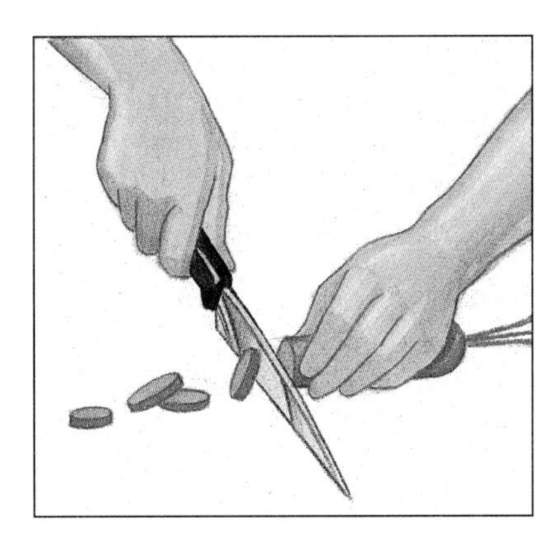

FIGURE 19.13
Two examples of how a wedge can be used: One function of the wedge is to multiply force.

THE SCREW

The simple model in Figure 19.14 uses a pencil and paper triangle to show you that a **screw** is simply an inclined plane wrapped around a rod. Like the wedge, the screw can do work by moving into or through an object. The screw's function is to multiply force. Because a screw is made of a very long inclined plane, it can have a very large mechanical advantage. It can enter hard materials such as wood and metal with only a small effort force. Tools such as clamps use the screw to hold objects in place. A corkscrew is a screw that can be used to move a load. A screw can also be used to hold the cap of a jar in place (Figure 19.15).

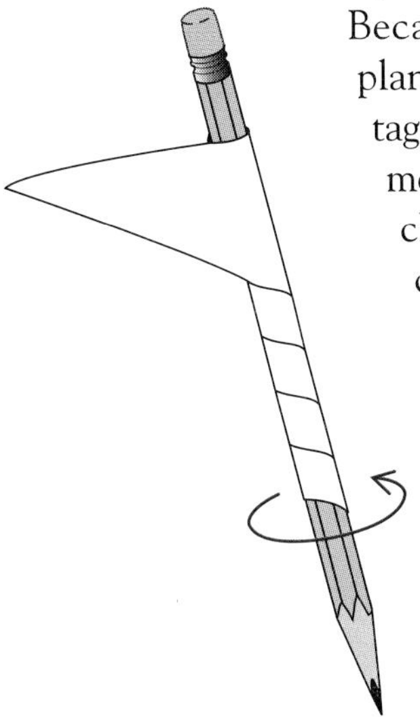

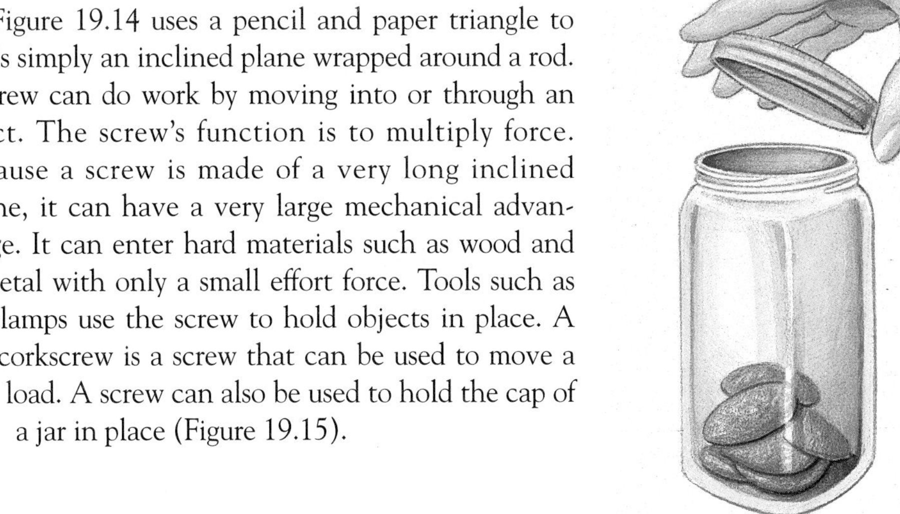

FIGURE 19.14 ◀
You can use a pencil and a paper triangle to make this simple model of a screw.

FIGURE 19.15
A screw cap provides a jar with an airtight seal.

THE LEVER

The children playing on the seesaw in Figure 19.16a are using a simple machine, the lever. The **lever** is a bar that is supported at one point. This point is called the **fulcrum**. The lever rotates freely around the fulcrum.

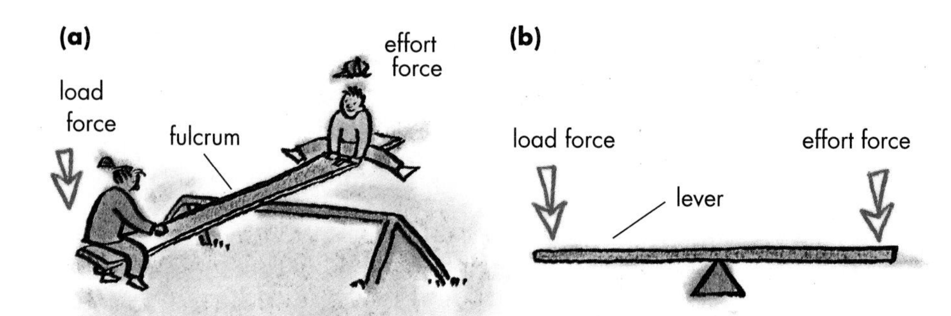

FIGURE 19.16 ▶
(a) A seesaw is an example of a lever.
(b) A lever with the fulcrum between the load force and the effort force.

FIGURE 19.17
A lever enables this man to raise one end of the trailer so that it can be placed on blocks. What would happen if he used a thin stick instead of this board?

Figure 19.16b shows a lever in which the effort and load forces are on opposite sides of the fulcrum. You can use levers arranged this way to help you lift heavy loads, if the fulcrum is closer to the load force than to the effort force. For example, the lever in Figure 19.17 is acting as a pry bar. By using it, an average-sized adult can produce a load force of 3600 N by using an effort force of only 600 N. The mechanical advantage of this lever can be found as follows:

$$\text{mechanical advantage} = \frac{\text{load force}}{\text{effort force}} = \frac{3600\text{ N}}{600\text{ N}} = 6$$

Not all levers are arranged like the pry bar shown in Figure 19.17. For example, a wheelbarrow enables a gardener to lift a load of soil using an effort force much less than the load force (Figure 19.18). In this case, the

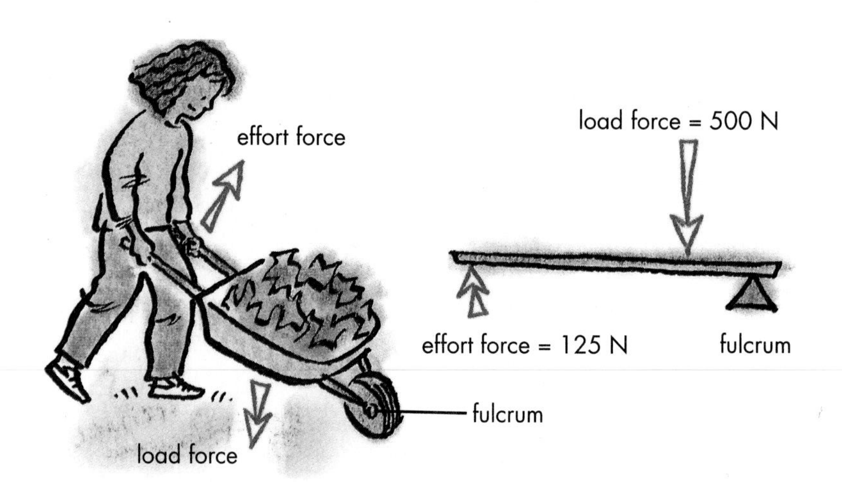

FIGURE 19.18 ▶
The wheelbarrow is an example of a lever in which the load force is between the fulcrum and the effort force.

fulcrum is at one end, the effort force is at the other end, and the load force is between the two. This arrangement multiplies force.

Levers can also be arranged so that the effort force is exerted between the load force and the fulcrum. A human forearm is an example of this arrangement (Figure 19.19a). To raise the arm holding the pizza, the arm muscle must exert an effort force much greater than the small load force exerted by the pizza. As a result, the mechanical advantage is less than 1.

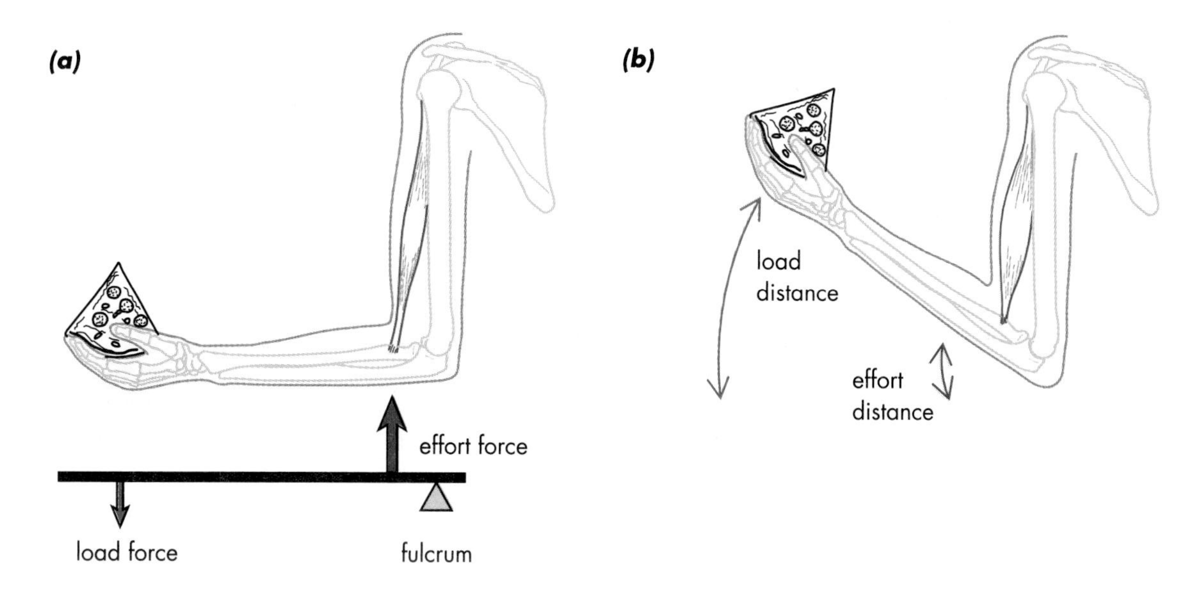

FIGURE 19.19
(a) A large effort is needed to hold up a small load force. (b) However, the load moves farther and faster than the muscle providing the effort force. Compare the load distance and effort distance for this lever with the load distance and effort distance for the wheelbarrow (Figure 19.18).

$$\text{mechanical advantage} = \frac{\text{load force}}{\text{effort force}} = \frac{2\ \text{N}}{10\ \text{N}} = 0.2$$

A lever arranged like this forearm does not multiply force, but multiplies distance and speed instead. Figure 19.19b shows that the hand and the pizza move much farther and faster than the muscle. Brooms and canoe paddles can also be used as levers that multiply distance and speed, if the effort force is exerted between the load and the fulcrum. Can you think of other examples?

Levers are often used in pairs. For example, a pair of scissors is really a pair of levers. Tongs are made up of two levers arranged like the forearm. Other paired levers include tweezers, nutcrackers, and pliers.

FIGURE 19.20
A screwdriver acts like a wheel-and-axle. The handle of a screwdriver is larger in diameter than the shaft. This allows you to turn the screwdriver with a small effort force. Which part of the screwdriver is the wheel? Which part is the axle?

THE WHEEL-AND-AXLE

Imagine trying to tighten a screw using a screwdriver without a handle. You would have to exert a very large force to turn the metal shaft with your fingers. With its handle attached, however, you can use the screwdriver to turn the screw with a much smaller effort force (Figure 19.20).

This is possible because the screwdriver's handle and shaft make up a simple machine called the **wheel-and-axle**. The wheel-and-axle consists of a large-diameter disk (the wheel), which is firmly attached to a small-diameter shaft (the axle).

The function of the screwdriver is to multiply force. Some other wheel-and-axle machines multiply distance and speed. Figure 19.21 shows that the wheel-and-axle is similar to a lever with the fulcrum located between the load force and the effort force.

Other examples of the wheel-and-axle are a pencil sharpener and a hand drill (Figure 19.22), which are both used to multiply force.

DID YOU KNOW

■ The ancient Incas, who lived in South America, had a highly advanced civilization. They discovered the wheel, but never used it for transportation. They used wheels only to make toys for children, while heavy loads were carried up the mountain roads by llamas or people. How would your life be different if the wheel had never been invented?

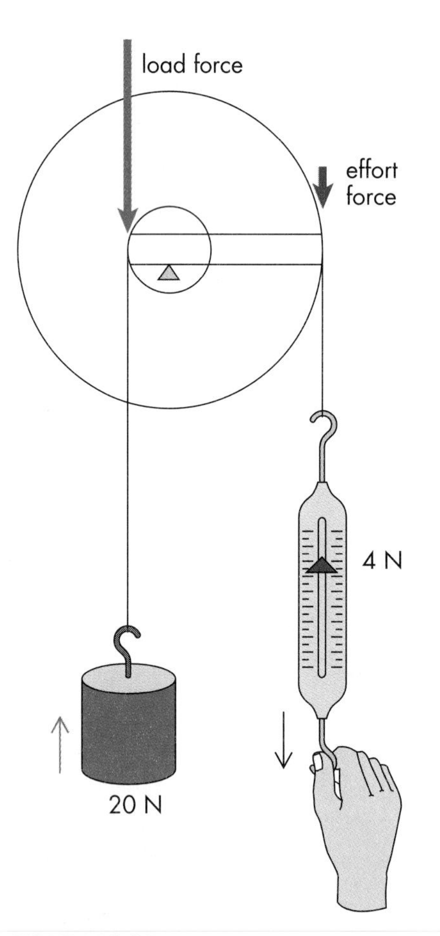

FIGURE 19.21
This wheel-and-axle machine is being used to multiply force. Can you see how it resembles a lever?

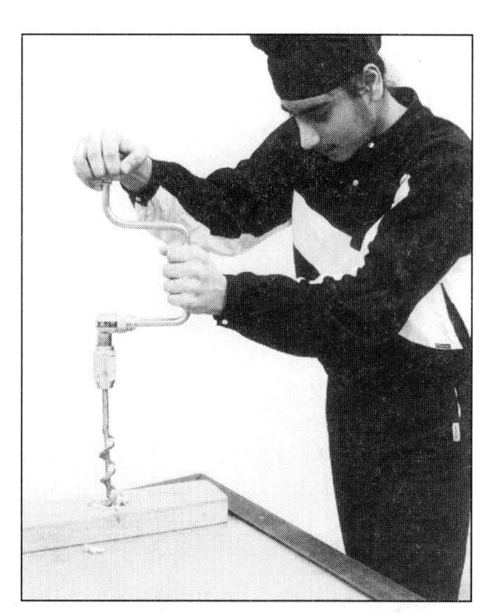

FIGURE 19.22
The handle of this hand drill moves in a complete circle as the student rotates it. What other type of simple machine do you see in this picture?

THE PULLEY

The **pulley** is a simple machine made from a rope or cable that is looped around a support, as shown in Figure 19.23. To reduce friction that might wear out the rope or cable, pulleys are made with smooth rope or cable and wheels that turn easily. The rope or cable sits on the wheel.

There are three types of pulley systems: fixed, movable, and a combination of the two. The simplest system is the single fixed pulley, shown in Figure 19.24a. "Fixed" means that the pulley does not move. This pulley does not multiply force. Its only function is to change the direction of the force. The other systems shown in Figure 19.24 are the single movable pulley and a four-pulley system. Their function is to multiply force.

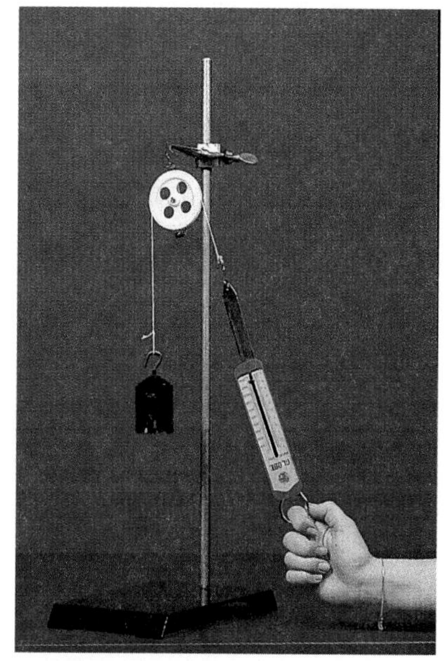

FIGURE 19.23
A pulley

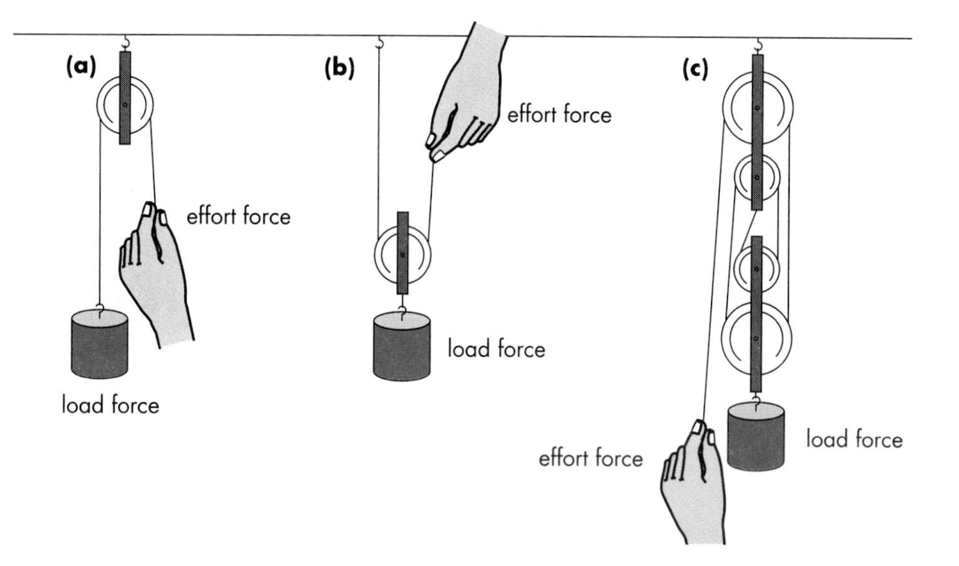

FIGURE 19.24 ◀
Three different pulley systems
(a) A single fixed pulley
(b) A single movable pulley
(c) A four-pulley system

ACTIVITY 19F / INVESTIGATING PULLEY SYSTEMS

In this activity, you will investigate different kinds of pulley systems.

MATERIALS

pulley systems

PROCEDURE

1. Your teacher will set up several pulley systems.
2. Design an investigation to find out as much as possible about each system.
3. In your notebook, set up a table to summarize all your findings. For each pulley system, include the following in your table:
   (a) a diagram
   (b) the effort force needed to lift a certain load force
   (c) the distance moved by the effort force when the load force is moved a certain distance
   (d) the mechanical advantage
   (e) the number of support strings (A support string is a string that pulls upward on the load, and so supports it. A single fixed pulley has one support string. A single movable pulley has two support strings. Figure 19.25 shows the number of support strings in different pulley systems.) ➡

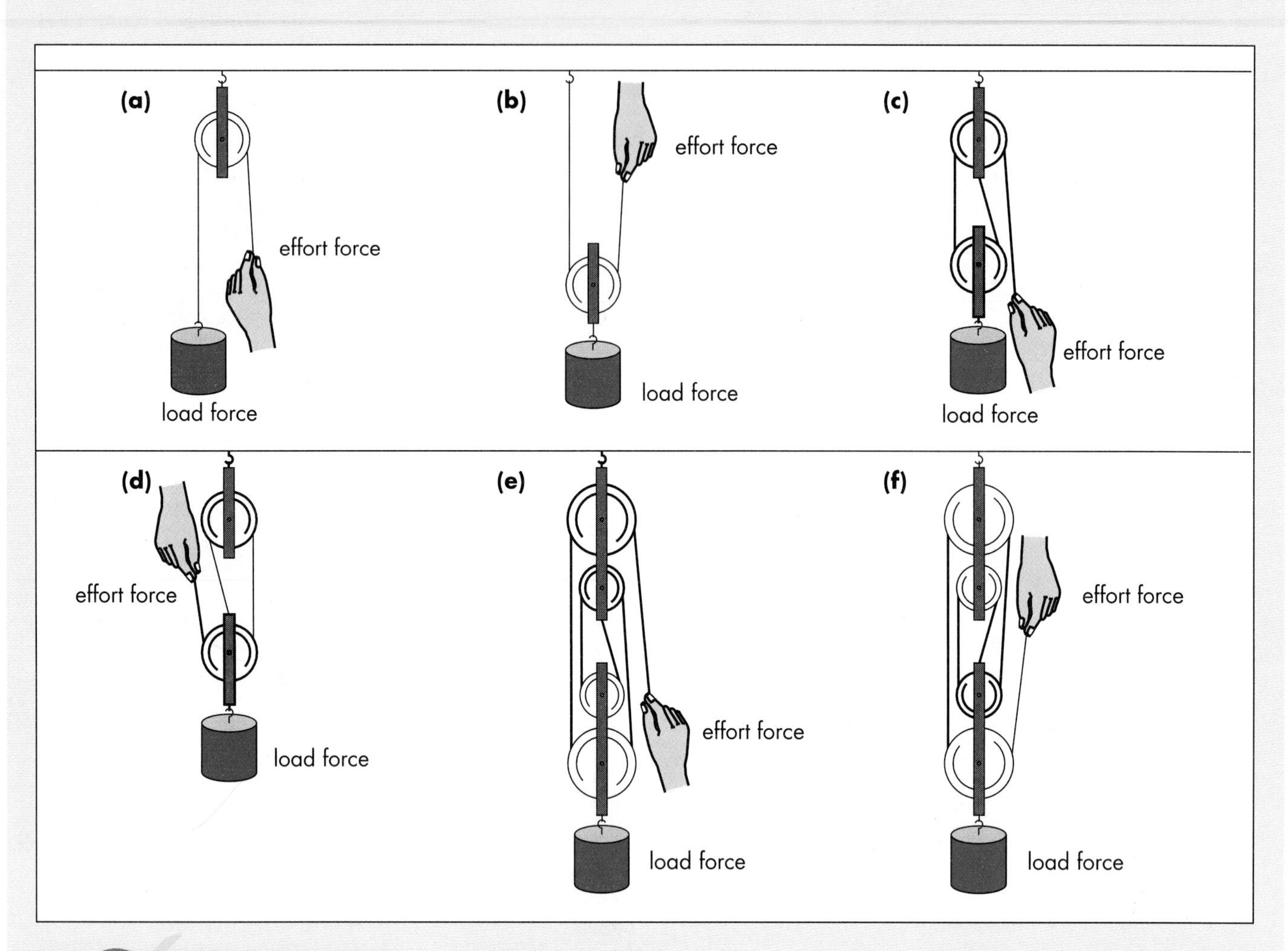

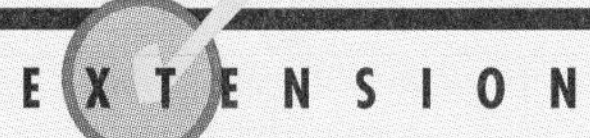

■ A compound machine is a device made of two or more simple machines. A bicycle is an excellent example of a compound machine. Trace the path of the force as it is transferred from the pedals of the bicycle to the rear wheel. At each stage, state what type of simple machine is involved in the transfer of force. Name all the other simple machines involved in the construction and operation of the bicycle. Consider the parts that move as well as the parts that can be adjusted. Also, state the function of each simple machine. Draw a diagram of each simple machine.

FIGURE 19.25
A string is a support string only if it is pulling upward.
(a) has one support string,
(b) and (c) have two,
(d) has three,
(e) has four, and
(f) has five support strings.
Try to identify them.

(f) the main function (Choose one of these functions: multiply force; multiply distance; change the direction of the force; transfer energy.)

DISCUSSION

1. For each pulley system, compare the number of support strings with the mechanical advantage. What do you conclude?
2. Discuss other patterns you observed in this activity. ❖

▶ REVIEW 19.4

1. (a) What functions can a machine perform?
   (b) Look back at Figure 19.11 Identify the machine function shown in each example in that figure.
2. (a) How do you know that an inclined plane is a machine?
   (b) Does an inclined plane reduce the amount of work needed to raise a load? Explain how you know.
3. List as many examples as you can of the following simple machines, and underline each example that you have used at home.
   (a) the inclined plane
   (b) the wedge
   (c) the screw
4. (a) Jody said: "When I used the screwdriver in the woodwork shop, I obtained a mechanical advantage." What did Jody mean?
   (b) Calculate the mechanical advantage of each simple machine in Figure 19.26.
5. For each of the devices in Figure 19.27:
   (a) Draw a simple diagram of the lever(s) involved.
   (b) On each diagram, label the fulcrum, load force, and effort force.
   (c) State whether the lever multiplies force or speed.
6. (a) How can the wheel-and-axle be used to multiply speed? Give two examples of this use (at least one from your own experience).
   (b) How can the wheel-and-axle be used to multiply force? Give two examples of this use (at least one from your own experience).
7. (a) How does a pulley resemble a lever?
   (b) How do the two differ?
   (c) Name two devices that use pulleys that you have used or have seen other people use.
8. Imagine that you are the owner of the Simple Machines Store. Your store has six floors, and each floor specializes in a different type of simple machine. Draw a poster to advertise your store, showing the products you sell.

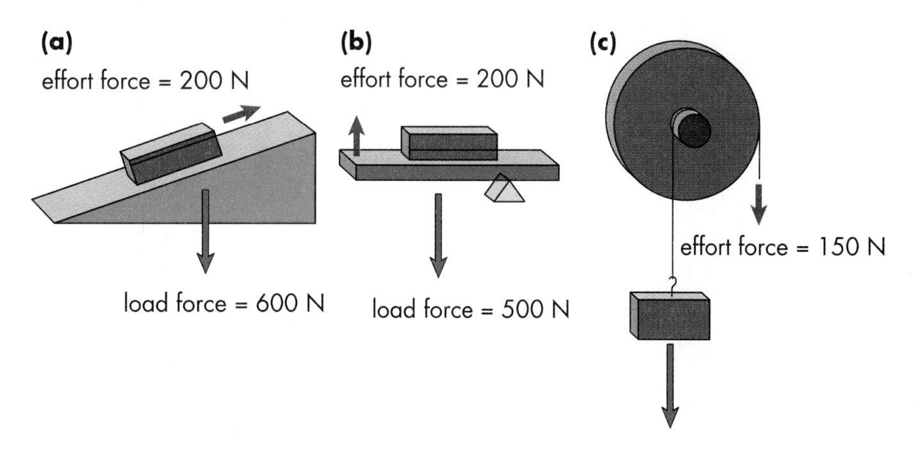

FIGURE 19.26
What is the mechanical advantage of each machine?

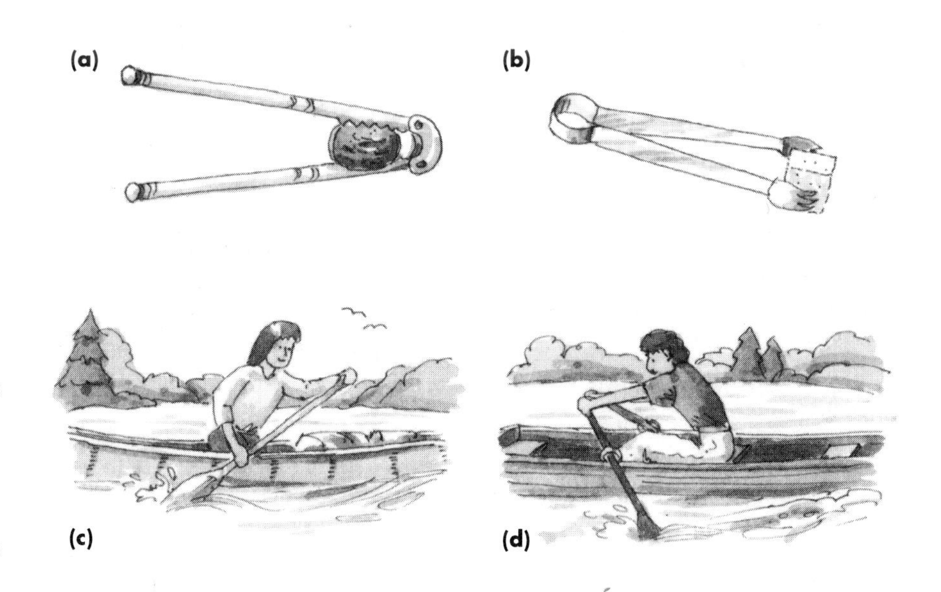

FIGURE 19.27
(a) Nutcracker. (b) Sugar tongs. (c) Paddling a canoe: Assume the fulcrum is at the top end of the paddle. (d) Rowing a boat: Assume the fulcrum is the point where the oar is attached to the boat.

COMPUTER APPLICATION

ROBOTS

Sparks fly as the welding gun touches the car body. When the spot weld is completed, the gun swings to the next spot. Then the next. Maybe 1,500 spot welds in all will be needed on this car body. Then the next car will roll into place on the assembly line.

It's hot and noisy, and the welding fumes are strong. This factory is an unpleasant place to work. Fortunately, the robot doing the welding doesn't mind.

Robots don't get bored or tired. They don't need vacations. They can work 24 hours every day, seven days a week. Today, computer-controlled robots are widely used throughout the automobile industry.

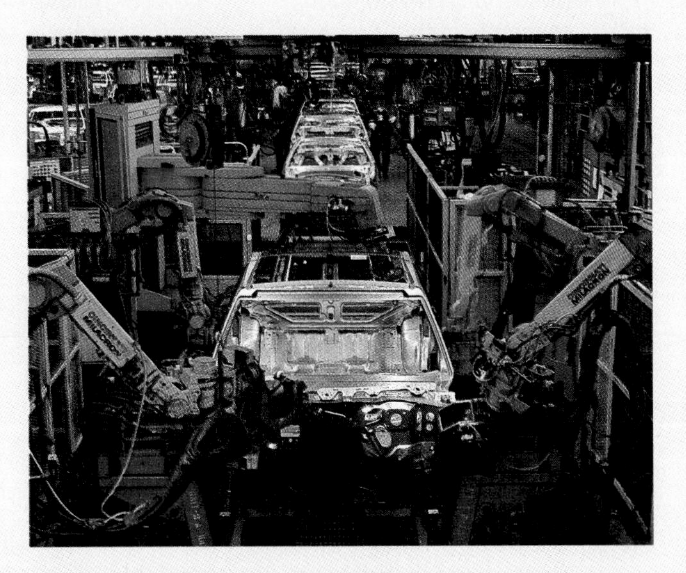

A robot at work helping to build cars.

Robots are also found in many other places—including places humans can't go. Robots have been used in space exploration and in the ocean depths. Robots can handle dangerous radioactive nuclear materials in power plants.

Japan is the leader in robot use. There, robots assemble electronic equipment, and even put together other robots.

Most robots are very simple machines. They consist of a movable arm that can grasp things. The work they do is easy and repetitive. For instance, a robot can screw a light bulb into a socket, but the parts must be lined up in exactly the right way. If the bulb is upside down, the robot will try to screw it in that way. If there is no bulb, the robot goes through the motions of screwing a bulb in anyway.

Robots are limited because they have no senses—they can't see, feel, hear, or handle objects the way we can. Nor can they make decisions or use common sense. Robots are just machines that do best at dull, repetitive jobs that bore us to tears.

Computer programmers are working to make "smarter" robots. These robots, with "artificial intelligence," won't be programmed to do exactly the same thing every time. They will be able to decide how to do a task. For instance, they may be able to "see" or "feel" the shape of an object, and then decide the best way to move their grasper to pick it up.

However, we are far from being able to develop a computer with an artificial brain anything like a human brain. While robot superbrains can do some things, like arithmetic, much faster than humans, we shouldn't underestimate our own abilities.

Imagine all the information you would have to program into a robot so it could do something as simple as recognize a dog (and know that it is not a cat or a stuffed toy). Think about how difficult it would be to design a robot that could tie a shoelace.

19.5 EFFICIENCY OF MACHINES

Moving parts are involved whenever simple machines are used. With an inclined plane, an object must be moved up the slope. In a pulley system, a wheel rotates on an axle. In these and all other machines, the surfaces of moving parts slide against each other, causing friction. Extra force must be used to overcome the friction.

To compare machines, we can calculate their efficiencies. The **efficiency** of a machine is a calculation of the work done by the machine (the work output) as a percentage of the work needed to operate it (the work input). Efficiency can be calculated in one of the two following ways:

1. $\text{efficiency} = \dfrac{\text{work output}}{\text{work input}} \times 100\%$

2. $\text{efficiency} = \dfrac{\text{load force} \times \text{load distance}}{\text{effort force} \times \text{effort distance}} \times 100\%$

As an example, think about the operation of a single movable pulley (Figure 19.28). As the wheel rotates on the axle, the surfaces of the moving parts slide against each other, causing friction. Extra force must be used to overcome this friction. Table 19.4 shows an example of this type of system. The calculations show that the work output is less than the work input, because of the extra force needed to overcome friction. You can calculate the efficiency of this pulley system using the first equation given above.

$$\text{efficiency} = \frac{\text{work output}}{\text{work input}} \times 100\%$$

$$\text{efficiency} = \frac{23\text{ J}}{25\text{ J}} \times 100\%$$

$$= 92\%$$

Thus, 92 percent of the work put into the machine is used to lift the load. The remaining 8 percent is used to overcome friction.

Imagine that someone has invented a machine with no friction between any of its moving parts. Its work output would be the same as its work input, and its efficiency would be 100 percent. In real machines,

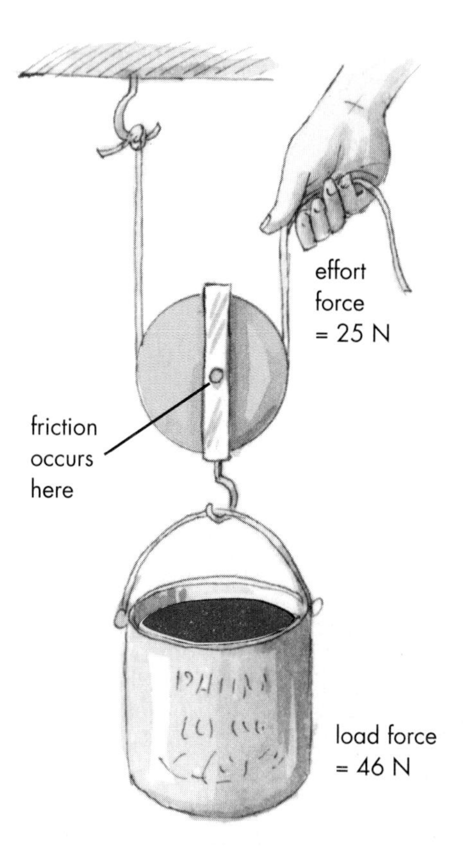

FIGURE 19.28
Because of friction between the wheel and the axle, the efficiency of this pulley is less than 100 percent. How can you tell when friction is present?

TABLE 19.4
Sample Work Input and Output of a Pulley System

Load force	46 N
Load distance	0.5 m
Work output	46 N × 0.5 m = 23 J
Effort force	25 N
Effort distance	1.0 m
Work input	25 N × 1.0 m = 25 J

however, work output is always less than work input. If machine parts are kept polished and oiled, friction can be reduced, but it cannot be eliminated completely. Because some work input is always needed to overcome friction, machine efficiency is always less than 100 percent.

Machines cannot create work. They cannot even transform all work input to useful work output. They can only make some tasks easier for you to do or more convenient by changing the forces or distances needed to do work.

REVIEW 19.5

1. A cook does 1,000 J of work on a hand-operated food grinder. The grinder does 750 J of work on the food. Calculate the efficiency of the food grinder.
2. (a) What is the best possible efficiency of any machine?
   (b) Can any real machine ever provide this efficiency? Explain your answer.
3. Soo Jin and Anita have the same mass. They are using special equipment to climb a rocky, vertical cliff. Soo Jin's climbing equipment has an efficiency of 80 percent and Anita's has an efficiency of 90 percent.
   (a) Who has to do more work to climb the same distance up the cliff? Why?
   (b) What do you think could cause Soo Jin's climbing equipment to have a lower efficiency than Anita's?
4. Calculate the efficiency of each simple machine shown in Figure 19.29.

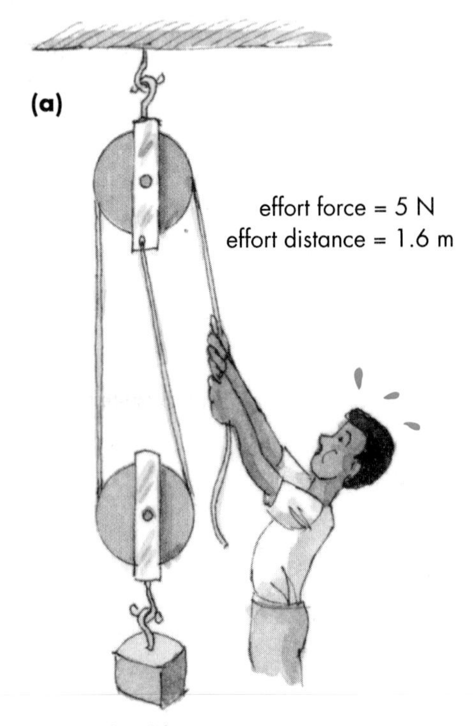

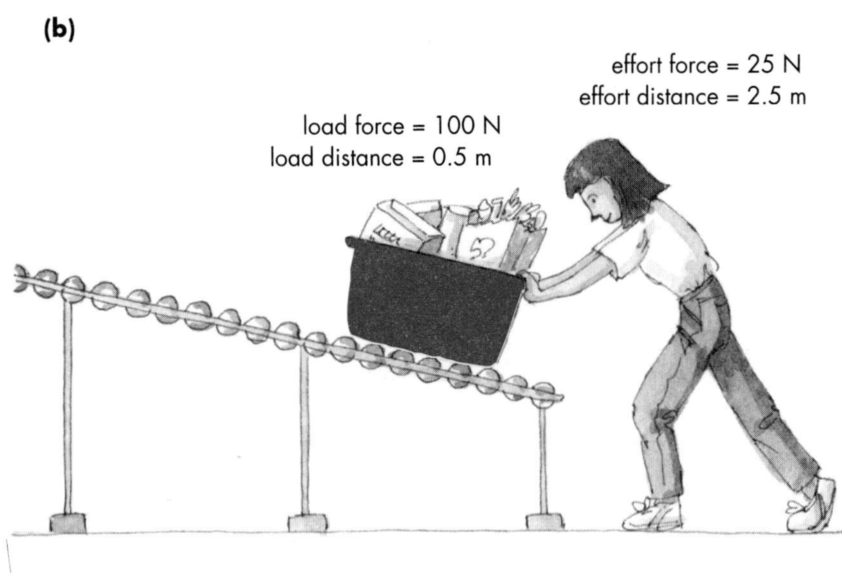

FIGURE 19.29
What is the efficiency of each machine shown here?

CHAPTER REVIEW

KEY IDEAS

- A force is a push or a pull; there are six force effects.
- Two types of force are friction and gravity.
- Force is measured in newtons (N).
- The force of gravity on each kilogram of mass on Earth is about 10 N.
- Work measures the transfer of energy to an object. You can calculate work by multiplying force and distance. Work is measured in joules (J).
- Machines help us do work more easily by performing at least one of five machine functions: transferring force, transforming energy, changing the direction of a force, multiplying force, or multiplying speed or distance.
- Every machine has a mechanical advantage that you can calculate by dividing the load force by the effort force.
- You can calculate the efficiency of a machine by using this equation:

$$\text{efficiency} = \frac{\text{work output}}{\text{work input}} \times 100\%$$

VOCABULARY

force
friction
gravity
newton (N)
work
joule (J)
machine
mechanical advantage
load force
effort force
simple machine
inclined plane
wedge
screw
lever
fulcrum
wheel-and-axle
pulley
efficiency

V1. In the middle of a page, write the phrase "Simple Machines." Then create a mind map using the ideas that you have learned in this chapter. Show how these ideas relate to each other by writing a word or phrase on the connecting lines. Add examples of machines from everyday life to your mind map.

V2. Imagine that you have just returned from a canoeing and camping trip. Write a story about your trip using as many words as possible from the vocabulary list.

CONNECTIONS

C1. Use examples of machines in your everyday life to explain the meaning of each of the following terms:
(a) force
(b) friction
(c) gravity

C2. (a) List the six force effects that may be observed when a force acts on an object.
(b) Which effect(s) can friction cause?

C3. Think about the following two situations. Skiers often apply wax to the bottom surface of their skis. Motorists often sprinkle sand on icy patches in the driveway.
(a) In which case is friction being reduced, and how? Explain why reducing friction is helpful in this case.
(b) In which case is friction being increased, and how? Explain why increasing friction is helpful in this case.

C4. A water pump raises 1,000 kg of water from a well 25 m deep.
(a) What is the force of gravity acting on the water?
(b) How much work does the pump do on the water in raising it?

C5. List the six simple machines. For each, give one example from this book, and a different one from your own experience.

C6. Mayan uses a lever to pry open the lid of a paint can. The load force is 100 N and the effort force needed is 20 N.
(a) What is the mechanical advantage of the lever?
(b) Is Mayan using the lever to multiply force or distance?

C7. Suppose you use your forearm to lift a ball.
(a) What kind of simple machine is your forearm in this case?
(b) What is the machine function of your forearm?

C8. A 55 kg passenger is standing on a moving sidewalk at an airport. The sidewalk is 140 m long, and it rises by 4 m from the beginning to the end.
(a) What is the force of gravity on the passenger?
(b) If the force of the sidewalk pushing forward on the passenger is 25 N, how much work does the sidewalk do in moving the passenger 140 m?
(c) Calculate the work output on the passenger.
(d) What is the mechanical advantage of the machine in this case?
(e) Calculate the efficiency of this machine.

EXPLORATIONS

E1. The huge pyramids built in Egypt thousands of years ago may have been constructed with the help of simple machines. Large stone blocks used to make the pyramids probably came from a long distance away.
(a) Describe what simple machines might have been used to take the stone blocks out of the ground, move them, and raise them to the layers of the pyramids.
(b) At the library, discover more about the building of the pyramids. Draw a picture showing the construction described in your research.

E2. The human body contains many examples of levers. Using Figure 19.30 as a guide, examine your own jaw. Locate the effort force, load, and fulcrum. Draw a simplified lever diagram showing this arrangement. Which of the following does your lower jaw resemble most closely in its function: a pry bar, a wheelbarrow, or a forearm? Explain.

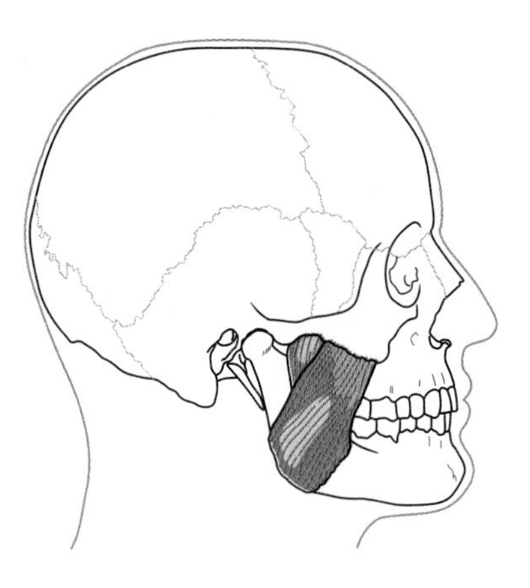

FIGURE 19.30
Your jaw is an example of a lever. What other parts of your body act as levers?

E3. Suppose you are applying for a job in a company that makes robots to handle dangerous materials. Your application includes the design of a robot that is made of at least two simple machines. The robot must be able to grasp and lift a small object (like an eraser). You should be able to control the robot by pulling strings, turning knobs, or some other easy method.
(a) Draw a diagram showing the robot design you would include with your job application.
(b) If possible, construct and test your design.

REFLECTIONS

R1. Look at the poster you created in Activity 19A. Describe how your knowledge and understanding of machines has changed after studying this chapter.

R2. (a) Which simple machine do you think was probably invented first? Give reasons for your answer.
(b) Which simple machine do you think was probably invented last? Explain.
(c) Which simple machine do you think is the most important one in use today? Give reasons for your answer.

THERMAL ENERGY AND HEAT

Can you think of any connections between the scene in the photograph and what you have studied in this unit? What forms of energy are being used here? What energy transformations do you think are happening? What possible energy resources are being used to operate the devices that you can see? Do you see any evidence of force or work? How many machines do you see?

Preparing food involves many forms of energy, such as thermal energy. Thermal energy and heat are the subjects of this chapter. Think of how often heat concerns you. Cooking involves adding heat to food. Many types of food must be stored in refrigerators and freezers, which are devices that remove heat. The clothes you wear, the heating systems in homes and schools, and your health all involve heat.

Because heat is part of your everyday life, and because you may have studied it in earlier grades, you will be able to build on the knowledge you already have about thermal energy and heat. In this chapter, you will observe the effects of gaining or losing heat, and you will learn more about them.

ACTIVITY 20A / THINKING ABOUT HEAT AND TEMPERATURE

In your notebook, name as many different situations as you can think of in which adding heat is useful. (For example, "cooking," which could include "cooking hamburgers," "cooking hot dogs," or cooking anything else.) Think of situations in your daily life, as well as situations in other parts of the world or even beyond the Earth.

Now name as many different situations as you can think of in which taking away heat is useful. Think of situations as you did in the first paragraph: in your daily life, in other parts of the world, or even beyond the Earth.

Your teacher may ask you to work with one or more other students. In your group, use everyone's ideas to make two lists. One list will include as many different situations as possible in which adding heat is useful. The other list will include as many different situations as possible in which taking away heat is useful. Record these lists in your notebook, leaving out any ideas that are the same as your own. After you have completed this chapter, you may want to add to your list. ❖

■ 20.1 TEMPERATURE, THERMAL ENERGY, AND HEAT

You may recall the terms "temperature," "thermal energy," and "heat" from your studies in other grades. What do these terms have in common? How do they differ? This section examines answers to these questions.

TEMPERATURE AND THE KINETIC MOLECULAR THEORY

Temperature is important in many ways (Figure 20.1). For example, have you ever had soggy french fries? If the temperature of the oil for cooking the fries is not high enough, the fries will be soggy. When you bake a cake, the oven temperature must be just right or the cake will be flat and hard. A car is designed so that a warning light comes on if the engine gets too hot. In the steel industry, the temperature in the furnace must be high enough to melt the iron and other substances used to make the steel.

Temperature is also important for your safety and comfort. Your senses tell you very quickly when you touch something that is too hot or too cold. You listen for the prediction of the day's high temperature so you can decide what to wear outdoors.

To understand the scientific meaning of temperature, you must combine your knowledge about energy with the theory of matter called the kinetic molecular theory. The **kinetic molecular theory** of matter can be summarized with the following statements:

1. All matter is made up of very small particles.
2. These particles are in constant motion, which means they have kinetic energy.

FIGURE 20.1
Why is temperature important in each of these examples?

3. There are empty spaces between the particles.
4. The particles and spaces are so small that they cannot be seen.
5. In a solid, the particles are so close together that they can only vibrate (move back and forth).
6. In a liquid, the particles are slightly farther apart and move slightly faster than those in a solid.
7. In a gas, the particles are far apart and move very quickly.
8. If heat is added to a substance, its particles gain kinetic energy and so move faster (Figure 20.2).

■ Have you ever seen coffee mugs that change appearance when hot water is added? You might find them for sale in planetariums and science museums, or in science specialty shops. They contain a thermosensitive material that changes color when its temperature changes. What other examples of thermosensitive materials have you seen?

(a)

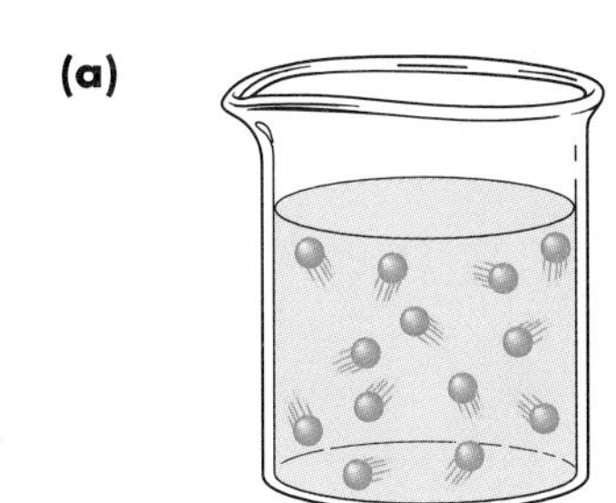

(b)

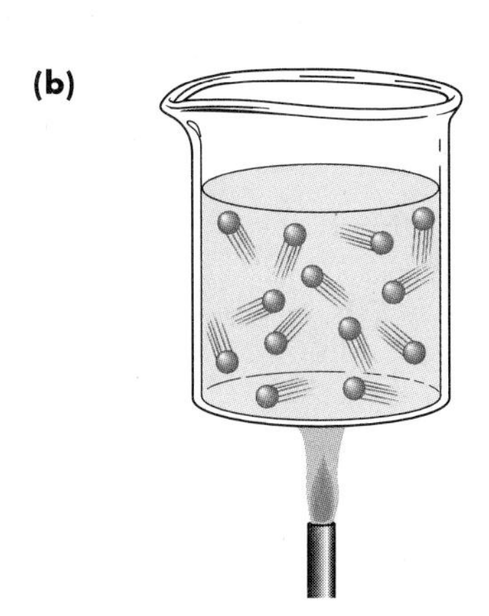

FIGURE 20.2 ◀
Which particles have more kinetic energy: those in (a) or those in (b)?

You can use the kinetic molecular theory to understand temperature. **Temperature** is a measure of the average kinetic energy of the particles in a solid, a liquid, or a gas. For example, look at the two cups of water in Figure 20.3. The water in one cup is at 20°C, which is room temperature. At any one instant, some of the molecules of water may be moving faster than others, but the average kinetic energy of all the molecules results in a temperature of 20°C. Similarly, some molecules in the 80°C cup may be moving faster than others, but the average kinetic energy of all the molecules results in a temperature of 80°C. The average kinetic energy of the molecules at 80°C is higher than the average kinetic energy at 20°C.

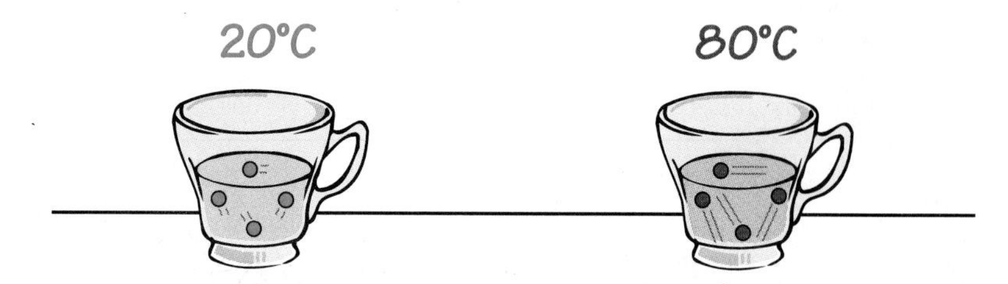

FIGURE 20.3 ▶
The temperature is higher in the cup at the right because the average kinetic energy of the molecules of water is greater.

ACTIVITY 20B / TEMPERATURES

GROUP ACTIVITY
CROSS-CURRICULAR

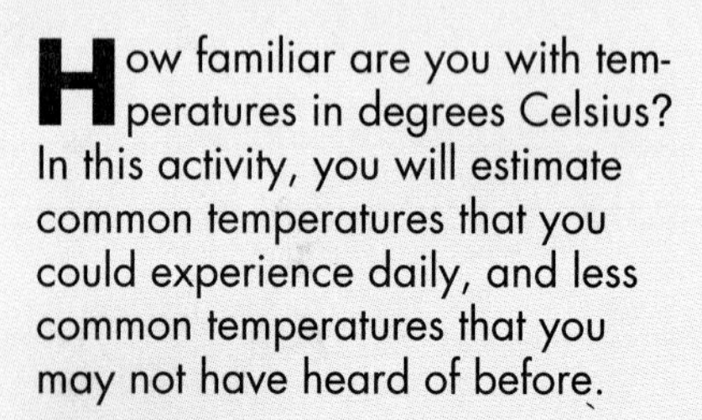

How familiar are you with temperatures in degrees Celsius? In this activity, you will estimate common temperatures that you could experience daily, and less common temperatures that you may not have heard of before.

MATERIALS

36 index cards provided by your teacher

PROCEDURE

1. In your group, sort cards 1 to 18 into four piles:
   - low temperatures
   - high temperatures
   - average everyday temperatures
   - temperature differences
2. Now arrange each pile of cards in order, from lowest to highest estimated temperatures or temperature differences.
3. Next match cards 19 to 36 with cards 1 to 18. If necessary, change the order of cards 1 to 18 until you have made your best estimate of what the number should be in each situation.
4. In your notebook record the temperature values for 1 to 18 in one list and the temperature difference values in another list.

DISCUSSION

1. What four or five temperatures do you think everyone should know?
2. (a) Do you think temperature differences between winter and summer are greater along the coast or in the interior of the U.S.?
   (b) Give possible reasons why the differences in temperature on the coast differ from those in the interior.
3. (a) Which temperature difference did you choose as zero Celsius degrees?
   (b) Give a possible reason why this value would be a zero. ❖

THERMAL ENERGY

Compare the two samples of water shown in Figure 20.4. One is small and one is large, but they are both at the same temperature. The molecules of the two samples have the same *average* energy because the temperatures are the same. But the *total* energy of the large sample is greater than that of the small sample, simply because there are more molecules in it. The total energy of all the particles (molecules or atoms) in an object is called the **thermal energy**. It was one of the many forms of energy mentioned in Section 18.1.

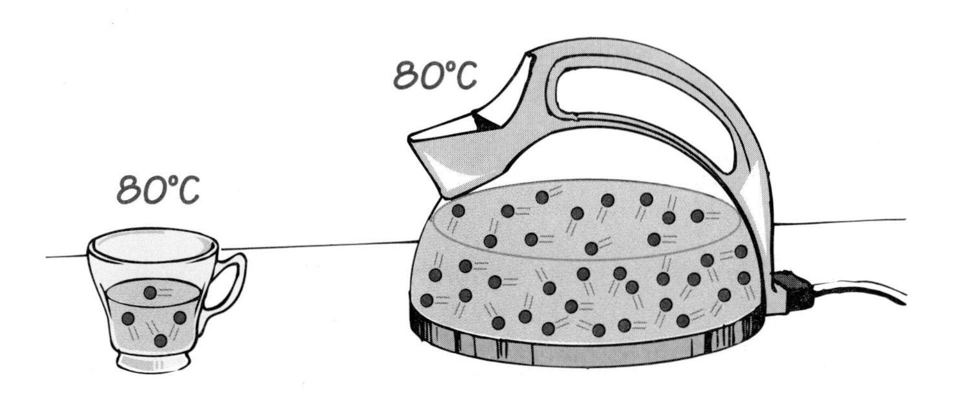

FIGURE 20.4 ◀
The water in the cup and kettle are at the same temperature. Which contains more thermal energy — the cup or the kettle?

HEAT

You may have noticed that any discussion of temperature and thermal energy involves one object or substance at a time. When discussing heat, however, think about two objects or substances. **Heat** is the amount of energy transferred from an object at a higher temperature to an object at a lower temperature. If you have ever touched a hot light bulb or the lid on a pot of boiling water, you have experienced heat transfer. Because your finger was at a lower temperature, the heat transferred from the bulb or lid to your finger.

The kinetic molecular theory provides a detailed explanation of heat. Imagine holding a mug into which you pour hot water. You soon notice that the outside of the mug feels hot. Some heat has transferred from the hot water, through the mug, and to your hand. According to the kinetic molecular theory, the hot particles of water have a high amount of kinetic energy. They collide with the inside surface of the mug, giving some of their kinetic energy to the particles that make up the mug. In turn, the particles of the mug transfer kinetic energy to other nearby mug particles. Soon, the particles at the outside surface of the mug have gained kinetic energy, and can transfer some of that kinetic energy to your hand (Figure 20.5). Do you remember the name given to heat transfer through solids?

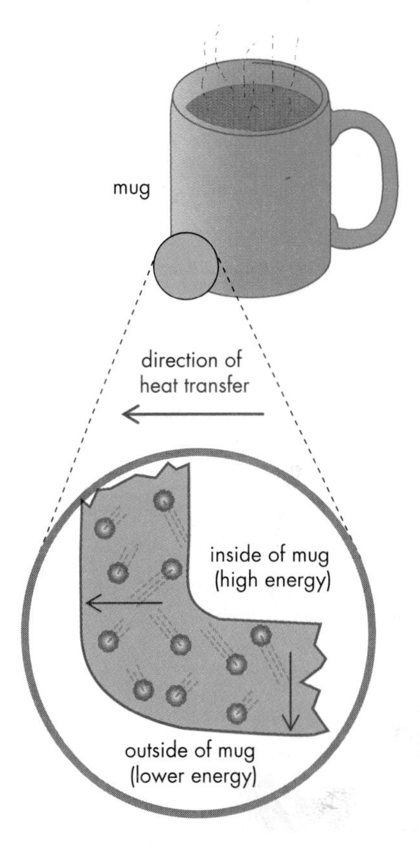

FIGURE 20.5
Heat transfers from an area of high energy (and temperature) to an area of lower energy (and temperature). As the mug becomes warm, where might the heat transfer to next?

▶ REVIEW 20.1

1. Describe why temperature is important
(a) in preserving food,
(b) in preparing food,
(c) to your health.

2. Suppose you had two cups containing the same amount of milk. What device would you use to compare the average kinetic energy of all the particles in each sample of milk? Why?

3. Describe the difference between
(a) thermal energy and temperature,
(b) thermal energy and heat.

4. Naomi fills a bathtub and a cup with water. The temperature of the water in the tub is 34°C and the temperature of the water in the cup is 40°C.
(a) In which sample do the water molecules have a higher average kinetic energy? Explain.
(b) Which sample do you think has the higher thermal energy? Why?
(c) If Naomi poured the water from the cup into the tub, would any heat transfer occur? Explain.

5. Mario has a thermometer containing a red liquid at room temperature. He places the thermometer into hot water. Use the kinetic molecular theory to explain how Mario's thermometer is able to indicate a higher temperature. Draw and label a diagram to go with your explanation.

20.2 HEAT CAPACITY AND HEAT TRANSFER

The map of the United States below in Figure 20.6 shows the average January temperature across the continent. You might ask, why do California and Virginia have about the same warmer temperatures when compared with Utah, Colorado, and Kansas? They really aren't much farther north. Why are the temperatures so high along the gulf coast?

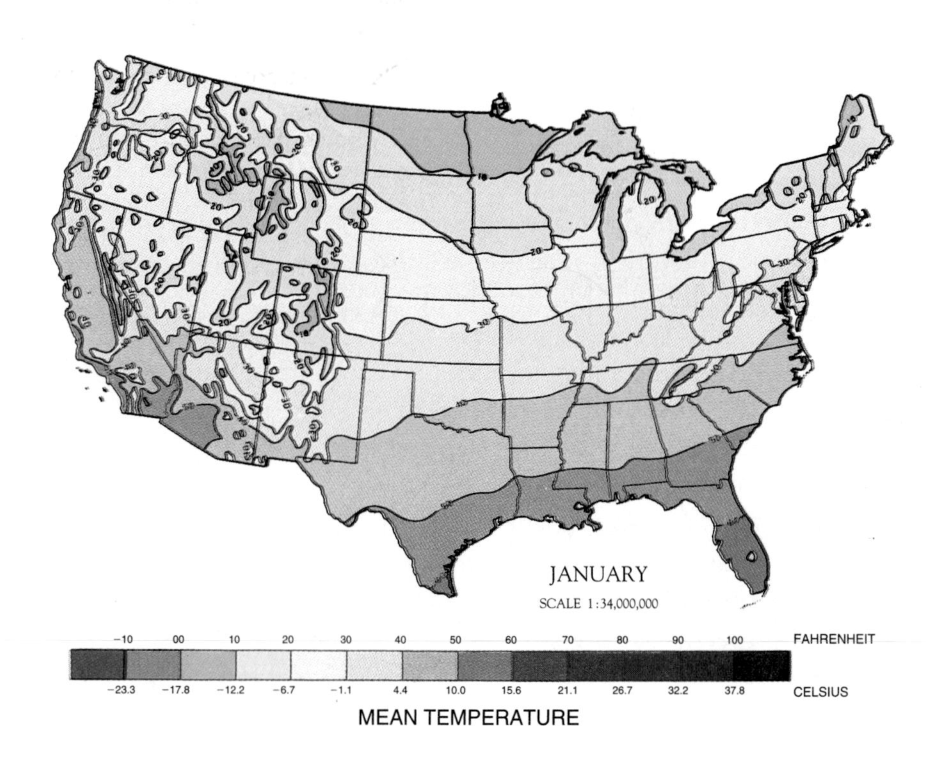

FIGURE 20.6 ▶ ***The average winter temperature at a location depends partly on how close the location is to a large body of water.***

One thing these warmer cities have in common is that they are near large bodies of water. As the weather gets colder in the fall and winter, the water takes a longer time to cool down than the land does. To put it another way, water takes a longer time to release or lose heat than the land does. Thus, we can say that water has a greater heat capacity than the land does. The large bodies of water keep the areas near them warmer in winter than those areas that are far from the water.

Water has a high heat capacity. Any substance that has a high heat capacity also requires a large amount of heat to raise its temperature. Different substances have different heat capacities.

ACTIVITY 20C / HEATING LIQUIDS

How do various liquids compare in the amount of heat needed for identical increases in temperature?

MATERIALS

safety goggles
apron
hot plate (or other safe source of constant heat)
balance
various liquids (e.g., water, vegetable oil, peanut oil, other non-flammable liquids)
glass beakers (one for each liquid)
stirring rod
thermometer
timer
beaker tongs or insulated gloves
graph paper

- **Avoid spilling any liquid on the heater or on yourself. Use gloves or tongs to hold a hot container. Do not allow the temperature of any liquid to rise above 70°C.**

PROCEDURE

1. Turn on the hot plate so that it can heat up and reach a constant temperature while you are carrying out the next three steps.
2. Copy Table 20.1 into your notebook. You will need one set of data for each liquid tested.
3. Put on your safety goggles and apron.
4. Use a balance to measure the mass of the beaker. Record this number in your notebook. Now pour enough of one liquid into the beaker so that the mass of the liquid and beaker together is equal to 100 g plus the mass of the beaker.
5. Place the beaker containing the liquid on the hot plate. As you stir the liquid with the stirring rod, have your partner observe the temperature of the liquid (Figure 20.7). For best results, be sure the thermometer does not rest on the bottom of the beaker. As soon as the temperature reaches 30°C, start taking readings of the temperature every 30 s according to your timer. Continue until the temperature reaches 70°C, then carefully remove the beaker from the hot plate using the tongs or gloves.
6. Repeat Steps 4 and 5 using each of the other liquids available.
7. Plot the data from your table onto a graph. Place time along the horizontal axis (x-axis), and temperature along the vertical axis (y-axis). Use a different colored line for each liquid. Label each line on your graph. ➡

Time (s)	0	30	60	90	120	150	...
Temperature (°C)	30						

TABLE 20.1 ***Sample Data Table for Activity 20C***

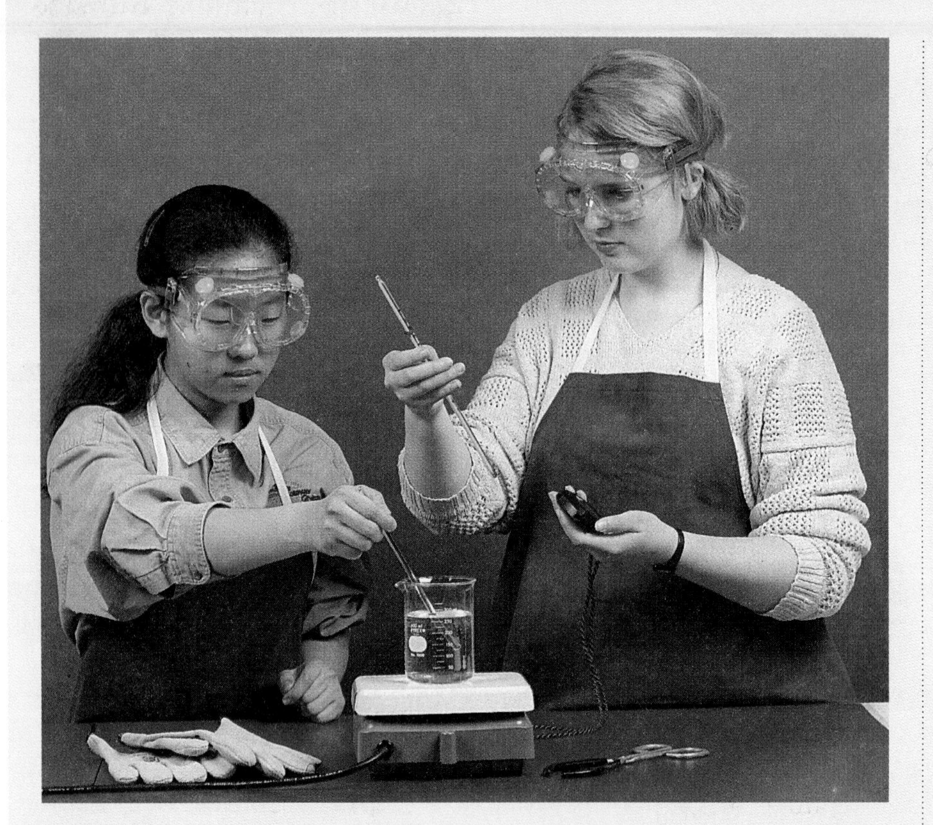

FIGURE 20.7
One student stirs the liquid while another student reads the thermometer.

DISCUSSION

1. (a) Which liquid took the longest time to reach 70°C?

(b) List the liquids in order of the time they took to be heated from 30°C to 70°C.

2. (a) Which liquid had the highest heat capacity? (That is, which liquid required the largest amount of heat to raise its temperature from 30°C to 70°C?)

(b) List the liquids used in this activity in order from greatest to least heat capacity.

3. In what ways was this activity a controlled experiment? ❖

EXTENSION

■ Predict the results in the following experiment. Suppose you use a single constant heat source (such as a hot plate that has been on for several minutes) to heat different masses of water. Each sample—100 g, 150 g, and 200 g—is in an identical beaker and starts at 20°C. You heat each sample until it reaches 60°C and record how much time each one takes. Which sample will reach 60°C most quickly? How much more time will be required for the slowest than for the fastest? If you have your teacher's permission, try this experiment and make a graph of your results.

THE SPECIAL CASE OF WATER

When you bite into a piece of hot apple pie (Figure 20.8), which do you think is more likely to burn your tongue—the crust or the filling? When the pie is removed from the oven, the crust and the filling are at the same temperature, but the filling remains hot for a longer time. Even the bottom crust cools off more quickly than the filling. The filling contains more water, so it cools more slowly. This is one example of the effect of high heat capacity.

Because of its high heat capacity, water is an ideal substance to use in some applications in technology. For example, a small amount of water can absorb a large amount of energy without having a large temperature change. Because of this property, water is often used in the cooling systems of car and truck engines, and other devices that produce heat from friction. Heat must be removed to prevent damage. The high heat capacity of water allows it to remove a large amount of heat. Only a small amount of water is required, so the cooling system does not have to be very large.

Water's high heat capacity also affects climate. Earlier in this section, you learned that large bodies of water, like a lake or ocean, help keep temperatures warmer in winter. In summer, the water takes a long time to gain heat, so regions near large bodies of water remain cooler than they would be away from the water.

FIGURE 20.8
Which releases heat more slowly, the filling of the pie or the crust?

SPECIFIC HEAT CAPACITIES

You have learned that different substances have different heat capacities. In order to compare heat capacities of various substances, you must use an equal mass of each substance. The amount used to compare heat capacities of substances is a mass of 1.0 kg. The **specific heat capacity** of a substance is the amount of heat transferred when the temperature of 1.0 kg of the substance changes by 1.0°C. (The temperature change can be either an increase or a decrease.) Like other energy, heat is measured in joules (J).

The word equation for specific heat capacity is:

$$\text{specific heat capacity} = \frac{\text{energy}}{\text{mass} \times \text{temperature change}}$$

Through experiments, scientists have determined that the value for the specific heat capacity of water is 4,200 J/(kg•°C). This means that it takes 4,200 J of energy to increase the temperature of 1.0 kg of water by 1.0°C. It also means that if the temperature of water drops by 1.0°C, it has released 4,200 J of energy.

If you know the amount of energy used, the mass of the substance, and the change in temperature, you can calculate the specific heat capacity of any substance. For example, suppose in an experiment your heater uses 8,000 J of energy to warm 100 g of vegetable oil from 30°C to 70°C. The oil's mass is 0.1 kg and its temperature changes by 40°C. So, you could use the equation above to calculate the specific heat capacity:

$$\text{specific heat capacity of vegetable oil} = \frac{8{,}000\ \text{J}}{0.1\ \text{kg} \times 40°\text{C}}$$

$$= 2{,}000\ \text{J/(kg•°C)}$$

The specific heat capacities of several common substances are listed in Table 20.2. You can see that water has a higher specific heat capacity than most substances, and that metals have low specific heat capacities.

Substance	Specific heat capacity [J/(kg•°C)]
Hydrogen gas	14,400
Helium gas	5,300
Water	4,200
Concrete	3,000
Ethanol	2,500
Ethylene glycol	2,200
Ice (at 0°C)	2,100
Steam (at 100°C)	2,100
Vegetable oil	2,000
Air	995
Aluminum	920
Glass	840
Sand	790
Iron	450
Copper	390
Brass	380
Silver	240
Lead	130

TABLE 20.2
Specific Heat Capacities of Common Substances at 25°C

▶ REVIEW 20.2

1. Use the kinetic molecular theory to explain each of the following statements.
   (a) The temperature of a pot of soup on the stove increases as the amount of heat transferred to it increases.
   (b) The heat required to raise the temperature of the soup in (a) by a specific amount increases as you add more soup to the pot. (Hint: The mass increases.)
   (c) Different liquids with the same mass require different amounts of heat to raise their temperatures by each degree Celsius.
2. Suppose you are at a beach on a clear, hot day.
   (a) Which becomes warm faster during the day, the sand or the water?
   (b) Which cools down more rapidly at night?
   (c) Explain your answer by referring to Table 20.2.
3. The graph in Figure 20.9 shows the results of an experiment in which 1.0 kg samples of three different liquids, A, B, and C, were heated.
   (a) Which liquid required the greatest amount of heat for a temperature change of 40°C?
   (b) Which liquid has the highest heat capacity?
   (c) For each line on the graph, divide the total amount of energy added by the total temperature change.
   (d) Use Table 20.2 to identify each liquid.
4. Mohan leaves a cup of hot tea on a table until the tea reaches room temperature. The thermal energy of the tea has decreased. Explain what happened.

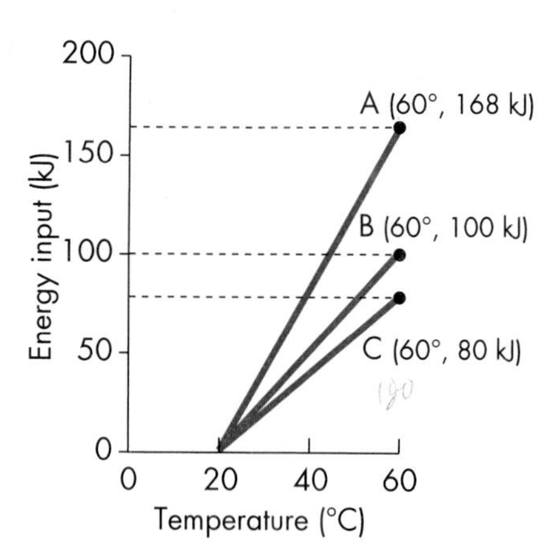

FIGURE 20.9
Graph for Review Question 3

EXTENSION

The equation for specific heat capacity can be rearranged to allow you to calculate the amount of energy required to heat a substance.

energy = mass × specific heat capacity × temperature change

This equation can also be written using symbols.

$$E = mc\Delta T$$

E stands for energy in joules. This is the amount of energy required to heat a sample of a substance or the amount of energy released when a substance cools. *E* equals the mass (*m*) of the sample in kilograms times the specific heat capacity (*c*) of the substance in J/(kg•°C) times the change in temperature (ΔT) in °C. ΔT is pronounced "delta T." The Greek letter *delta* is used to indicate "change in."

You can use the equation $E = mc\Delta T$ to solve a variety of problems. For example, one way to check the energy content of a food is to burn the food and measure how much energy it gives to another substance.

In a controlled experiment, a student burns several peanuts. The heat from the flames causes the temperature of 200 g of water to increase from 20°C to 25°C. Use the equation $E = mc\Delta T$ to determine how much energy was given to the water. How accurate do you think the results of this experiment would be? Why?

20.3 CHANGING STATES OF MATTER

You know that ice is different from water, and water is different from steam. In general, matter may exist in one of three states — solid, liquid, or gas. The properties of these states of matter are listed in Table 20.3.

If heat is added to or released from a substance, the state of the substance can change. For example, if heat is added to liquid water, the water can evaporate or boil, changing to a gas. If heat is removed from the water, the water can change to the solid state, ice.

TABLE 20.3 ◀
The States of Matter

State	Example	Properties
Solid	Stainless steel cutlery	• Has a fixed volume • Has a fixed shape • Cannot flow
Liquid	Milk	• Has a fixed volume • Takes shape of container • Can flow
Gas	Oxygen	• Takes volume of container • Takes shape of container • Can flow

The six possible changes of state are listed below and are illustrated in Figure 20.10.

- **Melting** is the change from a solid to a liquid.
- **Vaporization** is the change from a liquid to a gas. Slow vaporization is called **evaporation**, and fast vaporization is called **boiling**.
- **Condensation** is the change from a gas to a liquid.
- **Solidification** or **freezing** is the change from a liquid to a solid.

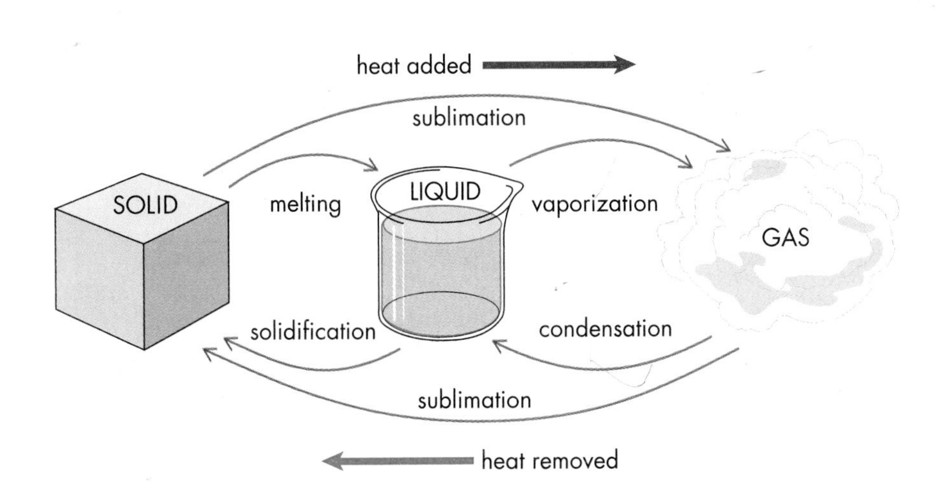

FIGURE 20.10 ◀
How many changes of state are shown here?

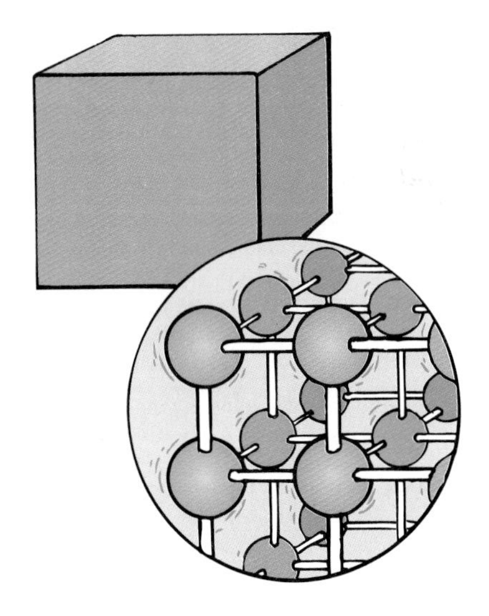

FIGURE 20.11
In a solid, the particles vibrate in place. What happens to the vibrations when the particles are moved further apart by heat?

- **Sublimation** is either the change from a solid to a gas or the change from a gas to a solid. Notice that sublimation is the name for two possible changes of state. An example of sublimation is the disappearance of frost on windows without melting.

CHANGES OF STATE AND THE KINETIC MOLECULAR THEORY

You can use the kinetic molecular theory to help you understand why the temperature of a substance remains constant, or nearly so, during a change of state. The kinetic molecular theory indicates that the particles of a substance are constantly moving. In a solid, the particles are constantly moving, but that motion is mainly vibration in one place (Figure 20.11).

As heat is added to the solid, its temperature rises until it reaches the melting point. What happens as more heat is added? You may be surprised to learn that the temperature does not rise. Instead, the heat causes the particles to become free of their set places. Eventually, when the particles are free to move around, the substance is in the liquid state. A graph of this change of state is called a **heating curve** (Figure 20.12). The flat part of the curve is called a plateau.

FIGURE 20.12 ▶
A heating curve for the melting of a solid with a melting point of 60°C. A plateau (the flat part of the curve) occurs during the time when heat causes the particles to become free of their set places. In what direction does heat flow when the solid is melting?

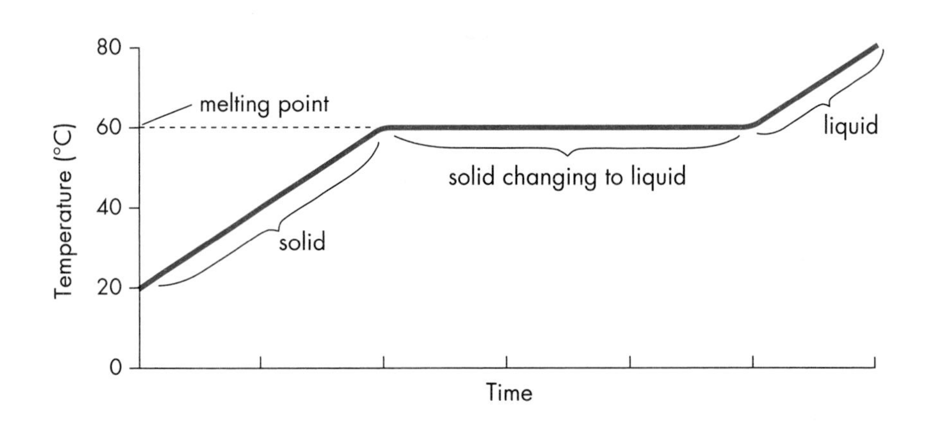

■ A fourth state of matter, called plasma, occurs only when temperatures are extremely high. For example, plasma exists in the sun. It also exists in devices called nuclear fusion reactors. How do you think plasma would differ from the other states of matter?

The other changes of state are similar to melting. Figure 20.13 illustrates the motion of particles in a liquid and a gas. As a liquid is heated, the particles of the liquid gain energy. The temperature rises until the liquid reaches its boiling point. The heat that is added helps to separate the particles, which are then free to move more easily. The temperature of the liquid during boiling remains constant so that the heating curve has a plateau at the boiling point (Figure 20.14).

When a substance cools, the reverse process occurs. The **cooling curve** shown in Figure 20.15 is for a substance that condenses at 120°C and solidifies at 60°C.

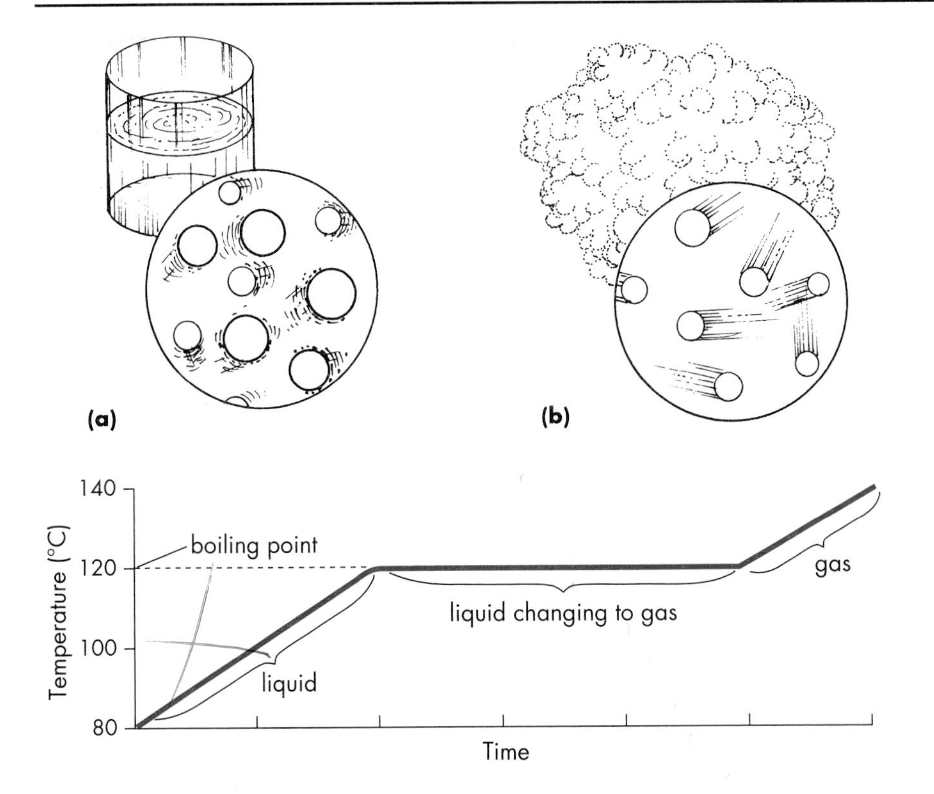

FIGURE 20.13 ◀
Particles in a liquid (a) are free to move around, but they are still close together as they flow around each other. Particles in a gas (b) move easily, and are separated by large spaces.

FIGURE 20.14 ◀
This is a heating curve for a liquid with a boiling point of 120°C. A plateau occurs when the heat added causes the liquid particles to become further apart.

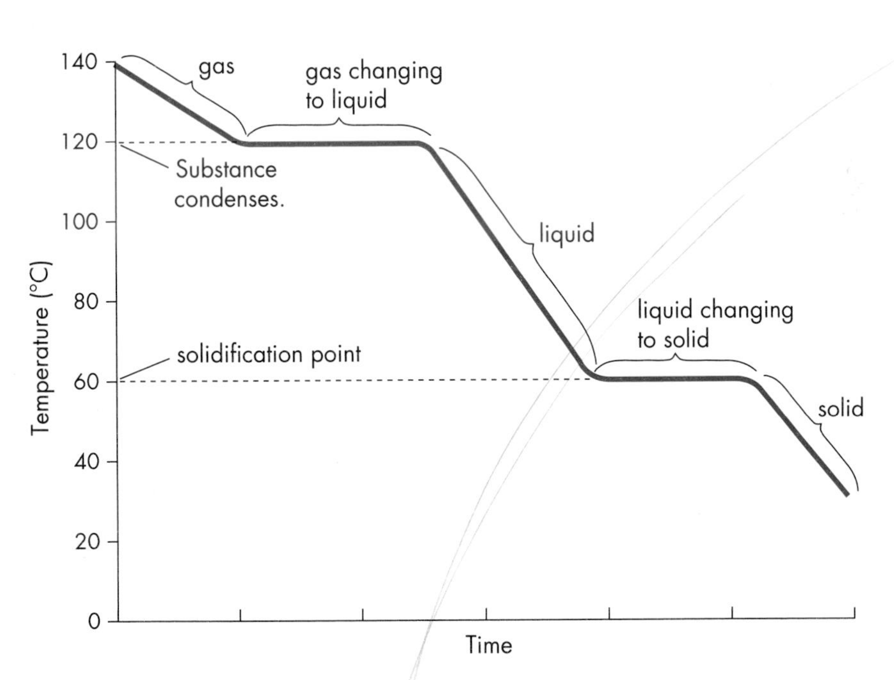

FIGURE 20.15 ◀
This graph shows a cooling curve for a substance that condenses at 120°C and freezes at 60°C. Can you explain why a plateau occurs at each of these temperatures? Are there any differences (other than the direction) between this cooling curve and the heating curves shown in Figures 20.12 and 20.14? In what direction does heat flow?

Different substances are made of different particles. Thus, the melting and boiling temperatures differ for different substances. Examples of melting and boiling temperatures for some substances are listed in Table 20.4. Through experiments, scientists have shown that the melting point and the freezing point of a single substance are the same.

EXTENSION

■ The amount of heat needed to change a substance from one state to another is called latent heat. The word "latent" means "hidden." Why do you think this heat is described as hidden?

Changes of state have several useful applications. You will learn about some of these in Section 20.4.

TABLE 20.4 ▶
Melting and Boiling Points of Common Substances

Substance	Melting point or freezing point (°C)	Boiling point (°C)
Aluminum	660	2,330
Copper	1,080	2,580
Hydrogen	–259	–253
Iron	1,535	2,800
Lead	327	1,750
Mercury	–39	357
Silver	691	2,190
Water	0	100

▶ REVIEW 20.3

1. What is the state of each of the following materials? (You may want to refer to Table 20.4.)
   (a) iron at room temperature
   (b) air
   (c) steam
   (d) mercury at room temperature
   (e) hydrogen at –255°C
   (f) lead at 326°C
2. In each of the following cases, name the change of state and indicate whether heat is added or removed during the change of state.
   (a) Molten (liquid) gold forms bars.
   (b) Unused ice cubes in a "frost-free" freezer gradually disappear.
   (c) Moisture forms on the outside of a glass filled with a cold drink.
   (d) Frost forms on a car's windshield during the night in winter.
   (e) Grapes turn into raisins.
3. During the fall, you are camping and you leave some water in a bucket overnight. The next morning, you notice a layer of ice on the top of the water. What do you think the temperature of the water is just beneath the ice?
4. Richard is making some popsicles from fruit juice in the freezer. He wants to eat them as soon as possible, so he puts a thermometer in one (Figure 20.16). Every 10 minutes he checks the temperature. At first everything is going well, then suddenly his popsicles stop getting colder. Later, however, they start getting colder again, and finally they reach the lowest temperature.
   (a) The readings Richard obtained every 10 minutes were: 20°C, 14°C, 8°C, 2°C, –2°C, –2°C, –2°C, –2°C, –4°C, –6°C, –8°C, –10°C, –10°C. Plot Richard's data on a graph. Use the horizontal axis (x-axis) for the time, and the vertical axis (y-axis) for the temperature.
   (b) At what temperature does a plateau appear on your graph? Explain what is happening at this time.

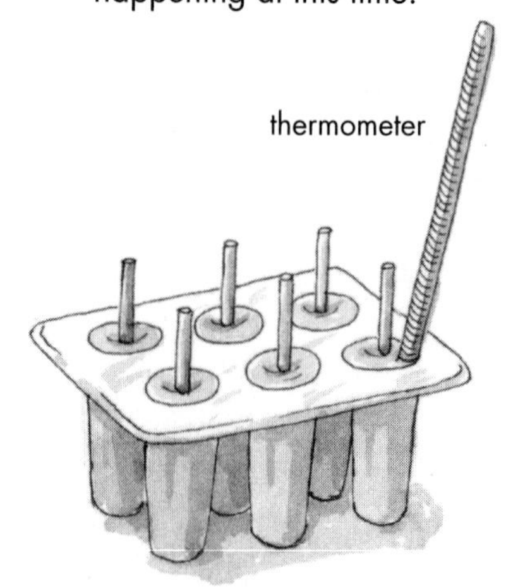

FIGURE 20.16
What happens to the temperature of Richard's popsicles (Review Question 4) as they form?

5. Charlene places a chemical with a melting point of 53°C in a test tube. She heats a sample of this chemical, starting at 20°C, until it reaches 60°C, then she allows it to cool back down to 20°C.

(a) What state of matter was the chemical in when Charlene started her experiment?

(b) Draw a temperature-time graph of Charlene's experiment, and label as much as possible on the graph.

6. Draw a heating curve for a substance that starts at room temperature and eventually reaches 200°C. The melting point of the substance is 80°C and the boiling point is 180°C. Label as much as you can on the graph.

■ 20.4 APPLICATIONS OF CHANGES OF STATE

You already know that there are many applications of science and technology in your daily life. But it may surprise you to learn that many of them relate to changes of state. These applications range from little things you observe in the kitchen to large changes in weather and climate throughout the world. As you read this section, look for ways in which changes of state relate to the main theme of this unit: energy.

ALTERING FREEZING AND BOILING POINTS

You know that the freezing point of pure water is 0°C. But this freezing point changes when other substances are mixed with water. For example, mixing antifreeze (ethylene glycol) with water lowers the freezing point. The antifreeze/water mixture is used in a car's radiator system. This is important in the winter when temperatures are cold enough to freeze water. Water expands when it freezes and could cause a car's radiator system or engine to crack.

Adding salt to water also lowers the freezing point. If salt is put on ice and snow, their melting point also becomes lower. This explains why salt is spread on roads and highways that are icy or snowy when the temperature is about –5°C to –10°C. The ice and snow can then melt at this temperature and the roads and highways become less slippery.

A change in air pressure affects boiling temperature. Normally, pure water boils at 100°C. However, if the air pressure pushing down on the water increases, the temperature of the water must be higher before boiling can occur. Water in a pressure cooker, for example, may have to reach 200°C before it boils. At such a high temperature, food cooks much more quickly than at 100°C.

If air pressure is reduced, as it is at high elevations in the mountains, the force of the air pushing down on the water decreases. So the water boils at a lower temperature. For a car travelling across mountain ranges, this can be a problem. As the water that is supposed to cool the engine becomes hotter, it may boil. This causes overheating of the car's engine.

■ A geyser is like a giant, erupting pressure cooker. Water seeps into a deep vertical hole. Heat from the Earth causes the water near the bottom of the hole, where the pressure is highest, to become very hot. When this water under pressure gets hot enough, it boils, and pushes its way to the top of the hole. This starts the eruption. Soon the pressure in the remaining water is lowered. Then the water boils rapidly, and a large eruption occurs.

CAREER

FROZEN FOOD INDUSTRY

Andrea Henderson works at a plant where fresh vegetables are turned into frozen ones. In an interview, she had this to say about her work.

I'm the quality control supervisor at the plant. It's my job to make sure that the frozen vegetables leave the plant looking and tasting good.

Take peas, for example. They arrive at the plant in wooden crates that hold about 1,000 kg of shelled peas, and they are dirty. They go through cleaning machines—washers and shakers that remove stones and anything else. Then they are blanched.

Blanching is a process that stops enzymes from working without actually cooking the vegetables. There are enzymes in all vegetables, and if the vegetables were frozen without first blanching, they would eventually develop unappealing flavors and odors. We blanch our vegetables either in steam tunnels or in boiling water.

Blanching also kills any bacteria that are on the vegetables. When the peas come out of the blancher they are clean, so it is very important that the production lines following the blanching are perfectly clean. I have a microbiologist working for me who checks that. There are some bacteria that can survive at freezing temperatures, and those are the ones we worry about.

The peas are then frozen. The freezers, which are about the size of a large room, shake the peas along a belt so they are always moving, and they don't freeze into one big lump. The peas must be taken very quickly from room temperature down to –18°C to make sure ice crystals do not develop inside them. The faster you freeze vegetables, the better their quality will be when they thaw.

The temperature of –18°C is the best temperature to use for freezing. Some vegetables come out of the freezer as low as –26°C. However, more fragile vegetables such as cauliflower would shatter at that temperature.

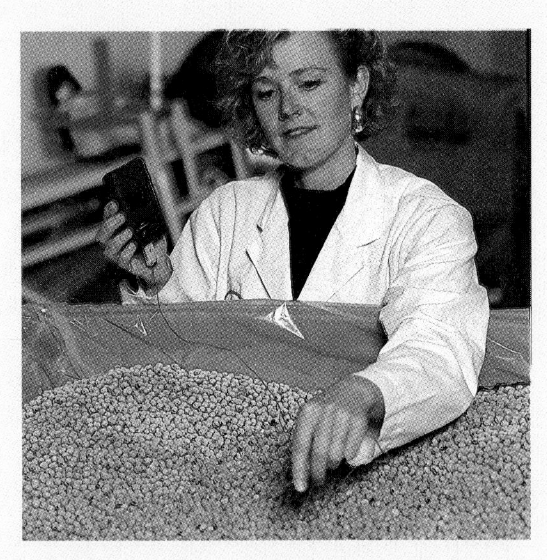

Andrea Henderson checks the temperature of frozen peas.

The crates of peas are kept in huge drive-in freezers. When we're ready to package them, we bring the peas out, inspect them again, and drop them into the bags. We ship them by truck to our customers.

On a good day we might process 100 tons of vegetables. I have to maintain the high quality of our products. To me, quality is number one. If we don't have high quality, we are not going to sell our vegetables.

I have always been interested in cooking and food. While I was at college, I worked part-time at a local food processing plant. That is when I became interested in food processing. When I completed my science degree, this plant had a job waiting for me. I am happy with the way it turned out.

What steps do you take at home to preserve the quality of the food you buy?

PROTECTING FRUIT CROPS

As a substance is freezing, it releases heat. Fruit farmers use this fact when a frost threatens to destroy their crop, especially during blossom season. If the weather forecast predicts an overnight temperature of about –2°C, the farmer can spray water onto the fruit plants when the temperature is about 1°C (Figure 20.17). As the temperature drops, the water freezes and releases heat to its surroundings. This helps to prevent the plants from freezing. This process does not work if the temperature drops too far below the freezing point of water.

FIGURE 20.17
These oranges were sprayed with water because the grower had heard a frost warning on the radio. Unfortunately, the frost was too severe, and the crop was damaged.

STEAM HEATING AND THE DANGERS OF CONDENSATION

Some older homes and buildings have heating systems that use radiators (Figure 20.18). Energy from burning fuel causes water to boil and change to steam. The steam moves through pipes to the radiators. Inside the radiators, the steam condenses back to water, releasing a large amount of energy. This energy provides heat to the room.

The energy released by condensing steam can be very dangerous. Steam burns are caused when steam from a kettle or pot with boiling water contacts skin. Always avoid placing your hand close to any source of steam (Figure 20.19).

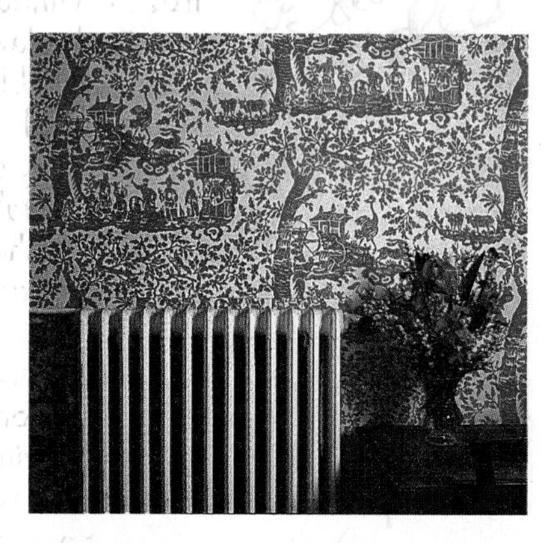

FIGURE 20.18
Steam releases heat when it condenses to a liquid in this radiator. What happens to the liquid?

Steam is condensing here.
Steam burns are most serious here where the steam has not yet started condensing.

FIGURE 20.19
Never place your hand just above the spout of a kettle with boiling water, where the steam has not yet condensed. If it condenses on your hand, it will cause a severe burn. Why does steam cause a more severe burn than boiling water?

▶ REVIEW 20.4

1. How do you think the freezing point of the mixture used to make a popsicle compares with the freezing point of water? Explain your answer.

2. Students on a hiking trip in the mountains discover that a much longer time is needed to boil potatoes at a high elevation than at sea level. Explain why this is so.

3. (a) As water changes to ice, does it absorb or release heat?
(b) Under what conditions might this help a fruit farmer?

4. Explain why a steam burn can be much more severe than a hot water burn.

5. Do you think it would be possible to cool your kitchen by leaving the refrigerator door open? Explain your reasoning.

CHAPTER REVIEW

KEY IDEAS

■ The statements of the kinetic molecular theory help you to understand and explain observations about thermal energy and heat.

■ Temperature is a way of measuring the average kinetic energy of the particles that make up a sample of a solid, liquid, or gas.

■ Thermal energy is the total energy of all the particles that make up a sample of a solid, liquid, or gas.

■ Heat is the amount of energy transferred from an object at a higher temperature to one at a lower temperature.

■ Specific heat capacity is the amount of heat required to change the temperature of 1 kg of a substance by 1.0°C.

■ There are three common states of matter and six possible changes of state. During the time a substance changes state, the temperature remains constant.

■ There are many applications of changes of state that you can observe or experience.

VOCABULARY

kinetic molecular theory
temperature
thermal energy
heat
specific heat capacity
melting
vaporization
evaporation
boiling
condensation
solidification
freezing
sublimation
heating curve
cooling curve

V1. Construct your own mind map for this chapter, including as many words as you can from the vocabulary list.

V2. Design a crossword puzzle using as many words as possible from the vocabulary list.

CONNECTIONS

C1. Use the kinetic molecular theory to explain what happens to the particles in a liquid when the liquid
(a) increases in temperature,
(b) changes into a gas,
(c) changes into a solid.

C2. In each case below, determine the specific heat capacity, and identify what the substance could be, using Table 20.2.
(a) It takes 1,920 J of energy to increase the temperature of a 100 g spoon from 20°C to 100°C.
(b) An empty container of mass 400 g releases 50,400 J of energy as it cools from 170°C to 20°C.

C3. The energy from the sun collected by solar panels is often stored in a liquid.
(a) Using the information in Table 20.2, which liquid do you think would be better for this purpose,

water or ethylene glycol (antifreeze)? Why?
(b) A mixture of these two liquids is used in solar collectors. Explain the advantage of this in the northern regions.

C4. For each situation below, name the first state of the matter involved, its final state, and the change of state that occurred.
(a) Ice cubes form in a freezer.
(b) Clothes are dried in a dryer.
(c) Ice disappears in a glass of lemonade.
(d) A bathroom mirror becomes foggy when you are taking a hot shower.
(e) Frost forms on the inside of windows on a cold winter night.

C5. Is it possible to add heat to a material without changing its temperature? Explain.

C6. On a graph, draw a heating and cooling curve for ice that starts at −10°C, melts, warms up to room temperature, cools, freezes, and then returns to its starting temperature. Label the axes on your graph, the states, changes of state, and important temperatures. Include a title box.

C7. Which is better for keeping food and drinks cool in a picnic cooler: water at 0°C or ice at 0°C? Explain why.

C8. Use the kinetic molecular theory to explain why steam can cause a serious burn.

EXPLORATIONS

E1. Think about how each of the following factors might affect how quickly water evaporates. In your notebook, write a hypothesis for each one.
(a) the surface area of water exposed to the air
(b) the depth of the water in the container
(c) the temperature of the water
(d) air movement across the water's surface
(e) the color of the water.

Design and perform a controlled experiment to determine whether or not your hypotheses are correct.

E2. In places that receive too little rainfall, a process called "seeding the clouds" has been used to try to solve the problem.
(a) What do you think is meant by "seeding the clouds"?
(b) Look up this process in a reference book. Describe what you discover on a poster, with a diagram showing the steps in the process.

E3. Cold water from a river, lake, or ocean may be circulated through a factory to cool machines. The heated water is then returned to the river, lake, or ocean it came from. This added heat often causes problems for the plants and animals living there. In some cases, this thermal pollution of bodies of water has been reduced by pumping the heated water from the factory through a large cooling tower. Much of the extra energy then goes into the air. However, the result is thermal pollution of the air rather than the water. Suggest how industries could reduce the amount of thermal pollution of water and air.

E4. Describe in detail how you would determine the amount of heat needed to warm the water you would use to fill your bathtub at home. To perform a sample calculation, estimate the temperatures of the water before and after you heat it. You can also use an estimate of the mass of the water in the tub.

REFLECTIONS

R1. Refer to the lists you made in Activity 20A. Now that you have completed this chapter, have you changed your mind about the ideas you listed? Use ideas you have learned in this chapter to add to your original lists.

R2. Write at least four questions about heat that you would like more complete answers to.

MANAGING OUR ENVIRONMENT

What do you think these boys will find in the water at the edge of this pond? Tadpoles? Insects? Fish? Whatever they find, it is part of a much larger picture, the area that makes up the valley floor around the pond. In this unit, you will learn about the way scientists describe large areas of the Earth's surface. Some of these areas may look like the area around the pond, or like the forest on the nearby mountainside, or like the distant mountain peak. You will find out that each of these large areas has characteristic kinds of plants and certain kinds of soil for the plants to grow in.

In this unit, you will also discover how much all the organisms in an area are affected by human beings. As an example, you will study the Pacific salmon and see how sensitive it is to changes in the streams, the rivers, and the Pacific Ocean, where it lives. The salmon is an important resource for humans. It is also an indicator of how well humans are sharing the Earth. If we cannot protect the organisms of the Earth, such as the salmon, how much longer will we be here ourselves?

UNIT VI

CHAPTER 21

THE LIVING PLANET

These photographs show three kinds of living things, or organisms, and their surroundings. Take a moment and look for clues in the photographs to help you match each organism to where it lives.

The biosphere—the part of the Earth where life can exist—is like a puzzle. Why do certain organisms live where they do? Why are certain species found together? How do parts of the biosphere change over time?

Perhaps you have wondered if there is a way to predict where different living things are found on Earth. In this chapter, you will search for patterns showing where different types of organisms live and look for the reasons for such patterns. By the end of the chapter, you will have a better understanding of how organisms interact with their surroundings, or environment. You will also have a better appreciation for the variety and beauty of nature—and why many scientists, world leaders, and citizens are concerned about protecting it.

ACTIVITY 21A / ASSIGNMENT: EARTH

Make a three-column chart in your learning journal. In the first column, list several kinds of places you can think of on Earth where you would expect to find wild animals and plants. For example, you might think of a tropical rain forest, a desert, and several other places. In the second column, for each place you have listed, write down several abiotic (nonliving) and several biotic (living) parts of the environment. (Abiotic parts of the environment could include soils, landforms, and temperatures. Biotic parts include plants and animals.) In the third column, write down what you would like to know about each place. For example, if you wrote "prairie" in your first column, you might list things such as flat landscape, prairie dogs, low rainfall, grass, and coyotes in your second column. In your third column, you might ask, "Where could I go to see a wild prairie, without any ranches or farms?" ❖

■ 21.1 THE PATTERNS OF LIFE

Imagine that you are a filmmaker and you have been asked to make a nature film. In order to show the wide variety of natural environments on Earth, you decide to begin with a selection of photographs taken by astronauts from space.

When you examine the astronauts' photographs, you quickly find wonderful examples of the variety of environments that you expected (Figure 21.1). You also find that many areas on Earth look very similar, even though they are far apart geographically. At the same time,

FIGURE 21.1 ◀
How does this view of the Earth, taken from space, help you identify the natural environments in which different types of plants and animals live?

you notice that one kind of environment often extends over a large distance.

How will you describe the features of the Earth in the script of your film? What patterns of life can you use to organize the information in your film? What information do you need about the Earth's surface?

ACTIVITY 21B / LOOKING FOR PATTERNS

An important part of film-making is organizing the visual images and spoken words to tell the right "story." For instance, if you want to produce a nature film that begins with the astronaut's view, you would need to organize information about the different environments seen from space. One way would be to classify each region and give it a useful name. Another way would be to make a map showing the size of different environments.

As you perform this activity, look for clues that might explain why certain organisms are found in certain places.

PART I Classifying Natural Areas

PROCEDURE

1. Prepare a table and record the following information for each of the environments shown in Figure 21.2:
   - vegetation (kind of plants, number of plants)
   - appearance of the landscape (sloping, hilly, flat, etc.)
   - weather when photograph was taken
   - other observations you think would be valuable
2. Using the information in your table, choose a name or term you could use in your film script to describe each environment. (Hint: Select a name that will help your audience remember the most striking feature in each.)

DISCUSSION

1. (a) What features are the same in the environments shown in Figure 21.2?
   (b) What features are different?
2. What feature(s) helped you to name each environment? Explain your reasoning.

(a)

(b)

FIGURE 21.2
Use the information in these photographs for Part I of Activity 21B.

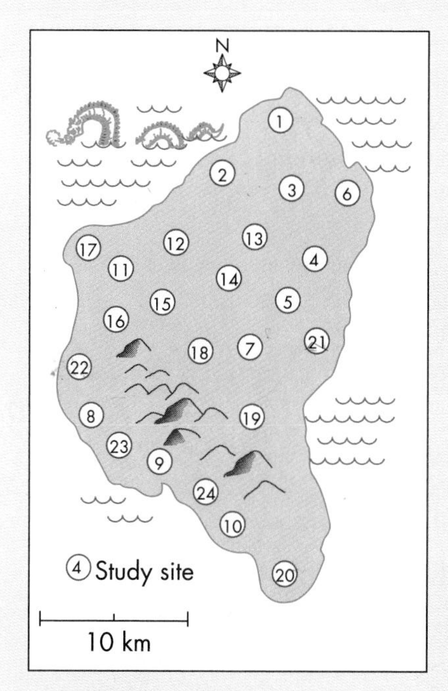

FIGURE 21.3
Noseeumland. Your enlarged copy of this map should fill one page of your notebook.

3. (a) Describe a place you have visited that has a natural environment different from those shown in Figure 21.2.
(b) What name would you use for this environment? Explain.

4. Why do you think names are a useful step in classifying environments?

PART II Mapping Living Things

PROCEDURE

One way to organize information about living things is to produce a map. Imagine that you are a member of a survey team sent to map the kinds of life on a newly discovered island called Noseeumland.

1. Copy the map of Noseeumland (Figure 21.3) into your notebook, or obtain a copy from your teacher. Each of the circled numbers represents a study site. Table 21.1 shows the observations made at each study site in one day.

2. Examine the data in Table 21.1. Decide which information will help you group similar sites. On your map, draw boundary lines around each group of similar sites.

3. Suggest a name for each of the groups you have chosen. Use the same type of naming system that you developed for Part I, adding names or using different ones if necessary.

Study site number	Plant life	Animal life	Landscape	Weather
1, 2, 6, 17, 20, 21	Short grasses	Antelope Prairie dogs Bison Horned larks Jackrabbits	Rolling hills	Clear and dry Windy
3, 11, 12, 13, 22	Tall grasses	White-tailed deer Cottontail rabbits Garter snakes	Flat plain	Clear and dry
4, 5, 7, 14, 15, 16	Cacti Small shrubs	Jackrabbits Rattlesnakes Kangaroo rats Ground squirrels Lizards Scorpions	Flat plain	Clear and dry
18, 19	Small trees Shrubs Open spaces	Wild Turkeys Elk	Gradual slope	Clear and dry
8, 9, 10, 23, 24	Tall trees Fern undergrowth	Mule deer Tree squirrels Black bears	Steep slope to ocean	Foggy and rainy

TABLE 21.1 Information Obtained for One Day, in Noseeumland

DISCUSSION

1. What characteristics did you use to group the study sites on Noseeumland into larger environments?
2. Describe how you think the boundaries between any two large environments of Noseeumland would appear if you were to walk from one area into another.
3. How could the larger environments you added to your map of Noseeumland be useful? Explain how you could use such a map, giving examples to support your answer.
4. How could a map like yours be used by a filmmaker to show patterns of life on Earth?
5. What do you think caused the patterns you mapped? ❖

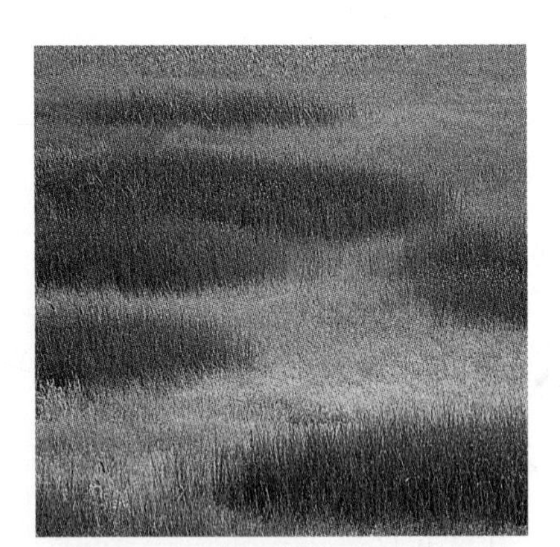

FIGURE 21.4
Identify the two biomes shown here. What differences would you expect to find in the kinds of animals found in these biomes?

MAPPING THE PATTERNS OF LIFE

If you did Activity 21B, the methods you used to map and name the areas of Noseeumland are the same as the methods used by scientists to map large areas of the Earth. Maps and names are useful tools for people who study the organisms found in different places. These maps also help people who plan the future uses for land.

A **biome** is a large geographical land area that has the same general kinds of plants and animals throughout. One example is the grassland biome, found in south central Canada and the central United States, also called prairie. The grassland biome is found in many other parts of the world. No matter where in the world you find a grassland biome, it will contain grasses that look about the same and grazing animals that feed on the grasses in similiar ways. The many different biomes on land make up part of the biosphere.

Figure 21.4 shows what you might see if you were to take a trip from a grassland biome to a desert biome. The most obvious difference is seen in the plants of the two biomes. A desert contains shrubs and cacti, whereas the grassland contains grasses. Plant species that are the most abundant or the largest in a location are called **dominant plant species**.

Ecologists may identify a biome and even name it for its dominant plant or plants, such as grassland or coniferous forest. Each biome contains one or more **ecosystems**. Recall that an ecosystem is a community of organisms that interact among themselves and with their environment. The dominant plant species are a very important part of that ecosystem because plants are the food producers in any ecosystem. Through photosynthesis, plants bring solar energy into the food web. Animals, as consumers, can only live where they can find their food. This means that certain plants must be growing in a place before the animals that feed on these plants will be found there. For this reason, plants are very useful in mapping biomes (Figure 21.5).

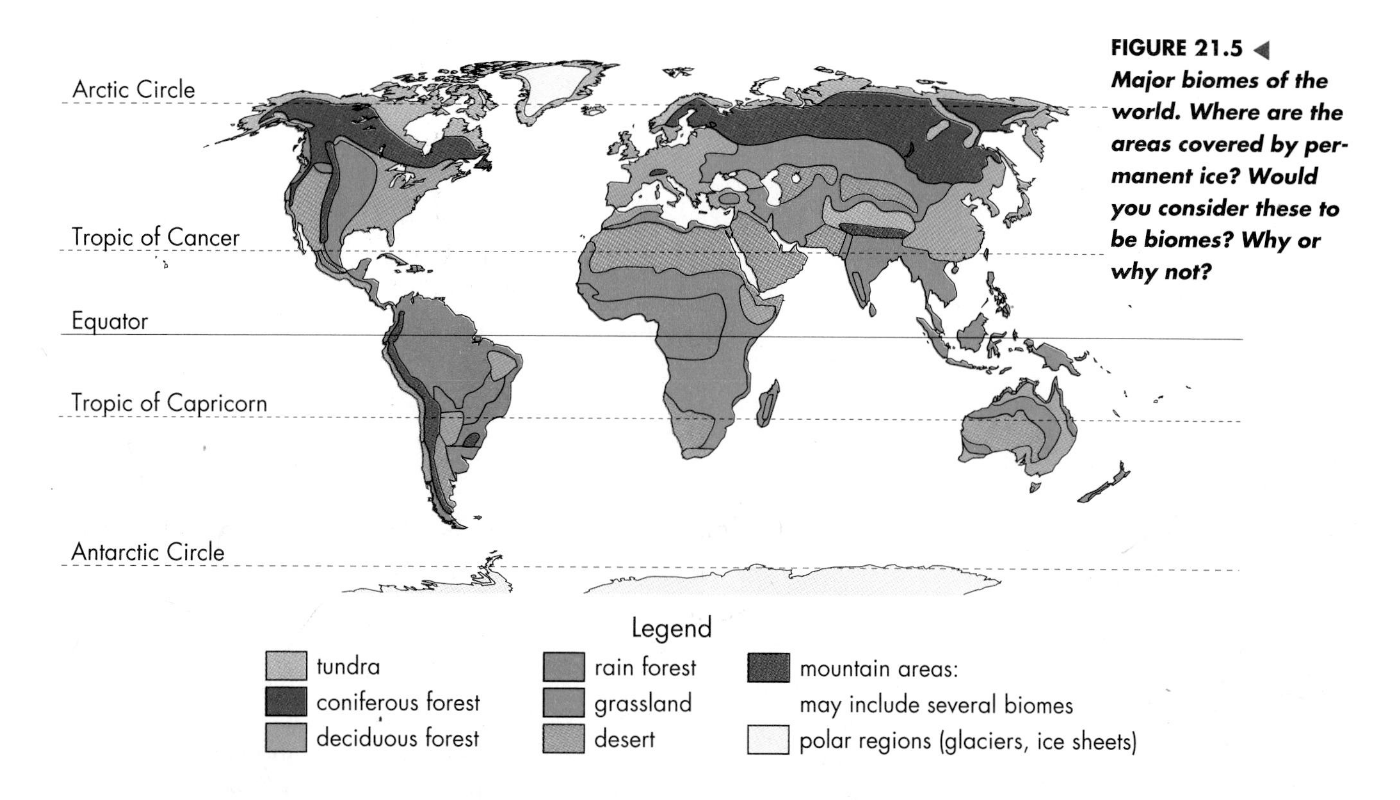

FIGURE 21.5 ◄ ***Major biomes of the world. Where are the areas covered by permanent ice? Would you consider these to be biomes? Why or why not?***

BIOMES AND CLIMATE

Plants may be the most visible way to identify a biome, but there is a more significant factor that makes each biome distinctive. **Climate** refers to the weather conditions that you would notice at a certain location over a long period of time. Weather conditions include the amount of rain or snowfall (precipitation) and the temperature.

The map in Figure 21.5 shows the major biomes on Earth. On a map this small, there is room to show only very large categories. These major biomes are often named after the plant life in the region. They can be further divided into smaller biomes that reflect regional conditions, particularly in climate. For example, the grassland biome can be divided into a temperate grassland biome and a tropical grassland biome. In the mountainous regions, there can be several biomes as you climb from valley floor to mountaintop.

Climate affects where certain kinds of plants and animals can survive. For example, plants such as banana trees or pineapple trees will grow well in a tropical country near the Equator but not in central North America because of the freezing winter temperatures.

Scientists use graphs called **climatographs** to help show what the climate of an area is like (Figure 21.6). Climatographs show the average monthly temperature and total monthly precipitation for one year. Since climate affects where living things can survive, climatographs can provide valuable clues to understanding the location of biomes.

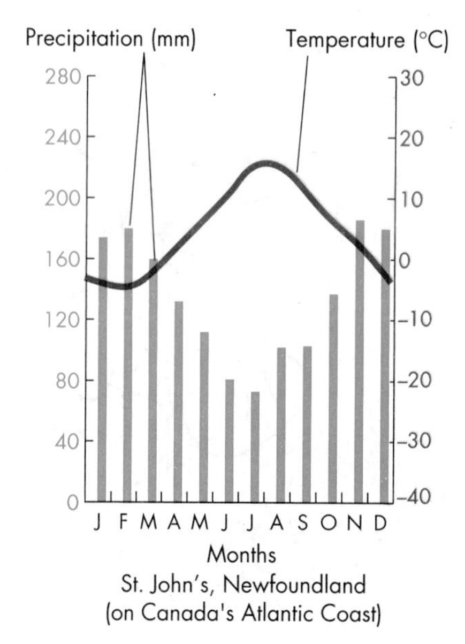

FIGURE 21.6
A climatograph for St. John's, Newfoundland. The curved line shows average monthly temperature; the bars show average total precipitation per month. You could draw a climatograph of your area; contact your nearest weather office for the data.

THE NATIONAL WILDLIFE REFUGE SYSTEM

It is likely that almost everyone is aware of our country's great National Parks. When we hear names like Yellowstone, the Grand Canyon, Mt. Rainier, Yosemite, and the Great Smokey Mountains, we immediately have a mental image of these parks. Unfortunately, the same cannot be said of the nation's National Wildlife Refuge system. Names such as Sabine, Aransas, Klamath Basin or Pelican Island National Wildlife Refuge are not likely to bring anything to mind. This discrepancy is not accidental.

The National Wildlife Refuge system is a collection of U.S. lands and waters preserved and maintained purely for wildlife, not for humans. The areas provide habitat for endangered species and hundreds of other species of birds, mammals, reptiles, amphibians, fish and plants. The refuges trace their origin to 1903, when President Theodore Roosevelt established the Pelican Island Federal Bird Reservation. A series of legislative acts are responsible for the expansion of the system. Today, the system is the world's largest system devoted to wildlife protection.

It is possible that you live near a National Wildlife Refuge and are not aware of its existence. There are 508 such refuges in the system with all 50 states represented. The system comprises 94 million acres, half the area of Alaska. By comparison, the National Parks system has 78 million acres. Many of the wildlife refuges are inacessible wetlands along the major north-south flyways used by large migratory birds such as ducks, geese, herons, eagles, and hawks. These refuges provide feeding and resting areas, and are important to maintaining populations of these animals. Many endangered species such as the bald eagle, ocelot, grizzly bear, and American crocodile find a haven here too.

While the National Parks feature geysers, glaciers, great mountains and other spectacular scenery that attract millions of visitors, the refuges regulate public access so as not to interfere with the birds, mammals and other wildlife that inhabit the area. Many of the refuges in the system do not have a visitor center or even paved roads. In spite of this, some public activities are permitted: wildlife observation, photography, hiking, and interpretive trails.

One example of a National Wildlife Refuge is the J. N. "Ding" Darling National Wildlife Refuge on Sanibel Island on the Gulf Coast of Florida. The "Ding" Darling refuge has a diversity of habitats including shallow bays, mangrove swamps, tangled sub-tropical forest, and sandy beaches. Among the wide variety of species that call the refuge home are alligators, herons, egrets, the rare roseate spoonbill, and the American crocodile, an endangered species much rarer than the alligator. Squadrons of brown pelicans soaring in formation along the beaches are a common sight. Because of financial support from local conservation groups, a $2 million visitor center is planned.

ACTIVITY 21C / HOW CLIMATE AFFECTS THE DISTRIBUTION OF BIOMES

In this activity, you will use climatographs to investigate how climate affects the location of certain biomes.

PART I Far Apart, but Similar

CROSS-CURRICULAR

PROCEDURE

1. Use an atlas to find out the distance between Seattle, Washington, and Iquitos, Peru. Then find the distance from each city to the Equator.

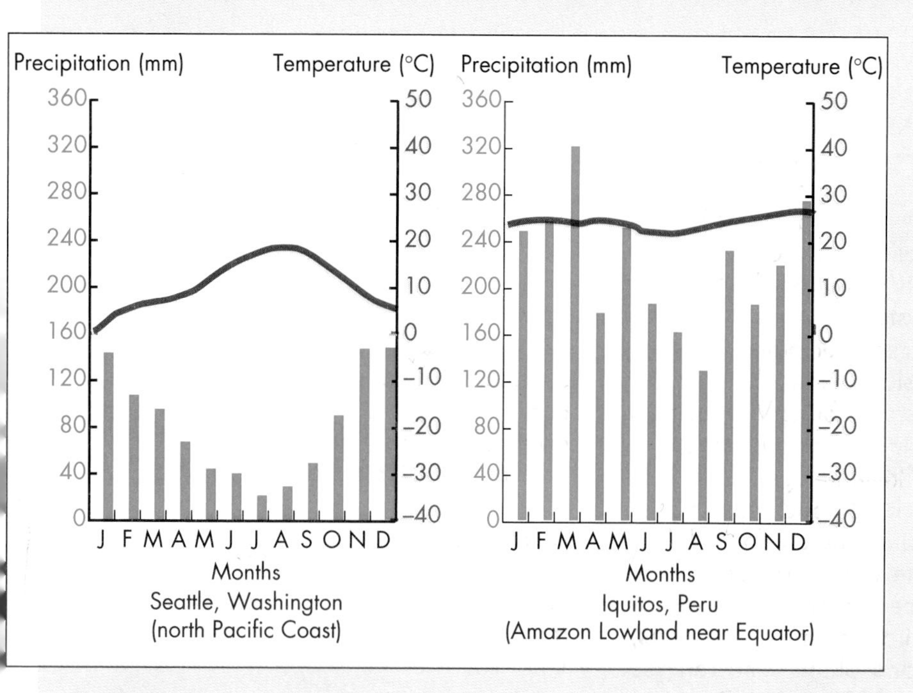

FIGURE 21.7
Climatographs for Part I of Activity 21C

2. Name the major biome each city is located in, using the map of biome distribution in Figure 21.5.
3. Compare the climatographs for the two cities (Figure 21.7). What are the similarities and differences? (Hint: Consider the yearly pattern, or climate, as a whole, rather than month by month.)

DISCUSSION

1. Why do you think the distances you calculated in Step 1 could result in a difference in the type of biome found at each location?
2. (a) What major biome are both these cities located in?
   (b) How might you divide this major biome into two smaller ones? What names would you give these smaller biomes?
3. If you were to bring plants from Iquitos to Seattle and plant them in your garden, do you expect they would survive? Why or why not?

Month	Jan.	Feb.	Mar.	Apr.	May	June	July	Aug.	Sept.	Oct.	Nov.	Dec.
Average monthly temperature (°C)	−3	0	4	9	13	17	21	20	15	9	2	−1
Average monthly precipitation (mm)	51	39	34	28	34	32	14	16	20	30	52	55

TABLE 21.2
Temperature and Precipitation Data for Spokane, Washington

PART II Close, but Not the Same

PROCEDURE

1. Using the data in Table 21.2, draw a climatograph for Spokane, Washington. Remember that on climatographs, a curved line is used to represent average monthly temperature, and bars are used to represent total monthly precipitation.
2. (a) Use an atlas to determine the distance between Spokane and Seattle.
   (b) Determine the latitude of Spokane and compare it with the latitude of Seattle.

3. Use your atlas to find the geographic feature or features that might explain the difference in climate between Seattle and Spokane.

DISCUSSION

1. Use your climatograph of Spokane to describe the annual pattern of weather, or climate, for this site.
2. Do you think latitude is important in producing the differences you observed between Spokane and Seattle? Explain your answer.
3. Would you expect to find the same type of plants in Seattle and Spokane? Why or why not? ❖

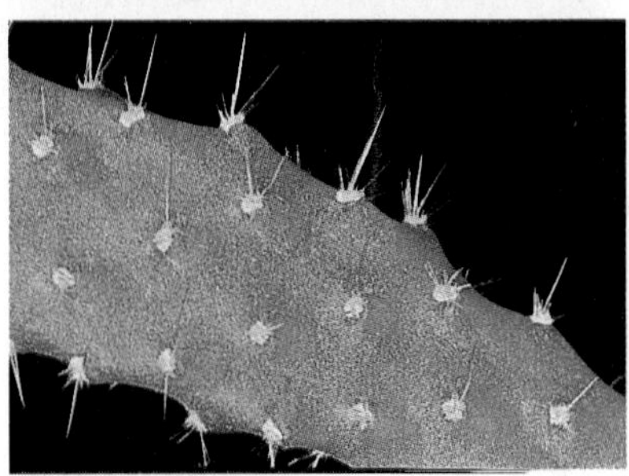

FIGURE 21.8
What features of these plants help them survive in the harsh desert climate?

A CLOSER LOOK AT THE EFFECT OF CLIMATE

The two most important features of climate are precipitation and temperature. How do you think they affect your everyday life? Precipitation may be important to activities you enjoy, like swimming or skating. Cities and farms need to be located where there is enough fresh water. Temperature affects your everyday life too. Just think of how the seasonal change from summer to winter temperatures affects what you wear and do outdoors.

How do precipitation and temperature affect the other living things in your area? How do patterns of precipitation and temperature affect the distribution of different biomes?

PATTERNS RELATED TO PRECIPITATION

Consider a desert biome. Deserts receive less than 25 cm of precipitation per year, and they often experience high temperatures that increase the amount of water that evaporates. If you observe desert organisms closely, you will see that each species has characteristics that help it survive with little or no water. These are **adaptations**—features that help an organism to survive in its particular environment.

For instance, why do you think desert plants grow far apart? Each plant needs a great deal of room in order for its roots to gather enough water to survive. Another adaptation of desert plants is the ability to store extra water in their stems or roots. Desert plants have smaller leaves to reduce the amount of evaporation, and they may have spines for protection against animals seeking food and water. Figure 21.8 shows two plants that have adapted to life in desert climates.

In contrast, rain forest biomes are found in areas where more than 150 cm of precipitation falls each year. Unlike desert plants, rain forest plants compete for sunlight rather than for water. Rapid growth to reach the sunlight is more important to rain forest plants than adaptations that store water or reduce evaporation (Figure 21.9).

Since plants bring food into their ecosystem, the more plants there are, the more food there will be in the ecosystem. There are far more plants in a rain forest than in a desert because conditions for plant growth are much better. Therefore, more animals can live in a rain forest than can live in a desert.

FIGURE 21.9 ◀
A great variety of species is found in the tropical rain forest biome.

FIGURE 21.10
During the cold winter months, many plants protect their new leaves inside tough buds. What is the advantage to a plant in having leaves ready to open when warmer weather arrives?

PATTERNS RELATED TO TEMPERATURE

Most plants can survive very warm weather as long as they take in more moisture than they lose through evaporation. For example, in hot tropical rain forests, millions of litres of water evaporate from the leaves of the trees each day. But the large amount of rainfall quickly provides more water to replace the water lost by evaporation.

In other biomes, low temperatures, especially those below the freezing point of water, present a different problem (Figure 21.10). Plants do not have the ability to keep themselves warm. In cold climates, their life functions slow down considerably. If the water inside the plant freezes, it can cause damage or death.

Plant species in biomes with cold climates have adaptations to overcome the problem of freezing. For example, before winter, many plants produce seeds that are resistant to freezing. Other plants store energy in their roots. They use the energy once the temperature rises again. Such plants are inactive, or **dormant**, in winter. There are other differences in adaptation found in the two main kinds of trees: deciduous and coniferous. Deciduous trees, like those in Figure 21.2a, drop their leaves during the cold months. With the loss of leaves, less water evaporates from the tree. Most coniferous trees do not lose their leaves. They have thin, wax-coated leaves called needles that do not become frost damaged (Figure 21.11). Needles have a much smaller surface area than leaves of deciduous trees, so needles do not lose water as easily.

FIGURE 21.11
What features do these leaves have that help the tree survive extremely low temperatures?

FIGURE 21.12
An adaptation that helps a species survive can be a behavior as well as a body part. What behaviors help these animals survive winter?

Animals also have adaptations that help them survive the conditions in a particular climate. Each species must be suited to the habitat where its food can be found. Look at Figure 21.12 and find the adaptations that make each of these animals well suited to the climate of its usual environment.

Now the definition of a biome can be expanded to include a characteristic climate as well as plants and animals.

▶ REVIEW 21.1

1. What is a biome?
2. What is a climatograph and how is it used?
3. Which is more important to consider when you compare climates—latitude or longitude? Give reasons for your answer.
4. Explain how the plant life of a biome determines the animal life that will be found there.
5. (a) State two examples of plant adaptations in a biome where there is a shortage of water.
   (b) State two examples of plant adaptations to low temperatures.
   (c) Would you expect to see these kinds of adaptations in the species of plants that live in tropical rain forests? Explain your reasoning.
6. Trin is a bird-watcher who lives in a forest in a northern region. She always sees more species of birds around her home in summer than in winter. One winter, Trin took a trip south to a desert region, hoping to see more birds. Even though the desert was quite warm, she actually saw fewer different species than she did at home in the snow. Explain Trin's observations, using the information you have gained about climate and adaptation.

■ 21.2 A TRAVELLER'S LOOK AT CALIFORNIA

Lang, Karen, Alishia, and Derek were among several high school students hired for a summer by a local conservation group. Their job was to design and create a set of booklets describing the natural areas in California (Figure 21.13). The booklets would be used by backpackers and hikers. The students hiked in each area, then talked to naturalists about what they observed. They often divided the task of writing the descriptions among themselves.

FIGURE 21.13
A view of California coastline. How many different biomes are in this state? What is the predominant biome?

ACTIVITY 21D / A TRAVELLER'S NOTEBOOK

CROSS • CURRICULAR

Divide a page in your notebook into four parts. In each part, put the name of one of the biomes described in the next section. As you read, make brief notes about the climate found in each biome. When you finish reading about all four biomes, go back over your notes and use a colored marker to highlight the features of climate that seem to affect the animals and plants the most. ❖

FIGURE 21.14 ▶
The alpine tundra biome. Where would you find this biome?

FIGURE 21.15
Lichens are a combination of algae and fungi. Lichens that grow on rock begin the process of forming soil by gradually breaking down the rock.

THE ALPINE TUNDRA, BY LANG

"The air in the mountains is cold and dry," Lang wrote (Figure 21.14). "The rocky landscape seems empty, but if you look carefully, you will see small living things. Most plants up here would not reach the tops of your shoes. They grow in any sheltered patch of soil and have bright flowers held close to the ground. Almost every boulder is covered in colourful lichens (Figure 21.15).

"This biome is called the alpine tundra biome. This treeless area high in the mountains is similar to the arctic tundra biome next to the Arctic Ocean. Both of these biomes have climates that are extremely cold. In many places, the ground is frozen all year long. Fewer species of plants and animals are able to survive in alpine and arctic tundra than in most other biomes. The photographs show two animals you might see in the tundra (Figure 21.16).

"One thing travellers notice is the lack of flying insects at very high altitudes in the mountains. High up in this biome, many insects are wingless since flying insects are blown around by the strong mountain winds. When the air is calm, however, you may see bees and flies on the alpine flowers."

(a)

(b)

FIGURE 21.16
Animals of the alpine tundra. What adaptations do you think help these animals survive?

THE CONIFEROUS FOREST, BY KAREN

Karen's booklet starts: "A large biome in the state is the coniferous forest biome, where coniferous trees are the dominant plants. You will enter this biome if you leave the alpine tundra and travel down most mountainsides. But first you will pass through the **tree line** (Figure 21.17). The tree line is the high-altitude limit of the coniferous forest. In order to grow, trees need soil that thaws completely for at least part

FIGURE 21.17
Locate the tree line in this photograph. What boundary does it form?

FIGURE 21.18
In the coniferous forest, dead trees decompose slowly. Other plants and trees start to grow on the decomposing wood. Plants and fungi that thrive in moist shade grow through the carpet of dead needles that covers the ground.

of the year. Above the tree line, the climate is too cold for trees to survive. (There is also a tree line in northern Canada, where the northern coniferous forest gives way to tundra.)

"The tree line is not a sharp boundary. In the zone between the alpine tundra biome and the coniferous forest biome, the trees grow as short, windblown shrubs. This zone is called a **transition zone** because it has similarities to both biomes. As you walk down the slope into the forest, you will notice first that the trees grow in small, widely separated stands and then that they become more densely packed and grow taller and straighter. Transition zones contain a mixture of organisms from both biomes. Many animals migrate across the transition zones, changing biomes as the seasons change.

"The fir, spruce, pine, and cedar towering above your head are home to many animals. Next take a look at the ground. It's carpeted with smaller plants that can live in the shade, such as mosses and ferns (Figure 21.18). They grow right through a layer of needles. The needles that fall from coniferous trees affect the soil in this biome. When the needles decay, they produce acid. Under the needles, the grey soil is acidic. This soil is low in nutrients because the dead plants do not decay very quickly in acidic soils. Also the acid makes nutrients wash out of the soil rapidly.

"One photograph shows a moose (Figure 21.19a). It's one of the largest animals in the coniferous forest. Moose eat vegetation found in the many bogs and marshes in the region. After feeling for roots under the water, the moose uses its lips to grab a mouthful. In winter, moose eat twigs, bark, and needles of conifers.

EXTENSION

■ Forest biomes are classified by the dominant tree species found in "old growth" areas. Find out what this term means. How much of California's forest is actually old growth?

(a)

(b)

FIGURE 21.19
Animals of the coniferous forest. How do each of these organisms depend on the plants in this biome?

"In spring and summer, the forest is full of birds and their songs. About 50 species of birds feed on the conifers. Some birds, like spruce grouse, eat needles. Others, like crossbills and siskins, eat the seeds from the cones. Many birds eat insects. The second photograph shows a red squirrel, which also eat seeds from the cones of the trees. (Figure 21.19b)."

THE DESERT, BY ALISHIA

"When you travel east from the coastal mountains in southern California," Alishia notes, "you will notice that there is less coniferous forest and more grassland. In the driest part of this region is the desert biome, the Mojave Desert. Shrubs adapted to the desert and cacti grow here (Figure 21.20). The dry climate is caused by the mountains to the west. My diagram shows what happens when moist air rises over the mountains from the oceans (Figure 21.21). As you can see, most of the rain and snow is on the ocean side of the peaks. By the time this air reaches the other side of the mountains, it is drier. In the areas on the east side of these peaks, long periods without rain are common. This dry side is called the rain shadow. There are deserts in the driest part of the rain shadow.

FIGURE 21.20
In this photograph, the desert biome occurs only in the distant valley bottom. Snowfall provides water for the forests of spruce and pine above the desert. Where does this biome occur in California?

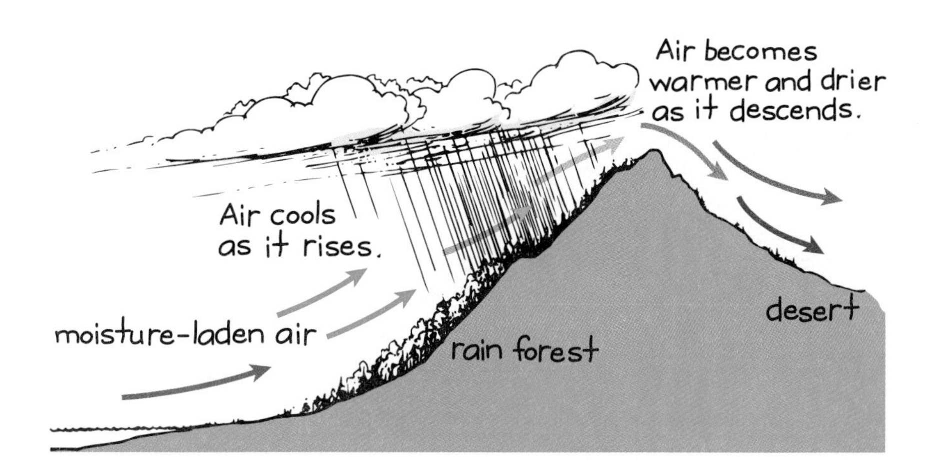
Air becomes
warmer and drier
as it descends.
Air cools
as it rises.
moisture-laden air
rain forest
desert

FIGURE 21.21
Explain what is happening in Alishia's diagram. Why are there more plants on the windward side of the mountain range than on the other side?

EXTENSION

■ How do reptiles survive in a desert? Prepare a presentation for your class on two different desert reptiles.

"Desert animals have had to adapt to survive in the dry climate. Some animals get their water from the food they eat and do not have to drink any water at all. Certain types of behavior also help them to conserve water. Most desert animals carry on their activities at night to avoid the drying heat of the sun. These animals are nocturnal. They spend the day in burrows or in holes in plants. Staying out of the sun and heat helps them to avoid losing water. The photographs show two animals that have adapted to the desert (Figure 21.22)."

FIGURE 21.22
How do these desert animals keep their bodies cool? Why is keeping cool important to their survival in this climate?

THE TEMPERATE RAIN FOREST, BY DEREK

Derek wrote: "When you travel north from San Francisco along the coast, you will be heading into mountainous terrain. The air starts to feel cool, although it is still very dry. The shade from the forest feels refreshing after the heat of the interior. As you reach the westward side of the mountains, you notice that the vegetation has changed dramatically. This biome, the temperate rain forest biome, is found only along the coast of California (Figure 21.23).

"The air in the rain forest is often damp, for this biome receives a great deal of precipitation. Redwood, spruce, hemlock, western red cedar, and Douglas fir are the main trees found in this forest. The climate is moderate, with higher temperatures in most months than in the interior. The moderate temperatures and the high rainfall allow trees to grow tall and close together. Their leaves and needles shade the forest floor, letting ferns and mosses grow. All year, you can see deer moving through the forest. If you come in winter, you will notice that it feels cold in the damp air. But you won't mind the cold if you are lucky enough to see elk."

FIGURE 21.23 ▶
A temperate rain forest. What evidence can you find in this photograph that the climate is usually moist and cool? What animals would you expect to find in this biome, and how do they depend on the plants here?

▶ REVIEW 21.2

1. (a) What other biome is similar to the alpine tundra?
   (b) What two environmental factors most affect plants and animals in both of these biomes?
   (c) Why is there a tree line?
2. (a) How is the soil in the coniferous forest biome different from the soil in the alpine tundra biome?
3. Suppose you are hiking up a trail from the bottom of a mountain in the temperate rain forest to its top in the alpine tundra. How do you think the increasing altitude affects each of the following? (Use the information in this chapter in your answers.)
   (a) the amount of precipitation
   (b) the average daily temperature
   (c) the strength of the wind
   (d) the type of soil
   (e) the number of species making up the ecosystem
4. During a long period of dryness, or drought, the amount of precipitation falling in an area is drastically reduced. Predict the effect of a long drought on
   (a) the kinds of plants and animals normally found in a temperate rain forest,
   (b) the kinds of plants and animals normally found in a desert.

■ 21.3 BIOMES AND CHANGE

So far, each area has been described as if it stayed exactly the same over time. However, as you can see in Figure 21.24, areas inhabited by living things change with time. Every living thing changes its own environment and that of its neighbors in the same ecosystem.

Some changes are small, such as the shading of a few square centimeters of soil by a tuft of grass. Other changes are immense, such as the tons of leaves dropped to the ground each year in a forest. Both small and large changes alter the environment of each organism as time passes.

(a)

(b)

FIGURE 21.24
These two photos show the same area years apart. What changes have occurred over the years?

PIONEER SPECIES

Think about what would happen if you prepared the soil for a flower bed near your home in the spring but did no planting. What sort of plants would begin to grow, if any? Would these same plants reappear in the following years?

Certain kinds of plants would naturally start to grow on your flower bed. These hardy plants are called **pioneer species**. They have the ability to grow quickly on bare ground as long as they receive a lot of bright sunlight (Figure 21.25). And they can survive harsh conditions. Pioneer species outgrow other plants in new areas, which is why some of these species are called weeds. What happens to these pioneers as time passes?

As the first-generation pioneer species grow, they shade the soil with their leaves. This small change to their environment slows the growth of the next generation. However, other plants, including shrubs and trees, can get started on shaded soil. Eventually, the shrubs and trees grow larger than the pioneer plants. The shade eventually prevents the pioneer species from growing at all.

FIGURE 21.25
The dandelion is an example of a pioneer species. It is able to sprout and grow almost anywhere as long as it has water and sunlight. Over time, how will this plant change its local environment?

SUCCESSION

The process of gradual change in the organisms that make up a community is called **succession**. Starting with bare ground, the change from pioneer plants to plants that prefer shade is the first step in succession. Succession occurs when living things change the abiotic and biotic parts of their habitat. Another set of species more suited to the newly altered environment pushes out the earlier species. As the plant community changes in succession, the animals change too.

Succession ends only when the changes caused by a particular community of organisms produce the conditions most favorable to the survival of those organisms and no others. This is the **climax community**. For example, the needles that drop from the trees of a coniferous forest make the soil acidic and low in nutrients. Coniferous tree seedlings

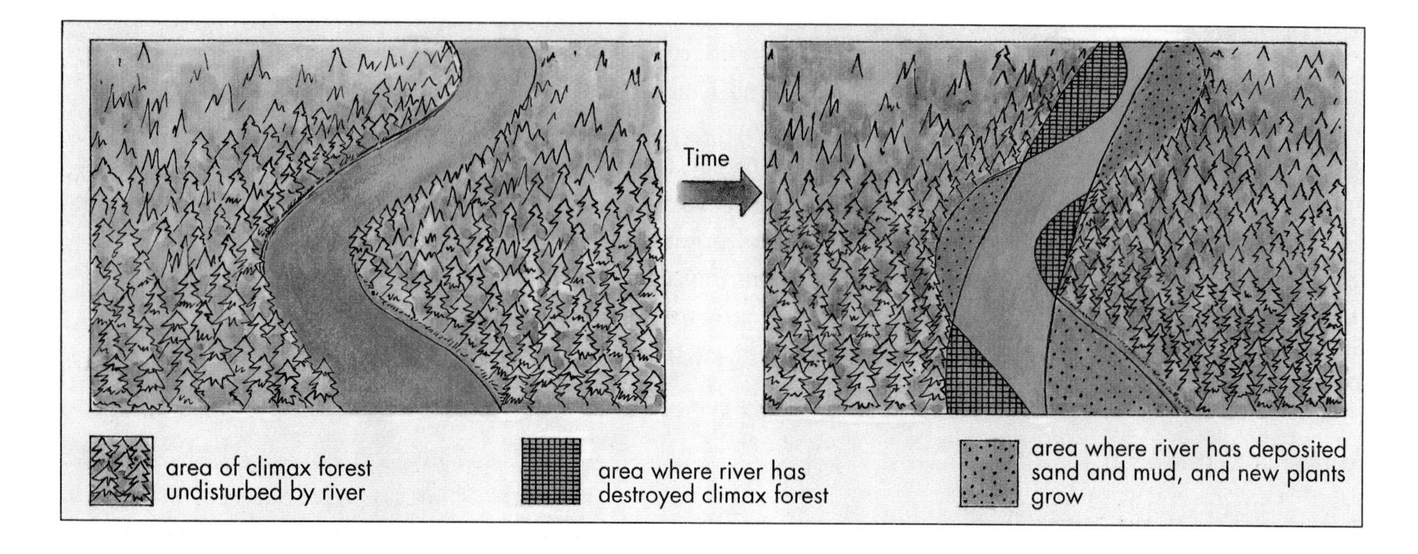

FIGURE 21.26
The path taken by a river changes over time as the soil at some bends is washed away and new sediments are deposited in other places. These changes affect the kinds of plants found along the river banks.

grow well in this soil. Other plants do not. The climax community of coniferous trees will remain because new generations will replace the older trees as they age and die.

A climax community will stay about the same unless the community is disturbed by catastrophes such as disease, forest fires, volcanic eruptions, floods, or human destruction of the habitat.

When ecologists identify and describe a biome, they pay particular attention to the plants in the climax community. The dominant plant species is the plant most commonly found there.

AN EXAMPLE OF SUCCESSION

How is it possible to find out what changes in succession are typical of an area? The best way is to find an area in which succession takes place over and over again. This happens along the banks of mountain river valleys.

These rivers often wind through the landscape. Because these rivers change their paths over the years, succession is constantly taking place along their twists and bends.

In Figure 21.26, you can see how a river's changing path has an effect on the plants along its banks. First the river erodes the shore along one side. As this happens, any plants growing there, including climax species, are destroyed. Eventually the river's path changes again, this time exposing the rocks and gravel of the riverbed to the air. Pioneer species start to grow, beginning succession.

This process takes place all along the river, continually affecting the types of plants found in the valley. If you walked beside this type of river, you would pass through plant communities that range from undisturbed climax species to new pioneers. Figure 21.27 shows the major stages in valley succession. You can think of these photographs as "snapshots" of succession.

FIGURE 21.27
Succession along a river.
(a) As the water changes its path through the valley, bare rocks and gravel are exposed to the air. The seeds of pioneer plants are able to take root and grow. The pioneer grasses and mosses form mats of soil as they trap and hold fine particles of sand.

(b) After several years, there will be enough soil to support the first trees, such as alders and willows. Their decomposing leaves further build and enrich the layer of soil.

(c) In a hundred years, there is a forest of trees that need plenty of sunlight, such as red alder and native crab apple. The willows were shaded by the bigger tree species, and did not survive. The soil is now deep and rich.

(d) A climax community of western hemlock, redwood, spruce, and red cedar has grown taller than the alders, which have not survived. There are no longer any deciduous trees. Mosses and ferns carpet the forest floor. Hemlocks outnumber the other coniferous trees, because hemlock seedlings are well adapted to growing in shade. From bare rock to climax community, this succession took several hundred years.

EXTENSION

■ A flowchart is one way to describe events. Draw a flowchart to organize the information in Figure 21.27. Show how each event in the valley succession leads to the next. Think about how each community of plants changes its environment.

▶ REVIEW 21.3

1. (a) What is succession?
   (b) Describe how a plant can cause changes to its environment. Use an example, such as a dandelion.
2. Do you think "pioneer" is a good word to describe the species found in the first stage of succession? Explain your answer.
3. Do you think there are any "weed biomes"? Explain why or why not.
4. Why do you think that species that eat dead plant and animal material (scavengers and decomposers) are more abundant in the climax stage of succession than in the pioneer stage?

DID YOU KNOW

■ Animals that were once rarely hunted in their biomes are now hunted illegally in order to supply markets in other countries. In North America, black bears are killed for their gall bladders, which are sold as medicine in Asia. Can you think of other examples? What do you think might be done about this problem?

■ 21.4 HOW HUMANS CHANGE BIOMES

How do people affect biomes? How do they affect the process of succession? Consider the grass in your schoolyard or in the neighborhood around your school. These lawns are "artificial" communities, made up of one kind of grass and soil organisms. If they were left alone, each lawn would eventually be replaced by the plants of the next stage of succession typical for your biome (Figure 21.28). The efforts of the school custodian or gardener have halted the process of succession.

Now think about the biomes over the Earth's surface. As the number of people living on the Earth has increased, so has the need for more land to grow food. Today, much of the land once covered by forest or grassland is used for agriculture.

FIGURE 21.28 ▶
What do you think has caused the difference in the appearance of the area on the left side and the area on the right side of this photograph?

A farm contains the plants and animals that are being raised for human use. A farm has little room for wild organisms to live. For example, wheat and other crops grow throughout the North American grassland where the bison used to graze (Figure 21.29). In parts of Africa, cattle graze where elephants and gazelles once roamed. The original plants and animals of the grassland biome are now found mostly in nature preserves and parks.

Human harvesting of wild plants and animals has also changed the biosphere. Rain forests, once considered wilderness, are being logged at an incredible rate. Many countries no longer have any wild forests. Fishing has changed the kinds of animals found in many biomes in the lakes and oceans.

The human ability to travel from continent to continent has changed biomes around the Earth. Hundreds of species, from plants to beetles, have accompanied human migrations. The species introduced may be hardy pioneer species or species with few predators in their new environment. The new species may change the pattern of succession for the species already present. For example, the starling, a small songbird originally from Europe, is now common throughout North America (Figure 21.30). A few pairs of starlings were deliberately released in a New York park in the 1800s by a homesick bird-watcher from England. It seemed a harmless thing to do. However, starlings are aggressive and force other birds, such as bluebirds or swallows, from the best nest sites. Starlings have now displaced many bird species throughout North America.

FIGURE 21.29
Modern agriculture has affected the living things found in most parts of the world. What do you think has happened to the number of different plant and animal species in this area?

IT IS YOUR HOME, TOO

Your connection with your own biome and with all the Earth's biosphere is not always easy to see, but it exists. Your home is where you live. On a hot summer's day, you probably enjoy the cool shade under a tree. Maybe you like fishing, or you might prefer to jog on grass rather than pavement. Take a deep breath. The air you breathe contains oxygen produced by plants. If it is winter, that oxygen was probably produced by plants in the tropics and carried to you by the winds that sweep the globe.

You can do your share for your biome as well. Many choices you make affect the entire biosphere. As a consumer, you buy foods and other items from around the world. To produce these goods for you and others to buy, people in different countries may have decided to cut down forests or grow crops for export instead of for food for their own people.

How else do you affect the biosphere? Water and air carry pollution throughout the world. Choosing a lifestyle that helps control pollution protects your biome and other biomes of the world as well. Exercising prevention is far easier than repairing any damage.

Figure 21.30
Species introduced from other continents, like this starling, compete with indigenous wildlife. The starlings were introduced to North America by people.

ACTIVITY 21E / LIVING TOGETHER

CROSS · CURRICULAR

Choose a biome from the map of the Earth (Figure 21.5) that you find interesting. Imagine that you are visiting this biome on your vacation. Write a letter home to a friend, describing the biome. What human activities are affecting this area? How much of the climax community is left? What organisms are endangered in this biome? What is being done to protect them? ❖

REVIEW 21.4

1. List three ways that human activities can affect the kinds of organisms found in an area.
2. Why can it be a problem to transport a species to an area where it is not normally found?
3. List several things that people living in a community can do to take care of their natural environment.
4. Write a limerick that compares looking after the biomes of the world with looking after your room.

CHAPTER REVIEW

KEY IDEAS

■ Biomes are large geographical land regions that have a characteristic climate and characteristic plants, animals, and soils. Biomes are useful in describing similarities within large regions on the Earth's land surface.

■ Precipitation and temperature are the most useful characteristics of climate in identifying biomes.

■ Transition zones are areas where one biome gradually ends and another begins.

■ Plant and animal species show a wide variety of adaptations to different climates.

■ Biomes in California include the temperate rain forest, the coniferous forest, the alpine tundra, the desert, and the grassland. In the mountain ranges, several biomes can occur in a small area.

■ Mountain ranges and moisture from the ocean are two major reasons for the diversity of biomes.

■ Succession occurs when organisms colonize an area, modify it, and are replaced by new species. Succession begins with pioneer species and continues in gradual stages to a climax community.

■ The dominant plant species is the most abundant or largest species in the climax community. The climax community is the characteristic community of a biome.

■ The climax community will continue to occupy an area if the climate is stable and there are no catastrophes or human interference.

■ With the increase in human population, human activities have a greater effect on the entire biosphere.

VOCABULARY

biome
dominant plant species
ecosystem
climate
climatograph
adaptation
dormant
tree line
transition zone
pioneer species
succession
climax community

V1. Construct a mind map that shows the connections between climate and biomes using as many of the terms in the vocabulary list as you can. You may also want to use other terms or the names of biomes.

V2. Copy the puzzle on the right into your notebook. Use the following clues to find the word that belongs on each line. What extra word, running up and down, have you also formed? Write a definition of this word. Compare your definition with that of a partner.

(a) the overall pattern of weather at a specific location
(b) the most abundant or largest species in a community
(c) large areas that have a characteristic climate and plant life
(d) the state in which many plants survive low temperatures
(e) a process of change in living communities over time
(f) hardy plants that can begin to grow almost anywhere
(g) the species that continue to live and reproduce in an area unless disturbed by change
(h) the zone between two biomes
(i) a community of living things interacting with their environment and each other

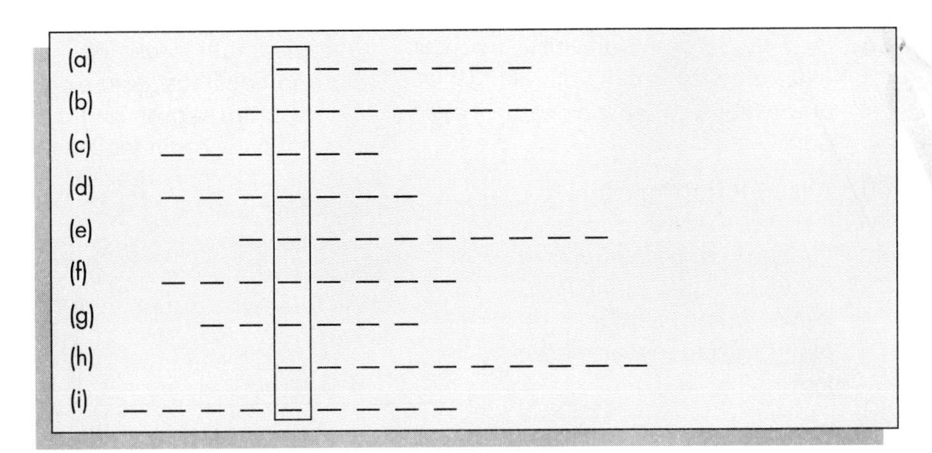

CONNECTIONS

C1. Compare any two biomes described in this chapter in terms of the following:
(a) climate
(b) characteristic climax community
(c) plant and animal adaptations

C2. Suppose ecologists from another planet visit Earth. They might decide that wheat fields form a biome on Earth.
(a) What evidence would the alien ecologists be using to make this decision?
(b) If you could communicate with them, how would you explain to them that there is no "wheat biome"?

C3. Different kinds of organisms have adaptations in their shape or behavior that help them survive in their particular environment. Figure 21.31 shows the ear sizes for three fox species. Large ears improve hearing and act as radiators to help an animal lose extra heat. Small ears help prevent heat loss. Use these clues to match the species of fox in Figure 21.31 to the latitude where they would most likely be found: near the Arctic Circle, near the northern border of the U.S., or near the Equator. Explain your reasoning.

FIGURE 21.31
Use this illustration to answer Question C3.

C4. Describe how you affect the other organisms in a natural area when you walk along a path.

C5. After a forest fire, many trees sprout new growth directly from buried roots or the undamaged portions of trunks. Are these pioneer plants? Explain your reasoning.

C6. Imagine you are travelling across the U.S. How would you know when you enter a new biome? How sudden is the change?

C7. Imagine that you are walking along the bank of a river. The forest around you abruptly changes from coniferous forest (with tall hemlock and spruce) to deciduous forest (with a mixture of broad-leafed oak and maple). As you continue your walk, you encounter the coniferous forest again. Give one explanation of how the patch of deciduous forest came to be there. Do you expect the patch of deciduous forest to still be there when your grandchildren walk along the same path? Why or why not?

C8. Your Schoolground Improvement Committee wants to plant several young trees. As a member of this group, you have been asked to decide which kinds of trees to order from the nursery. What do you need to know in order to choose trees that will survive and produce more trees in the area?

C9. The tropical rain forests of India and Southeast Asia experience a dry season and a season of heavy rain (called the monsoon). Why are the climax tree species of this biome deciduous? During which season do you think these trees would shed their leaves?

EXPLORATIONS

E1. Make a climatograph for your community. You could collect weather information for the past year from the daily newspaper or local weather station.

E2. (a) Start a notebook to record the kinds of living things you see in your community. Include observations such as the date you first see a particular bird arrive in spring, or when certain wildflowers bloom.
(b) Join a wildlife or naturalist organization and help record data on plant and animal populations, such as an annual bird census or a search for endangered species.

E3. Volunteer for a community or school project that helps improve or preserve a natural area.

E4. Research what your biome was like 25 to 35 million years ago. What plants made up the climax community? What animals lived there? Describe the climate.

E5. How does logging affect the process of succession? Find out what happens after an area of forest has been logged by clearcutting. What determines how long it takes for the climax community to occur again? Compare the succession after clearcutting with the succession that occurs after another logging method, such as selection logging. What factors besides the method of logging are important to the process of succession?

REFLECTIONS

R1. Many researchers and environmentalists have expressed concern that air pollution may be causing a gradual warming trend in the Earth's atmosphere. Computer models predict that if global warming occurs, there may be a change in the pattern of precipitation around the world. For example, the temperate grassland biome could receive less precipitation and the biomes in the tropics more precipitation. Look at the map in Figure 21.5. How might the major biomes you see be changed if global warming occurs?

R2. Look back at your chart from Activity 21A. Note any answers to your questions that you obtained from this chapter. Write down any new questions you have. How might you look for the answers?

CHAPTER 22

SOIL: THE VITAL SURFACE

Step off the pavement and onto some soil. You may not be aware of it, but below your feet lies a world full of life. Living in the soil are many organisms, such as worms, ants, and bacteria. These living things are as much a part of the Earth's biomes as the plants and animals that live on the surface. In fact, without organisms in the soil, the living things that you see on the surface would be unable to survive.

In this chapter, you will find out how you depend on soil both for the growth of plants and for the recycling of materials throughout the biosphere.

ACTIVITY 22A / WHAT DO YOU KNOW ABOUT SOIL?

What do you know about soil? In your learning journal, write down your answer to each of the following questions. Check your answers after you finish this chapter.

1. Which organisms depend on soil? How do you know?
2. If you cut some grass and leave the clippings on the lawn, what will happen to them?
3. In what ways do you think soil organisms interact with surface organisms?
4. Do you think that soil is important to your survival? Why? ❖

■ 22.1 WHAT IS SOIL?

Suppose you dig up a piece of soil the size of this book, and hold it in your hands (Figure 22.1). You would be holding more individual organisms than live on any square kilometer of the Earth's surface. These organisms belong to a greater number of species as well. The soil contains a rich variety of living things. The soil is one of the places where life can exist—it is part of the biosphere.

FIGURE 22.1 ◀
How many living things do you think are in this piece of soil?

ACTIVITY 22B / LIFE UNDERFOOT

What kinds of organisms live in the soil? How do organisms in the soil interact with one another and with their environment?

PART 1 A Soil Food Web

When you are looking at any ecosystem, one of the most important things to consider is how each species interacts with other species.

PROCEDURE

Draw a two-column chart in your notebook. In the left-hand column, list the organisms shown in Figure 22.2. In the right-hand column, describe the role or roles of each organism using the following terms: producer, consumer, herbivore, carnivore, decomposer, scavenger. (You may want to refer to the Glossary to review the meaning of these terms before you begin.) If you are not sure, put a question mark beside the organism's name, then continue. When you have finished your chart, use class or library resources to find out the role of any organisms you were unsure of before.

DISCUSSION

1. What information did you use to help you decide on the role of each organism?

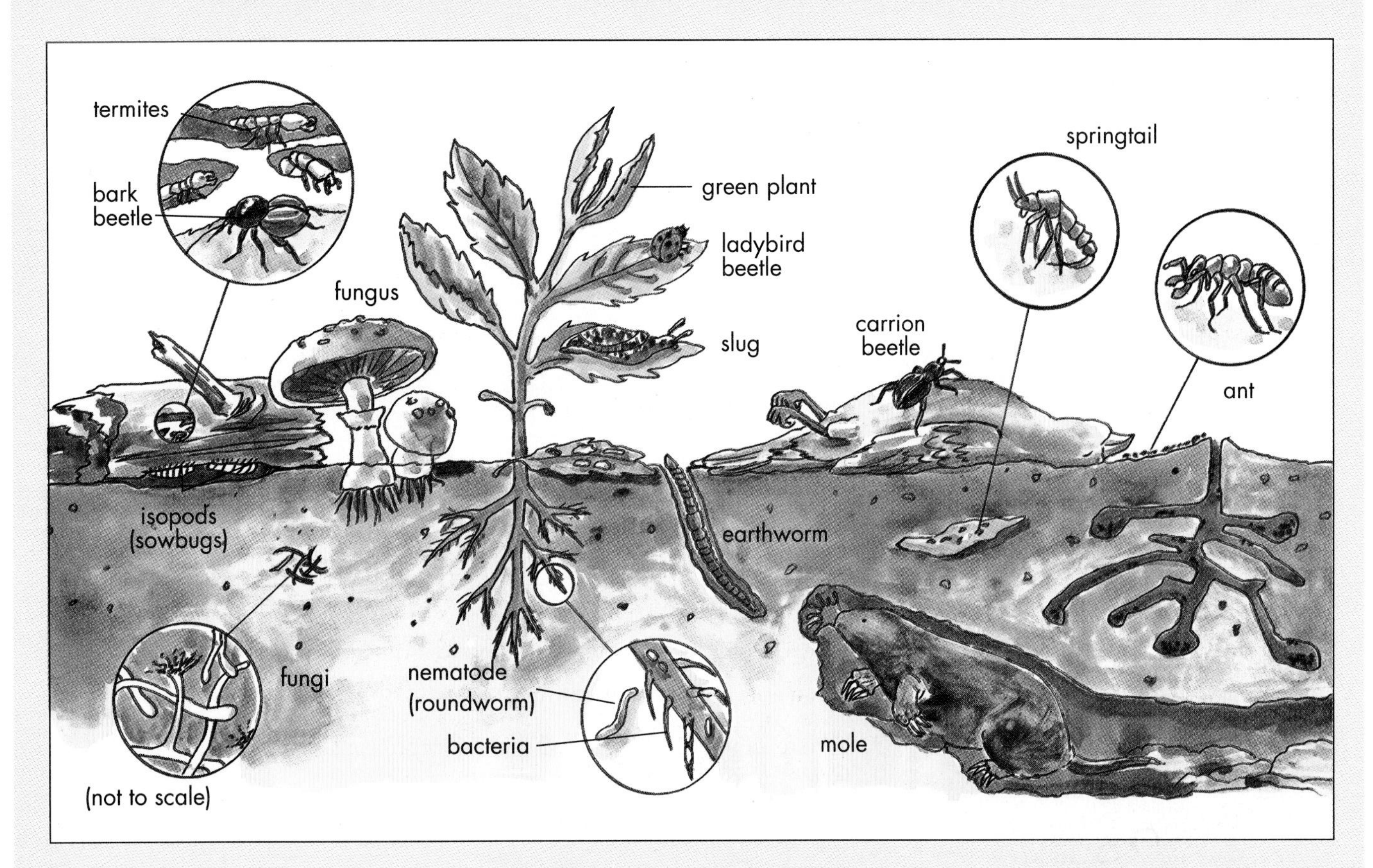

FIGURE 22.2
These are just a few of the organisms that live in the soil. What others can you think of?

2. What was the most abundant kind of organism? How do you know?
3. How does food energy enter the soil food web?
4. What would happen to this community of organisms if the decomposers were removed? Explain your answer.

MATERIALS

Part II

gloves
a container suitable for a terrarium (see Figure 22.3)
aquarium gravel or marbles
charcoal (optional)
soil
soil organisms (e.g., ants, sowbugs, earthworms, mealworms)
plant material (e.g., leaves, sticks, small plants)
plastic wrap, if needed

■ **When handling soil organisms, wear gloves and follow your teacher's instructions.**

PART II Preparing a Terrarium

You can observe many of the larger soil organisms in a terrarium in your classroom.

PROCEDURE

1. Work in small groups to prepare a terrarium, using Figure 22.3 as a guide. When selecting soil organisms to put in the soil avoid larger predators such as centipedes, millipedes, and carnivorous beetle larvae. Wear gloves when handling soil organisms.
2. Observe your terrarium for a few weeks, or longer. Add water to keep the soil moist.
3. Treat any living things in your terrarium with care. Return them to their natural environment at the conclusion of your investigation.

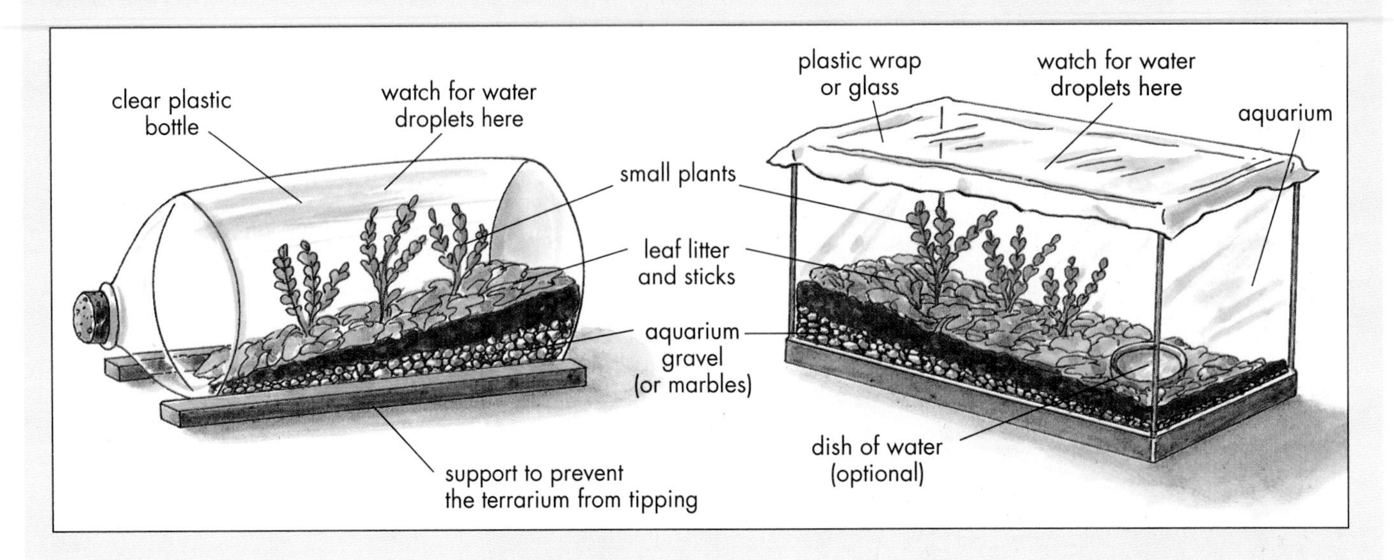

FIGURE 22.3
Terrariums. If water droplets appear on the inner surface, loosen the lid for a few minutes. (A thin layer of charcoal between the gravel and soil is recommended if you plan to keep the terrarium over several weeks. The charcoal absorbs odors.)

DISCUSSION

1. Why do you wear gloves when handling soil organisms?
2. What does the word "terrarium" mean?
3. Where in your terrarium did the soil organisms spend most of their time? Suggest a reason for this behavior.
4. What did the organisms in the soil use for food?
5. What would you need to do in order to keep your soil organisms alive and well in the terrarium for a long period of time? ❖

A JOURNEY BENEATH THE SURFACE

Imagine that you are able to shrink yourself to the size of an ant. Soil contains channels and other openings throughout. You would be able to move through these openings just as you move through the halls and rooms of your school. As you can see from Figure 22.4, the size of an ant is perfect for a firsthand look under the surface.

Each **pore**, or tiny space between the particles of soil, contains air. The oxygen in this air is important for the roots of plants and for animals that live in the soil. Drops of water cling to the sides of each pore, slowly moving downward. This water contains dissolved substances such as minerals. As the water drops get near plant roots, the water and minerals are pulled into the small root hairs.

Look closely at a water drop on your underground journey. You may see tiny shrimp-like creatures swimming. Clear, waxy-looking worms, called nematodes, may poke their heads through the water drops hunting for food.

Like an ant, you continue your journey downward. You are surrounded by pieces of rock, ranging in size from grains smaller than sand to large pebbles. Between these grains you see small clumps of dark,

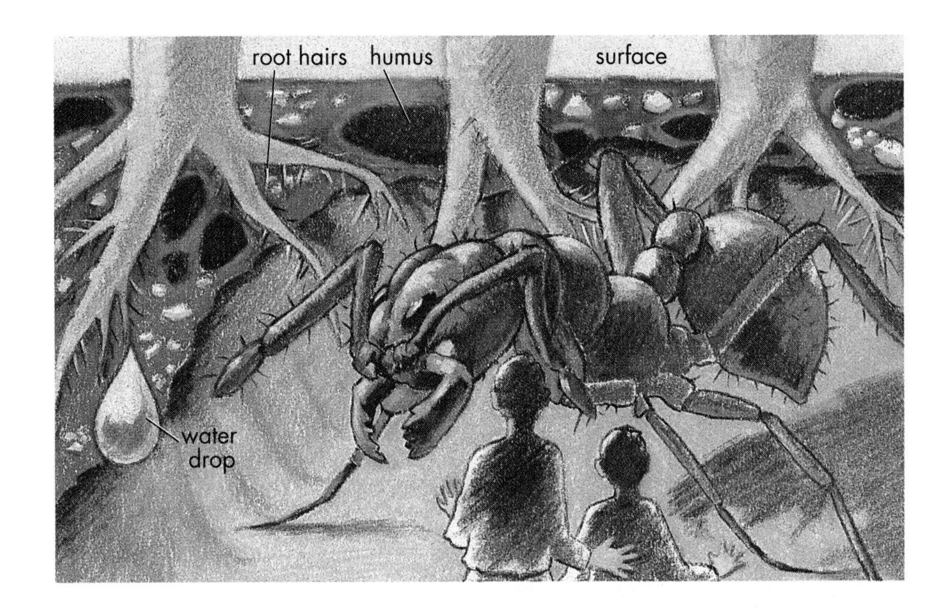

FIGURE 22.4
An imaginary field trip can help you understand more about soil and the living things found underground. What other organisms and things would you find?

moist material. This is **humus**, decaying plant and animal matter in the soil. Each clump of humus is full of life. Tiny decomposers, such as bacteria and fungi, break down the humus. Other organisms, such as springtails, prey on the decomposers. Humus helps hold moisture in the soil.

If you look up, you will see that the network of plant roots overhead is the liveliest part of the underground world. Tiny organisms coat the roots as shingles cover a roof. Some organisms feed on dead plant material. Some animals are predators that hunt other animals.

Next to you there's a sudden rumble. An earthworm eats its way past, swallowing whole particles of soil, moving like a whale through the ocean. Any material the worm cannot digest passes out of its body to enrich the soil.

The soil is a rich environment for living things. Plant and animal wastes fall constantly on the soil, supplying abundant food for organisms in the soil. The air temperature and moisture levels a few centimeters beneath the surface rarely change, making soil a stable environment for the most delicate organisms. Plants growing on the surface protect the soil from erosion and shade it from the drying heat of the sun. As organisms live and die in the soil, they help produce more soil.

SOIL FORMATION

The soil itself is composed both of tiny particles of rock and of **organic matter**. Organic matter includes wastes from living organisms as well as parts of living things that have died. Fallen leaves and animal bodies, for example, contribute to organic matter. As organic matter in the soil decays, it becomes humus.

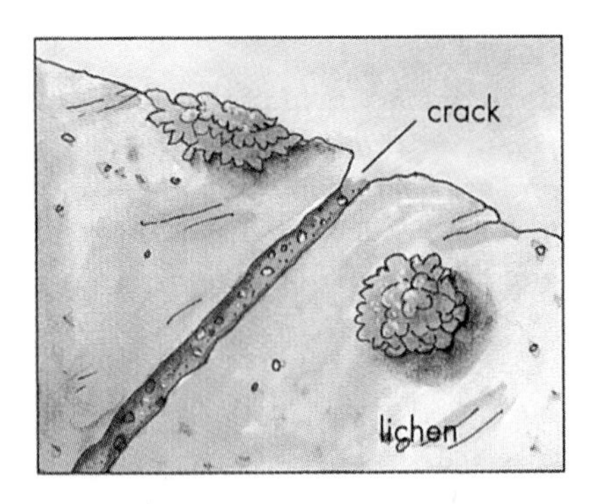

(a) Particles of dirt and sand collect in cracks in the rock.

(b) Clumps of lichens and mosses grow in the cracks. When they die, they decompose into humus and combine with the sand and dirt to form a pocket of soil.

(c) The layer of soil becomes thicker as more humus is formed and more dirt and sand collect. Flowering plant seeds germinate and grow.

FIGURE 22.5
As lichens and mosses grow on rock, they release chemicals that speed up the weathering process. How else do these organisms contribute to the formation of soil? How might water and ice speed up the process?

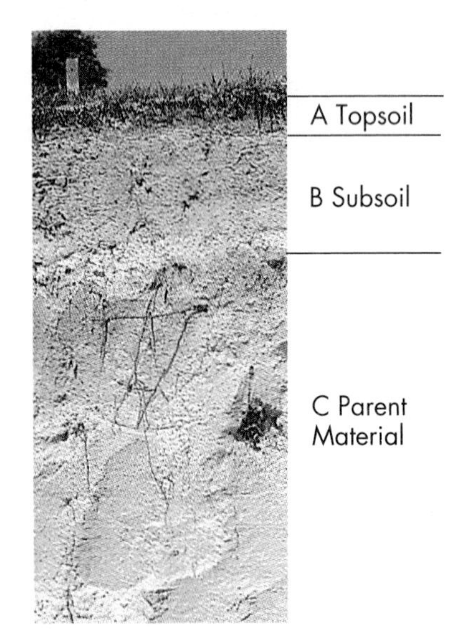

FIGURE 22.6
A soil profile describes the features of a particular soil, and can be used to study how the soil developed. The topsoil contains pieces of plant material and humus along with the sand and clay. The subsoil can be sandy or clay, depending on the type of soil. It contains roots and a few rocks or pebbles. The parent material is usually coarser than the subsoil.

How does rock get broken down and mixed with organic matter? Rock may be broken into smaller particles through weathering. Water freezing in cracks in a rock, for example, may help break the rock into smaller pieces. Plants also help break down rocks. Succession begins on bare rock or particles of rock when pioneer species, such as lichens and mosses, start to grow. Figure 22.5 illustrates how these organisms help break the rock into smaller particles. After they die, their bodies add organic matter to the rock particles. This process is the beginning of soil.

As succession continues, small plants begin to grow. The soil beneath plants is enriched and becomes thicker. Eventually, distinctive layers form. These layers can be shown in a diagram called a **soil profile** (Figure 22.6). Soil profiles vary from place to place. One reason for this is that the underlying rock that makes up the soil can be different from place to place. Another reason is that the amount and type of organic matter varies, depending on the plant and animal life.

Soil scientists have developed a classification system for soil profiles that helps them compare different soils. This system is used to produce maps that show where different types of soils are found. The photographs in Figure 22.7 show two different biomes in the state of California. Although a biome has generally similar soils throughout, there may be a lot of variation within a biome. For example, plants may be at a different stage of succession in different parts of a biome, or local conditions such as moisture may change the kinds of plants and soils.

Soil is produced mainly through the action of plants on the underlying rock and weathered material. Since each biome has its own kinds of plants, each biome also has a distinctive soil type.

(a)

(b)

FIGURE 22.7
What is the connection between the plants, climate and soil for each of these biomes?

ACTIVITY 22C / WHY ARE THERE DIFFERENT TYPES OF SOIL?

In this activity, you will look at the connection between plants, climate, and soil.

PROCEDURE

1. Draw a two-column chart in your notebook with the headings "Abiotic Factors Affecting Soil Formation" and "Biotic Factors Affecting Soil Formation." Use the information in Figure 22.8 on the next page to complete this chart for the coniferous forest biome.
2. Repeat step 1 for the desert biome.

DISCUSSION

1. Why do you think more soil organisms can live in the forest soil than in the desert soil? Suggest at least two reasons.
2. (a) Which soil contains the least amount of humus?
   (b) Why do you think this is so?
   (c) What would you expect to happen in a soil that had a low amount of humus?
3. (a) Why is the soil under coniferous trees more acidic?
   (b) Is this harmful to the trees? Explain your answer.
4. (a) Which soil holds the most moisture?
   (b) Why is this so?

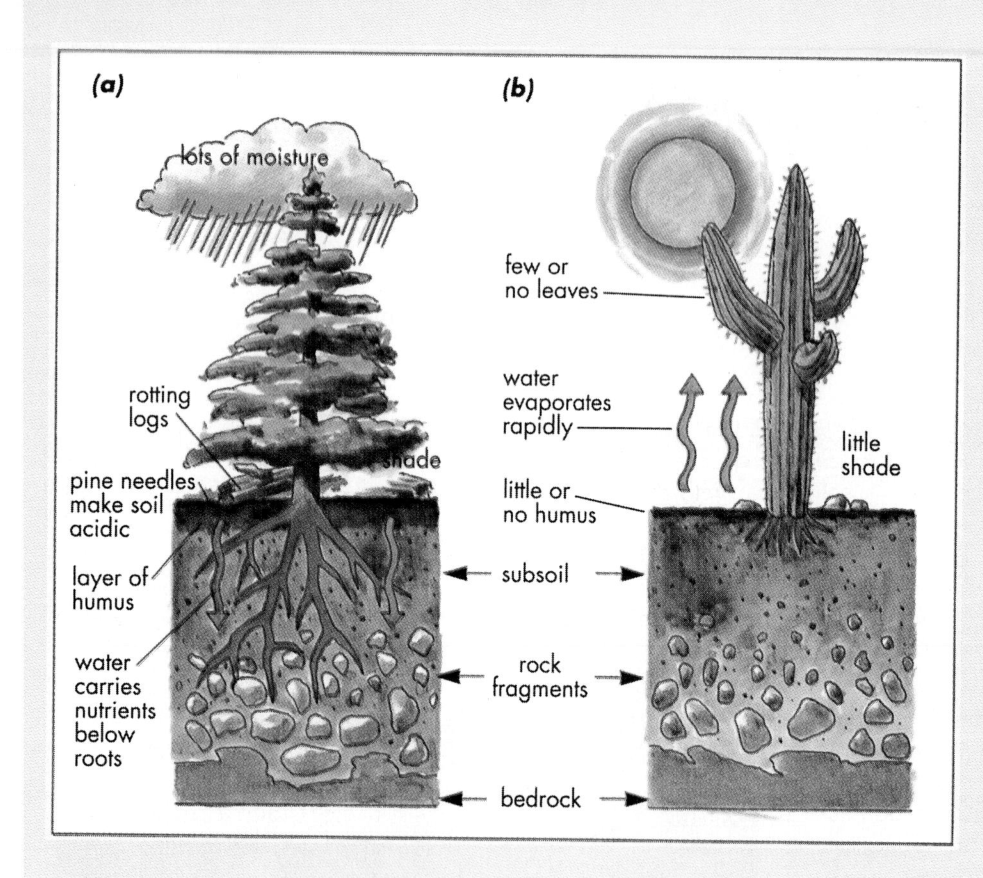

5. (a) Describe the major differences between desert soil and coniferous forest soil.

(b) What is responsible for the differences between desert soil and coniferous forest soil?

6. Suppose you carefully removed a cubic meter of soil from a desert and buried it in a coniferous forest. Predict what might happen to this soil after a long period of time. Explain your reasoning. ❖

FIGURE 22.8 ◀
(a) Soil in a coniferous forest biome. (b) Soil in a desert biome. In which biome might you find more soil organisms? Suggest reasons for your answer.

INTERDEPENDENCE

The soil and its organisms are closely related to the plant and animal life above the ground. They are both part of an ecosystem that is interconnected in many ways. For example, the beaver depends on the trees in a forest, which in turn depend on a suitable type of soil and on organisms living in the soil. The pond created by the beaver also creates a distinctive habitat within the forest for certain plants and animals (Figure 22.9).

FIGURE 22.9 ▶
Beavers depend on forest plants for food and building materials. In turn, the plants depend on the soil and its organisms in order to survive.

PROFILE

SOIL SCIENTIST GERRY DAVIS

Gerry Davis gets muddy a lot in her job. "Sometimes, when I'm up to my knees in a peat bog and it's raining, I wonder why I ever went into this profession!"

Davis is a pedologist—a soil scientist. She is particularly interested in the quality of forest soils.

One part of her job is measuring how much soil is disturbed by logging equipment. The rich surface layer of the soil is important for holding nutrients and water. How should logging be done to protect that soil layer? And how can the soil be restored after disturbance has taken place?

"All our research has to answer questions," Davis explains. The questions come from foresters working in the woods. The answers Davis provides help improve the productivity of the forest. "A healthy soil resource is essential for a healthy forest."

Soil nutrition is an important area of research. For instance, Davis finds out how different tree species respond to fertilizers applied to the soil. This research is important because many trees in North America lack nitrogen, which can be supplied by fertilizers.

Another problem that currently concerns Davis is a root-rot fungus that is killing trees in the area. When loggers cut trees and leave the stumps, the root-rot fungus feeds on the roots of the stumps. It then spreads to other healthy trees. Foresters are considering a new kind of logging to cope with this problem, called pushover logging. The whole tree, including the stump, is pushed over and removed from the ground. Davis needs to find out how this new logging method will disturb the soil.

Soil is just one part of the forest ecosystem. One thing Davis likes about her job is that she works with many other scientists in forest-related fields, such as ecology and silviculture—the science that deals with the development of the forest. "In the past, individual researchers might work mostly on their own," she says. "But that's not really how an ecosystem works. If you change one thing in the forest ecosystem, you affect another. So I love to learn about other fields, and how my field relates to them."

Her job is a mix of office work and fieldwork. It helps that Davis likes physical exercise, because she digs a lot of soil pits to get a better look at the soil. "Most soil scientists have bad backs," she says.

When she is not digging in the mud for her job, Davis is often outdoors anyway, whitewater canoeing, hiking, bird-watching—or even gardening in the dirt.

Soil scientists find work in many different places, such as old mine sites, archaeological digs, and golf courses. What might a soil scientist be hired to do at each of those locations?

▶ REVIEW 22.1

1. In the center of a page in your notebook, write "Soil." Make a mind map that includes the biotic and abiotic components mentioned in this section.
2. (a) What is a soil profile?
   (b) Describe one use of a soil profile.
   (c) Which layer of a soil profile do you think is most like the rock from which the soil was made? Explain why you think this is so.
3. Organisms can affect other organisms in several ways. For each characteristic of earthworms listed below, describe how you think the earthworm might affect other organisms living in or on soil.
   (a) Earthworms consume dead leaves and other plant material, producing wastes rich in organic material.
   (b) As earthworms move through soil, they create a network of interconnected tunnels and spaces in the soil.
   (c) Earthworms carry small soil particles to the surface from areas lower down.
   (d) Earthworms reproduce quickly.
4. Humus in soil helps plants to grow.
   (a) What is humus?
   (b) What is the source of humus in soil?

■ 22.2 HOW SOIL SUPPORTS THE GROWTH OF PLANTS

All living things need certain substances in order to live and grow. These necessary substances, called **nutrients**, are taken into the organism's cells and used by the organism. You, like other animals, obtain nutrients from the food you eat, but plants must obtain most of their nutrients directly from the soil. The table in Figure 22.10 lists three nutrients that plants need in large amounts.

FIGURE 22.10
The effects of some nutrients on plant growth. How could you tell if a plant you were observing needed one of these nutrients?

Nutrient	Symbol	Used in form of	Effect on plants
Nitrogen	N	Nitrate	Helps plant produce chlorophyll, which adds a rich green color to leaves, important for building plant parts such as flowers and seeds 
Phosphorus	P	Phosphate	Root development and growth 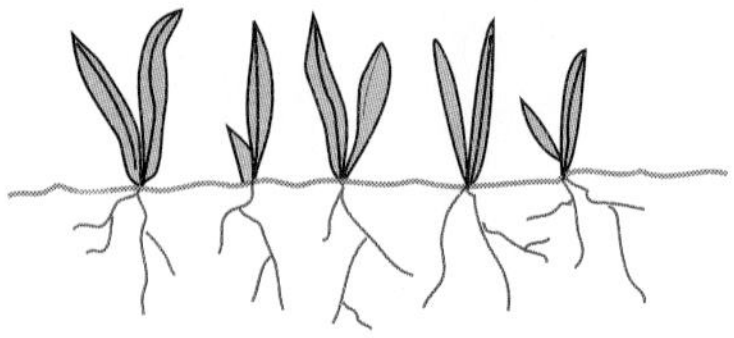
Potassium	K	Potassium	Needed for health 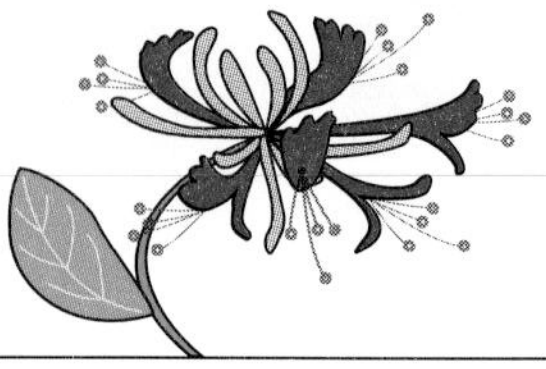

THE ROLE OF ORGANISMS IN THE SOIL

If there were no organisms in the soil to break down leaves and needles in a forest, would these pile up until they buried the forest? The answer is no, because long before the trees could be buried, they would have died from a lack of nutrients. All organisms need nutrients to live. Humans, like other animals, need nutrients to form all parts of their bodies. Plants need nutrients to form their leaves, stems, roots, and other parts. Organisms obtain nutrients from food or from the soil they grow in. The most important role of organisms in the soil is to recycle nutrients, making them available again to plants.

Think of a leaf falling to the ground in autumn. The dead leaf is no longer useful to the tree. But the leaf is like a locked treasure chest full of valuable nutrients. Soil organisms are the "key" to that chest.

The most common organisms living in the soil are decomposers or scavengers, such as fungi, bacteria, and nematodes. When they consume dead plant or animal material, these organisms use some of the nutrients in this material for themselves. The nutrients they do not use are released into the soil through their body wastes (Figure 22.11). Once in the soil, these nutrients can be absorbed by plants through their root hairs. Many gardeners use this ability of soil organisms to recycle nutrients from plant waste. They allow plant waste (from the garden and from food) to decay in a compost pile. They can then add this compost to the soil to supply nutrients and humus.

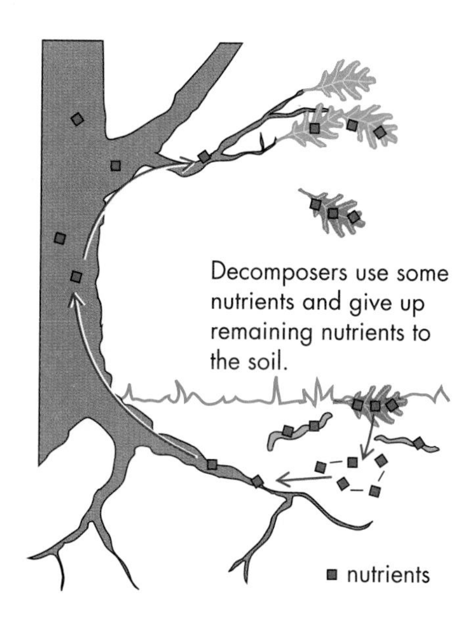

Decomposers use some nutrients and give up remaining nutrients to the soil.
nutrients

FIGURE 22.11
Follow the passage of nutrients from the tree, though the soil, and back into the tree. Could this cycle continue without the soil organisms?

ACTIVITY 22D / A COMPOSTER FOR YOUR SCHOOL

Composting makes use of nature's recycling crew—the organisms in the soil—to break down plant material into compost, a crumbly humus-rich substance that can be used to enrich soil for plants.

PROCEDURE

As a class, design a composting project to recycle lunchroom waste. If your school already has a composting project, your class can prepare a report on how well it is working. If your school does not have such a project under way, your class can make recommendations on how best to begin a school or classroom composting project. Use the following questions to help you carry out the project.

(a) What kinds of lunchroom waste can be composted?

(b) How much of these wastes does your school (or class) produce?

(c) What composter design would work best in your school, keeping in mind the climate and the amount of waste you want to put into the composter? (See Figure 22.12 on the next page.)

(d) How will responsibility for constructing and maintaining the composter be shared among the members of your class?

(e) What safety rules must be followed?

(f) What will be done with the final compost?

FIGURE 22.12
There are several types of manufactured composters, but a simple homemade version works just as well. What advantages are there to composting garden and food wastes instead of dicarding them?

FIGURE 22.13
These seedlings will die if the soil becomes too dry or too wet. A good soil for these plants stays moist but drains extra water away quickly.

WATER, NUTRIENTS, AND PLANT GROWTH

Compost is a dark, odorless material that is an excellent source of humus. Humus itself is not a nutrient for plants, since it is not taken into the plant's cells, but the decaying organic matter contains nutrients that plants need. Humus improves soil for plant growth in other ways. It loosens soil, making it easier for air and roots to penetrate. Humus also holds water, preventing the soil from drying out. In general, the more humus there is in soil, the better plants will grow.

The most useful way to find out whether a particular soil is good for growing plants is to observe how water moves through the soil. A good soil for most plants is one that holds some water in the spaces between soil particles. The water should drain slowly enough to allow roots to absorb the water they need without drowning (Figure 22.13). Too much water will fill in the air spaces in the soil, depriving both roots and soil organisms of oxygen.

Water is important to plants. It has a variety of functions within the plant, just as it does within your own body. Much of the plant is made of water, and you may have seen how some plants collapse or wilt if they do not receive the water they need.

Water is also essential for dissolving nutrients in the soil. As water moves through the soil, it "picks up" nutrients and carries them along. Plant roots can only absorb nutrients that have been dissolved in water. This ability of water to carry dissolved substances can also cause a problem for plants. Think of how you use water in a sink to wash your hands. The water dissolves the dirt and carries it down the drain. In a similar way, water dissolves nutrients near the surface of the soil and carries them downward through the soil. This movement is called **leaching**. If water moves too quickly through the soil, the nutrients are washed below the plant roots before the plants have a chance to absorb the nutrients they need.

ACTIVITY 22E / OBSERVING LEACHING IN ACTION

How does water's ability to dissolve and carry substances affect where a substance is found in soil? Figure 22.14 shows an experiment done by some students using the following procedure.

The students added the following materials to their bottle: one handful of aquarium gravel, two handfuls of sand, one handful of soil (Figure 22.14a). They poured water onto the soil at the top until water began to drip out of the bottom. When water no longer dripped out, the students poured 5 mL of red food coloring on top of the soil. Then they slowly poured water onto the soil, once more stopping when water began to drip out of the bottle. In Figure 22.14b, you can see what happened.

DISCUSSION

1. What did each layer of material the students put into the bottle represent?
2. What substance represented nutrients dissolved in water?
3. Why do you think most nutrients are present in the upper layer of soil?
4. A soil with a lot of humus will hold water longer than a soil with little humus. Based on the results from this experiment, explain which kind of soil would be best for plant growth.
5. Water dissolves many substances, not just nutrients. Predict what might happen to organisms living in the soil beneath a lawn if a strong poison was sprayed on the grass to kill insects or weeds. Explain your reasoning. ❖

(a) Before adding food coloring

(b) After adding red food coloring

FIGURE 22.14 ▶
This experiment was designed to find out what happens when a substance is leached (moved by water) through different layers of soil. The solution obtained by leaching is called the leachate.

▶ REVIEW 22.2

1. (a) What is a nutrient?
   (b) How do you obtain the nutrients you need?
   (c) How do plants obtain the nutrients they need?

2. (a) List the three main nutrients that plants need.
   (b) For each of the nutrients you listed in (a), describe how it affects plant growth.

3. What is the connection between organisms in the soil and the nutrients needed by plants?

4. In tropical rain forests around the world, it is a common practice for people to burn the trees and other plants in order to clear the soil for agriculture. At first the soil is extremely rich in nutrients and humus, and crops grow very well. However, the soil soon becomes poor in nutrients and the farms must be abandoned. The problem is not lack of water, because these areas receive a large amount of rain. Explain why the soil loses its ability to support plant growth.

5. Mr. Morgan wanted to grow a lawn around his desert home. He knew his grass would need daily watering, but he was willing to pay for it.
   (a) What should Mr. Morgan add to the soil before he plants his grass? Explain why this "ingredient" will help the plants grow in this soil.
   (b) What will happen to any nutrients in the soil as Mr. Morgan continues to water his grass every day? What will happen to the grass?

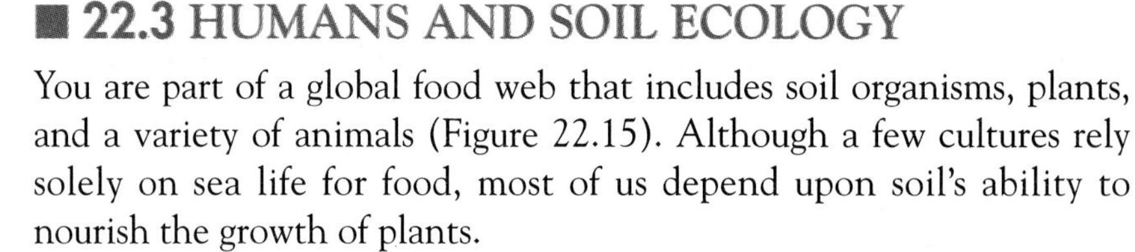

■ 22.3 HUMANS AND SOIL ECOLOGY

You are part of a global food web that includes soil organisms, plants, and a variety of animals (Figure 22.15). Although a few cultures rely solely on sea life for food, most of us depend upon soil's ability to nourish the growth of plants.

FIGURE 22.15
What other foods began as plants growing in soil?

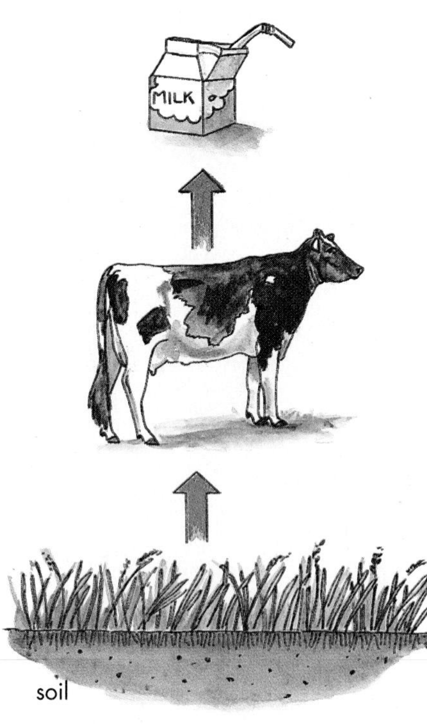

Almost everyone on this planet depends on soil for their food. So soil is an essential resource. Like any resource, there is a limited amount of soil that is good for growing food. To make matters worse, soil takes several hundred years to form but can be destroyed or lost much more quickly.

SOIL FOR AGRICULTURE

If you had to choose somewhere in the world to grow plants for food, what would you look for? You would look for a favorable climate. You would need a temperature range that suited the plants you wanted to grow. You would need enough precipitation to supply your plants with water throughout the growing season. You would also need a deep, nutrient-rich layer of soil.

Does this description sound familiar? It should, since these are the conditions found in the biomes that have abundant plant life. The best soils for farming were generally formed in grassland and forest biomes. Throughout the world, these were the first areas turned into farmland (Figure 22.16).

The human demand for food plants has greatly increased in recent years as the number of people on Earth has increased. Some researchers predict that the human population will double by the year 2030, reaching 10 billion. Even without this population growth, it has been necessary to find ways to increase food production. Several ways have been developed to help food crops grow where they normally would not survive.

FIGURE 22.16 ◀
What kinds of crops grow well in the grassland biome?

FIGURE 22.17
Water is continually removed from this soil by putting it into a series of canals. Otherwise, the soil would be flooded and crop plants would be unable to grow.

HELPING PLANTS GROW: WORKING WITH WATER

In the past, most farms had areas that were not used for growing crop plants because the soil was too wet or too dry. Many of these areas are increasingly being used for crops.

A wet soil can be drained by adding ditches or laying drainage tiles underneath the soil. With a great enough effort, even land that was once underwater can be turned into farmland. Figure 22.17 shows an area that was once a marsh. The river was diverted, and pumps were installed to control the water level. This former marsh now supplies vegetables and fruits for people to eat.

When wetlands such as floodplains and marshes are drained, they are no longer suitable for the plants and animals that naturally live there. As wetlands were drained in the prairies, for example, hunters and naturalists noticed a sharp decline in the number of waterfowl. People have only recently begun to appreciate the importance of wetlands in holding back floodwater. Wetlands are also vital for filtering and cleaning the water supplies that we depend on. Many wetland areas are now protected, so they cannot be drained.

The opposite situation—dry soil—can be improved by adding humus in the form of compost or animal manures. Figure 22.18 shows several methods used to help keep water in a soil.

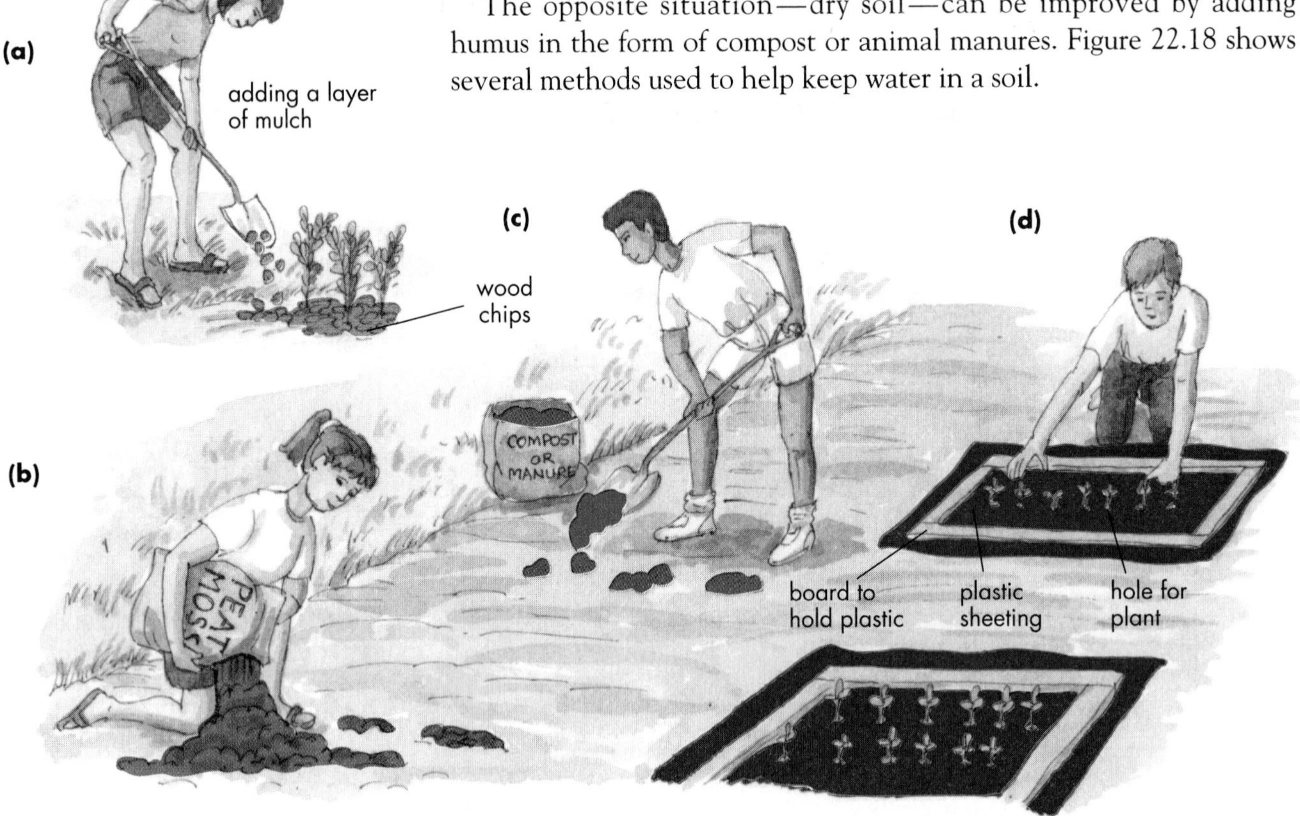

FIGURE 22.18
These methods help keep moisture in soil. Which methods reduce evaporation? Which methods reduce the amount of water that drains out of the soil?

Sometimes, however, an area does not have enough water in the form of precipitation. In such cases, water can be added by **irrigation**. Irrigation is a technique of providing water to the land in order to grow plants.

HELPING PLANTS GROW: ADDING NUTRIENTS

Leaves drop to the soil, organisms in the soil recycle the nutrients in the dead leaves, and plants absorb these nutrients from the soil. Most agriculture breaks this cycle. As you can see in Figure 22.19, the plant material is used for food by people or other animals that are often far away from the soil where the plants are grown. Year by year, the soil nutrients are taken up by the crop plants, which are harvested and carried away. These nutrients have to be replaced in order for plants to continue to grow.

Nutrients can be added to the soil in the form of **fertilizers**. For example, decayed animal wastes and compost as well as manufactured fertilizers are used. Next time you are in a store that sells fertilizer, look for a series of numbers on the front of each bag. These are the amounts of each of the major plant nutrients in the fertilizer. For example, a fertilizer labelled 10-20-30 contains 10 percent nitrate (the usable form of nitrogen), 20 percent phosphate (the usable form of phosphorus), and 30 percent potassium (see Figure 22.20).

EXTENSION

■ Design an experiment to test the effect of a store-bought fertilizer on house plants or strips of sod. Be sure to include a control in your experiment.

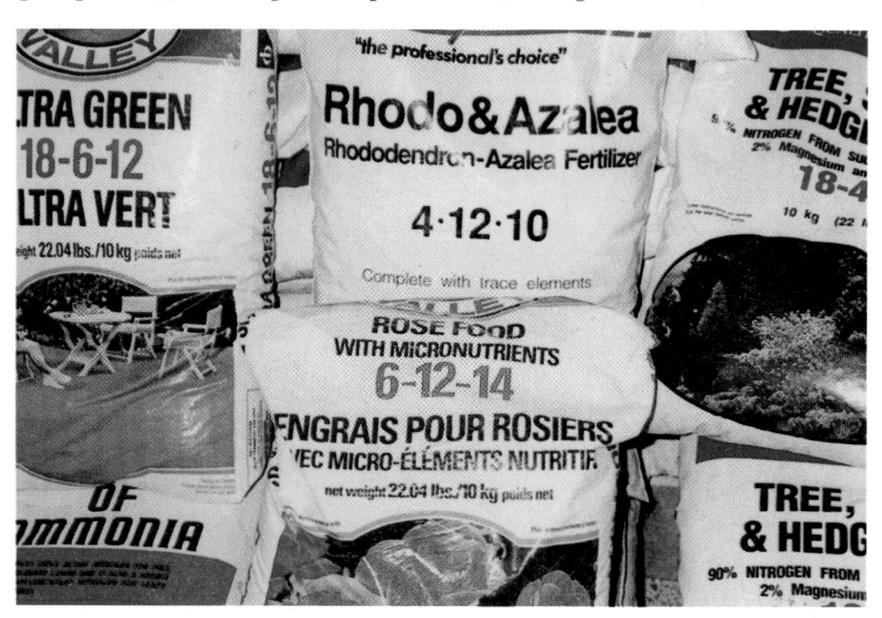

FIGURE 22.20
Using numbers for percentages of nitrate, phosphate, and potassium, you can choose the fertilizer best suited to the needs of certain plants at specific stages of their growth. For example, a fertilizer for encouraging lawns to grow thick green leaves in spring has the formula 19-3-3. A fertilizer designed to promote deep root growth in lawn grasses in the fall has the formula 5-10-5.

FIGURE 22.19
In what form are the nutrients taken from the soil? What organisms use these nutrients? What happens to the amount of nutrients left in the soil?

■ With a partner, prepare for a debate about what can happen when a fertilizer or a pesticide enters a waterway. One person will argue on behalf of aquatic organisms such as fish. The other person will argue on behalf of consumers who buy food.

The nutrients supplied by fertilizers, like those supplied by recycling in the soil, dissolve in water. The nutrients can then be absorbed by the plant roots. Any fertilizer that is not used by the plants will be leached by water down through the soil. If the leached fertilizer reaches the water table, it may be carried into waterways such as streams, rivers, and lakes. Although the fertilizer does increase crop yield, excess fertilizer that reaches waterways can increase the growth of algae and water plants. When these organisms die, decomposers in the water begin to break them down. Unfortunately, the increased number of decomposers uses up the oxygen in the water, and there is not enough for fish and aquatic animals.

HELPING PLANTS GROW: CONTROLLING PESTS

Pesticides are substances that kill plant or animal pests. Pesticides are used to prevent damage to crop plants. Pesticides can be helpful or harmful, and they vary according to how long they keep their toxic effects. Some pesticides can remain poisonous for many years. Like fertilizers, some pesticides can leach down into soil. If they do, they may kill organisms that the plants depend on for their survival. If pesticides reach rivers and lakes, they affect all other ecosystems as well.

THE NATURAL TOUCH

It is possible to use methods that will increase food production without causing damage to the environment. Some of these methods have been used throughout the world for centuries. As you read about the following methods in Table 22.1, think of which ones you could use in a home garden.

TABLE 22.1
Methods of Farming Without Pesticides

Method	Technique
Organic farming	Remove pest insects by hand or use natural predators to control pests. Use compost to improve soil.
Crop rotation	Use legumes such as clover to enrich soil with nitrates. Plow entire plants back into soil to add humus. The following year, grow and harvest crop plants in enriched soil.
Companion planting	Use combinations of plants to improve the soil. For example, oak trees improve the soil for corn and provide corn with shelter from wind.
Zero tillage	Do not till the soil (that is, no turning over of soil to loosen clumps and remove weeds) since this tillage causes soil to dry out and erode more quickly. Experiments show that many crops grow better in undisturbed soil, even though there are more weeds.

A FRAGILE RESOURCE

Mud slides and floods can sweep away centuries' worth of soil in the blink of an eye (Figure 22.21). Succession begins again, as lichens and mosses join the weathering of bare rock to start a new layer of soil. These natural events usually affect small areas at a time. In the biosphere, there is always enough soil somewhere to support plant growth.

FIGURE 22.21
What kinds of plants will start to grow where the soil has been removed by erosion?

Is there enough soil? Researchers estimate that an average of 2.5 cm of soil is being lost in farming areas such as the North American prairies every 16 years. A few centimeters may not seem like much, since these prairie soils are often several meters deep. But that 2.5 cm of soil took more than 300 years to form under the grassland biome. The soil resource is being eroded by wind and water faster than it can be replaced. And the natural plant community that built the soil is no longer there to replace it.

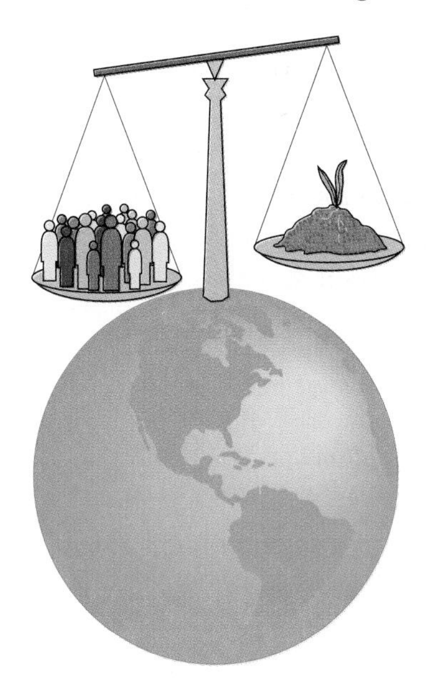

FIGURE 22.22
On a global scale, the issue of soil management becomes one of survival.

This slow but steady loss of soil is taking place even though farmers use techniques such as planting trees as windbreaks to reduce erosion. Imagine how much more soil is lost when bare soil is left exposed to wind or rain. And what of the soil buried under roads and cities? This kind of loss adds up very quickly. For example, most of the soil suitable for growing peaches in the U.S. is now covered by either highways or suburbs.

The loss of soil is not happening only on farmland or at the edges of expanding cities. The clearcutting of large areas of forests removes trees, which are not only the producers in the forest ecosystem but also the soil's protection. Unshaded, the soil heats up and dries out in the sun. Rain strikes the bare surface with more force, washing the soil and its nutrients away. On steep slopes, bare rock is exposed and succession must begin again.

A few decades ago, people could not imagine that soil for farming would become a scarce resource. But human population growth combined with the loss of soil has changed that way of thinking forever (Figure 22.22). Soil is a fragile, limited, and vitally important resource. It is part of your environment. Understanding and caring for the soil is part of caring for the biosphere on which all life depends.

■ Over 2000 years ago, the golden age of Greek civilization ended not because of conquest or war but because of damage to the country's most precious resource—soil. Trees were cut from every hillside to build cities and ships, as well as for fuel. Without trees to prevent erosion and enrich the soil, the land was not able to support farming. The famine that followed killed many people. Even today, large areas of Greece remain bleak and barren.

REVIEW 22.3

1. What kind of soil is usually good for agriculture? Why?
2. (a) List two ways you could improve a dry soil for agriculture.
   (b) What benefit is there to each of these methods?
   (c) What problems, if any, can result from each of these methods?
3. (a) Why is a very wet soil not suitable for most plants?
   (b) How can a wet soil be improved for agriculture?
   (c) What problems, if any, can result?
4. (a) Why is it necessary to add fertilizer to soil that is used to grow food plants year after year?
   (b) What problems can result from fertilizer use?
   (c) How can these problems be prevented?
5. One farming technique designed to improve soil quality is to grow clover and then plow it into the soil before planting the same field with a different crop.
   (a) What natural process is this technique copying?
   (b) How is the soil improved by the growth and plowing under of the clover plants?
6. (a) Describe two ways in which human activities cause a loss of soil.
   (b) Why is this loss a concern?

CHAPTER REVIEW

KEY IDEAS

■ Soils are formed by a combination of the weathering of rock and the decay of organic matter.

■ Soil supports the growth of plants and has many organisms living in it.

■ Plants have a great effect on the characteristics of a particular soil, as seen in the soil profile. Plants provide organic matter and protect soil from moisture loss and erosion.

■ In general, biomes have characteristic soils developed by and suited to their climax plant communities.

■ Organisms living in the soil consume dead plant and animal material, releasing nutrients into the soil.

■ Knowing how water moves through soil is important for predicting the soil's ability to support plant growth.

■ Methods to help crops grow include draining wetlands, irrigating dry soils, adding fertilizers, and using pesticides. These methods can also result in undesirable changes to the environment.

■ The soil is a fragile but vital resource, which is being lost by erosion and some human activities.

VOCABULARY

pore
humus
organic matter
soil profile
nutrient
leaching
irrigation
fertilizer
pesticide

V1. Make a mind map that shows how soil is affected by the activities of living things. Use as many words as possible from the vocabulary list, along with other words.

V2. For each of the terms in the vocabulary list, write a brief definition in your own words.

CONNECTIONS

C1. (a) What kind of organisms are the producers in an ecosystem?
(b) How do organisms in the soil obtain food from these producers?
(c) How is the soil a part of an ecosystem that includes the living and non-living things above the surface? Give an example.

C2. As an experiment, a gardener removed all the plants from one area of her lawn, and kept removing any that sprouted from seeds. In a nearby spot, she left the plants alone. After a couple of months, she examined the soil in the two areas very carefully. Predict three ways in which the soil without plants will be different from the soil with plants.

C3. (a) What do you think would happen if some kind of disease or poison got into a farmer's field and killed all the organisms that release usable nitrogen into the soil?
(b) What might happen if all the earthworms were killed?
(c) What might happen if all the decomposers of plant materials were killed?
(d) What do you think would have to be added to this field in order to grow plants if all the organisms living in the soil were killed?

C4. How would you solve the following problems in a garden? Explain your answers.
(a) Puddles of water stay on the surface long after rainfall ends.
(b) Soil dries quickly, is very pale, and cracks.
(c) Plants are small and grow very slowly.

C5. Commercially available potting soil has been treated to remove weed seeds and plant-eating pests. The treatment also kills all soil organisms. If this soil contains no living things, how can it be useful to plants?

EXPLORATIONS

E1. To raise funds for your school, your class can start popular seeds such as petunias or tomatoes indoors. Have the young plants ready for sale in the spring.

E2. The foods available in most grocery stores today are quite different from the foods available even 25 years ago. Much of the change has come as a result of new farming and food preserving methods. Ask older family members and friends what foods they used to find in the store and when these foods were available. Compare this with what is available to you now.

E3. Farmers in the U.S. are facing tough economic times. Many are forced to sell their farms. Find out why this tragedy is happening. Your best source of information is to speak to a farmer or farm association, but many magazines available in the public library will have good articles on this topic.

E4. Investigate the problem of salting in irrigated fields. Find out what can be done about it.

E5. The headline of a local paper reads: "Charter Fishing Boat Owner Sues Local Farmers Over Loss of Business." Explain how fishing could be affected by farming practices.

REFLECTIONS

R1. Look back at your answers to Activity 22A and make any changes you wish. Has your opinion about the importance of soil changed? If so, how?

R2. Phosphates from detergent, certain industrial wastes, and sewage from livestock operations have all been shown to affect the plant life and therefore the ecosystems of lakes and streams.
(a) Find out what happens when a body of water is "over-fertilized" in this way.
(b) What do you think would happen if the source of the pollution were stopped?
(c) How could you help repair the damage to the water ecosystem following an "overdose" of fertilizer?

THE EARTH'S OCEANS

You pause along the shoreline, and the water washes over your bare feet. Each time the waves come in, perhaps imperceptibly, the water is a little higher. If you were to stay here for several hours, the water would be at your knees. The cycle of high and low tides is one of many cosmic and terrestrial cycles that governs our lives. The sun tells us when to get up and when to go to bed. The moon seems to drag the Earth's waters along with it each time it circles the Earth.

From space, our planetary home looks as if it is mostly water. Water covers seventy percent of our planet's surface. Most of it is stored in the oceans. But, most of us live not far from lakes, rivers, and streams. And frequently we experience water from the sky—rain.

Where does rain come from? Where does it go? Why are most lakes, rivers, and streams fresh water, while the oceans are salt water?

In this chapter, you will learn about the great patterns that govern the Earth's waters, their composition, and their physical properties.

■ 23.1 THE OCEANS: AN OVERVIEW

Most of us think of the Earth as having several oceans, but in fact, they are all one ocean, one vast body of water. And like a giant hour hand on a clock, the moon sweeps around the Earth, creating high tides beneath it.

EXPLORING THE OCEANS

For thousands of years man has had a strong attachment to the ocean. Explorers sailed the oceans in search of new lands and riches. Artists and poets have sought to describe and interpret the ocean, and people all over the world have harvested the sea for food. It wasn't until the late nineteenth century, however, that oceanography developed as a science.

In December 1872 the H.M.S. Challenger left Portsmouth, England on a scientific expedition that would last almost three and a half years and cover 68,890 nautical miles (Figure 23.1). The expedition was headed by Wyville Thomson and included a group of five scientists. Their first scientific endeavors took place about two months later when they established a research station just off the west coast of the island of Tenerife (Figure 23.2). In the words of J. Y. Buchanan, the

FIGURE 23.1
Wyville Thomson's HMS Challenger.

chemist on the expedition, "it may be taken that the Science of Oceanography was born at sea in Lat. 25° 45′ N, Long 20° 14′ W, on 15th February 1873." Buchanan further points out that, at the time of the expedition, there was no word for the scientific work they were to undertake. After the expedition, it was given the name **oceanography.**

Over the course of the expedition, the scientists collected and described 715 genera and 4,417 species which were previously unknown. They made soundings with rope and weights and found the Mariana trench to be 9,000 meters deep, proving the ocean was much deeper than previously thought.

The resulting report from the expedition was published in 50 volumes from 1880 to 1895. By then, other nations mounted expeditions of their own, firmly establishing the science of oceanography.

EXTENSION

- Imagine you were Wyville Thomson. Who would you have taken on the expedition? What kinds of scientists would your research require?

THE GEOGRAPHY OF THE EARTH'S OCEANS

Oceans cover approximately seventy percent of the Earth's surface. The Earth's oceans are actually one continuous body of water, but geographers and cartographers recognize four major oceans: the Pacific, the Atlantic, the Indian, and the Arctic (Figure 23.3).

Notice the vastness of the Pacific Ocean. The Pacific is the largest ocean; it is twice as large as the Atlantic. At a point north of the equator, the distance across the Pacific Ocean is nearly half the circumference of the Earth. The Indian Ocean is about ten percent smaller than the Atlantic Ocean. The Arctic Ocean is the smallest. Table 23.1 lists the area of each ocean.

FIGURE 23.2
Journey of the HMS Challenger.

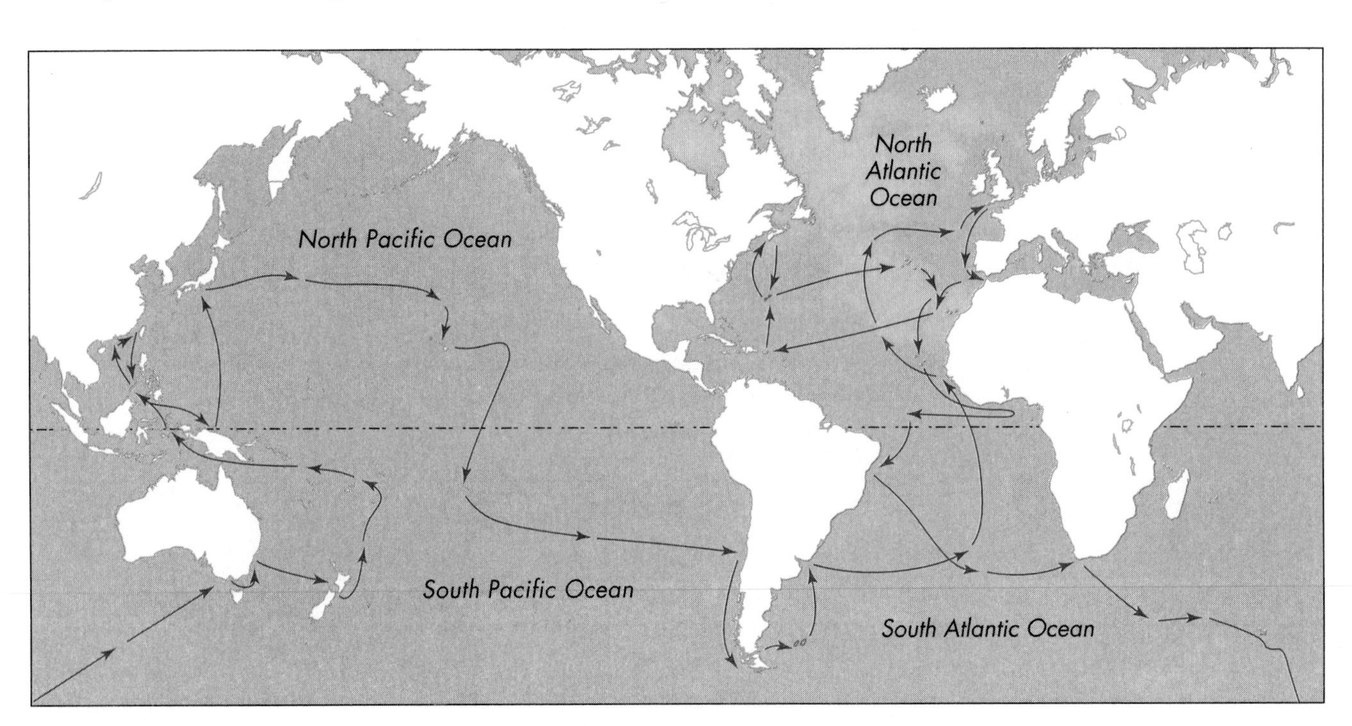

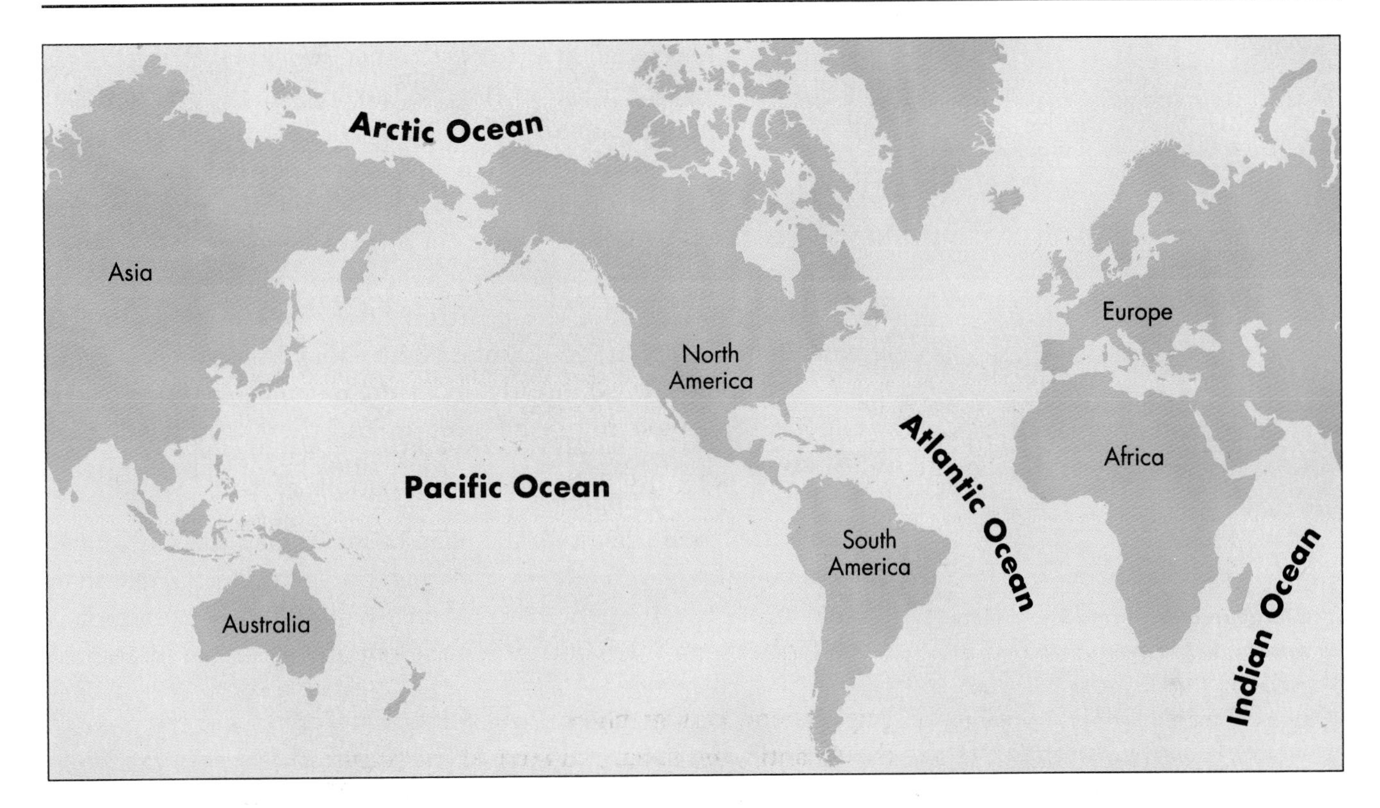

FIGURE 23.3
The four major oceans.

Because the Pacific and Atlantic Oceans range into both the northern and southern hemispheres, they are frequently referred to as the North Pacific, South Pacific, North Atlantic, and South Atlantic, with the equator being the dividing line. The Arctic Ocean is the area above the continental land masses in the region of the Earth known as the Arctic Circle.

OCEANS, MOON, AND SUN

If you have ever been to the beach, you have heard of high tides and low tides. At **high tide** the waves come farther up the beach. At **low tide** the sea is farther out.

Body of Water	Area
Pacific Ocean	165,200,000 square kilometers
Atlantic Ocean	82,400,000 square kilometers
Indian Ocean	73,400,000 square kilometers
Arctic Ocean	12,900,000 square kilometers
All Oceans and Seas	361,000,000 square kilometers

TABLE 23.1 ◀
Areas of the Four Major Oceans

Were it not for the moon and the sun, there would be no tides. This rising and falling of the ocean is caused mainly by the gravitational pull of the moon on the Earth. The ocean is at its highest at a point directly beneath the moon. There is also a corresponding surge on the other side of the Earth, away from the moon. These two surges are the high tides. The ocean's low tides occur midway between the two high tides (Figure 23.4). The high tide moves across the ocean surface as the Earth rotates under the position of the moon. If the moon remained stationary, it would take twenty-four hours for the same point on the Earth to be directly under the moon again; but because the moon is also moving in its orbit around the Earth, it takes 24 hours and 50 minutes. Therefore, high tides occur approximately every 12 hours and 25 minutes.

While the moon's gravity is the main cause of tides, the sun's gravitational pull also plays a role. Even though the sun is much larger than the moon, its effect on the Earth is only half as strong because it is much farther from the Earth. When the sun and moon are positioned

FIGURE 23.4
The location of high and low tides in relation to the moon.

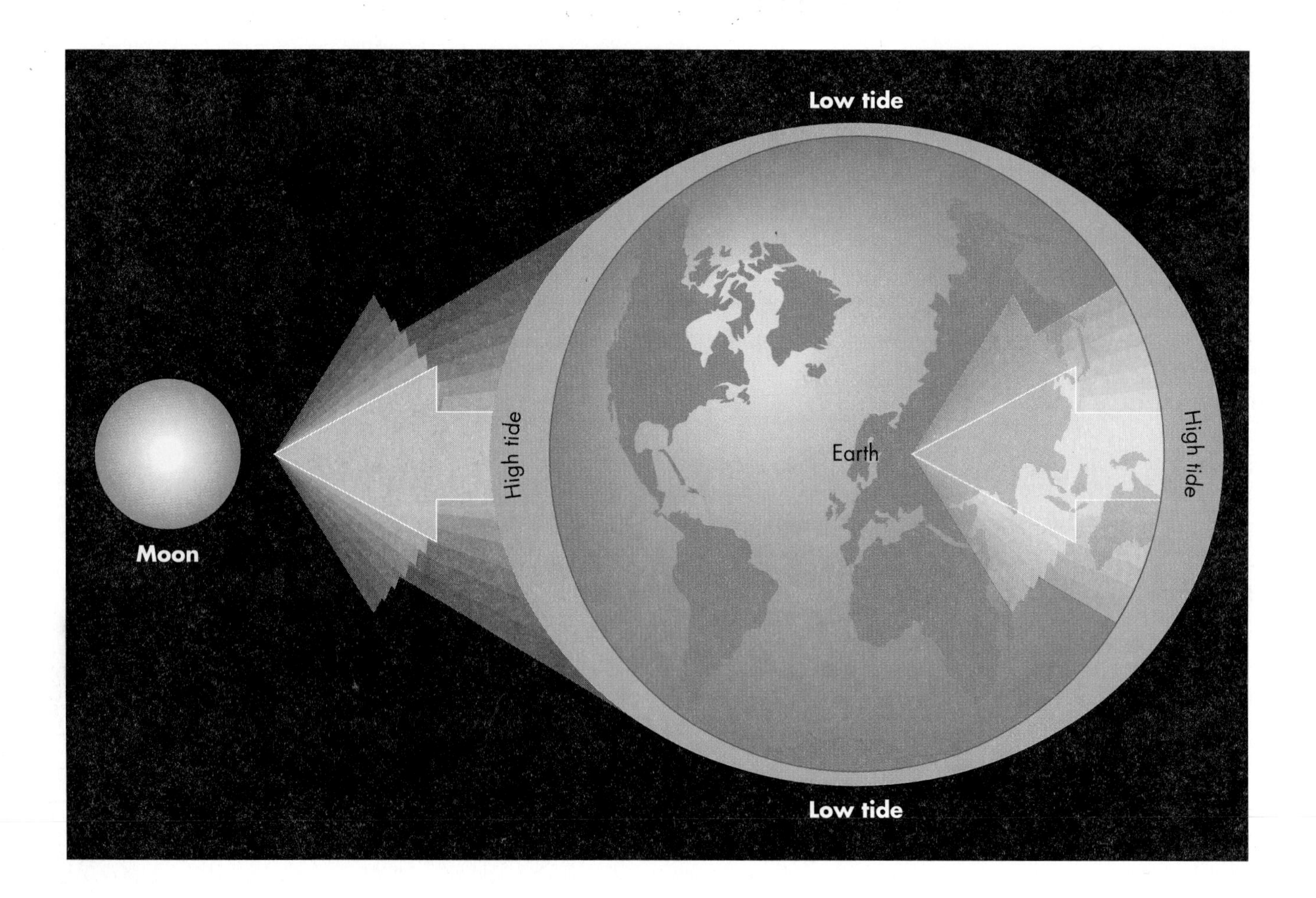

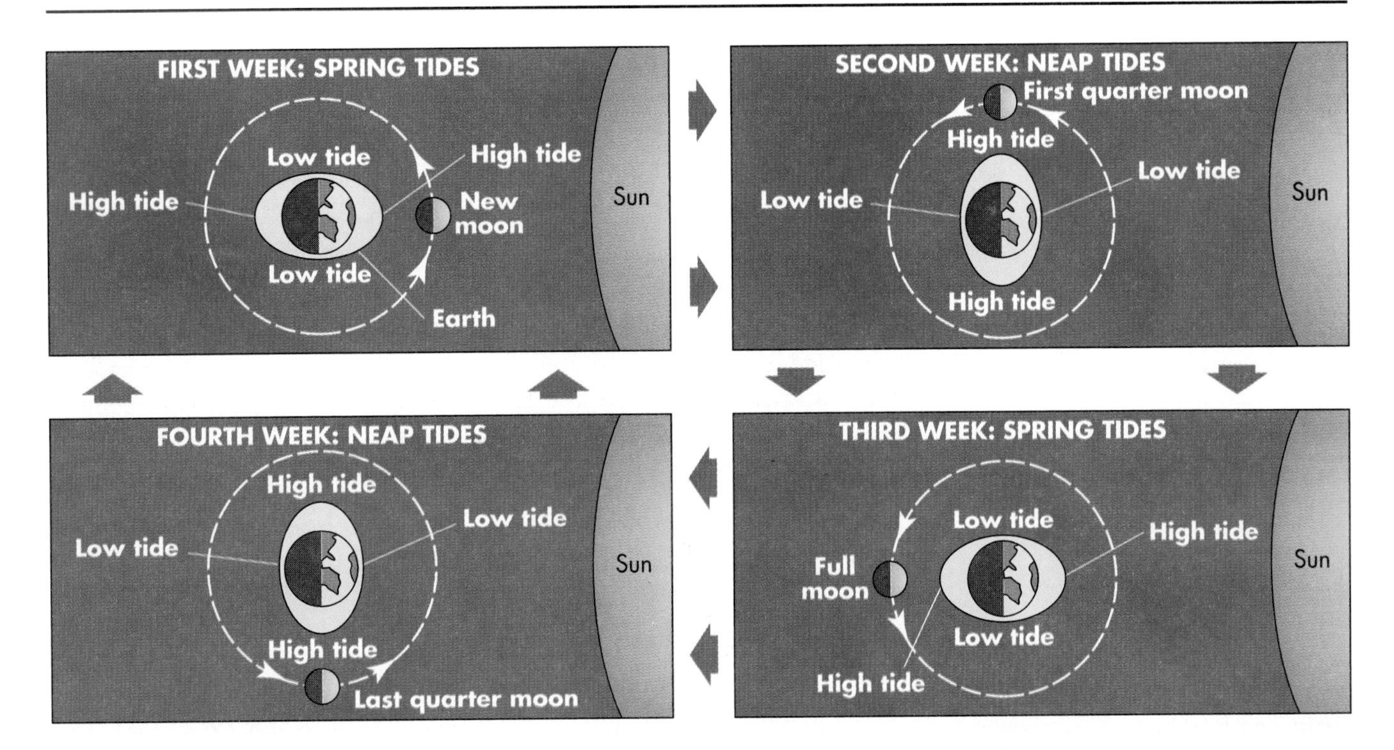

in a direct line with the Earth, the combined gravitational pull creates higher than normal high tides. These are known as **Spring tides**. When the sun and moon are at 90 degrees to each other, the high tides are lower than normal. These tides are called **Neap tides** (Figure 23.5). Spring tides and Neap tides each occur twice a month.

Although the tidal surge is only about 0.9 meters high out in the open ocean, it rises to two to three meters as it approaches the shore. When it is funneled into a bay or narrow inlet, the difference can be even more dramatic. In the Bay of Fundy in Nova Scotia, for example, the difference between high tide and low tide can be as much as 18 meters (Figure 23.6).

FIGURE 23.5
The Neap and Spring tides in relation to the sun and the moon.

FIGURE 23.6
High and low tides in the Bay of Fundy, Nova Scotia.

▶ REVIEW 23.1

1. In what year did the science of oceanography begin?
2. Name the four major oceans. Which is the largest and which is the smallest?
3. How often do high tides occur? How often do Spring tides occur?
4. Describe the difference between the Spring tides and the Neap tides.
5. Why don't high tides occur every twelve hours?

■ 23.2 OCEANS, CLIMATE, AND WEATHER

The atmosphere and the oceans together form one large system. The atmosphere affects the oceans in many ways, and conversely, the oceans affect the atmosphere in just as many ways.

THE SUN AND THE WATER CYCLE

One interaction between the oceans and the atmosphere is the continual exchange of water. This process is called the **water cycle**. Like any cycle, it has no beginning or end, so let's jump into the cycle just above the ocean's surface on a hot, dry day. Look at Figure 23.7. The ocean's

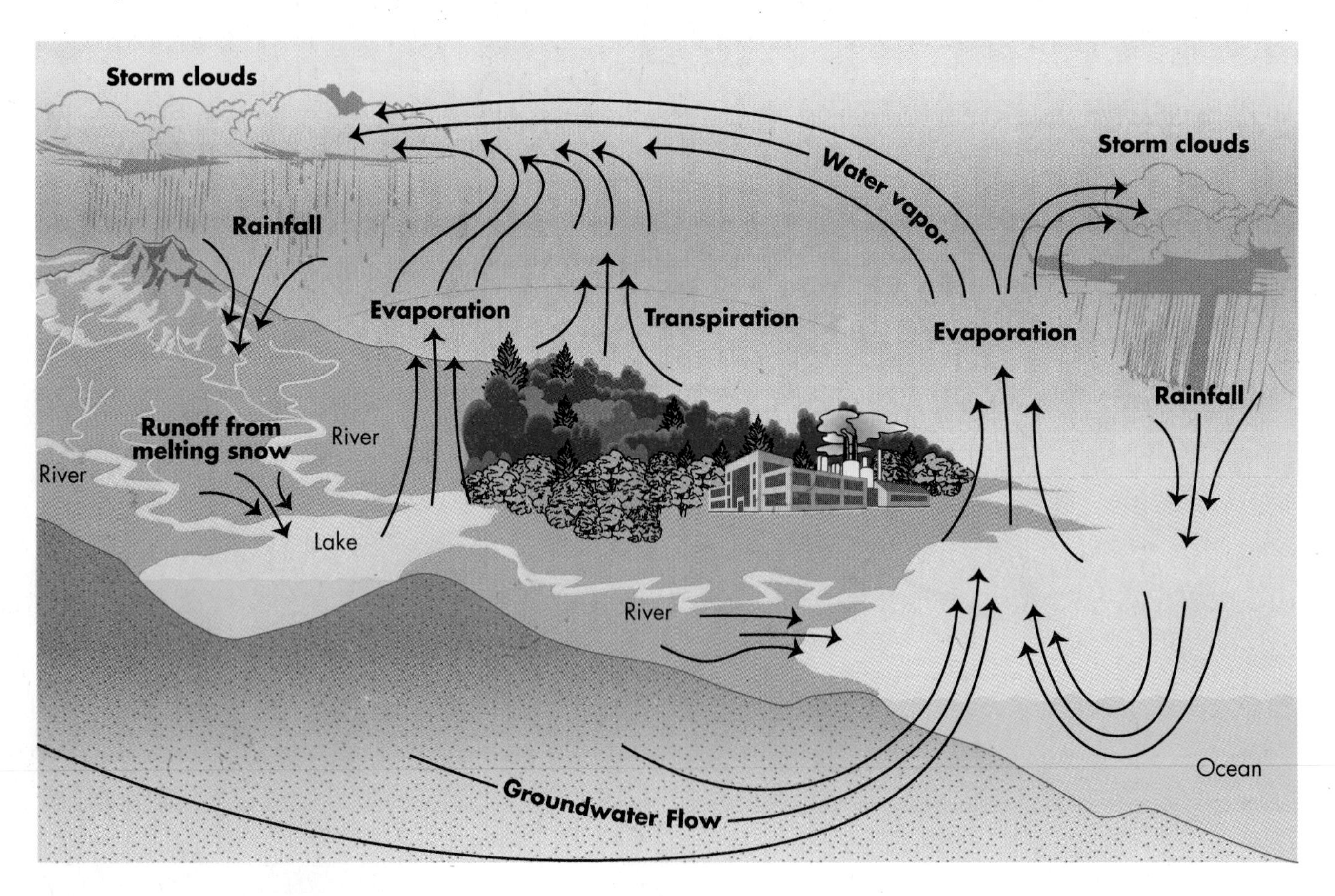

FIGURE 23.7
The flow of water through the water cycle.

surface waters are beginning to **evaporate**, the process by which liquid water is changed to vapor. It travels up into the atmosphere, cools, and condenses to form water droplets which become clouds. **Condensation** is the reverse of evaporation; water vapor is changed to liquid water. The winds may carry the clouds to another region before **precipitation**, rain or snow, falls from the clouds. As it reaches the ground it can go four places. It can evaporate back up into the atmosphere. It can soak into the ground and be absorbed by the **groundwater**, water that flows beneath the Earth's surface. It can become **runoff**, draining into surrounding bodies of water, such as lakes and rivers. Finally, it can be absorbed by vegetation and then transpired back into the atmosphere. **Transpiration** is the escape of moisture through pores in a plant. The rainfall that becomes runoff or groundwater eventually finds its way back to the ocean, only to evaporate again and return to the clouds to become rain or snow (Figure 23.8).

The water cycle is balanced; the volume of water entering the atmosphere is equal to the amount leaving it. What drives the entire process? The sun's energy powers the process by heating and evaporation.

DID YOU KNOW

■ It takes a molecule of water about 3,600 years from the time it evaporates to the time it returns to the ocean through precipitation, groundwater seepage, and river flow.

DID YOU KNOW

■ 99.9% of the Earth's heat energy comes from the sun. Where does the remaining 0.1% come from? From the interior of the Earth, from sources such as natural radioactive decay.

FIGURE 23.8 ◀
Water evaporating out of a tropical rain forest.

ACTIVITY 23A / THE WATER CYCLE IN YOUR REGION

Work with a group of classmates to design a poster about the role your region plays in the water cycle. Your region will be defined by the area served by your local water supply.

MATERIALS

felt-tip markers
poster board

PROCEDURE

1. List the bodies of water in your area: groundwater, lakes, rivers, streams, ocean. ➡

■ Scientists have found that the oceans lose 3.6×10^{13} cubic meters of water each year. They also know that 3.6×10^{13} cubic meters of water falls on land each year. The water cycle is in perfect balance. Imagine how you would measure this. Where, when, and how often would you perform measurements?

2. List other components of the water cycle in your area: sun, rain, snow, runoff from melting snow, vegetation, and others.
3. Use felt-tip markers and poster board to draw a diagram of your region's water cycle. Connect components with arrows of two colors: arrows indicating energy flow and arrows indicating exchange of water.
4. Display your group's poster with those completed by other groups. After studying all the posters, make a list of questions about your region's water cycle.

DISCUSSION

1. How has human habitation altered these natural pathways in the water cycle?
2. How has your water company and your sewage system altered the cycle?

THE SUN AND THE WINDS

Weather and climate are not the same. When we talk about the **weather**, we are talking about daily, or even hourly, changes in meteorological conditions. But **climate** is the averaging of these conditions over long periods of time for an entire region. Every region has changing wind conditions, and these can also be averaged over time. When this is done, certain global patterns emerge. Figure 23.9 shows the major global wind patterns.

What causes wind patterns? Let's begin at the equator. The equatorial region receives more solar energy than other latitudes, because the sunlight is more direct. Warm air builds up in this region, it rises, and then it begins moving northeast. At about latitude 30°, the air begins to cool and sinks back to the Earth's surface. As it reaches the surface it spreads out in two directions. Part of it flows toward the equator in a southwesterly direction, called the **trade winds**. The remaining descending air moves toward the poles in a northeasterly direction, called the **westerlies**. Their northeasterly flow is blocked at about 60° latitude by the **polar easterlies**. This extreme cold air near the pole is not distributed out of the polar region, but it is locked in a circular pattern around the pole from east to west. This entire pattern in the northern hemisphere is repeated in the southern hemisphere, with directions reversed.

Why does the warm air rising from the equator move northeast toward the pole instead of north? The answer to this mystery lies in the **Coriolis effect**. In Activity 23B, you will construct a model of the Coriolis effect and begin to untangle the mystery.

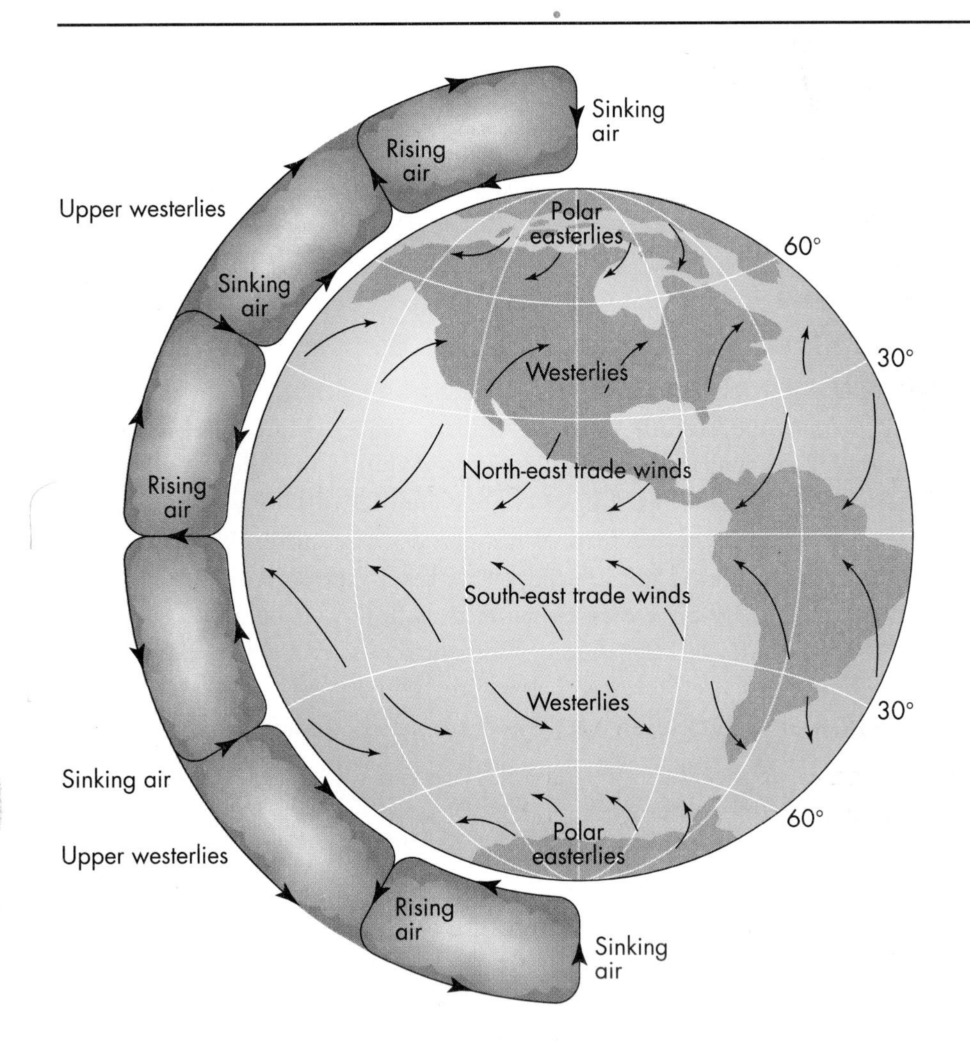

FIGURE 23.9 ◀
The major global wind patterns. Why do the winds north and south of the equator go in the same direction?

ACTIVITY 23B / A DEMONSTRATION OF THE CORIOLIS EFFECT

Work with one of your classmates to demonstrate the Coriolis effect.

MATERIALS

a slate globe
chalk dust
water pistol
tap water
drawing paper
drawing pencil

PROCEDURE

1. Moisten the slate globe and apply chalk dust evenly to the surface.
2. Fill the water pistol with water.
3. One student rotates the ball clockwise about the poles in a slow and steady motion.
4. At the same time, the other student aims the water pistol at the equator, fires a steady stream and slowly moves, in a straight line, toward the north pole. After the ball has rotated a fraction of a turn, aim and fire again.
5. Repeat Steps 3 and 4, but this time start the pistol's aim at the equator and move it toward the south pole.

DISCUSSION

Draw, on paper, an illustration of the paths created by the water on the ball, with arrows indicating the direction of movement. Use this model to explain the effect of the Earth's rotation on the movement of the atmosphere.

OCEAN CURRENTS AND WINDS

Ocean currents flow like rivers of water through the ocean, moving at different speeds than the surrounding water. Mariners have long known that they can shorten their travel times by following the currents across the ocean. Currents are hundreds of kilometers wide and several hundred meters deep. What causes these currents?

Surface currents are caused by winds at the Earth's surface. But there are also **deep water currents** that flow well below the surface. Colder, saltier, and therefore denser water near the poles sinks toward the bottom of the ocean and flows in the direction of warmer waters. These deep water currents flow in the opposite direction to the surface currents above them. When the winds blow the warmer water away, the colder water is able to move toward the surface. This process is known as **upwelling**. Upwelling is good for fish and other marine life because it brings nutrients to the surface, providing food.

DID YOU KNOW

■ How long does it take surface water to sink and return to the surface again? About 700 years in the Atlantic and about 1,500 years in the Pacific.

Oceanographers measure the ocean currents by using special buoys. These buoys contain instruments or meters that record the speed and direction of the current. Present day meters may sometimes also be suspended from research ships (Figure 23.10).

Space satellites are also used to measure currents. These satellites carry sensors that detect the temperature differences at the ocean surface. The warmer waters show up red and the cooler waters are blue (Figure 23.11).

How do wind patterns affect the oceans? The wind striking the ocean surface causes waves and moves the surface water in patterns corresponding to the wind patterns (Figure 23.12). If you look at the North Pacific just west of the U.S., you will see that the westerlies cause the N. Pacific Current, driving it toward the northwest coast of the continent, where it is blocked and diverted southward and driven westward by the trade winds.

FIGURE 23.10
Research vessel measuring currents.

But not only do the winds drive the currents, but conversely, the currents alter the atmosphere and determine the climate of entire regions. Ocean currents distribute heat around the globe by moving large masses of warm water from the equatorial region to the cooler latitudes. In the North Pacific, for example, the North Equatorial Current flows westward until it approaches Asia, where it is diverted

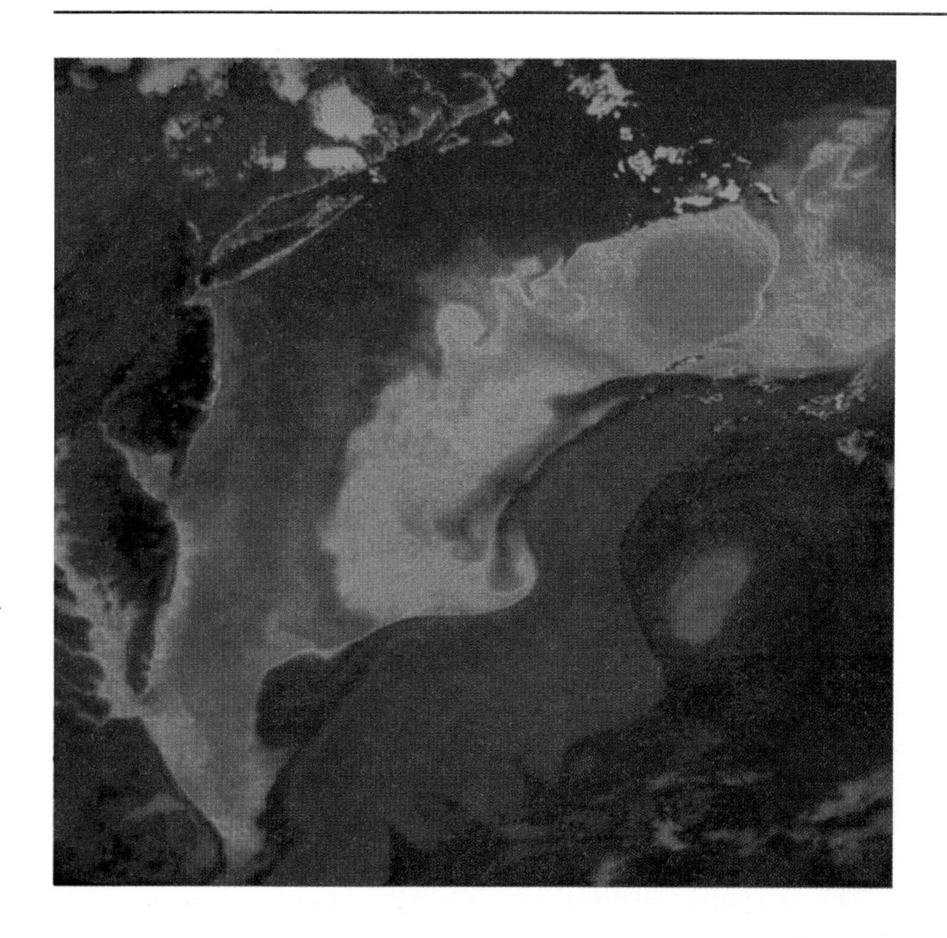

FIGURE 23.11 ◀
Satellite photo showing different water temperatures. What might cause the differences in temperature seen here?

FIGURE 23.12
Major surface currents.

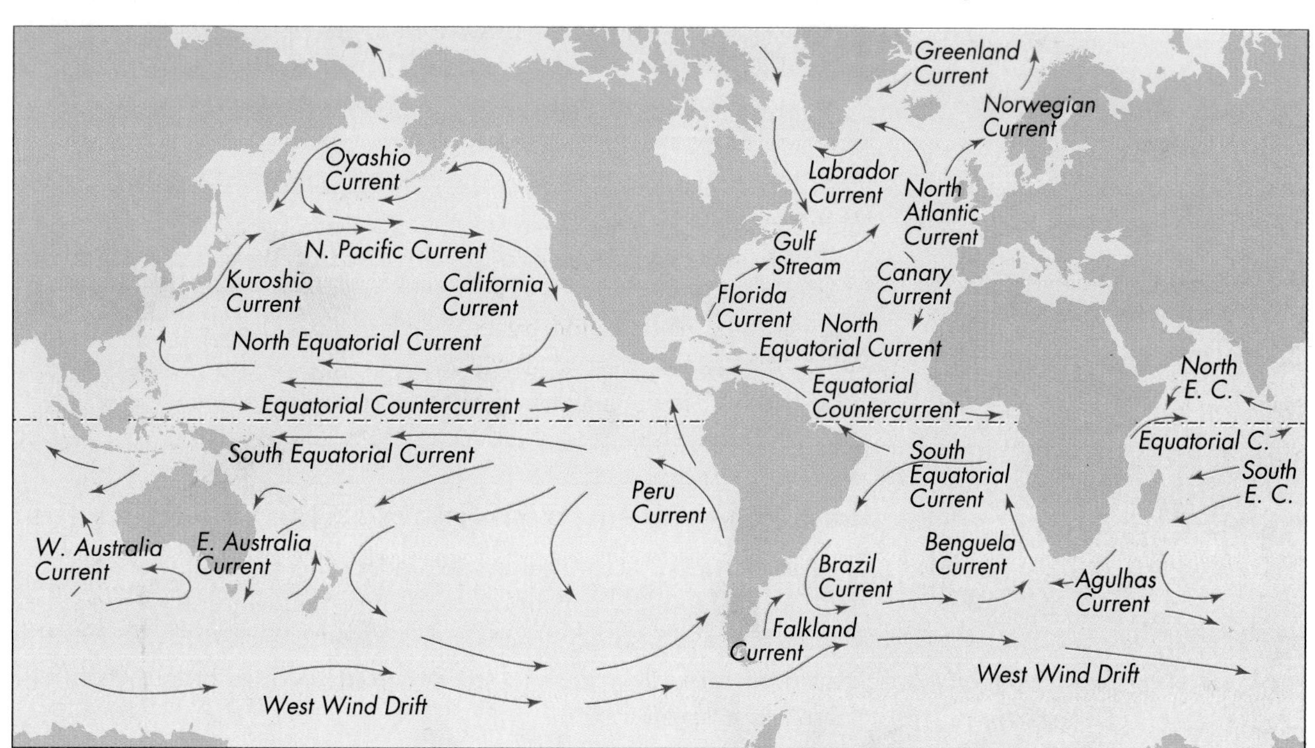

DID YOU KNOW

■ Without the oceans, the Earth's temperatures would soar to more than 100° C during the day and plunge below −100° C at night. During the winter, temperatures in the polar regions would drop to −200° C.

northward and northeastward along the coastline, carrying warm water with it. At about 45° latitude, it turns eastward back across the Pacific. As it approaches the northwest coast of North America, the warmer air moderates the temperatures of the Pacific Northwest. It then turns southward, carrying now cooler northern water toward the equator.

The Gulf Stream, as an example, carries warmer waters from the equator northward along the southeast coast of North America, then northeastward across the Atlantic Ocean to Europe. As a result, winter weather in Great Britain is milder than one would otherwise expect (Figure 23.13).

FIGURE 23.13 ▶
View of Ashleam Bay, Achill Island, Ireland.

▶ REVIEW 23.2

1. Describe the water cycle.
2. What is the role of the sun in the water cycle?
3. Describe the difference between weather and climate.
4. Do the polar easterlies mix with the westerlies? Explain your answer.
5. Describe the winds' effect on ocean currents and the ocean currents' effect on climates.

■ 23.3 A MORE DETAILED LOOK AT OCEANS

The properties of ocean water vary under different conditions. In this section, you will learn about ocean salinity, temperature, pressure, and depth. You will also learn how these relate to each other and to the global patterns described in the previous section. Finally, you will learn about ocean waves.

PROPERTIES OF OCEAN WATER

The composition of ocean water can be affected by the atmosphere. Ocean water is approximately 96.5 percent pure water. Dissolved salts make up the remaining 3.5 percent. The measure of the ocean's saltiness is called **salinity**. Salinity is expressed in parts per thousand. The average salinity of the ocean is 35 parts per thousand. More than 75 percent of these dissolved salts are sodium chloride (Figure 23.14).

The salinity of the ocean in any one area is dependent, in part, on climate. In warm dry regions, the water will have a higher concentration of salts. In the Mediterranean, for example, the air is hot and dry. This leads to evaporation of surface water, leading to surface water with a higher salinity. When the salinity is higher, the **density** of the surface water increases, causing the surface water to descend to lower depths. This can affect the circulation of the deep water currents.

The temperature of the ocean's waters varies by region, by depth, and by season. Because the sun's rays strike the Earth more directly at the equator, surface waters there are warmest all year long. Waters at the poles are coldest, because the sun's rays strike less directly there.

Temperatures also vary by depth. Surface water is warmer than deeper water. The sun's rays penetrate about 100 meters deep into the ocean. Because warmer water is less dense, it floats on top of the denser water instead of mixing with it. There is a zone between warm surface water and the cold deep water where the temperature changes abruptly. This is called the **thermocline** (Figure 23.15).

Surface water at the equator can be as warm as 30° C. In polar regions all year and in middle latitudes in the winter, the waters are as

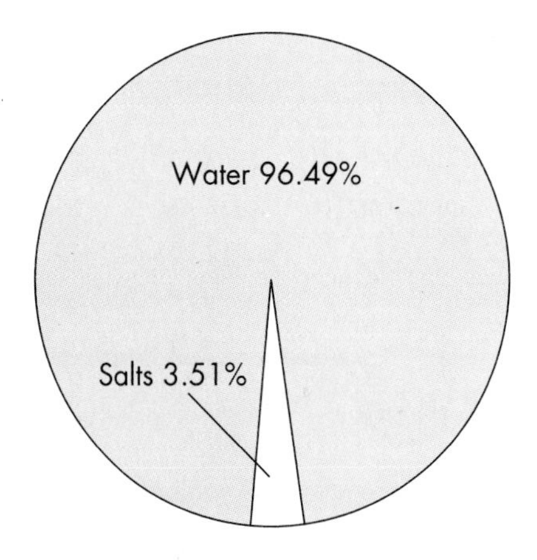

Salts		
	sodium chloride:	2.72%
	magnesium chloride:	0.38%
	magnesium sulfate:	0.17%
	calcium sulfate:	0.13%
	potassium sulfate:	0.09%
	calcium carbonate:	0.01%
	magnesium bromide:	0.01%

FIGURE 23.14
Salt content of ocean water.

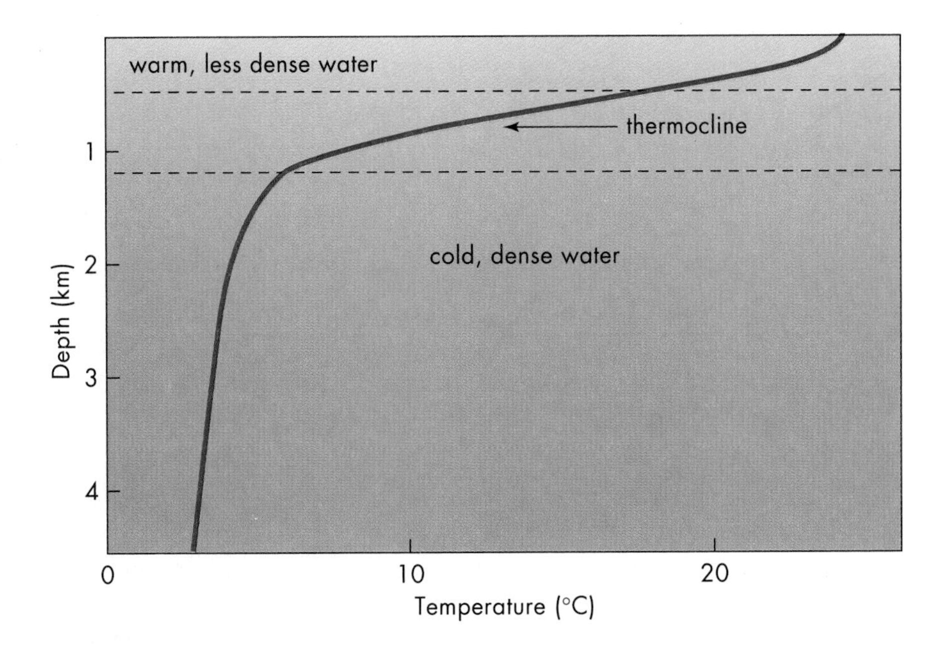

FIGURE 23.15 ◀
Water temperature changes abruptly at the thermocline.

cold as −2° C. Sea water freezes at −2° C instead of 0° C because of its salinity. Polar ice rarely reaches more than 3 meters in depth. Water below that point is protected by the ice from the frigid polar air, so remains above the freezing point.

We have already seen that cooler and saltier water is denser and therefore sinks to lower depths, where it joins the deep water currents. As the water becomes deeper, the water pressure becomes greater. Gravity presses on the Earth's surface and pressure is exerted by the water as it presses down on the ocean floor. But water is not easily compressed; therefore, its density is not greatly increased by pressure. The water density in the deepest parts of the ocean is only seven percent greater than it is at the surface.

Water pressure is measured in **atmospheres**. The pressure of the atmosphere at sea level is one atmosphere. The pressure increases by one atmosphere every 10 meters. At the ocean's deepest point, the pressure can reach 1,100 atmospheres.

DID YOU KNOW

■ Where is all the water? 97% of the Earth's water is in the oceans. Another 2% is locked in glaciers. This means that all the water that flows underneath and on top of the Earth's surface constitutes only 1% of the Earth's water (Figure 23.16).

(a)

(b)

FIGURE 23.16 ◀
Two percent of the Earth's water is locked in glaciers. What might happen to sea level if these were to melt?

ACTIVITY 23C / DEPTH AND PRESSURE

In this activity, you will explore the relationship between water depth and pressure.

MATERIALS

plastic soft-drink bottle
nail
masking tape
tap water
graduated beaker
watch or clock with second hand

PROCEDURE

1. With the nail, make three holes in a vertical line up the side of the soft-drink bottle.
2. Cover each hole with a piece of tape.
3. Fill the bottle with water to a level well above the top hole.
4. Place the graduated beaker beneath the top hole, remove the tape and allow it to fill the beaker for 10 seconds. Record the amount of water in the beaker.
5. Repeat the process in Step 4 for the second and third holes.
6. Compare results for the three trials.

DISCUSSION

1. Construct a bar graph for the results of the three trials.
2. Describe and explain the results.
3. What is the relationship between pressure and depth?

FIGURE 23.17 ◀
Materials required for Activity 23C.

WINDS, WAVES, AND TSUNAMIS

Most waves are caused by surface winds. The speed of the wind at the ocean's surface and the length of time it blows are the main factors that create high waves.

A wave is a periodic motion of the ocean surface. Notice in Figure 23.18 that waves have high points called **crests** and low points called

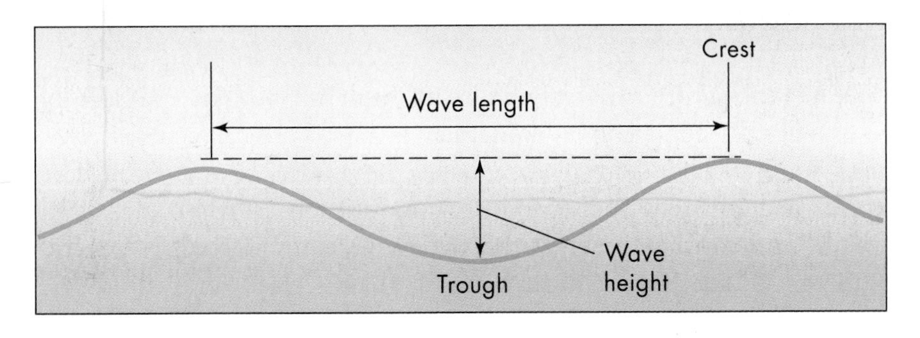

FIGURE 23.18 ◀
Parts of a wave.

troughs. The distance between crests is called **wavelength**. The vertical distance between crest and trough is the **wave height**. In most instances, the water in a wave does not actually advance forward. The wave moves forward but the water does not (Figure 23.19). The water particles move in a circular pattern during the passing of a wave so that during one wave period the water particle returns to its original starting point.

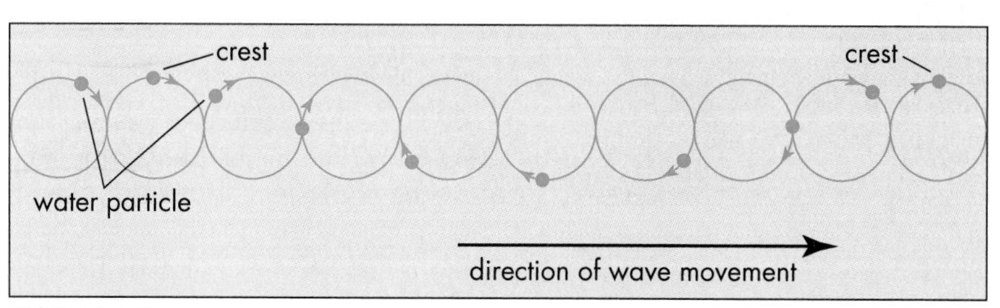

DEEP WATER

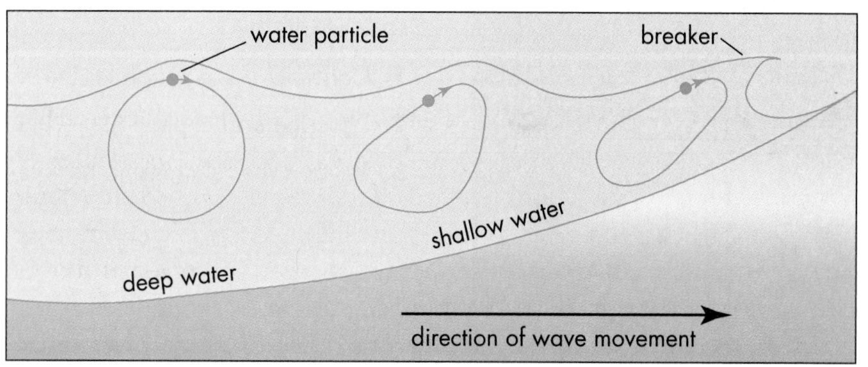

SHALLOW WATER

FIGURE 23.19 ▶
In deep water, the particles do not move forward. In shallow water, they advance with the wave.

When a wave approaches the shore, however, things change. As the wave approaches the shore, water is increasingly shallower and the ocean floor interferes with the wave's motion. The bottom part of the wave is slowed down and the crest topples forward, creating what are known as breakers. In this instance, the water particles do advance with the wave (Figure 23.20).

In the open ocean, the average wave is a few meters high. The average wavelength in the open ocean is 60 to 120 meters. When the height of a wave is greater, its corresponding wavelength is also greater. Under extreme wind conditions, waves may be 30 meters high.

But wind is not the only cause of high waves. An example of extreme wave conditions is **tsunamis**. Tsunamis are caused by sudden undersea movements from earthquakes, landslides, or volcanoes. The waves generated by such conditions may reach 20 to 30 meters in height and can move across the Pacific Ocean at speeds of several hundred kilometers per hour. When tsunamis crash onto the shore,

FIGURE 23.20 ◄
As a wave approaches the shore, the crest topples, forming breakers.

they cause widespread destruction (Figure 23.21). Because of this, a tsunami warning network has been set up to alert people living in the Pacific region.

STORMS

Storms can affect the oceans. A violent storm can create waves that are capable of altering sea level and can even alter temperatures of surface waters. Likewise, oceans can affect storms. Surface temperatures of the ocean can alter storms. For example, a hurricane moving across the Atlantic Ocean will increase in intensity when it moves over the warmer waters of the Gulf of Mexico.

The temperature of a patch of ocean can even cause violent storms. (Figure 23.22) For example, some years an **El Niño**, a large patch of warmer ocean water, forms in the western Pacific Ocean and slowly moves eastward to South America. It spawns large storms and heavy rains in many parts of the world and drought in other places. California, as a result, experienced much rainier and stormier winter weather than normal during a recent El Niño, resulting in beach erosion, mudslides, and property destruction.

(a)

(b)

FIGURE 23.21
Before (a) and after (b) sequence of tsunami damage at the Scotch Cap Lighthouse on Unimak Island, Alaska.

FIGURE 23.22 ◄
Storms created by an El Niño can create destructive waves such as those shown here

SCRIPPS INSTITUTION OF OCEANOGRAPHY

The Scripps Institution of Oceanography in California is one of the premier oceanographic research centers in the world. Scripps is a part of the University of California, San Diego, and operates from its 180-acre campus on the Pacific coast in La Jolla. Its graduate degree program is the top-rated program in oceanography in the country.

Scripps, however, is much more than a graduate school. It recently launched its newest research vessel, the 274-foot R/V *Roger Revelle,* bringing the number of ships in its fleet to four and making it the largest academic research fleet in the country. Scripps conducts an ambitious program of research projects, currently numbering more than 300, covering such areas as marine biotechnology, the Antarctic Ozone Hole, oil spills, beach erosion, tropical rain cycles, thermostat theory, and many others.

Scripps recently received a major grant to establish the International Research Institute for climate prediction. The Institute's goal will be to provide early prediction of El Niños and other climate changes, and to alert those parts of the world expected to be affected. Scripps also plans to integrate the El Niño data researchers collect into the forecasting of global climatic changes.

Another project of note developed by Scripps is a technology called acoustic daylight. This process uses the sound generated by bubbles in the ocean. When the bubbles come in contact with objects, the sound they produce changes. By detecting and recording these sound changes, researchers are able to create a pictorial image of the object and view it on a monitor. This new technology has many applications such as navigation, salvage, and even the detection of mines and submarines.

Another important part of the Scripps Institution of Oceanography is the Birch Aquarium. This is Scripps' main public education center offering educational programs to people of all ages with special emphasis on school-age children. Students can work in the classroom/laboratories or explore the local tide pools and beaches. The aquarium complex includes more than 30 aquarium tanks displaying marine life from a variety of ocean habitats. The museum also features a simulated trip on a deep-diving submersible. The Birch Aquarium's continuing mission is to interpret the oceanographic research conducted at Scripps, to provide ocean science education to the public, and to promote ocean conservation.

▶ REVIEW 23.3

1. How can the hot, dry air affect deep water currents?
2. Which water has the highest salinity, surface water or deep water? Explain you answer.
3. Describe the difference between water above the thermocline and water below the thermocline.
4. What is the pressure at 50 m below sea level? At 200 m below sea level? And at 500 m below sea level?

CHAPTER REVIEW

KEY IDEAS

- The tides are caused primarily by the moon and secondarily by the sun.
- High tides occur along a line connecting the Earth and moon; low tides occur at points midway between high tides.
- Spring tides occur when the sun, Earth, and moon are in alignment. Neap tides occur when the sun, Earth, and moon form a right angle.
- The atmosphere and the ocean affect each other in many ways.
- Water is exchanged between the ocean and atmosphere through the water cycle.
- The sun evaporates water which rises into the atmosphere.
- Water condenses out of the atmosphere and returns to the surface through precipitation.
- Precipitation returns to the oceans through groundwater and runoff.
- Precipitation returns to the atmosphere through evaporation and transpiration.
- The global wind patterns are caused when the sun heats equatorial air and the Earth rotates, causing the movement.
- Global wind patterns create the global ocean circulation patterns.
- Rising air moves northeast instead of north due to the Coriolis effect.
- Surface currents carry warm water to otherwise cooler regions, warming the air and causing an unexpectedly warm climate.
- The ocean's waters are salty, 3.51%. The deeper, the saltier, because water with a higher salinity is denser and sinks beneath less salty, less dense water.
- Surface waters are warmer than deep waters. They are sharply divided by the thermocline.
- As depth increases, pressure increases dramatically, but density increases only slightly.
- Most waves are caused by surface winds.
- The wave height is determined by the speed and duration of surface winds.
- Tsunamis are caused by earthquakes, landslides, and volcanoes.
- Patches of warm water in the ocean can increase the intensity of a storm or even cause a storm.

VOCABULARY

oceanography
high tide
low tide
Spring tide
Neap tide
water cycle
evaporate
condensation
precipitation
groundwater
runoff
transpiration
weather
climate
trade winds
westerlies
polar easterlies
Coriolis effect
surface currents
deep water currents
upwelling
salinity
density
thermocline
atmospheres
crests
troughs
wavelength
wave height
tsunami
El Niño

V1. Construct a mind map that shows the influences between the atmosphere and the ocean using as many terms in the vocabulary list as you can.

V2. For each term in the vocabulary list, write a brief definition in your own words.

CONNECTIONS

C1. Compare the sun's role in the water cycle and its role in generating the global wind patterns.

C2. Describe all the ways in which the atmosphere affects the oceans and all the ways the oceans affect the atmosphere.

C3. List the paths that precipitation can follow after it falls to the surface.

C4. Describe the factors responsible for the fact that ocean currents flow in large loops.

C5. The next time your area experiences an unexpectedly warm day in winter, or an unexpectedly cool day in summer, determine the reason for this change in weather. Did the ocean currents have anything to do with it?

EXPLORATIONS

E1. What global winds are in your area? When it rains, where did the water for the rain originate?

E2. Go outside and locate the moon. It may be in your daytime sky or out at night. Estimate its position relative to the zenith point. From this, determine what longitude on Earth is experiencing a high tide at that moment.

E3. Investigate the most recent El Niño and its effects on the California coast. Study newspaper reports to find the exact location of the El Niño, and trace the path of storms it generated.

E4. Write a report on the tsunami warning network around the Pacific region.

REFLECTIONS

R1. Locate a tide table for the ocean nearest you. Study the rising and setting times of the moon, the times of high and low tides, and the dates of Spring and Neap tides for one month listed in the table. Analyze the mathematical relationship between these times. The *Farmer's Almanac* is an excellent source for such tables.

R2. Construct a diagram that traces the path of one water molecule. It can begin its journey in one of the oceans, or in a stream nearby. Be certain to pick an actual geographical location for your starting point. Trace its journey as far as you can. Can you estimate how long its journey might last?

Appendix A

SI UNITS OF MEASUREMENT

The International System of Units—the metric system—is usually referred to as the SI. SI comes from the French name *Le Système international d'unités*.

Table A.1 shows some common SI and non-SI units. It includes all the units used in this book. Sometimes units are combined to measure other quantities. For example, speed can be measured in kilometers per hour (km/h).

Table A.2 shows the most commonly used SI prefixes. You can see that all the prefixes are related to each other by multiples of 10. The SI prefixes may be combined with the base units in Table A.1—for example, centimeter (cm), kilojoule (kJ), milligram (mg).

TABLE A.1 ◀ ***Some SI Symbols and Units, and Units Permitted with SI***

Symbol	Unit	Quantity
a	year	time
d	day	time
g	gram	mass
h	hour	time
ha	hectare	area
J	joule	energy, work
K	kelvin	temperature
L	liter	volume
m	meter	length
min	minute	time
N	newton	force
s	second	time
t	metric ton	mass
°C	degree Celsius	temperature

Note: one tonne (1 t) = one megagram (1 Mg)
one hectare (1 ha) = one square hectometer (1 hm^2)

TABLE A.2 ***SI Prefixes***

Prefix	Symbol	Factor by which the base unit is multiplied	Example
mega	M	10^6 = 1 000 000	10^6 m = 1 Mm
kilo	k	10^3 = 1 000	10^3 m = 1 km
hecto	h	10^2 = 100	
deca	da	10^1 = 10	
		10^0 = 1	10^0 m = 1 m
deci	d	10^{-1} = 0.1	
centi	c	10^{-2} = 0.01	10^{-2} m = 1 cm
milli	m	10^{-3} = 0.001	10^{-3} m = 1 mm

Appendix B

PERIODIC TABLE OF ELEMENTS

Metals
Non-metals

1 **H** Hydrogen								
3 **Li** Lithium	4 **Be** Beryllium							
11 **Na** Sodium	12 **Mg** Magnesium							
19 **K** Potassium	20 **Ca** Calcium	21 **Sc** Scandium	22 **Ti** Titanium	23 **V** Vanadium	24 **Cr** Chromium	25 **Mn** Manganese	26 **Fe** Iron	27 **Co** Cobalt
37 **Rb** Rubidium	38 **Sr** Strontium	39 **Y** Yttrium	40 **Zr** Zirconium	41 **Nb** Niobium	42 **Mo** Molybdenum	43 **Tc** Technetium	44 **Ru** Ruthenium	45 **Rh** Rhodium
55 **Cs** Cesium	56 **Ba** Barium	57 – 71 **La–Lu** * (see below)	72 **Hf** Hafnium	73 **Ta** Tantalum	74 **W** Tungsten	75 **Re** Rhenium	76 **Os** Osmium	77 **Ir** Iridium
87 **Fr** Francium	88 **Ra** Radium	89 – 103 **Ac–Lr** ** (see below)	104 **Rf** Rutherfordium	105 **Db** Dubnium	106 **Sg** Seaborgium	107 **Bh** Bohrium	108 **Hs** Hassium	109 **Mt** Meitnerium

*LANTHANIDE SERIES	57 **La** Lanthanum	58 **Ce** Cerium	59 **Pr** Praseodymium	60 **Nd** Neodymium	61 **Pm** Promethium	62 **Sm** Samarium	63 **Eu** Europium
**ACTINIDE SERIES	89 **Ac** Actinium	90 **Th** Thorium	91 **Pa** Protactinium	92 **U** Uranium	93 **Np** Neptunium	94 **Pu** Plutonium	95 **Am** Americium

								2 **He** Helium
			5 **B** Boron	6 **C** Carbon	7 **N** Nitrogen	8 **O** Oxygen	9 **F** Fluorine	10 **Ne** Neon
			13 **Al** Aluminum	14 **Si** Silicon	15 **P** Phosphorus	16 **S** Sulfur	17 **Cl** Chlorine	18 **Ar** Argon
28 **Ni** Nickel	29 **Cu** Copper	30 **Zn** Zinc	31 **Ga** Gallium	32 **Ge** Germanium	33 **As** Arsenic	34 **Se** Selenium	35 **Br** Bromine	36 **Kr** Krypton
46 **Pd** Palladium	47 **Ag** Silver	48 **Cd** Cadmium	49 **In** Indium	50 **Sn** Tin	51 **Sb** Antimony	52 **Te** Tellurium	53 **I** Iodine	54 **Xe** Xenon
78 **Pt** Platinum	79 **Au** Gold	80 **Hg** Mercury	81 **Tl** Thallium	82 **Pb** Lead	83 **Bi** Bismuth	84 **Po** Polonium	85 **At** Astatine	86 **Rn** Radon

64 **Gd** Gadolinium	65 **Tb** Terbium	66 **Dy** Dysprosium	67 **Ho** Holmium	68 **Er** Erbium	69 **Tm** Thulium	70 **Yb** Ytterbium	71 **Lu** Lutetium
96 **Cm** Curium	97 **Bk** Berkelium	98 **Cf** Californium	99 **Es** Einsteinium	100 **Fm** Fermium	101 **Md** Mendelevium	102 **No** Nobelium	103 **Lr** Lawrencium

Appendix C

LIST OF ELEMENTS

Element	Symbol	Element	Symbol	Element	Symbol
Actinium	Ac	Holmium	Ho	Rhenium	Re
Aluminum	Al	Hydrogen	H	Rhodium	Rh
Americium	Am	Indium	In	Rubidium	Rb
Antimony	Sb	Iodine	I	Ruthenium	Ru
Argon	Ar	Iridium	Ir	Samarium	Sm
Arsenic	As	Iron	Fe	Scandium	Sc
Astatine	At	Krypton	Kr	Selenium	Se
Barium	Ba	Lanthanum	La	Silicon	Si
Berkelium	Bk	Lawrencium	Lw	Silver	Ag
Beryllium	Be	Lead	Pb	Sodium	Na
Bismuth	Bi	Lithium	Li	Strontium	Sr
Boron	B	Lutetium	Lu	Sulfur	S
Bromine	Br	Magnesium	Mg	Tantalum	Ta
Cadmium	Cd	Manganese	Mn	Technetium	Tc
Calcium	Ca	Mendelevium	Md	Tellurium	Te
Californium	Cf	Mercury	Hg	Terbium	Tb
Carbon	C	Molybdenum	Mo	Thalium	Tl
Cerium	Ce	Neodymium	Nd	Thorium	Th
Cesium	Cs	Neon	Ne	Thulium	Tm
Chlorine	Cl	Neptunium	Np	Tin	Sn
Chromium	Cr	Nickel	Ni	Titanium	Ti
Cobalt	Co	Niobium	Nb	Tungsten	W
Copper	Cu	Nitrogen	N	Uranium	U
Curium	Cm	Nobelium	No	Vanadium	V
Dysprosium	Dy	Osmium	Os	Xenon	Xe
Einsteinium	Es	Oxygen	O	Ytterbium	Yb
Erbium	Er	Palladium	Pd	Yttrium	Y
Europium	Eu	Phosphorus	P	Zinc	Zn
Fermium	Fm	Platinum	Pt	Zirconium	Zr
Fluorine	F	Plutonium	Pu		
Francium	Fr	Polonium	Po	**Recently Named Elements**	
Gadolinium	Gd	Potassium	K	Bohrium	Bh
Gallium	Ga	Praseodymium	Pr	Dubnium	Db
Germanium	Ge	Promethium	Pm	Hassium	Hs
Gold	Au	Protactinium	Pa	Meitnerium	Mt
Hafnium	Hf	Radium	Ra	Rutherfordium	Rf
Helium	He	Radon	Rn	Seaborgium	Sg

Appendix D

FISH ANATOMY AND DISSECTION

In this activity you will have a chance to examine the internal (inside) and external (outside) structure of a fish. You will learn more about how a fish swims, breathes, digests food, and reproduces. The term "dissect" means "to separate into parts and examine." Before beginning your examination of fish anatomy, take a moment to become familiar with the dissecting equipment you will be using. Adopting a scientific approach to using this equipment will result in a dissection that will show all the structures, undamaged, and in their proper position. If you are using a preserved fish, it should be soaked in water overnight and rinsed thoroughly under a tap before beginning the dissection.

Note: Part IX, on cleaning-up procedures, should be followed at the end of every dissection lab.

MATERIALS

safety goggles
apron
disposable gloves
pencil and drawing paper
fresh or preserved fish
dissection kit
dissecting tray
paper towels
plastic bag with tag
magnifying lens or dissecting microscope
watchglasses or Petri dishes

CAUTION! CAUTION!

- **Dissection instruments are extremely sharp. Read and follow the dissection safety rules, number 26 in the Safety Rules section at the beginning of this book.**
- **Put on your safety goggles, apron, and gloves whenever you are handling the fish or the dissecting instruments.**

PART I External Features

PROCEDURE

1. Your teacher will provide you with an outline drawing of a fish. Refer to Figure D.1, and using a pencil, carefully draw your fish and neatly print the following labels on the diagram:
   (a) head
   (b) tail fin (caudal fin)
   (c) mouth
   (d) gill cover
   (e) dorsal fin or fins (the number may change depending on the species of fish you are dissecting) located along the fish's back, or dorsal surface

FIGURE D.1 ▶
External features of the fish

(f) pectoral fins (paired) located just posterior to (behind) the gill covers
(g) pelvic fins (paired) located ventral to (below) the pectoral fins
(h) anus, the external opening of the digestive system located posterior to the pelvic fins
(i) anal fin (one), if present in the species of fish you are examining, located posterior to the anus

2. Put on your safety goggles, apron, and gloves.

3. Note the coloring of your fish. Indicate on your drawing where its lighter and darker areas are located.

4. Move your fingers over the scales covering the body, from anterior (front) to posterior (back) and back the other way. Describe how they look and feel. Use forceps to remove a scale, then look at it under a magnifying lens or dissecting microscope. Note the shape of the scale and any lines that can be seen. (The lines on a fish scale can be counted to give the fish's age; this is similar to the way that rings on a tree stump can be counted to give the tree's age.) Make a sketch of a scale, labelling the anterior and posterior ends and showing any lines that you can see.

5. Locate the gill covers. Use forceps to gently raise one of the gill covers to observe the gills under it. Look at the gills with a magnifying glass or microscope.

6. Find the nostrils, a pair of small pits located at the anterior end of the fish. The nostrils of a fish do not join to form a common breathing passage as they do in humans. Fish nostrils are used only for the sense of smell; these pits contain sensitive cells capable of detecting the presence of even a few molecules of certain chemicals in the water. Insert the pointed end of the probe into a nostril. Gently move it forward to determine the depth of the nostril. Do not push too deep.

7. Locate the line running along the side of the fish from the edge of the gill to the tail fin. This is called the lateral line. The cells of the lateral line (sensory receptors) help the fish to detect vibrations in the water, such as those that might be caused by the motion of other fish. The sensitivity of this lateral line enables fish in groups or "schools" to move and change direction as a group. Label the lateral line on your diagram.

8. Describe the location of the fish's eyes. Are there eyelids?

DISCUSSION

1. (a) How do you think the body shape of your fish is suited to its environment? Explain.
(b) What do you think is the function of each of the different fins?

2. Explain how you think the differences in color from the top to the bottom of a fish help it avoid enemies
(a) from above (b) from below.

3. (a) Are the nostrils of a fish used for the same purposes as the nostrils of humans? Describe any similarities and differences.
(b) How do you think a fish's nostrils might be affected by the addition of industrial wastes to rivers and streams?

4. What is the function of the lateral line?

PART II Respiratory System

Fish need to exchange carbon dioxide and oxygen just as we do. In humans, this exchange of gases takes place in the lungs (see Chapter 8). In fish, gas exchange takes place in the gills. Whereas our lungs obtain oxygen from air, the gills of fish are capable of removing dissolved oxygen from water.

PROCEDURE

1. Insert the tip of a blunt probe through the mouth of your fish and out under its gill cover.

This is the path taken by water circulating over the gills. Leaving the probe in place, carefully remove the gill cover with your scissors. (Note: Be careful not to damage the delicate gills underneath.) Each gill consists of a backward C-shaped arch of cartilage, with bumps called gill rakers on the inner edge and a double row of thin-walled projections called gill filaments. Count the number of gills on the side that you are examining.

2. Look at the gill filaments. Blood is pumped through these filaments in order to pick up oxygen from the water and to release carbon dioxide.
3. Touch one gill arch. It will feel firm. The comb-like bumps along the inner edge of the arch are the gill rakers mentioned above. Their function is the same as that of the hairs of your nose: to filter out particles and keep them from passing farther into the respiratory system.
4. Remove one complete gill and float it in a Petri dish containing water. Draw the gill for your fish, labelling the gill arch, gill rakers, and gill filaments.
5. Sketch the anterior end of a fish and use arrows to show the movement of water through the respiratory system. Include the mouth and gills in your sketch.

DISCUSSION

1. Why do you think the gill rakers are on the anterior surface of the gills?
2. The gill filaments are highly folded. Why is this necessary?
3. Most fish are able to move their gill covers, opening and closing them rhythmically. Why do you think they do this?
4. Why do you think fish have covers over their gills?

PART III Opening the Body Cavity

PROCEDURE

1. Under the skin of a fish lie the muscles that propel it through the water. In order to examine the muscles of your fish, you need to peel away an area of skin. Select an area along the side of the specimen, then use your scalpel to make a shallow cut through the scales and skin to outline the area (Figure D.2).
2. Pull the skin back very carefully, using forceps. The exposed wavy segments of muscle are attached to the ribs.
3. Insert the point of your scissors into the anus of your fish. Be careful how you make the cut, since the internal organs are rather loose, and close to the body wall. Make your first cut slowly and not too deeply along the middle of the ventral surface (Figure D.3).

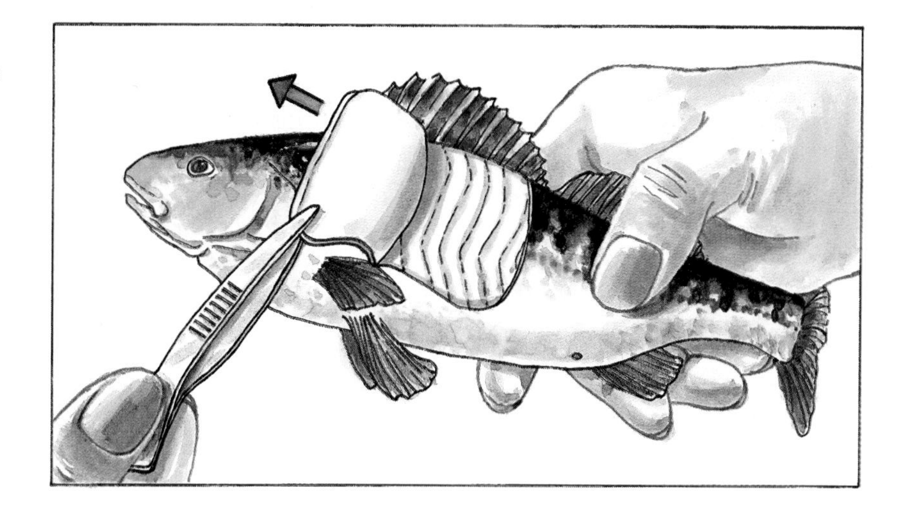

FIGURE D.2 ▶
When the skin is pulled back, the muscles are visible.

4. The second cut, as shown in Figure D.3, goes from the anus up to the lateral line. Gently pull up the body wall as you cut. This will help you adjust the depth of your scissors. As you near the lateral line, you will see a thin-walled sac inside the fish. This is the air bladder, or swim bladder. (Note: Try not to puncture it.)
5. Continue your cut just below the lateral line to the front of your fish. This is the third cut shown in Figure D.3.
6. Cut along the gill cover (fourth cut) to remove the entire section of the body wall.
7. Compare the internal organs of your fish with those shown in Figure D.4 and with the specimens your classmates have prepared.

DISCUSSION

1. What do you think is the function of the ribs?
2. Describe the motion of a fish through water. How do you think the structure of the fish's skeleton helps it move this way?
3. Why do you think the fish has a fluid in its body cavity?

PART IV Reproductive Organs

PROCEDURE

1. Locate the reproductive organs, or gonads. These will be testes if the fish is male, and ovaries if the fish is female. Depending on the condition of each specimen at the time of its death, the gonads will vary in size. Close to breeding time, the gonads (especially the ovaries of the females) can fill most of the body cavity and even push out the body wall.
2. Make a sharp cut with a scalpel through the middle of the gonad and examine the structure. If possible, examine fish of both sexes. Describe any differences in appearance or texture between the male and female gonads.

DISCUSSION

1. Why do you think the size of the gonads varies so greatly among the fish?
2. How many male and female fish were there in the specimens provided for your class? Calculate the percentage of the total population made up by each sex. Do you think these values reflect the situation in the wild? Explain.

PART V Digestive System

The digestive system of a fish is similar to your own. Refer to Chapter 7, Digestion, for more information about the functions of the parts of the human digestive system.

PROCEDURE

1. Examine the mouth of your fish. Use a blunt probe to help open it. Carefully feel inside the mouth with your finger and note the location of

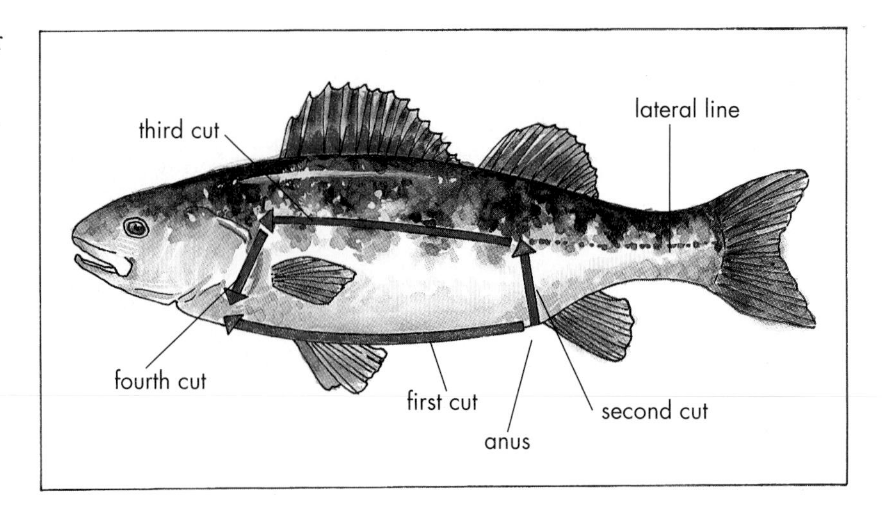

FIGURE D.3 ▶
Opening the body cavity. Be careful not to damage the internal organs. Do not insert your scissors too deeply.

the teeth. Does your fish have a tongue? (Note: Do not puncture your glove.)

2. Place a blunt probe in the fish's mouth and push it gently but firmly deeper into the fish. Try to make the probe enter the esophagus. Follow the esophagus to the stomach. The stomach is S-shaped and usually larger in diameter and thicker than the esophagus.

3. The liver is a large mass of tissue that lies in front of the stomach in the body cavity. In order to examine the rest of the digestive system, the liver should be carefully removed. (Be careful not to remove the heart at the same time. See Figure D.4.) Place the liver in a Petri dish, and add enough water to keep the liver moist. Now try to locate the gall bladder. This small, usually greenish-colored sac will probably be attached to the underside of the liver.

4. The stomach leads to the intestine. Follow the intestine to the anus.

5. Compare the positions of the organs of the digestive system of your fish with the positions shown in Figure D.4.

6. Carefully remove the stomach of the fish, and place it in another Petri dish with some water. Use scissors to cut open the stomach, then examine its contents. Compare the stomach contents of your fish with those of your classmates' fish. (Note: This is an interesting investigation to do on freshly caught fish, especially if you are familiar with the area where the fish lived.)

DISCUSSION

1. List the parts of a fish's digestive system and the functions of each part in the process of digestion. (Chapter 7, Digestion, contains helpful information.)

2. Describe the stomach contents of your fish and of the fish dissected by your classmates. Suggest what might be the diet of the particular fish you are studying. (This type of examination is routinely done by Department of Fisheries and Oceans biologists to determine the eating habits of fish.)

PART VI Circulatory System

The circulatory system of a fish is similar to yours. The major difference lies in the structure of the heart. Chapter 9 has useful information on the circulatory system.

PROCEDURE

1. Using the scissors, continue your cut along the ventral surface (underside) of your fish forward until you reach the beginning of the gills. Cut upward to expose the heart. (See Figure D.4.)

2. Note the color and texture of the fish heart. Examine the cavity in which the heart lies.

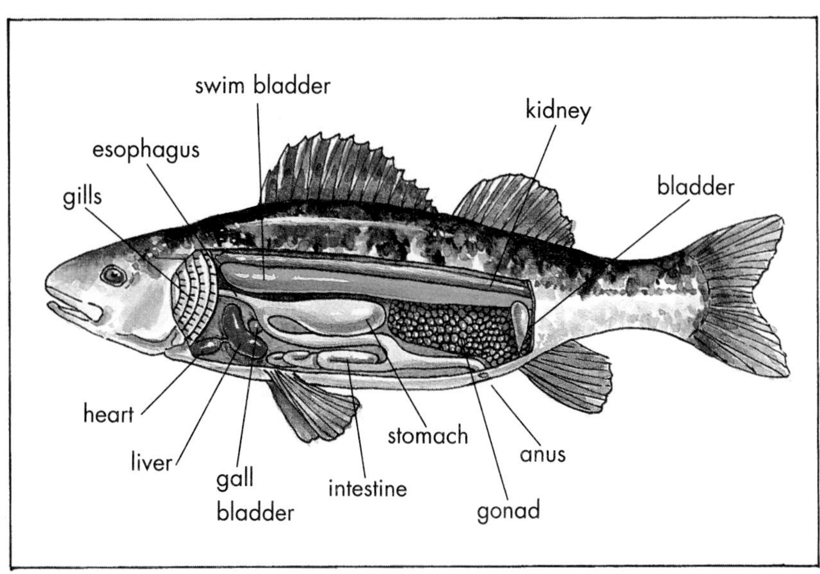

FIGURE D.4 ▶
Internal organs of the fish.

3. If possible with your specimen, locate tubes connected to the heart.

DISCUSSION

1. (a) What do you think is the function of the tubes connected to the heart?
   (b) What are these tubes called? (Refer to Chapter 9, Circulation and Excretion.)

2. Compare the number of chambers in the fish heart with the number in the human heart. Discuss the differences.

PART VII Excretory System

PROCEDURE

1. Locate the kidneys of the fish. They lie in a long dark-red line along the backbone and are covered by a transparent membrane. These organs remove wastes from the blood and produce urine. (Refer to Chapter 9 for information about the function of the kidneys in humans.)

2. You may be able to locate tiny tubes connecting the ends of each kidney with a small urinary bladder located just above the anus. The bladder empties into the same opening as the reproductive system, which is separate from and posterior to the anus.

DISCUSSION

1. Fish die if their kidneys are damaged. The position of the kidneys in the body of a fish helps to protect them from injury. Name the organs or parts that might help protect the kidneys.

PART VIII The Air Bladder or Swim Bladder

Bony fish (the major group of fish other than sharks and their relatives) possess one organ that is unique to them, the swim bladder. This thin-walled, air-filled sac enables fish to remain at a particular depth in the water, neither floating to the surface nor sinking to the bottom. Changing the volume of the air in the bladder allows the fish to change the depth at which it can remain motionless. A similar technique is used in the ballast tank of a submarine.

PROCEDURE

1. Locate the swim bladder along the dorsal surface of the body cavity just below the kidneys. Describe its shape and appearance.

DISCUSSION

1. What might happen to a fish if its swim bladder were located in its belly? Explain.

2. Sharks do not have swim bladders.
   (a) What do you think happens to sharks when they stop swimming?
   (b) Do you think sharks require more or less energy to swim than fish with swim bladders? Explain.

PART IX Cleaning up

Follow these steps at the end of every dissection lab.

PROCEDURE

1. If your fish is to be used again, wrap it in a moist paper towel, and place it in a plastic bag to be given to your teacher for storage. Be sure your name is on a tag attached to the bag or written clearly in pencil on a piece of paper inside the bag.

2. Dispose of any waste material in the container provided by your teacher, not in a sink.

3. Clean and dry your dissection tools carefully to keep them sharp and rust-free.

4. Wash your hands, fingernails, and lower arms thoroughly before leaving the lab.

GLOSSARY

A

abiotic: Non-living parts of the environment, such as water, rocks, and mud.

absolute magnitude: A number that is a measure of the actual amount of light given off by a star at a standard distance. The smaller the absolute magnitude, the greater is the amount of light given off by the star. The absolute magnitude is not the same as apparent magnitude.

absorption: The movement of molecules into cells. For example, nutrients are absorbed by the cells lining the small intestine.

acid: A type of compound that tastes sour, is soluble in water, and reacts with bases.

acid-base indicator: A substance that is one color in an acid and a different color in a base.

adaptation: A feature that helps an organism survive in a particular environment.

aerobic: With oxygen. While performing an aerobic activity, you continue to supply your body cells with the oxygen they need.

alloy: A solidified solution of two or more metals. An alloy is made by melting two or more metals and then mixing them. Brass is an alloy made of copper and zinc.

alpine tundra: A biome found at high altitudes, with a cold climate and a seasonal shortage of sunlight. Plant life includes small alpine flowers.

alternative energy resource: An energy resource used as an alternative to fossil fuels.

alveoli (singular: alveolus): Tiny air sacs, found in the lungs, where the exchange of gases takes place.

amino acid: A type of molecule that, when joined with other amino acid molecules, makes up a protein. In other words, amino acids are the building blocks of proteins.

anaerobic: Without oxygen. When you perform an anaerobic activity, you force your body to work so hard (over a short period of time) that you cannot supply your cells with oxygen as fast as they are using it up.

antibody: A type of blood protein that recognizes and tries to destroy foreign matter (such as bacteria) in the body.

anticline: An arch-like upfold of rock layers.

anus: The opening at the end of the digestive system through which feces are eliminated.

apparent magnitude: A number that is a measure of how bright a star looks to astronomers on Earth. The smaller the apparent magnitude, the brighter the star looks. The apparent magnitude is different from the absolute magnitude.

artery: A thick-walled blood vessel that carries blood away from the heart.

asteroid: A rocky body that orbits the sun. Most asteroids are found in a region between Mars and Jupiter.

asteroid belt: The region between the orbits of Mars and Jupiter where many asteroids are found.

astrology: The belief that objects in the sky influence people's lives. Astrology is not a branch of science.

astronomer: A person who studies the position and movement of objects and energy in the universe, and what they are made of.

astronomy: The branch of science that studies the position and movement of all objects and energy in the universe.

atherosclerosis: A disease in which cholesterol and other fats stick to the inner surface of arteries, reducing the amount of blood moving through the artery. Also called "hardening of the arteries."

atmospheres: Used to measure water pressure. The pressure at sea level is one atmosphere and increases by one atmosphere every ten meters.

atom: The smallest particle of an element that has the properties of that element.

atrium (plural: atria): One of the two chambers at the top of the heart. The right atrium receives deoxygenated blood from the body. The left atrium receives oxygenated blood from the lungs.

axis: The imaginary line around which a spinning object rotates.

B

barred spiral galaxy: A special type of spiral galaxy that has a bar-shaped region passing through it.

Basal Metabolic Rate (BMR): The rate at which a person's body uses energy to run basic life processes, such as those carried out by the digestive, circulatory, and respiratory systems.

base: A type of compound that tastes bitter, is soluble in water, feels slippery, and reacts with acids.

bias: A preference for one thing over another based on preconceived ideas.

Big Bang theory: The theory that the universe began with a huge explosion, and that the universe will continue to expand outward from that explosion.

bile: A liquid produced by the liver, stored in the gall bladder, and used in the small intestine to break apart large drops of fat so they can be acted upon by enzymes more rapidly.

biomass fuel: An energy resource made from plants and animal waste products.

biome: A large area of the Earth that has characteristic plants, climate, animals, and soil. The plants are part of a climax community.

biosphere: The total area of the Earth where living things are found.

biotic: All the organisms in the environment.

black hole: The extremely dense center of a star, which is much smaller in size than the star from which it was produced. The matter in a black hole is so dense that nothing, not even light energy, can escape from it.

blood protein: A protein carried in the plasma of the blood. Blood proteins include antibodies, hormones, and clot-forming proteins, as well as other kinds.

blood vessel: A tube that carries blood through the body. There are three major types of blood vessels: arteries, veins, and capillaries.

body mass: The amount of matter in a person's body. Body mass is often referred to as weight.

boiling: A fast change of state from liquid to gas.

breathing rate: The number of times you breathe in a minute.

bright nebula: A nebula that either gives off its own light energy or reflects light energy from a nearby star.

bronchi (singular: bronchus)**:** The two branches of the trachea. Each bronchus carries air into one lung.

by-products: Products other than the main ones in a chemical reaction.

C

capillary: A tiny, thin-walled blood vessel that carries blood close to body cells for the exchange of supplies and wastes. Capillaries connect arteries to veins.

carbohydrates: A group of nutrients made up of carbon, hydrogen, and oxygen atoms arranged in various ways into sugar molecules. Most kinds of carbohydrates contain two or more sugar molecules joined together. Carbohydrates are the major source of energy for your body.

cardiorespiratory fitness: A type of fitness describing how well a person's circulatory and respiratory systems are working.

catalyst: A substance that speeds up a chemical reaction without being changed itself.

cell: The basic unit of all living things.

cellular respiration: A process that takes place inside body cells. This process uses oxygen and glucose and releases energy, which the cells need to perform their jobs. Carbon dioxide and water are also produced.

chemical change: A change in matter in which one or more new substances are produced.

chemical digestion: The action of digestive enzymes in breaking down the chemical bonds that hold large molecules of food together.

chemical energy: The energy that is stored in chemical compounds.

chemical formula: The combination of chemical symbols that represents a particular compound.

chemical reaction: A change in matter in which one or more new substances are produced.

chest cavity: A sealed compartment in the body that holds the lungs and heart. The walls of the chest cavity are formed by the diaphragm and the bones and muscles of the rib cage. The chest cavity is also called the thoracic cavity.

cholesterol: A fat-like substance found in foods that come from animal products. Your body makes cholesterol as well as obtaining it from various types of food.

chromosphere: The region of the sun just above the photosphere of the sun. The gases there are hotter than the photosphere. Solar flares and solar prominences come out of the chromosphere.

cilia: Tiny hair-like projections of cells. Cilia are able to move back and forth. In the air passages, cilia move mucus and any trapped matter away from the lungs.

cinder cone: A small volcano formed when pieces of cinder-like rock erupt from an opening in the Earth and build up into a mound.

circulatory system: The system of blood vessels, blood, and the heart responsible for the transport of substances through the body.

climate: The average pattern of weather occurring at a place over a period of years. For example, the pattern would include temperature, wind velocity, rain and snowfall, and relative humidity.

climatograph: A graph showing average monthly temperature (as a line) and total monthly precipitation (as bars) over a year at one place on the Earth's surface.

climax community: A stable group of two or more species that is able to survive and reproduce indefinitely in the same habitat. The climax community is the final stage of succession unless the area is disturbed.

clot-forming blood proteins: A set of proteins that form a hard plug (clot) wherever a blood vessel has been damaged.

coal: A solid fossil fuel composed mainly of carbon. Coal formed from plants that lived in swamps millions of years ago.

combining capacity: The ability of an element to combine with other elements.

combustion: A chemical reaction in which oxygen is one of the reactants and in which heat is produced.

comet: A chunk of ice with some small, solid particles mixed in. When the comet comes close to the sun, the ice evaporates and glows. Comets travel in very long, oval orbits around the sun.

community: A group of interacting populations of two or more species.

complete protein: Any protein that contains enough of all eight essential amino acids to supply your body's needs.

complex carbohydrate: A carbohydrate made up of several or many sugar molecules joined together. Cellulose, glycogen, and starch are examples of complex carbohydrates.

composite cone: A steep-sided volcano formed from alternating layers of cinder-like rock and hardened lava.

compound: A pure substance that is made up of two or more elements that are chemically combined.

concentrated: Containing a great deal of dissolved solute.

concentration: The amount of solute present in a specific amount of solution.

condensation: The change of state from a gas to a liquid.

conductor: A substance that allows heat or electricity to move through it easily.

conifer: A plant that produces seed-bearing cones, such as a fir, spruce, or pine tree.

constellation: A group of stars that forms a shape or pattern.

constipation: A condition in which acts of elimination (bowel movements) are unusually infrequent, with feces that are dry and hard.

consumer: An organism that obtains its food by eating other organisms.

continental shelf: A gently sloping area that extends under water from the coastline and ends in a steep drop to deep water.

controlled experiment: An experiment in which only one variable is changed at a time, while all the other possible variables remain the same; also called a fair test.

controlled variable: A factor in an experiment that is kept constant or unchanged.

convection current: The moving stream or current that forms in a fluid when it is heated.

cooling curve: A graph of temperature and time that shows what happens to the temperature of a substance as it releases heat: the temperature remains constant during a change of state.

core: The hot center of the Earth's interior.

Coriolis effect: The deflection of air currents to the right in the northern hemisphere and to the left in the southern hemisphere as a result of the Earth's rotation.

corona: The outer layer of the sun, where the gases are very hot.

corrosion: The eating away of a metal as a result of a chemical reaction.

cosmology: The study of the origin of and changes in the universe.

crests: The high point of an ocean wave.

crust: The Earth's outer layer of rock.

cycle: A series of actions or events that occur in the same order every time.

D

dark nebula: A nebula that consists mostly of dust. A dark nebula blocks out light energy from stars and bright nebulas behind it.

deciduous plant: A plant that loses its leaves during cold or dry seasons, or at a particular stage of growth.

decomposer: An organism that feeds on material that used to be living.

deep water currents: Swift-moving water that flows in a continuous direction well below the ocean's surface.

defining operationally: Describing an object or event according to observable or measurable characteristics.

density: The mass of a substance per unit volume. When the salinity of the ocean's water is higher, the density of the surface water increases, causing the surface water to descend to lower depths.

deoxygenated blood: Blood in which the red blood cells have given up their oxygen to body cells and so are no longer carrying large amounts of oxygen.

desert: A biome found in areas that receive less than 25 cm of precipitation per year but that are not as cold as tundra areas. Plant life includes cacti and shrubs.

diaphragm: A large, dome-shaped sheet of muscle that forms the bottom of the chest cavity. Movement of this muscle helps in breathing.

diarrhea: A condition in which acts of elimination (bowel movements) are unusually frequent, with feces that are watery.

diet: The types of food that a person regularly eats and drinks.

dietary fiber: Complex carbohydrates that cannot be digested. They cannot, therefore, be used as nutrients because they cannot enter your cells. Dietary fiber is still necessary for good health.

digestion: The overall process (which is made up of many smaller processes) by which the food you eat is broken down into molecules small enough to enter and be used by your body cells.

digestive enzyme: An enzyme that breaks apart large food molecules into smaller components.

digestive system: A series of organs and tubes that carry out digestion. The human digestive system includes the esophagus, stomach, and intestines, as well as other organs.

dilute: Containing only a small amount of dissolved solute.

direct energy use: The use of energy involving only one change of the form of energy.

disease: A condition in which any structure or part of an organism does not function correctly. Diseases have many causes, including birth defects, poor diet, infections, and poisons in the environment. The terms "illness," "condition," and "disorder" are sometimes used instead of "disease."

dissolve: To completely mix the particles of two or more substances, forming a solution.

dominant plant species: The most abundant or obvious plant species found in a location.

dormant: A state in which an organism is alive but not active or growing. For example, many plants survive the cold of winter by storing food and becoming inactive.

ductile: Capable of being stretched out.

duration: In an exercise program, duration refers to the length of time that you do a physical activity.

E

Earth-centered universe: The ancient idea that the Earth was standing still and all the other bodies in the universe moved around it.

earthquake: A shaking of the Earth's surface caused by large sections of rock sliding past each other.

ecology: The study of interactions in the environment.

ecosystem: A network of interactions linking living (biotic) things with non-living (abiotic) things.

efficiency: A calculation of the work done by a machine (the work output) as a percentage of the work needed to operate it (the work input). The equation is:

$$\text{efficiency} = \frac{\text{work output} \times 100\%}{\text{work input}}$$

efficient: Describing something or someone that can perform a particular task with a minimum amount of time or energy. An efficient body cell, for example, needs less energy to perform a task than a less efficient cell.

effort force: The force needed to move an object when you use a machine.

elastic energy: The energy stored in an object whose shape is changed by stretching, compressing, or twisting.

electrical energy: The energy of moving electrical particles.

electromagnetic spectrum: A broad band of energy that consists of radio waves, microwaves, infrared rays, visible light, ultraviolet rays, X rays, and gamma rays.

element: A pure substance that cannot be broken down into any other pure substances. An element contains only one type of atom.

elimination: The act of removing solid waste materials from the body. For example, in humans, elimination involves the large intestine, rectum, and anus.

elliptical galaxy: A galaxy that is oval in shape.

El Niño: A large patch of unusually warm surface water that forms in the western Pacific Ocean and moves slowly eastward to South America.

emulsion: A suspension of one liquid in another liquid that has been treated to keep it from separating. For example, homogenized milk is an emulsion of fat droplets in the watery part of milk.

endothermic reaction: A chemical reaction that absorbs energy from its surroundings.

energy: The ability to make things move.

energy conservation: Using energy carefully in order to reduce wasteful use.

energy resource: Something in nature that can be used to produce useful energy. Coal, wood, petroleum, and falling water are examples of energy resources.

energy transformation: A change of energy from one form to another.

enriched: An enriched food is one in which the amounts of some nutrients normally present in the food are increased by food processing.

environment: Everything in an organism's surroundings: physical habitat, air, water, and all other organisms, including those of the same species.

enzyme: A protein that speeds up the rate of a chemical reaction inside an organism. A type of catalyst.

epiglottis: A flap of tissue that closes the opening to the trachea during swallowing. This prevents food or liquid from passing into the lungs.

erosion: The process that carries away rock, particles of rock, and soil through the action of water, ice, wind, or gravity.

esophagus: The tube leading from the throat to the stomach.

essential amino acid: An amino acid that your body cannot make, or cannot make enough of to supply your needs. You must obtain these amino acids from the food you eat. Eight of the 20 amino acids your body needs are essential amino acids.

evaporation: A slow change of state from a liquid to a gas.

excretion: The process of removing excess water, salts, and the waste products of cells from the body. Excretion occurs through the lungs and skin, and in the urine produced by the kidneys.

excretory system: The organs of the body responsible for removing excess water, salts, and the waste products of cells. They include the lungs, skin, kidneys, and other associated organs.

exothermic reaction: A chemical reaction that releases energy to its surroundings.

extrusive igneous rock: Rock formed from lava that has cooled and hardened on the Earth's surface.

F

fair test: An experiment in which only one variable is changed at a time, while all the other possible variables remain the same; also called a controlled experiment.

fats: A group of nutrients made up of carbon atoms, some oxygen atoms, and many hydrogen atoms. These atoms are combined in various ways to form different types of fatty acids. Fatty acids are then joined together with other molecules to form fats. Fats supply your body with energy.

fatty acid: A type of molecule that, when joined together with other molecules (including other fatty acids), makes up a fat. In other words, fatty acids are the building blocks of fats.

fault: A crack or break in the Earth's lithosphere, along which the sections of rock have shifted or moved.

feces: The undigestible wastes that have been made drier and more solid in the large intestine. They are stored in the rectum and eliminated through the anus.

fertilizer: Any substance that is added to the soil to provide nutrients to plants, especially nitrogen (as nitrate), phosphorus (as phosphate), and potassium.

flexibility: The amount of movement a person has at his or her movable joints.

fold: Rock layers that have been squeezed into a wave-like curve.

food additive: Any substance added to food during food processing. Some additives are used to keep food from spoiling; others are used to improve taste or appearance.

food energy: The energy that your body cells obtain from food.

food poisoning: Illness caused by micro-organisms living in food.

food processing: Anything done to plant or animal materials between the time they are harvested or slaughtered and the time they are eaten. Cooking and preserving (by drying, freezing, or canning, for example) are types of food processing.

force: A push or a pull.

fortified: A food that has been fortified has had nutrients added to it (during food processing) that are not normally present in the food.

fossil fuel: An energy resource formed from plants and animals that died millions of years ago. Coal, petroleum, and natural gas are examples of fossil fuels.

freezing: The change of state from a liquid to a solid; also called solidification.

frequency: The number of times an event occurs within a certain period of time. In an exercise program, frequency refers to how often you do a physical activity—for example, how many times per week.

friction: A force that resists motion whenever one material rubs against another.

fulcrum: The support point around which a lever can rotate.

function: Purpose or job. For example, the function of your eyelids is to protect your eyes.

G

galaxy: A huge collection of gases, dust, and hundreds of billions of stars. Our solar system belongs to the Milky Way Galaxy.

gall bladder: A small balloon-like organ that stores the bile produced by the liver until it is needed for fat digestion in the small intestine.

gas: A state of matter that has a fixed mass but does not have a fixed volume or a fixed shape.

gas giants: The four largest planets (Saturn, Jupiter, Uranus, and Neptune), whose atmospheres consist mainly of the gases hydrogen and helium.

geothermal energy: Energy that comes from the Earth's interior.

gland: An organ or tissue that makes and releases (secretes) one or more substances needed by another part of your body. For example, the pancreas secretes digestive enzymes, and mucous tissue secretes mucus.

globular star cluster: A group of about a million stars forming a shape like a globe. Globular clusters are outside the main part of a galaxy. The stars in a globular cluster are closer together than the stars in an open cluster.

glycogen: A complex carbohydrate found in the bodies of animals.

grasslands: A biome where the dominant plant species are grasses.

gravitational energy: The energy that an object has when it is above the surface of the Earth. Any object that can drop, roll, or slide down has gravitational energy.

gravity: A force of attraction between all objects.

groundwater: Water which flows beneath the Earth's surface.

H

habitat: The physical space where a certain species lives.

hazardous chemical: A chemical that is dangerous to human health or to the environment.

heart attack: A failure of the blood supply to the muscles of the heart. The heart muscles cannot work properly and may be damaged or die.

heart rate: The number of times the heart beats in a minute.

heat: The energy transferred from an object at a higher temperature to an object at a lower temperature.

heating curve: A graph of temperature and time that shows what happens to the temperature of a substance as it is heated: the temperature remains constant during a change of state.

hemoglobin: A protein contained in red blood cells that picks up oxygen in areas of high concentration and releases the oxygen in areas of low concentration.

high blood pressure: A condition in which a person's blood pressure is higher than usual most of the time.

high tide: The tide when the water is at its greatest elevation. Caused mainly by the gravitational pull of the moon on the Earth. The ocean is at its highest at a point directly beneath the moon. There is a corresponding surge on the other side of the Earth, away from the moon. These two surges are the high tides.

hormone: A type of blood protein that acts as a chemical messenger. Hormones are produced in one part of the body (a gland) and often affect other parts.

humus: Dead plant or animal matter in the soil that has been almost completely decomposed. Humus absorbs water and so improves the soil's ability to support plant growth.

hydrocarbon: A chemical compound made of the elements of hydrogen and carbon. The fossil fuels petroleum and natural gas are examples of hydrocarbons.

hydroelectricity: Electricity generated from falling water.

I

ideal body mass: A range of body masses that describes the best (most healthy) body mass for a person. A person's ideal body mass depends on many factors, including the individual's sex, age, height, and bone structure.

igneous: Used to describe a type of rock formed when molten rock cools and hardens.

immune system: The parts of the body involved in defending against disease-causing organisms.

For example, antibodies and white blood cells help destroy bacteria and other foreign matter.

inclined plane: A simple machine that is a sloping surface along which a load is moved.

incomplete protein: Any protein that does not contain all eight essential amino acids in sufficient amounts to supply your body's needs.

indirect energy use: The use of energy involving more than one transformation of energy from one form to another.

inner planets: The four planets of the solar system that are closest to the sun (Mercury, Venus, Earth, and Mars); also called the terrestrial planets.

intensity: In an exercise program, intensity refers to how hard your body has to work to do the activity.

intergalactic distances: Distances between galaxies.

interplanetary distances: Distances between the planets in our solar system.

interstellar distances: Distances between the stars. These distances are often measured in light-years.

intestines: The small intestine and the large intestine, sometimes called the bowels.

intrusive igneous rock: Rock formed from magma that has cooled and hardened beneath the Earth's surface.

irregular galaxy: A galaxy that has no distinct shape.

irrigation: The technology of bringing water to the land in order to promote the growth of plants.

issue: A topic about which people have different points of view.

J

joint: A place where two or more bones come together in your body.

joule (J): The SI unit of work. A joule is the same as a newton meter. Joules are the units used to measure energy.

K

kidneys: A pair of organs that form part of the excretory system. The kidneys filter blood to retain useful substances and remove wastes in the form of urine.

kinetic energy: The energy of motion.

kinetic molecular theory: The theory that all matter is made up of small particles that are constantly in motion. When heat is added to a substance, its particles gain kinetic energy, and so move faster.

L

landform: The naturally occurring surface features of the Earth, such as mountains, plains, valleys, rivers, and lakes.

large intestine: The tube leading from the small intestine to the rectum. The large intestine removes water from undigestible wastes, making feces.

lava: Molten rock that has risen through the crust to the Earth's surface.

law of conservation of energy: A scientific law stating that energy can neither be created nor destroyed; it can only change form.

law of conservation of mass: A scientific law that states that in a chemical reaction, the total mass of the reactants is always equal to the total mass of the products.

leaching: The dissolving and movement of a material through the soil. For example, nutrients are leached from top soil layers by water and move down through the soil.

lever: A simple machine made up of a bar that is supported at one point. Levers are sometimes used in pairs. For example, scissors are made of a pair of levers.

life cycle: The stages an organism goes through in life.

light energy: The form of energy that you can see.

light-year: The distance that light energy can travel in space in one year. This distance is about 9,460,000,000,000 km.

liquid: A state of matter that has a fixed mass and volume but does not have a fixed shape. A liquid takes the shape of the container in which it is placed.

lithosphere: The rigid outer layer of the Earth made up of the crust and the upper mantle.

liver: A large organ, located above the stomach, that has several functions in the body. These include controlling the storage and release of substances such as nutrients, the breakdown and removal of dangerous substances such as alcohol, and the production of bile.

load force: The force needed to move an object without using a machine.

low tide: The farthest ebb of the tide. The ocean's low tides occur midway between the two high tides.

lungs: A pair of spongy organs that receive air during inhalation. The lungs are made up of clusters of tiny air sacs, called alveoli, surrounded by blood vessels.

M

machine: A device that helps people do work more easily.

magma: Molten rock below the Earth's surface.

magnetic energy: The form of energy that causes some kinds of metal to attract or repel some metal objects. (Also known as magnetism.)

magnetic reversal: A sudden change in the Earth's magnetic poles that causes the North and South Magnetic Poles to reverse their positions.

magnitude: The brightness of a star; see also apparent magnitude and absolute magnitude.

malleable: Capable of being flattened into thin sheets.

malnutrition: A condition that results from eating improper amounts or an improper balance of nutrients. Overnourishment (eating too much of one or more types of nutrients) or undernourishment (eating too few nutrients) can both lead to malnutrition.

mantle: The thick layer of the Earth found between the crust and outer core.

mass: The amount of matter in an object. Mass can be measured in milligrams, grams, kilograms, or tons.

matter: Anything that has mass and occupies space, including all solids, liquids, and gases.

mechanical advantage: The amount by which a machine can multiply a force. Mechanical advantage can be calculated by using the equation:

$$\text{mechanical advantage} = \frac{\text{load force}}{\text{effort force}}$$

mechanical digestion: The physical breaking apart of food into smaller pieces.

mechanical energy: The energy in any set of moving parts.

mechanical mixture: A mixture in which two or more parts can be seen with the unaided eye.

melting: The change of state from a solid to a liquid.

metabolism: All of an organism's basic life processes (such as those carried out by the digestive, circulatory, and respiratory systems).

metal: An element that is shiny, ductile, and malleable and is a good conductor of heat and electricity.

metamorphic: Used to describe a type of rock formed when rock is altered by heat and pressure beneath the Earth's surface.

meteor: A lump of rock or metal that falls from space toward the Earth's surface. It enters the Earth's atmosphere at high speed and burns up. We see a meteor as a streak of light in the night sky.

meteorite: A meteor that is large enough to reach the Earth's surface before totally burning up.

micro-organism: A tiny living thing that can only be seen with the help of a microscope.

mid-ocean ridge: An elongated ridge that runs along the floor of all major oceans. Lava flows from the center of the ridge as the sea floor spreads apart.

migrate: To go to another place to live when the seasons change.

minerals: A group of nutrients that are elements needed by your body in small amounts.

mixture: Matter that has two or more kinds of particles and thus may have different properties in different samples.

molecule: A particle that is made up of two or more atoms combined together.

moon: An object that revolves around a planet.

muscular endurance: The ability of a muscle or muscles to work (and therefore perform activities) for a long period of time.

muscular strength: The amount of force (push or pull) that a muscle is able to exert at any one time.

N

nasal cavity: The space within the nostrils and behind the eyes. Air travels through your nasal cavity when you breathe through your nose.

natural gas: A fossil fuel found in the form of a gas. Natural gas is composed mainly of compounds of carbon and hydrogen.

neap tides: Lower than normal high tides that occur twice a month. Neap tides occur when the sun and moon are at 90 degrees to each other.

nebula: A spread-out cloud of interstellar dust or gas.

neutral: Neither an acid nor a base.

neutralization: The chemical reaction between an acid and a base.

newton (N): The SI unit for measuring force.

non-essential amino acid: An amino acid that your body can make from the atoms in the food you eat. You do not have to eat foods containing these amino acids.

non-metal: An element that has few or none of the properties of metals.

non-renewable energy resource: A resource that takes an extremely long time to be renewed. Once it is used up, it is considered to be gone forever. Coal, petroleum, and natural gas are examples of non-renewable energy resources.

northern coniferous forest: A biome found in the Northern Hemisphere, characterized by severe winters. Plant life includes coniferous trees, mosses, and ferns.

nuclear energy: The energy stored in the central part of an atom (the nucleus).

nutrient: A substance that organisms require for life, growth, and reproduction.

nutrition: The study of nutrients and how they affect people's health.

nutritionist: A person who studies nutrients and how they affect people's health.

O

observatory: A large building where telescopes are placed permanently.

oceanography: The scientific study of the oceans, their properties, and motions.

open star cluster: A group of about 10 to 120,000 stars that are found in the main part of the Milky Way Galaxy. It is called an open cluster because the stars are not as close together as they are in a globular cluster.

orbit: The path of one body in space as it revolves around another body.

organ: A body part that performs a specific function and is made up of several types of tissues. For example, the stomach and the heart are organs.

organic matter: Matter that is or was once part of living organisms.

organism: A living thing.

oscillating theory: The theory that the expanding universe will reach a maximum size, then start to contract again.

outer planets: The five known planets of the solar system that are farthest from the sun (Saturn, Jupiter, Uranus, Neptune, and Pluto).

oxygenated blood: Blood in which the red blood cells are carrying large amounts of oxygen.

P

pancreas: An organ located near the stomach that produces digestive enzymes used in the small intestine.

particle model: A model that suggests that matter is made up of tiny particles. This model explains the properties of solids, liquids, and gases.

peat: A fossil fuel that is produced when plants in a swamp die and then decay. Peat that is under a lot of pressure may eventually form into coal.

peristalsis: A series of muscular contractions that help move food along the tubes of the digestive system.

pesticide: Any substance used to control populations of pest organisms, such as crop-eating insects or weeds.

petroleum: A liquid fossil fuel that is made up mostly of compounds of carbon and hydrogen.

physical change: A change in matter in which no new substance is produced.

physical fitness: A condition describing how well various parts of your body work together and how much energy you require to make those parts work.

pioneer species: A tough plant with the ability to survive and grow on bare soil or in a crack between rocks. The species found growing in the first stage of succession.

planet: A large object that revolves around the sun. We do not know if there are planets revolving around any other stars.

plasma: The liquid component of blood. Plasma consists of water, dissolved substances, and blood proteins.

plate: A section of the Earth's rigid lithosphere that moves over the less rigid mantle beneath.

platelet: A small, cell-like fragment found in the blood. Platelets contain the chemicals that start blood clotting. They break open and release these chemicals wherever a blood vessel is damaged.

polar easterlies: Extremely cold air from the polar region that is locked in a circular pattern around the poles from east to west.

population: The number of individuals of a species that live together in an area.

pore: Small spaces between particles of soil that contain water and air. (The term "porosity" refers to the volume that the pores occupy in a soil sample.)

potential energy: Stored energy.

precipitation: Water which falls to Earth from the clouds in the form of rain, snow, sleet, or hail.

preservative: Any substance that is added to food to prevent it from spoiling or to keep it fresh (unspoiled) for a longer period of time.

preserve: To process food in some way in order to keep it safe to eat.

primary consumer: An organism that eats producers.

process: A set of actions done in a specific order.

process skill: A procedure, such as observing or measuring, that is used in scientific research.

producer: An organism that can produce its own food from materials in the abiotic part of the environment. Green plants and other organisms with chlorophyll are producers.

product: Any substance that is produced in a chemical reaction.

property: A characteristic used to describe matter.

proteins: A group of nutrients that consist of molecules made up of chains of smaller molecules called amino acids. Proteins contain atoms of nitrogen, hydrogen, oxygen, and carbon. Proteins help our body cells work properly and assist in the building and repair of these cells.

pulley: A simple machine made from a rope or cable that is looped around a wheel.

pulsar: A dense, rotating object in the universe that emits bursts (pulses) of radio waves.

pulse: The rhythmic surge of blood passing a certain place in an artery. A pulse occurs after each beat of the heart.

pure substance: A substance that has the same properties in any sample you choose. There are two kinds of pure substances: elements and compounds.

Q

quasar: A star-like object in the universe that has a high mass and emits radio waves.

qualitative property: A property that supplies information about the appearance, smell, taste, feel, or sound of an object.

quantitative property: A property that answers the question "How much?" or is the result of exact measurement.

R

radio telescope: A device that receives radio waves from space.

reactant: Any substance that you start with in a chemical reaction.

reaction rate: The speed of a chemical reaction.

reactive: Able to react easily.

rectum: The final section of the large intestine, which stores feces until they can be eliminated through the anus.

red blood cell: A small, disk-shaped blood cell that contains hemoglobin, a protein that combines with oxygen. This allows red blood cells to carry oxygen in the blood.

red giant: A stage in the life cycle of low-mass and medium-mass stars. After a star's energy runs out, it cools down and swells up into a red giant.

red shift: A change in the color of light from the blue end of the spectrum to the red end of the spectrum. This movement along the spectrum is called a shift. A red shift occurs when a source of light is travelling away from the observer. It provides evidence that the universe is expanding.

red supergiant: A stage in the life cycle of high-mass stars. After a star's source of energy is used up, the star swells into a red supergiant.

refining: A process used to separate substances from a mixture. This process is used to obtain useful products from petroleum.

reflecting telescope: An instrument that uses a curved mirror to gather light so we can view distant objects. It is different from a refracting telescope.

refracting telescope: An instrument that uses a lens to gather light so we can view distant objects. It is different from a reflecting telescope.

renewable energy resource: A resource that can be renewed within a normal human lifetime (about 75 years) or less. Trees are considered to be a renewable resource.

residual air: The amount of air that stays inside the lungs even after as much air as possible has been exhaled.

respiration: Breathing—inhaling and exhaling air.

respiratory system: The series of tubes and organs involved in breathing. The function of breathing is to bring oxygen into the body and remove carbon dioxide. The human respiratory system includes the trachea, lungs, and other body parts.

revolution: The motion of an object around another.

rib cage: The rib bones and the muscle that connects them.

rotation: The spinning of an object around an imaginary line called an axis.

runoff: Precipitation that falls to Earth and drains into surrounding bodies of water, such as streams, lakes, rivers, and oceans.

S

salinity: A measure of the salt content of water.

saliva: A watery fluid in the mouth that moistens food and contains enzymes that start to digest carbohydrates.

salt: A compound formed by the reaction of an acid and a base.

saturated fat: A type of fat that contains all the hydrogen atoms it possibly can. In other words, all the spaces in the fat molecule into which hydrogen atoms could be added are already full. Saturated fats come from animal products, such as meat, cheese, and milk.

scavenger: An organism that habitually feeds on dead organic material.

science: Both a body of knowledge and a process for gaining more knowledge about the world in which we live.

scientific law: A general statement that sums up the conclusions of many experiments.

screw: A simple machine made up of an inclined plane wrapped around a rod. A screw does work by moving into or through an object, such as a piece of wood.

sea floor spreading: The moving apart of two plates as lava flows out at a mid-ocean ridge.

second-hand smoke: Unfiltered smoke from the burning end of a cigarette, cigar, or pipe, along with the exhaled smoke that enters the environment when a person smokes.

secondary consumer: An organism in a food chain that eats primary consumers.

sedimentary: Used to describe a type of rock formed when small particles of rock, such as mud and sand, are deposited in layers and pressed together.

shield volcano: A large volcano with very gently sloping sides, formed of hardened basalt lava.

simple carbohydrate: A small carbohydrate made up of one or just a few sugar molecules joined together.

simple machines: The six basic devices that make work easier and more convenient for us to do. They are the inclined plane, wedge, screw, lever, pulley, and wheel-and-axle.

small intestine: The long tube that leads from the stomach to the large intestine. Proteins, carbohydrates, and fats are digested in the first 25 cm and nutrients are absorbed in the remaining length.

society: A group of individuals who must work together.

soil: A thin layer of material at the surface of the Earth that sustains plants. Soil consists of weathered rock and organic material.

soil profile: A diagram showing the different layers within the soil at one place. Soil profiles are studied in order to classify different types of soil and help to define biomes.

solar energy: Energy from the sun.

solar flare: A burst of energy from the sun's photosphere.

solar prominence: A large sheet of glowing gases that bursts outward from the sun's chromosphere.

solar system: The sun, the planets that travel around the sun, and all the other objects that travel around the planets and the sun.

solid: A state of matter that has a fixed mass, volume, and shape.

solidification: The change of state from a liquid to a solid. Also called freezing.

soluble: Able to be dissolved.

solute: The material that dissolves in a solution. For example, in a solution of sugar and water, sugar is the solute.

solution: A mixture that appears to be all one substance.

solvent: The material in which a solute dissolves to produce a solution. For example, in a solution of sugar and water, water is the solvent.

sonar: A technique used to measure the depth of the ocean by bouncing sound waves from the ocean floor back to the water's surface.

sound energy: The form of energy that you can hear. It travels as vibrations through materials and through air.

space probe: A spacecraft with many instruments on board that is sent from Earth to explore our solar system.

specific heat capacity: The amount of heat transferred when the temperature of 1.0 kg of a substance changes by 1.0°C. The equation is:

$$\text{specific heat capacity} = \frac{\text{energy}}{\text{mass x temperature change}}$$

spectral types: A system of classifying stars by their spectra, which you can observe by using a spectroscope. There are seven color types.

spectroscope: An instrument that splits light energy into a series of colors called a spectrum.

spectrum: A series of colors that you observe when light is split up by instruments such as a spectroscope.

spiral galaxy: A galaxy that has a spiral, disk shape.

spring tides: Higher than normal high tides that occur twice a month. Spring tides occur when the sun and the moon are positioned in a direct line with the Earth, exerting a combined gravitational pull on the ocean.

star cluster: A group of stars that are fairly close together and move together.

starch: A complex carbohydrate found in plants.

states of matter: Solid, liquid, and gas.

subduction: The downward movement of one plate as it slides underneath the edge of another and sinks into the mantle.

sublimation: The change of state from a solid to a gas or a gas to a solid.

succession: The process of gradual change that occurs when organisms colonize a habitat, modify it, and are forced out by a new species better adapted to the now-altered environment. Succession begins with pioneer species and ends with the development of a climax community.

sugar: A type of molecule that serves as the building block for complex carbohydrates. There are several types of sugars, which, either singly or joined together into much larger molecules, make up carbohydrates.

sun-centered solar system: Our solar system, in which all the planets revolve around the sun.

sunspot: A darker region on the surface of the sun.

supernova: A huge explosion that occurs near the end of the life cycle of a large star.

surface currents: Water movement caused by winds at the Earth's surface.

suspension. A cloudy mixture that consists of pieces of a solid or droplets of liquid scattered throughout the mixture. The parts of a suspension will separate if the suspension is left to stand.

swallowing: A reflex action that moves food or liquid from the mouth to the esophagus. During swallowing, the trachea is closed so that food or liquid cannot enter the lungs.

syncline: A trough-like downfold of rock layers.

system: A group of organs that work together to perform a function. For example, the digestive system and the circulatory system are two of your body systems.

T

tars: Tars are made up of several substances that condense out of smoke as it cools. For example, tars from cigarette smoke contain substances that cause cancer and substances that paralyse the cilia of cells.

technology: The use of scientific knowledge to solve a practical problem. Technology refers to the tools, machines, and processes that help people use the natural world.

telescope: An instrument used by astronomers to view the universe.

temperate rain forest: A biome found in areas that receive plentiful rain and have moderate to cool temperatures. Plant life includes very large tree species, ferns, and mosses.

temperature: A measure of the average energy of the particles of a substance. The SI unit for temperature is degrees Celsius (°C).

terrestrial planets: The Earth and the planets that resemble the Earth. They are also called the inner planets.

theory of continental drift: A theory proposing that Earth's continents are not stationary. Evidence suggests that in the past, a single huge land mass broke into pieces and drifted apart to form today's continents.

theory of plate tectonics: A theory that describes the movement of plates on the Earth's surface and links this movement to the formation of mountains, earthquakes, and volcanoes.

thermal energy: The total energy of all the particles (molecules or atoms) in an object.

thermocline: The zone between warm surface water and cold deep water where the temperature changes abruptly.

tidal energy: Energy that we obtain from ocean tides. The tides are caused by the gravitational pull of the moon and sun on the oceans.

tides: Surges in the ocean's water level due to the gravitational pull of the moon, and, to a lesser extent, the sun.

tissue: A group of similar cells that work together to perform a specific function. For example, muscle tissue and nervous tissue are two types of tissue.

trachea: The tube that carries air from the throat to the lungs. Rings of cartilage prevent the trachea from collapsing. The opening of the trachea is covered by the epiglottis when you swallow.

trade winds: A wind pattern that forms as cooled air reaches the Earth's surface at approximately 30° latitude and flows toward the equator in a southwesterly direction.

training effect: A result of exercise that improves the strength and endurance of your heart muscle. To have a training effect, a physical activity must increase your heart rate to within a certain safe range.

transform fault: A break in the Earth's lithosphere where sideways movement of plates occurs.

transition zone: The area along the boundary of neighboring biomes, where organisms from both biomes live.

transpiration: The escape of moisture through pores in a plant.

tree line: The upper or northern limit of tree growth in the northern coniferous forest. Trees do not grow above the tree line on mountain ranges or north of the tree line in northern Canada.

trench: A long, deep, narrow depression in the ocean floor along the edge of a continent or a chain of islands.

triangulation: A method of measuring distances indirectly by drawing a scale diagram of a triangle.

troughs: The low point of an ocean wave.

tsunami: A large, destructive, fast-moving wave caused by an earthquake on the ocean floor.

turbine: A wheel-shaped device with blades. A liquid or a gas moving through the turbine pushes against the blades, causing the turbine to spin. One way to produce electricity is to connect a spinning turbine to an electrical generator.

U

universe: Everything that exists, including all matter, all energy, and all the space between the matter.

unsaturated fat: A type of fat that still has space for one or more hydrogen atoms. Unsaturated fats come from plant products, such as corn and nuts.

upwelling: The process of colder, deeper water moving toward the ocean's surface, when winds blow the warmer surface water away. Upwelling is good for fish and other marine life because it brings nutrients to the surface, providing food.

urethra: The tube that leads from the urinary bladder so that urine can be removed from the body.

urinary bladder: A balloon-like organ that stores urine until it can be released from the body.

urine: A mixture of water, urea, and other wastes produced by the kidneys for removal from the body.

V

valve: Flaps of tissue found in the circulatory system that act to prevent the backward flow of blood.

vaporization: The change of state from a liquid to a gas. Slow vaporization is called evaporation. Fast vaporization is called boiling.

variable: A factor that could affect the results of an experiment.

vein: An elastic-walled blood vessel that carries blood to the heart.

ventricle: One of the two chambers at the bottom of the heart. The right ventricle sends deoxygenated blood to the lungs. The left ventricle sends oxygenated blood to the body.

vital capacity: The maximum amount of air that a person can move in and out of his or her lungs in one breath.

vitamins: A group of nutrients, most of which help enzymes do their work in your body.

volcano: A mountain formed of igneous rock, with an opening, or vent, through which steam, ashes, and lava erupt.

volume: The amount of space that matter occupies. Volume can be measured in milliliters or liters.

W

water: A compound made up of molecules containing two hydrogen atoms and one oxygen atom. Some people consider water to be the most important nutrient for your body.

water cycle: The cycling of water throughout the planet through the processes of evaporation and condensation.

water table: The level of water in the ground.

wave height: The vertical distance between a wave crest and a wave trough.

wavelength: The distance between wave crests.

weather: The daily, or even hourly, changes in meteorological conditions with respect to heat or cold, wetness or dryness, calm or storm, and clearness or cloudiness.

weathering: The process that breaks down rock into smaller particles (such as gravel, sand, mud, or clay) or solutions.

wedge: A simple machine shaped like an inclined plane. A wedge does work by being pushed into the object.

westerlies: Winds from the west that move toward the poles in a northeasterly direction.

wheel-and-axle: A simple machine made up of a large-diameter disk (the wheel) attached to a small-diameter shaft (the axle).

white blood cell: A blood cell that forms part of the body's immune system. Some types of white blood cells detect and consume foreign particles such as bacteria.

wind energy: Energy that we obtain from moving air. Wind energy comes from solar energy.

word equation: An equation that uses words to show the reactants that react to give the products of a particular chemical reaction.

work: A measure of the transfer of energy to an object. Work, measured in joules, is equal to force, measured in newtons, times distance, measured in meters.

Y

year: The length of time for the Earth to complete one revolution around the sun.

Z

zodiac constellations: The constellations through which the planets appear to move.

INDEX

D

E

F

T

U

V

W

PHOTO CREDITS

Front Matter: p. vi, far left photo courtesy of Bureau of Reclamation; p. vi (welder) © Nancy Bundt; p. vii, bottom left photo courtesy of Chevron Corporation; p. vii, top right © Mike Beedell, Ottawa, Canada; p. xiii photo courtesy of Hawaii Visitors Bureau.

Unit Openers: Unit I Maria Taglienti/Image Bank; Unit IV p. 262 Royal Observatory, Edinburgh/ AATB/Science Photo Library; Unit V photo courtesy of American Petroleum Institute; Unit VI Al Harvey.

Sidebars: Computer and Science W. Martin Jr. /COMSTOCK; Career and Profile COMSTOCK.

Chapter 1: p.1, 1.1 David Keevil; p.15 courtesy The Environmental Children's Organization.

Chapter 2: p. 20 Louis Doqville and Gisele Dincl, Lasalle, Quebec, Canada; 2.2 Al Harvey; 2.3 Mark Tomalty/Masterfile; 2.6 © Nancy Bundt; 2.7 David Keevil; 2.8 © Paul Couvrette, Ottawa, Canada; 2.10 Greg Kinch; 2.11 © 1995, PhotoDisc, Inc.; 2.12 Wiley Photo Files; p. 32 Teth Breyere; 2.15a The Bettman Archives; 2.15b A. Penner; 2.19 Richard Megna/Fundamental Photographs; 2.21 Mason Moffit/Masterfile; 2.22 David Keevil; 2.23 Jeremy Jones; 2.24a NASA; 2.24b Skip Dean/Image Bank; 2.24c Bill Ivy; 2.24d L. T. Webster; 2.24e Alvis Upitis/Image Bank; 2.25 Canapress Photo Service.

Chapter 3: p. 42 Al Harvey, p. 43 Canadian Forces Photograph-DND; 3.2, 3.3 David Keevil; 3.4, p. 48 photo courtesy of Andy Cooperman; Buschlen/Mowatt Fine Arts Ltd.; 3.10 Martyn Chillmaid/Oxford Scientific Films; 3.11 David Keevil.

Chapter 4: p. 60 David Brownell/ Image Bank; 4.1 Environment Canada; 4.2 David Keevil; 4.3 photo courtesy of Alcoa-Aluminum Co. of America; 4.4 © Esther Mugar and David Ryan, San Francisco, CA; 4.5 Jeremy Jones; 4.7 David Keevil; 4.11 L. T. Webster; p. 74 American Automobile Association; 4.14 © Berkus Design Studio; 4.17 Greg Kinch; 4.18 Mary Evans Picture Library; 4.19 Al Harvey.

Chapter 5: p. 80 Steve Niedorf/ Image Bank; 5.5 David Keevil; 5.7a, b Jeremy Jones; 5.8, 5.9, 5.10 David Keevil; 5.11 © William P. McElligott, Ottawa, Canada.

Chapter 6: p. 98 © Bob Daemmrich, Stock Boston; 6.4 top and bottom, 6.5 © Jeff Isaac Greenberg; 6.8, 6.9, 6.10 David Keevil; 6.13 Ian MacLeod/ Unicef; p. 110 W. Durward; 6.14, 6.15 David Keevil; 6.18a Lisa Dennis/ Image Bank; 6.18b Vanessa Vick/ Photo Researchers; 6.18c Patrick J. LaCroix/The Image Bank; 6.19a, b Al Harvey; 6.20, 6.23 David Keevil; 6.24 Daily Food Guide Pyramid, courtesy of National Dairy Council ®; 6.25 Bob Semple; 6.26 Tom Abrahams/ Viewpoints West; 6.27 photo courtesy of Giant Bicycle, Inc., © John Tomac.

Chapter 7: p. 142 Wiley Photo Files.

Chapter 8: p. 150 NASA; 8.2 left © Jeff Isaac Greenberg, right Co Rentmeester/Image Bank; 8.3 © Larry Goldstein, Tony Stone Images, Inc.; 8.5 Al Harvey; p. 158 photo courtesy of Malaysian Tourist Information Center; 8.11 Prof. P Motta/Dept. of Anatomy/University "La Sapienza," Rome/Science Photo Library; 8.15 Al Harvey.

Chapter 9: p. 170 © Billy E. Barnes, Stock Boston; 9.3a Professors P. M. Motta & S. Correr/Science Photo Library; 9.3b, c Francis Leroy, Biocosmos/Science Photo Library; 9.8 Wiley Photo Files; 9.11, 9.15 J. Czerneda; p. 183 Polar/RBI Inc.; 9.17 Wiley Photo Files.

Chapter 10: p. 192 © Harold Davis, NY, NY; 10.1 Al Harvey; 10.2 top Al Harvey; bottom photo courtesy of Florida Dept. of Commerce/Division of Tourism; 10.3 © Bob Daemmrich, Stock Boston; 10.4 Al Harvey; 10.5 David Keevil; 10.6 Brian Milne/First Light; 10.9 © Rudy Molacek, G. W. Magazine, Inc.; 10.12 Greg Kinch; p. 206 Vivian Bowers; 10.13b R. Walker/H. Armstrong Roberts/ Comstock; 10.14 photo courtesy of UNRWA; 10.15 right photo courtesy of Hawaii Visitors Bureau; 10.16 Greg Kinch.

Chapter 11: p. 218 photo courtesy of Arkansas Dept. of Parks & Tourism, © Robyn Horn; inset 1993 Dr. Gerald Newlands; 11.1, 11.2 V. Last/ Geographical Visual Aids; p. 223 J. Souther; 11.7 Black Star/Victor Englebert; 11.8 U.S. Geological Survey; 11.10 photo courtesy of Hawaii Visitors Bureau; 11.11 James Mason/First Light; 11.13 Mike McNamee/Science Photo Library; 11.14 U.S. Geological Survey; p. 230 photo courtesy of LA City Fire Dept.; 11.17 Canapress; 11.18, 11.19 Wiley Photo Files; 11.21 V. Last/ Geographical Visual Aids.

Chapter 12: p. 239 Public Archives of Canada/War Mitchell; 12.3,12.11 David Keevil.

Chapter 13: p 264 photo courtesy of NASA; 13.5, 13.6, 13.7, 13.8, 13.9, 13.10, 13.11a, b NASA; 13.12 photo courtesy of NASA; 13.14 NASA; p. 279 Terence Dickinson; 13.16, 13.17a, b, c, d, 13.18a, b, 13.19a, b, 13.20, 13.21, 13.23a NASA; 13.23b Hertzburg Institute of Astrophysics; 13.24 U.S. Geological Survey; 13.25 Rev. Ronald Royer/Science Photo Library; 13.26 NASA.

Chapter 14: p. 291 John Sandford/Science Photo Library; p. 305 © 1993, W. T. Sullivan, III; 14.18 © Jeffrey Muir Hamilton, Stock Boston; 14.20 Al Harvey; 14.21 (all) Dept. of Physics, Imperial College/Science Photo Library.

Chapter 15: p. 318 NASA; 15.3 Scala, Art Resource, NY; 15.4 Wiley Photo Files; 15.5 David Keevil; 15. 9 © Robert Coello, photo courtesy of Hawaii Visitors Bureau; 15.10 NASA; 15.11 National Optical Astronomy Observatory; 15.12 NAIC, Cornell University; 15.13 Chuck O'Rear/First Light; 15.14 NASA; 15.16 Dennis Di Cicco/Science Photo Library; 15.17a NASA; 15.17b Dr. Jean Lorre/Science Photo Library; 15.17c NASA; 15.17d Science Photo Library; 15.18a photo courtesy of NASA; 15.18b Royal Observatory/Science Photo Library; 15.18c, 15.19, 15.20 NASA; p. 335 H. D. Hogg; 15.22, 15.23a, NASA; 15.23b Kim Gordon/Science Photo Library; 15.23c NASA; 15.24 NRAO/Science Photo Library.

Chapter 16: p. 343 Nasa/ JPL/Science Photo Library; 16.3 Dr. Rudolph Schild/Science Photo Library; p. 355 Royal Greenwich Observatory/ Science Photo Library; 16.13 NASA.

Chapter 17: p. 362 photo courtesy of Chevron Corporation; 17.2 © Tony Bell, photo courtesy of Aspen Skiing Company; 17.3 © 1995, PhotoDisc, Inc.; 17.4 photo courtesy of USDA; 17.5 © Jack K. Blonk, Laval, Quebec, Canada; 17.7 Gary Cralle/Image Bank; 17.8 © 1995, PhotoDisc, Inc.; 17.9 Stephan McBrady/Photo Edit; 17.11, 17.12 Al Harvey; 17.13 Al Hirsch; 17.17 Al Harvey; 17.18 Wiley Photo Files; 17.19 Reuters/Bettman; 17.20, 17.21 Al Harvey; p. 378 Jim Wiese; 17.27a Al Harvey; 17.27b © Wendt Worldwide; 17.28 K. Svensson, Viewpoints West; 17.29 © Lawrence Migdale, Tony Stone Images, Inc.

Chapter 18: p. 385 Peter Turner/Image Bank; 18.2 J. Tapochaner/Image Finders; 18.3 © 1995, PhotoDisc, Inc.; 18.6 Mary Kate Denny/Photo Edit; p. 395 Wiley Photo Files; 18.11 Al Harvey; 18.15 John Mead/Science Photo Library; 18.17 © 1995, PhotoDisc, Inc.; 18.18 John Mead/Science Photo Library; 18.19 NASA; 18.20 Al Harvey.

Chapter 19: p. 410 © 1995, PhotoDisc, Inc.; 19.1a NASA/Science Photo Library; 19.1b photo courtesy of NASA; 19.1d © Bob Daemmrich, Stock Boston; 19.2 © 1995, PhotoDisc, Inc.; 19.4 David Keevil; 19.7 Jay Freis/Image Bank; 19.22 Wiley Photo Files; 19.23 David Keevil; p. 430 Ford Motor Co.

Chapter 20: 20.1a © 1995, PhotoDisc, Inc.; 20.1b photo courtesy of Chevrolet Motors Corporation; 20.1c Al Sutterwhite/Image Bank; 20.7 David Keevil; p. 450 Greg Kinch; 20.17 Canapress; 20.18 Lise Dennis/Image Bank.

Chapter 21: 21.1 Canada Centre for Remote Sensing, Natural Resources Canada; 21.2a Crystal Images Photography; 21.2b © Richard Jackson, Leavenworth, WA; 21.4a Ken Straiton/First Light; p. 462 © 1995, PhotoDisc, Inc.; 21.8a V. Last/ Geographical Visual Aids; 21.8b Crystal Images Photography; 21.9 © 1995, PhotoDisc, Inc.; 21.10 Crystal Images Photography; 21.11 © 1995, PhotoDisc, Inc.; 21.12a © Richard Jackson, Leavenworth, WA; 21.12b John Serrao/Photo Researchers Inc.; 21.12c © 1995, PhotoDisc, Inc.; 21.14 © Barbara Douglas; 21.15 © Jack K. Blonk, Laval, Quebec, Canada; 21.17 © Barbara Douglas; 21.18 FBM Photo/First Light; 21.19a, b © Richard Jackson, Leavenworth, WA; 21.20 Al Harvey; 21.22a Tom McHugh/Photo Researchers Inc.; 21.22b © Richard Jackson, Leavenworth, WA; 21.24a, b Pacific Forestry Centre; 21.25, 21.27a Crystal Images Photography; 21.27b, c, d Al Harvey; 21.28 Crystal Images Photography; 21.29 Agriculture Canada; 21.30 © Larry Parsons, Bowmanville, Canada.

Chapter 22: p. 481 top left Crystal Images Photography, top right, bottom left, bottom right © Leslie Saul, San Francisco, CA; 22.6 V. Last/Geographical Visual Aids; 22.9 Al Harvey; p. 489 Greg Kinch; 22.13, 22.14a, b Crystal Images Photography; 22.16, 22.17 V. Last/Geographical Visual Aids; 22.20 W. Durward; 22.21 © 1995, PhotoDisc, Inc.

Chapter 23: p.502 © 1995, PhotoDisc, Inc.; 23.1 © North Wind Picture Archives; 23.6 (both) © John Elk III, Stock Boston; 23.8 © Joel Bozman; 23.10 photo courtesy of NOAA/Dept. of Commerce; 23.11 photo courtesy of NOAA, National Environmental Satellite Service, National Climatic Data Center, Asheville, NC; 23.13 © Jamey Stillings, Tony Stone Images, Inc.; 23.16b © Barbara Douglas; 23.17 © Sholeh Ashtiani; 23.20 © Barbara Douglas; 23.21 (both) U.S. Coast Guard, photo provided by National Oceanic & Atmospheric Administration, National Geophysical Data Center; 23.22 © 1995, PhotoDisc, Inc.; p. 520 photo courtesy of Scripps Institution of Oceanography, University of California, San Diego.